Business Statistics for Competitive Advantage with Excel 2013

Cynthia Fraser

Business Statistics for Competitive Advantage with Excel 2013

Basics, Model Building, Simulation and Cases

Third Edition

Cynthia Fraser
McIntire School of Commerce
University of Virginia
Charlottesville, VA, USA

ISBN 978-1-4614-7380-0 ISBN 978-1-4614-7381-7 (eBook)
DOI 10.1007/978-1-4614-7381-7
Springer New York Heidelberg Dordrecht London

Library of Congress Control Number: 2013938375

Printed on acid-free paper

Springer is part of Springer Science+Business Media (www.springer.com)

Contents

Preface

Exceptional managers know that they can create competitive advantages by basing decisions on performance response under alternative scenarios. To create these advantages, managers need to understand how to use statistics to provide information on performance response under alternative scenarios. Statistics are created to make better decisions. Statistics are essential and relevant. Statistics must be easily and quickly produced using widely available software, e.g., Excel. Then results must be translated into general business language and illustrated with compelling graphics to make them understandable and usable by decision makers. This book helps students master this process of using statistics to create competitive advantages as decision makers.

Statistics are essential, relevant, easy to produce, easy to understand, valuable, and a powerful source of competitive advantage.

The Examples, Assignments, and Cases Used to Illustrate Statistics for Decision Making Come from Business Problems

McIntire Corporate Sponsors and partners, such as Rolls-Royce, Procter & Gamble, and Dell, and the industries that they do business in, provide many realistic examples. The book also features a number of examples of global business problems, including those from important emerging markets in China, India, and Chile. Students are excited when statistics are used to study real and important business problems. This makes it easy to see how they will use statistics to create competitive advantages in their internships and careers.

Learning Is Hands On With Excel and Shortcuts

Each type of analysis is introduced with one or more examples. Following this is an example of how to create the statistics in Excel, and what the numbers mean in English.

Included in Excel sections are screenshots which allow students to easily master Excel. Featured are a number of popular Excel shortcuts, which are, themselves, a competitive advantage.

Powerful PivotTables and PivotCharts are introduced early and used throughout the book. Results are illustrated with graphics from Excel.

In each chapter, assignments or cases are included to allow students to practice using statistics for decision making and competitive advantage. Beginning in Chap. 11, Harvard Business School cases are suggested which provide additional opportunities to use statistics to advantage.

Focus Is On What Statistics Mean to Decision Makers and How to Communicate Results

From the beginning, results are translated into English. In Chap. 7, results are condensed and summarized in PowerPoint presentations and memos, the standards of communication in businesses. Later chapters include example memos for students to use as templates, making communication of statistics for decision making an easy skill to master.

Instructors, give your students the powerful skills that they can use to create competitive advantages as decision makers. Students, be prepared to discover that statistics are a powerful competitive advantage. Your mastery of the essential skills of creating and communicating statistics for improved decision making will enhance your career and make numbers fun.

New in the Third Edition

The third edition includes more explanations of Monte Carlo simulation. Chapter 4 introduces simulation as a tool for inference and forecasting based on decision makers' assumptions. Chapter 6 demonstrates the use of Monte Carlo simulation to integrate naïve forecasts of performance components. Chapter 13 illustrates use of simulation to integrate nonlinear regression forecasts of performance components for forecasting amidst uncertainties.

The use of naïve models to fit trends for longer range forecasts is introduced in Chap. 6 and extended in Chap. 13, where nonlinear trends and indicators are combined for forecasting. In some situations, a longer-term forecast is desired without the immediate need to explain variation, and naïve forecasts offer a relatively quick solution.

The financial and economic events of 2008–2010 changed business dramatically. Examples have been updated to illustrate how the impacts of recent changes can be acknowledged to build powerful, valid models.

Global examples now include analysis of several emerging markets multinationals in Chile. Such markets create unique opportunities for global business, and the third edition moves beyond the BRICs to explore these.

Acknowledgements

The first and second editions of *Statistics for Decision Making & Competitive Advantage* were used in the Integrated Core Curriculum at The McIntire School, University of Virginia, and I thank the many bright, motivated, and enthusiastic students who provided comments and suggestions. I am grateful to Klaus Mehner for particularly insightful comments.

Cynthia Fraser
Charlottesville, VA

Chapter 1
Statistics for Decision Making and Competitive Advantage

In the increasingly competitive global arena of business in the twenty-first century, the select few business graduates distinguish themselves by enhanced decision making backed by statistics. Statistics are useful when they are applied to improve decision making. No longer is the production of statistics confined to quantitative analysis and market research divisions in firms. Managers in each of the functional areas of business use statistics daily to improve decision making. Excel and other statistical software live in our laptops, providing immediate access to statistical tools which can be used to improve decision making.

1.1 Statistical Competences Translate into Competitive Advantages

The majority of business graduates can create descriptive statistics and use Excel. Fewer have mastered the ability to frame a decision problem so that information needs can be identified and satisfied with statistical analysis. Fewer can build powerful and valid models to identify performance drivers, compare decision alternative scenarios, and forecast future performance. Fewer can translate statistical results into general business English that is easily understood by everyone in a decision making team. Fewer have the ability to illustrate memos with compelling and informative graphics. Each of these competences provides competitive advantage to those few who have mastery. This text will help you to attain these competences and the competitive advantages which they promise.

Most examples in the text are taken from real businesses and concern real decision problems. A number of examples focus on decision making in global markets. By reading about how executives and managers successfully use statistics to increase information and improve decision making in a variety of mini-case applications, you will be able to frame a variety of decision problems in your firm, whether small or multi-national. The end-of-chapter assignments will give you practice framing diverse problems, practicing statistical analyses, and translating results into easily understood reports or presentations.

Many examples in the text feature bottom line conclusions. From the statistical results, you read what managers would conclude with those results. These conclusions and implications are written in general business English, rather than statistical jargon, so that anyone on a decision team will understand. Assignments ask you to feature bottom line conclusions and general business English.

Translation of statistical results into general business English is necessary to insure their effective use. If decision makers, our audience for statistical results, don't understand the conclusions and implications from statistical analysis, the information created by analysis will not be used. An appendix is devoted to writing memos that your audience will read and understand, and to effective PowerPoint slide designs for effective presentation of results. Memos and PowerPoints are predominant forms of communication in businesses. Decision making is

C. Fraser, *Business Statistics for Competitive Advantage with Excel 2013: Basics, Model Building, Simulation and Cases*, DOI 10.1007/978-1-4614-7381-7_1, © Springer Science+Business Media New York 2013

compressed and information must be distilled, well written and illustrated. Decision makers read memos. Use memos to make the most of your analyses, conclusions and recommendations.

In the majority of examples, analysis includes graphics. Seeing data provides an information dimension beyond numbers in tables. To understand well a market or population, you need to see it, and its shape and dispersion. To become a master modeler, you need to be able to see how change in one variable is driving a change in another. Graphics are essential to solid model building and analysis. Graphics are also essential to effective translation of results. Effective memos and PowerPoint slides feature key graphics which help your audience digest and remember results. We feature PivotTables and PivotCharts in Chap. 7. These are routinely used in business to efficiently organize and display data. When you are at home in the language of PivotTables and PivotCharts, you will have a competitive advantage. Practice using PivotTables and PivotCharts to organize financial analyses and market data. Form the habit of looking at data and results whenever you are considering decision alternatives.

1.2 The Path Toward Statistical Competence and Competitive Advantage

This text assumes basic statistical knowledge, and reviews basics quickly. Basics form the foundation for essential model building. Chapters 2 and 3 present a concentrated introduction to data and their descriptive statistics, samples and inference. Learn how to efficiently describe data and how to infer population characteristics from samples.

Model building with simple regression begins in Chap. 5 and occupies the focus of the remaining chapters. To be competitive, business graduates must have competence in model building and forecasting. A model building mentality, focused on performance drivers and their synergies is a competitive advantage. Practice thinking of decision variables as drivers of performance. Practice thinking that performance is driven by decision variables. Performance will improve if this linkage becomes second-nature.

The approach to model building is steeped in logic and begins with logic and experience. Models must make sense in order to be useful. When you understand how decision variables drive performance under alternate scenarios, you can make better decisions, enhancing performance. Model building is an art that begins with logic.

Model building chapters include nonlinear regression. Nearly all aspects of business performance behave in nonlinear ways. We see diminishing or increasing changes in performance in response to changes in drivers. It is useful to begin model building with the simplifying assumption of constant response, but it is essential to be able to grow beyond simple linear models to realistic models which reflect nonconstant response. Visualize the changing pattern of response when you consider decision alternatives and the ways they drive performance.

1.3 Use Excel for Competitive Advantage

This text features widely available Excel software, including many commonly used shortcuts. Excel is powerful, comprehensive, and user friendly. Appendices with screenshots follow each chapter to make software interactions simple. Recreate the chapter examples by following the steps in the Excel sections. This will give you confidence using the software. Then forge ahead and generalize your analyses by working through end of chapter assignments. The more often you use the statistical tools and software, the easier analysis becomes.

1.4 Statistical Competence Is Powerful and Yours

Statistics and their potential to alter decisions and improve performance are important to you. With more and better information from statistical analysis, you will be equipped to make superior decisions and outperform the competition. You will find that the competitive advantages from statistical competence are powerful and yours.

Chapter 2
Describing Your Data

This chapter introduces *descriptive* statistics, center, spread, and distribution shape, which are almost always included with any statistical analysis to characterize a dataset. The particular descriptive statistics used depend on the *scale* that has been used to assign numbers to represent the characteristics of entities being studied. When the distribution of continuous data is bell shaped, we have convenient properties that make description easier. Chapter 2 looks at dataset types and their description.

2.1 Describe Data with Summary Statistics and Histograms

We use numbers to measure aspects of businesses, customers and competitors. These measured aspects are *data*. Data become meaningful when we use statistics to describe patterns within particular *samples* or collections of businesses, customers, competitors, or other entities.

Example 2.1 Yankees' Salaries: Is It a Winning Offer?

Suppose that the Yankees want to sign a promising rookie. They expect to offer $1M, and they want to be sure they are neither paying too much nor too little. What would the General Manager need to know to decide whether or not this is the right offer?

He might first look at how much the other Yankees earn. Their 2005 salaries are in Table 2.1:

Table 2.1 Yankees' salaries (in $MM) in alphabetical order

Crosby	$.3	Johnson	$16.0	Posada	$11.0	Sierra	$1.5
Flaherty	.8	Martinez	2.8	Rivera	10.5	Sturtze	.9
Giambi	1.34	Matsui	8.0	Rodriguez	21.7	Williams	12.4
Gordon	3.8	Mussina	19.0	Rodriguez F	3.2	Womack	2.0
Jeter	19.6	Phillips	.3	Sheffield	13.0		

What should he do with this data?

Data are more useful if they are ordered by the aspect of interest. In this case, the Manager would re-sort the data by salary (Table 2.2):

Table 2.2 Yankees sorted by salary (in $MM)

Rodriguez	$21.7	Williams	$12.4	Rodriguez F	$3.2	Sturtze	$.9
Jeter	19.6	Posada	11.0	Martinez	2.8	Flaherty	.8
Mussina	19.0	Rivera	10.5	Womack	2.0	Crosby	.3
Johnson	16.0	Matsui	8.0	Sierra	1.5	Phillips	.3
Sheffield	13.0	Gordon	3.8	Giambi	1.3		

C. Fraser, *Business Statistics for Competitive Advantage with Excel 2013: Basics, Model Building, Simulation and Cases*,
DOI 10.1007/978-1-4614-7381-7_2, © Springer Science+Business Media New York 2013

Now he can see that the lowest Yankee salary, the *minimum*, is $300,000, and the highest salary, the *maximum*, is $21.7M. The difference between the maximum and the minimum is the *range* in salaries, which is $21.4M, in this example. From these statistics, we know that the salary offer of $1M falls in the lower portion of this range. Additionally, however, he needs to know just how unusual the extreme salaries are to better assess the offer.

He'd like to know whether or not the rookie would be in the better paid half of the Team. This could affect morale of other players with lower salaries. The *median*, or middle, salary is $3.8M. The lower paid half of the team earns between $300,000 and $3.8M, and the higher paid half of the team earns between $3.8M and $21.7M. Thus, the rookie would be in the bottom half. The Manager needs to know more to fully assess the offer.

Often, a *histogram* and a *cumulative distribution plot* are used to visually assess data, as shown in Figs. 2.1 and 2.2. A histogram illustrates central tendency, dispersion, and symmetry.

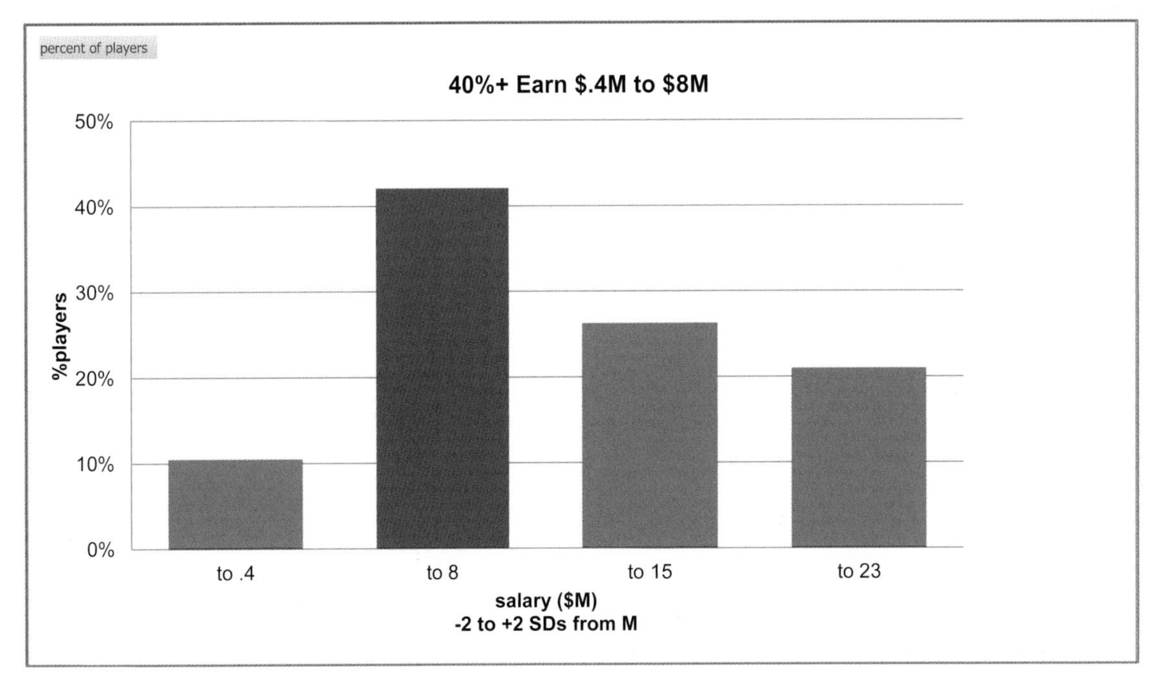

Fig. 2.1 Histogram of Yankee salaries

The histogram of team salaries shows us that a large proportion, more than 40 %, earn more than $400,000, but less than the average, or *mean*, salary of $8M.

The cumulative distribution makes it easy to see the median, or 50th percentile, which is one measure of central tendency. It is also easy to find the *interquartile range*, the range of values that the middle 50 % of the datapoints occupy, providing a measure of the data dispersion.

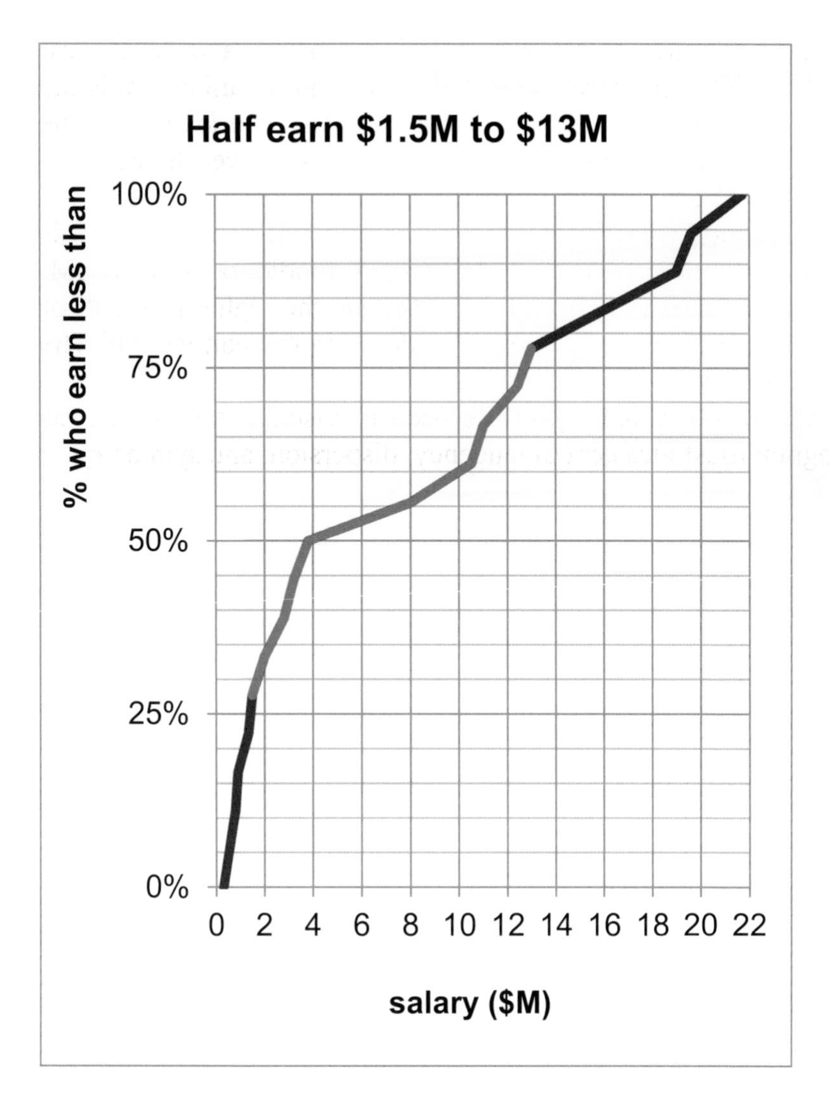

Salary ($MM)	
25 %	1.4
Median	4
75 %	13

Fig. 2.2 Cumulative distribution of salaries

The cumulative distribution reveals that the *Interquartile Range,* between the 25th percentile and the 75th percentile, is more than $10M. A quarter earns less than $1.4M, the 25th percentile, about half earn between $1.5M and $13M, and a quarter earns more than $13M, the 75th percentile. Half of the players have salaries below the *median* of $4M and half have salaries above $4M.

Example 2.2 Executive Compensation: Is the Board's Offer on Target?

The Board of a large corporation is pondering the total compensation package of the CEO, which includes salary, stock ownership, and fringe benefits. Last year, the CEO earned $2,000,000. For comparison, The Board consulted Forbes' summary of the total compensation of the 500 largest corporations. The histogram, cumulative frequency distribution and descriptive statistics are shown in Figs. 2.3 and 2.4.

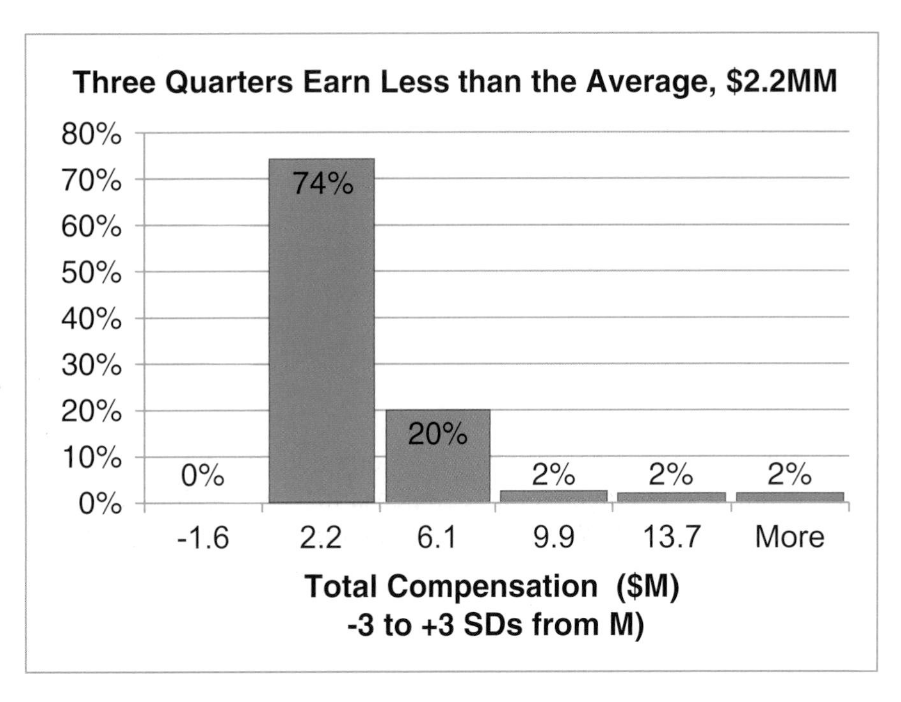

Total compensation (−1 to +3 sds from M)	*% Executives*
2	74 %
6	20 %
10	2 %
14	2 %
More	2 %

Fig. 2.3 Histogram of executive compensation

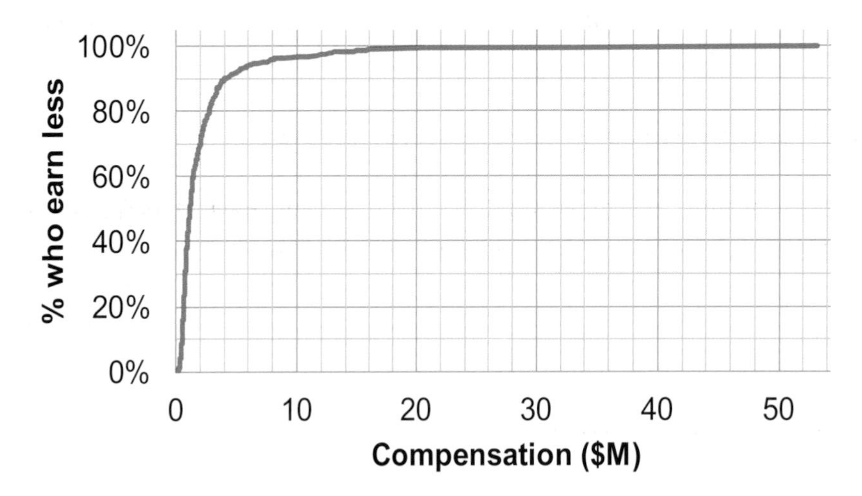

Total compensation ($M)	
M	2.2
SD	3.8
75th percentile	2.3
Median	1.1
25th percentile	.7

Fig. 2.4 Cumulative distribution of total compensation

The average executive compensation in this sample of large corporations is $2.2M. Half the sample of 447 executives earns $1.1M (the median) or less. One quarter earns less than $.7M, the middle half, or *interquartile range,* earns between $.7M and $2.3M, and one quarter earns more than $2.3M.

2.2 Round Descriptive Statistics

In the examples above, statistics in the output from statistical packages are presented with many decimal points of accuracy. The Yankee manager in Example 2.1 and The Board considering executive compensation in Example 2.2 will most likely be negotiating in hundred thousands. It would be distracting and unnecessary to report descriptive statistics with significant digits more than two or three. In the **Yankees** example, the average salary is $8,000,000 (*not* $7,797,000). In the **Executive Compensation** example, average total compensation is $2,200,000 (not $2,215,262.66). It is deceptive to present results with many significant digits, creating an illusion of precision. In addition to being honest, statistics in two or three significant digits are much easier for decision makers to process and remember. If more significant digits don't affect a decision, round to fewer and make your statistics easier to process and remember.

2.3 Share the Story that Your Graphics Illustrate

Use your graphics to support the conclusion you have reached from your analysis. Choose a "bottom line" title that shares with your audience what it is that they should be able to see. Often this title should relate specifically to your reasons for analyzing data. In the executive compensation example, The Board is considering a $2M offer. The chart titles capture Board interest by highlighting this critical value. The "bottom line," that a $2M offer is relatively high, when compared with similar firms, makes the illustrations relevant.

Many have the unfortunate and unimaginative habit of choosing chart titles which name the type of chart. "Histogram of executive salaries" tells the audience little, beyond the obvious realization that they must form their own, independent conclusions from the analysis. Choose a "bottom line" title so that decision makers can take away your conclusion from the analysis. Develop the good habit of titling your graphics to enhance their relevance and interest.

2.4 Data Is Measured with Quantitative or Categorical Scales

If the numbers in a dataset represent amount, or magnitude of an aspect, **and** if differences between adjacent numbers are equivalent, the data are *quantitative* or *continuous*. Data measured in dollars (i.e., revenues, costs, prices and profits) or percents (i.e., market share, rate of return, and exam scores) are continuous. Quantitative numbers can be added, subtracted, divided or multiplied to produce meaningful results.

With quantitative data, report central tendency with the *mean, M:*

$$\mu = \frac{\sum x_i}{N}$$ for describing a *population* and

$$\overline{X} = \frac{\sum x_i}{N} \quad \text{for describing a } \textit{sample} \text{ from a population,}$$

where x_i are data point values, and

N is the number of data points that we are describing.

The *median* can also be used to assess central tendency, and the *range, variance,* and *standard deviation* can be used to assess dispersion.

The *variance* is the average squared difference between each of the data points and the mean:

$$\sigma^2 = \frac{\sum (x_i - \mu)^2}{N} \quad \text{for a population and}$$

$$s^2 = \frac{\sum (x_i - \overline{X})^2}{(N-1)} \quad \text{for a sample from a population.}$$

The *standard deviation SD,* σ for a population and s for a sample, is the square root of the variance, which gives us a measure of dispersion in the more easily interpreted, original units, rather than squared units.

To assess distribution symmetry, assess its skewness:

$$\text{Skewness} = \frac{n}{(n-1)(n-2)} \sum \left(\frac{x_i - \overline{x}}{s} \right)^3.$$

Skewness of zero indicates a symmetric distribution, and skewness between −1 and +1 is evidence of an approximately symmetric distribution.

If numbers in a dataset are arbitrary and used to distinguish categories, the data are *nominal,* or *categorical.* Football jersey numbers and your student ID are nominal. A larger number doesn't mean that a player is better or a student is older or smarter. Categorical numbers can be tabulated to identify the most popular number, occurring most frequently, the *mode,* to report central tendency. Categorical numbers cannot be added, subtracted, divided or multiplied.

Quantitative measures convey the more information, including direction and magnitude, while categorical measures convey the less, sometimes direction, and sometimes, merely category membership. One, more informative type of categorical data are *ordinal* scales that used to rank order data, or to convey direction, but not magnitude. With ordinal data, an element (which could be a business, a person, a country) with the most or best is coded as '1', second place as '2', etc. With ordinal numbers, or rankings, data can sorted, but not added, subtracted, divided or multiplied. As with other categorical data, the mode represents the central tendency of ordinal data.

When focus is on membership in a particular category, the *proportion* of sample elements in the category is a continuous measure of central tendency. Proportions are quantitative and can be added, subtracted, divided or multiplied, though they are bounded by zero, below, and by one, above.

2.5 Continuous Data Are Sometimes Normal

Continuous variables are often *Normally distributed*, and their histograms resemble symmetric, bell shaped curves, with the majority of data points clustered around the mean. Most elements are "average" with values near the mean; fewer elements are unusual and far from the mean.

Skewness reflects lack of symmetry. Normally distributed data have skewness of zero, and approximately Normal data have skewness between −1 and +1.

If continuous data are Normally distributed, we need only the mean and standard deviation to describe this data and description is simplified.

Example 2.3 Normal SAT Scores

Standardized tests, such as SAT, capitalize on Normality. Math and verbal SATs are both specifically constructed to produce Normally distributed scores with *mean* M = 500 and *standard deviation* SD = 100 over the population of students (Fig. 2.5):

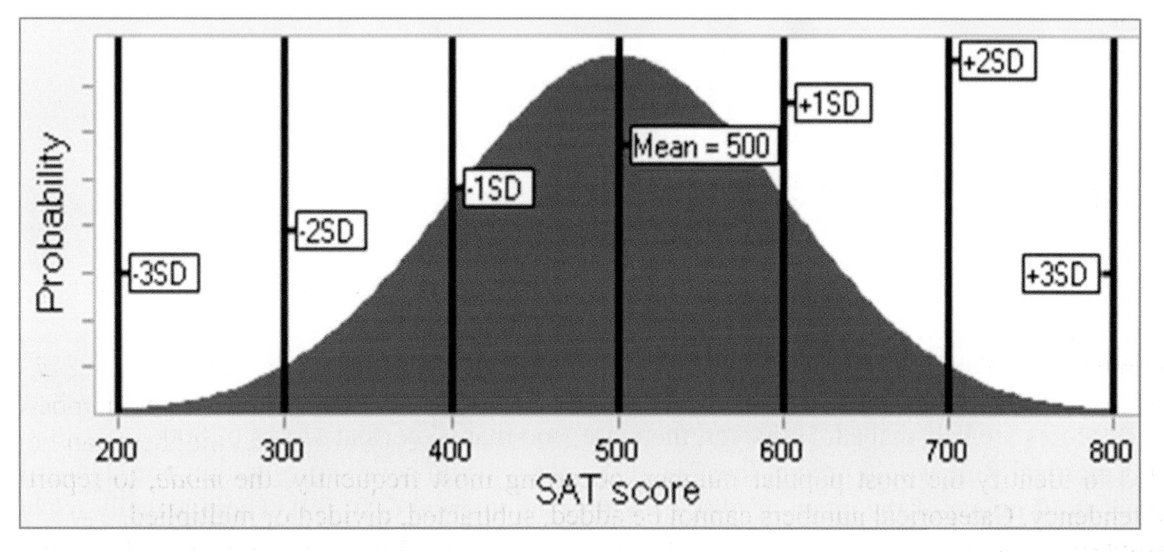

Fig. 2.5 Normally distributed SAT scores

2.6 The Empirical Rule Simplifies Description

Normally distributed data have a very useful property described by the *Empirical Rule*:

- 2/3 of the data lie within one standard deviation of the mean
- 95 % of the data lie within two standard deviations of the mean

This is a powerful rule! *If data are Normally distributed, data can described with just two statistics: the mean and the standard deviation.*

Returning to SAT scores, if we know that the average score is 500 and the standard deviation is 100, we also know that

- 2/3 of SAT scores will fall within 100 points of the mean of 500, or between 400 and 600,
- 95 % of SAT scores will fall within 200 points of the mean of 500, or between 300 and 700.

Example 2.4 Class of '10 SATs: This Class Is Normal & Exceptional

Descriptive statistics and a histograms of Math SATs of a third year class of business students reveal an interquartile range from 640 to 730, with mean of 690 and standard deviation of 70, as shown in Fig. 2.6. Skewness is −.5, indicating approximate symmetry, an approximately Normal distribution.

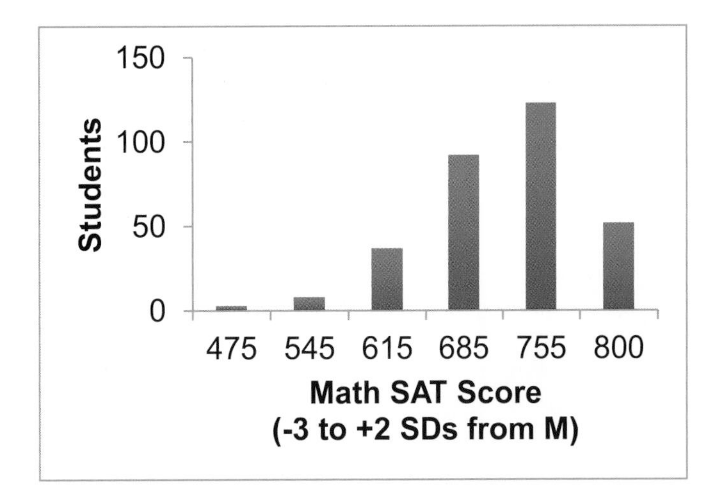

95 %	800
90 %	790
75 %	730
Median	690
25 %	640
10 %	600
5 %	560
Mean	690
sd	70
Skewness	−.5

Fig. 2.6 Histograms and descriptive statistics of class '06 math SATs

Class '10 scores are bell shaped. However, there are "too many" perfect scores of 800.

The Empirical Rule would predict that 2/3 of the class would have scores within one standard deviation of 70 points of the mean of 690, or within the interval 620–760. There actually 67 % (=37 % + 30 %).

The Empirical Rule would also predict that only 2–1/2 % of the class would have scores more than two standard deviations below or above the mean of 690: scores below 550 and above 830. We find that 4 % actually do have scores below 530, though none score above 830 (since a perfect SAT score is 800). This class of business students has Math SATs that are nearly Normal, but not exactly Normal.

To summarize Class '10 students' SAT scores, report:

- Class '10 students' Math SAT scores are approximately normally distributed with *mean* of 690 and *standard deviation* of 70.
- Relative to the larger population of all SAT takers, the smaller *standard deviation* in Class '10 students' Math SAT scores, 70 versus 100, indicates that Class '06 students are a more homogeneous group than the more varied population.

2.7 Outliers Can Distort the Picture

Outliers are extreme elements, considered unusual when compared with other sample elements. Because they are extraordinary, they can distort descriptive statistics.

Revisiting the **Executive Compensation** example, why is the *mean*, $2.2M, so much larger than the *median*, $1.1M? There is a group of eight *outliers*, shown as *MORE* than three standard deviations above the mean in Fig. 2.3, who are compensated extraordinarily well. Each collects a compensation package of more than $14M, a compensation level that is more than three standard deviations greater than the mean.

When we exclude these eight outliers, eleven additional outliers emerge. This cycle repeats, since the distribution is highly skewed. When we remove outliers, the new mean is adjusted, making other executives appear to be more extreme. After removing about ten percent, or the 44 best compensated executives, we see a clearer picture of what "typical" compensation is, shown in Fig. 2.7:

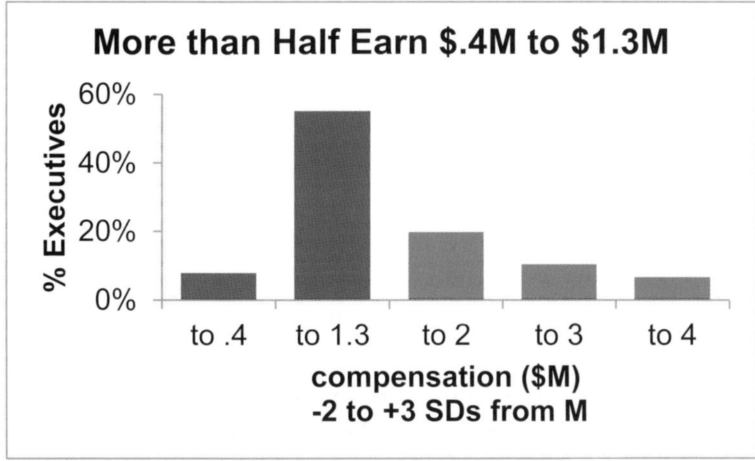

Total compensation ($MM) −1 to +3 SDs from M	*% Executives*
≤.4	8 %
.5 to 1.3	55 %
1.4 to 2.3	20 %
2.4 to 3.2	10 %
3.3 to 4.1	7 %

Fig. 2.7 Histogram and descriptive statistics with 44 outliers excluded

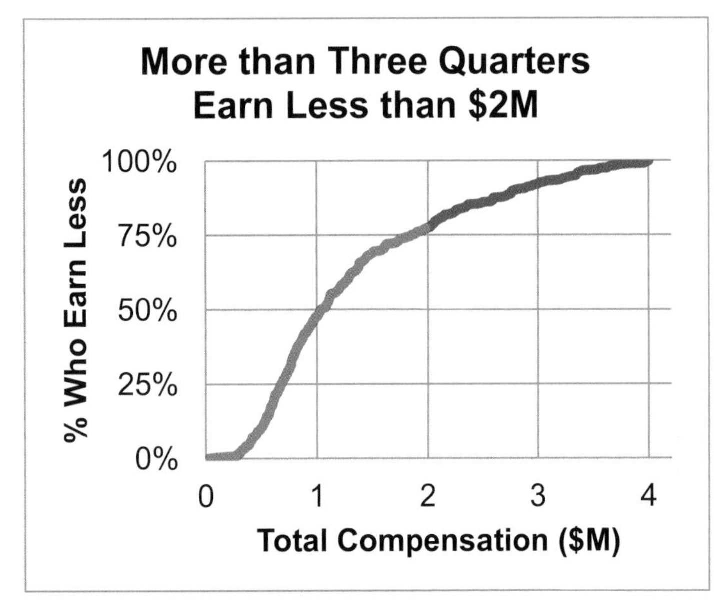

Total compensation ($M)	
M	1.3
SD	.9
75th percentile	1.8
Median	1.0
25th percentile	.7

Fig. 2.8 Cumulative distribution of total compensation

Ignoring the 44 outliers, the average compensation is about $1.3M, and the *median* compensation is about $1M, shown in Fig. 2.8. The *mean* and *median* are closer. With this more representative description of executive compensation in large corporations, The Board has an indication that the $2M package is well above average. More than three quarters of executives earn less. Because extraordinary executives exist, the original distribution of compensation is *skewed*, with relatively few exceptional executives being exceptionally well compensated.

2.8 Central Tendency, Dispersion and Skewness Describe Data

The baseball salaries and executive compensation examples focused on two measures of *central tendency*: the *mean*, or average, and the *median*, or middle. Both examples also refer to a measure of *dispersion* or variability: the *range* separating the minimum and maximum. *Skewness* reflects distribution symmetry. SATs were approximately symmetric and Normal; Executive compensation values were skewed, until outliers were removed. To describe data, we need statistics to assess central tendency, dispersion, and skewness. The statistics we choose depends on the *scale* which has been used to code the data we are analyzing.

2.9 Describe Categorical Variables Graphically: Column and PivotCharts

Numbers representing category membership in nominal, or categorical, data are described by tabulating their frequencies. The most popular category is the *mode*. Visually, we show our tabulations with a *Pareto* chart, which orders categories by their popularity.

Example 2.5 Who Is Honest & Ethical?

Figure 2.9 shows a column chart of results of a survey of 1,014 adults by Gallup:

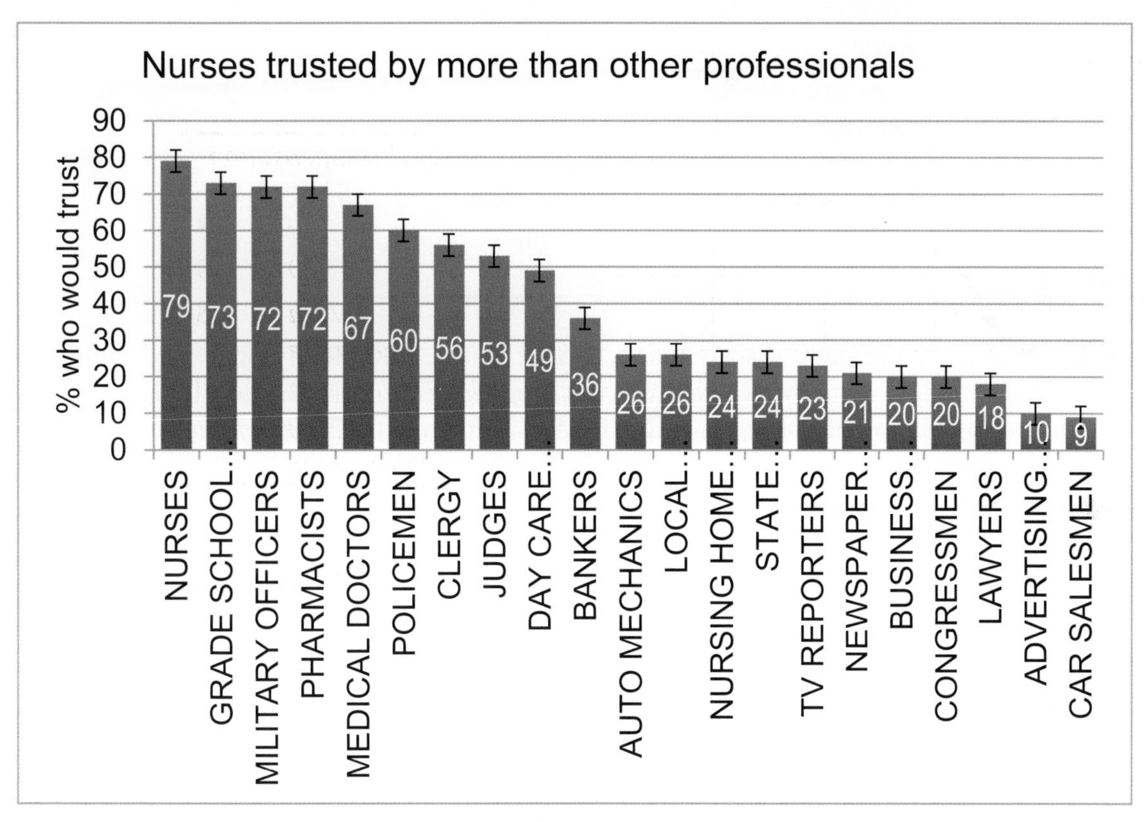

Fig. 2.9 Pareto charts of the percents who judge professions honest

More Americans trust and respect nurses (79 %, the *modal* response) than people in other professions, including doctors, clergy and teachers. Though a small minority judge business executives (20 %) and advertising professionals (10 %) as honest and ethical, most do not judge people in those fields to be honest (which highlights the importance of ethical business behavior in the future).

2.10 Descriptive Statistics Depend on the Data and Rely on Your Packaging

Descriptive statistics, graphics, central tendency and dispersion, depend upon the type of scale used to measure data characteristics (i.e., quantitative or categorical).

Table 2.3 summarizes the descriptive statistics (graph, central tendency, dispersion, shape) used for both types of data:

Table 2.3 Descriptive statistics (central tendency, disperson, graphics) for two types of data

	Quantitative	Categorical
Central tendency	*Mean*	*Mode*
	Median	*Proportion*
Dispersion	*Range*	
	Standard deviation	
Symmetry	*Skewness*	
Graphics	*Histogram*	*Pareto chart*
	Cumulative distribution	*Pie chart*
		Column chart

If continuous data are normally distributed, a dataset can be completely described with just the mean and standard deviation. We know from the *Empirical Rule* that 2/3 of the data will lie within one standard deviation of the mean and that 95 % of the data will lie within two standard deviations of the mean.

Effective results are those which are remembered and used to improve decision making. Your presentation of results will influence whether or not decision makers remember and use your results. Round statistics to two or three significant digits to make them honest, digestible, and memorable. Title your graphics with the "bottom line," to guide and facilitate decision makers' conclusions.

Excel 2.1 Produce Descriptive Statistics

Executive Compensation

We will describe executive compensation packages by producing descriptive statistics, a histogram and cumulative distribution.

First, freeze the top row of **Excel 2.1 Executive Compensation.xls** so that column labels are visible when you are at the bottom of the dataset. Select the first cell, **A1,** and then use Excel shortcuts **Alt WFR**. (The shortcuts, activated with **Alt** select the vie**W** menu, the **F**reeze panes menu, and then freeze **R**ows.)

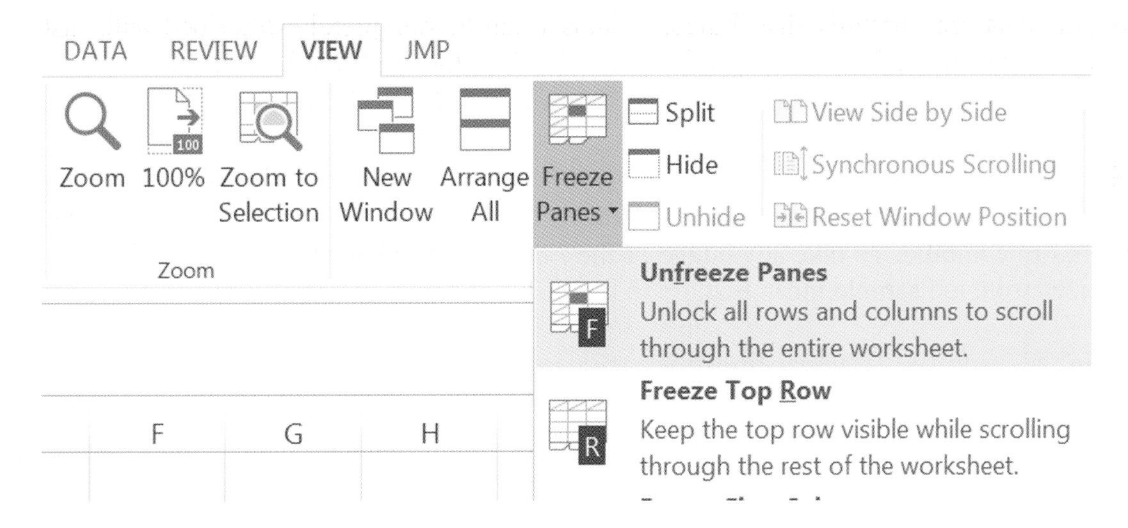

Use shortcuts to move to the end of the file where we will add descriptive statistics. **Cntl+down arrow** scrolls through all cells containing data in the same column and stops at the last filled cell.

Descriptive statistics. In the first empty cell in the column, below the data, use shortcuts to find the sample mean: Alt **MUA**.

Use the:

STDEV.S(*array*) function to find the standard deviation,

PERCENTILE.INC(*array*, .75) and **PERCENTILE.INC(*array*, .25)** to find the 75th and 25th percentile values,

MEDIAN(*array*) function to find the median, and

SKEW(*array*) function to find skewness:

B454 =SKEW(B2:B448)

	A	B	C
1		Total Compensation (MM$)	
448		53.1	
449	M	2.2	
450	SD	3.8	
451	75%	2.3	
452	median	1.1	
453	25%	0.72	
454	skew	7.4	

Set up Histogram Bins. To make a histogram of salaries, Excel needs to know what ranges of values to combine. To take advantage of the *Empirical Rule,* create *bins*, or categories, using differences from the sample mean that are in widths of standard deviations.

Excel uses bin values to set the upper limit for each category. Start with a bin with upper limit equal to the mean, which will include compensation values that are at less than or equal to the mean. Specify the cell containing the mean rather than typing in the mean value:

= B449

This will be the first bin, since subtracting one standard deviation from the mean produces a negative number, and none of the executives earns negative salary dollars.

C449 =B449

	A	B	C	D
1		Total Compensation (MM$)		
448		53.1	total compensation (MM$) bin	
449	M	2.2	2.2	≤ M

In each of the three cells below this first bin, add one SD to the cell above, creating bins with upper limits of

M + 1SD,
M + 2SD and
M + 3SD.

(Lock the cell reference to the SD in B450 by pressing fn4.)

T.IN... | =C451+$B$450

	A	B	C	D
1		Total Compensation (MM$)		
448		53.1	total compensation (MM$) bin	
449	M	2.2	2.2	≤M
450	SD	3.8	6.1	M< ≤M+SD
451	75%	2.3	9.9	M+SD< ≤M+2SD
452	median	1.1	=C451+$B$450	M+2SD< ≤M+3SD

Excel 2.2 Sort to Produce Descriptives Without Outliers

Outliers are executives whose total compensation is more than three standard deviations greater than the mean.

To easily identify and remove outliers, sort the rows from lowest to highest *total compensation ($M)*:

Select *total compensation* data in column **B** (but not statistics below the data), then use shortcuts to sort:

Alt ASA, Continue with the current selection, Sort. (**A** selects the dat**A** menu, **S** selects the Sort menu, and **A** specifies **A**scending.)

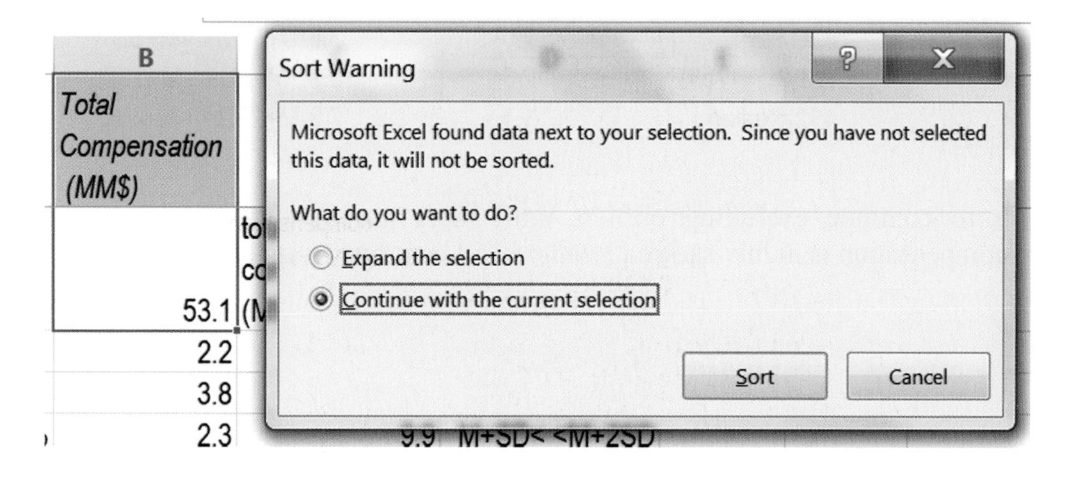

Scroll up from the end of **B** to identify the rows with outlier values more than 13.7:

	A	B
1		*Total Compensation (MM$)*
440		13.1
441		14.7
442		14.9
443		15.7
444		15.9
445		16.2
446		20.7
447		32.6
448		53.1

Recalculate the mean and standard deviation, including only rows with *total compensation* less than 14 million, by changing the end of the array in both Excel functions, which will update your bin upper limits:

B450 =STDEV.S(B2:B440)

	A	B	C	D
1		*Total Compensation (MM$)*		
448		53.1	total compensation (MM$) bin	
449	M	1.8	1.8	≤M
450	SD	2.0	3.8	M< ≤M+SD
451	75%	2.3	5.9	M+SD< ≤M+2SD
452	median	1.1	7.9	M+2SD< ≤M+3SD

Repeat this process to continue excluding outliers until there are no outliers. Since the distribution of total compensation is highly skewed, *outliers* will continue to appear.

Update the 75 % percentile, median, 25th percentile, and skewness.

B454 =SKEW(B2:B404)

	A	B	C	D
1		*Total Compensation (MM$)*		
447		32.6		
448		53.1	total compensation (MM$) bin	
449	M	1.3	1.3	≤M
450	SD	0.9	2.3	M< ≤M+SD
451	75%	1.8	3.2	M+SD< ≤M+2SD
452	median	1.0	4.1	M+2SD< ≤M+3SD
453	25%	0.68		
454	ske	1.1		

Including only executives whose total compensation is less than $4.1 million, the descriptive statistics are more representative.

Excel 2.3 Make a Histogram

Excluding outliers, there are now executives whose compensation is more than one standard deviation beneath the mean. Add a histogram bin M−SD:

T.IN... =B449-B450

	A	B	C	D
1		*Total Compensation (MM$)*		
447		32.6		
448		53.1	total compensation (MM$) bin	
449	M	1.3	=B449-B450	≤M-SD
450	SD	0.9	1.3	M-SD< ≤ M

To see the distribution of *Total Compensation*, use shortcuts to request a histogram:

Alt AY3 H Enter (**Alt AY3** selects the dat**A** menu and the data Anal**Y**sis menu.)

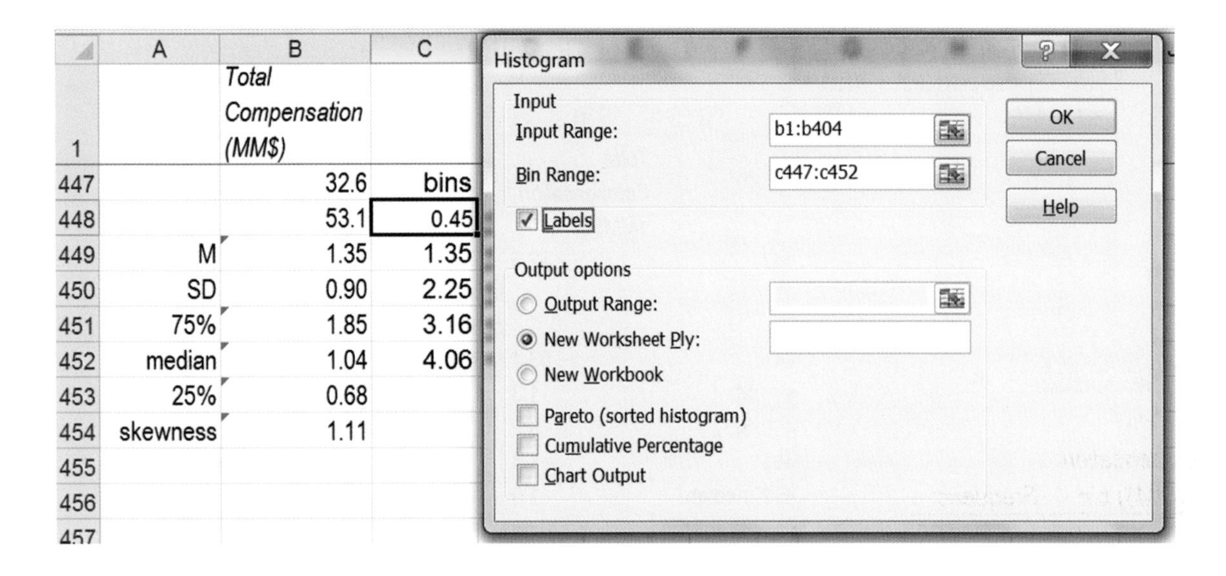

For **Input Range**, enter the *total compensation* cells, including the label, excluding outliers; for **Bin Range**, enter the *total compensation bin* cells, including the label, and choose **L**abels, **OK**.

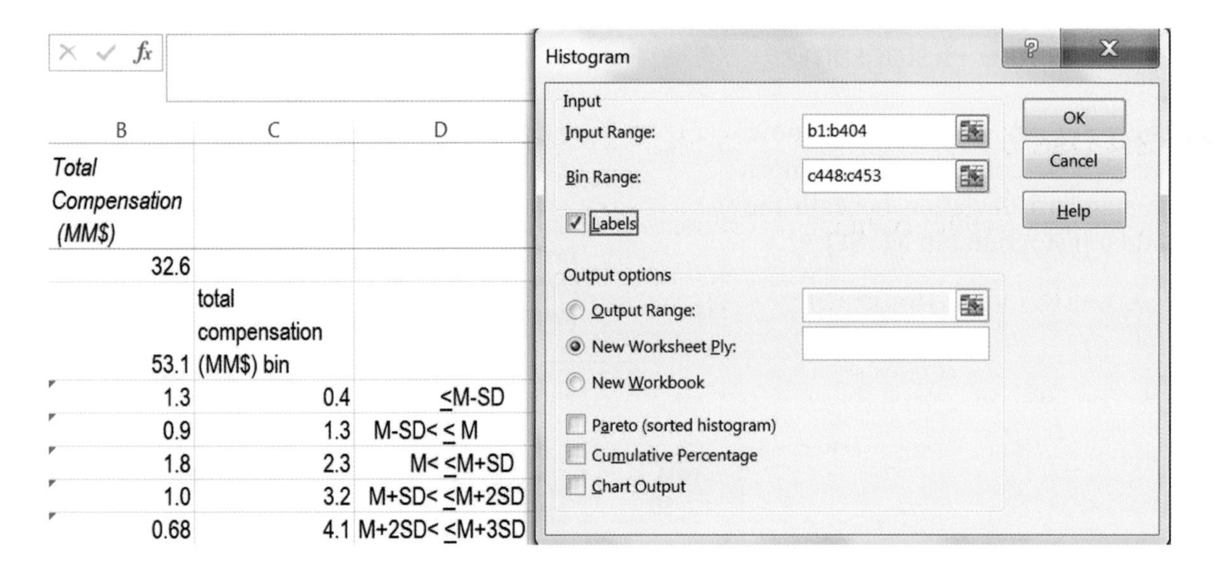

Select the bins cells with more than two significant digits and use shortcuts **Alt H 9** to reduce the unnecessary decimals. (**H** selects the **H**ome menu and **9** selects the reduce decimals function of the Number menu.)

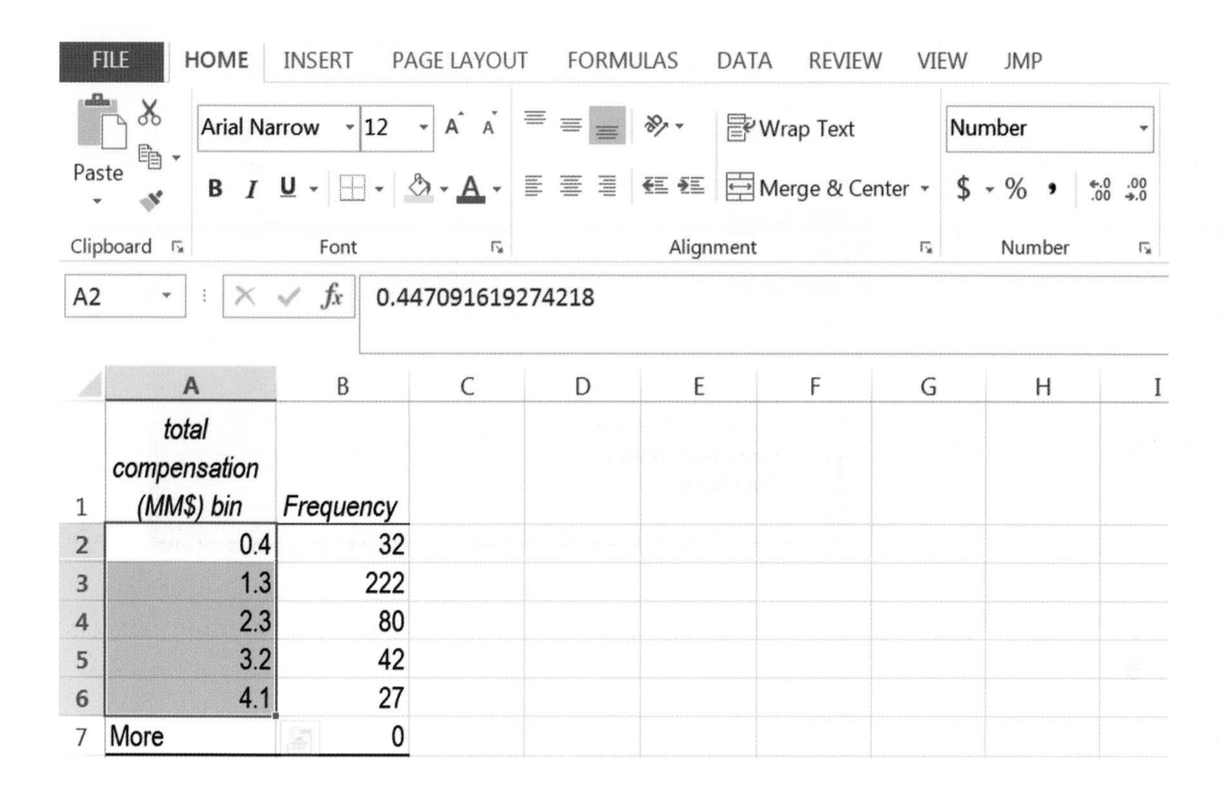

Excel 2.4 Use a PivotTable to Plot the Distribution

Select the histogram table, excluding the More row, and use shortcuts to produce a PivotTable:

Alt N V Enter (Shortcuts for i**N**sert pi**V**ot.)

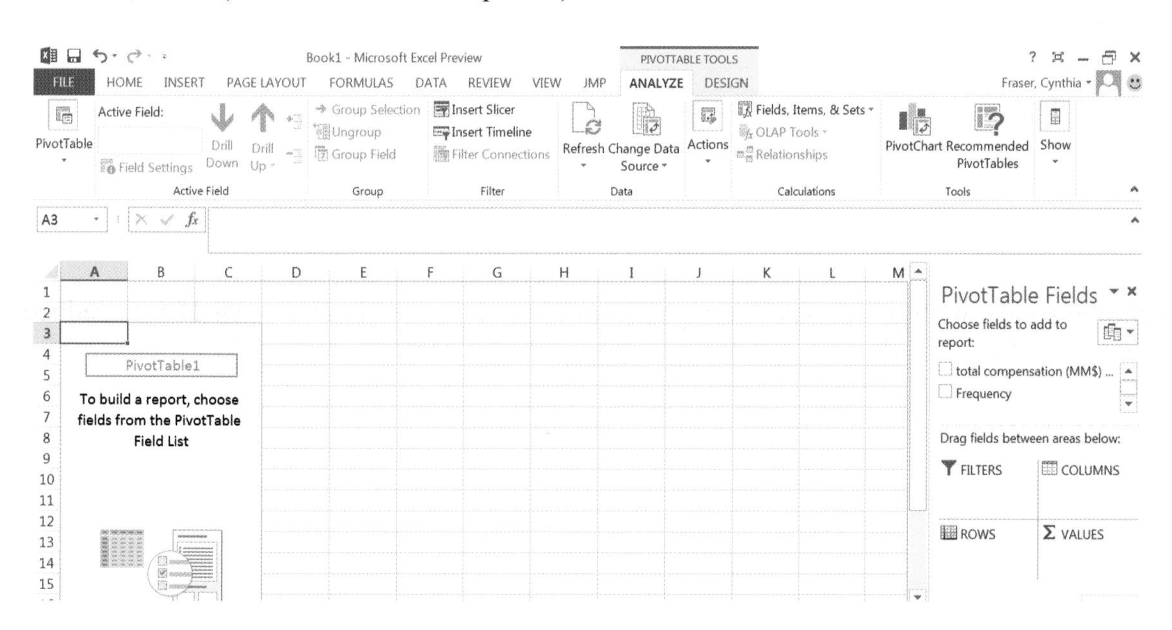

Drag *total compensation bins* to ROWS and drag *frequency* to the ∑ VALUES:

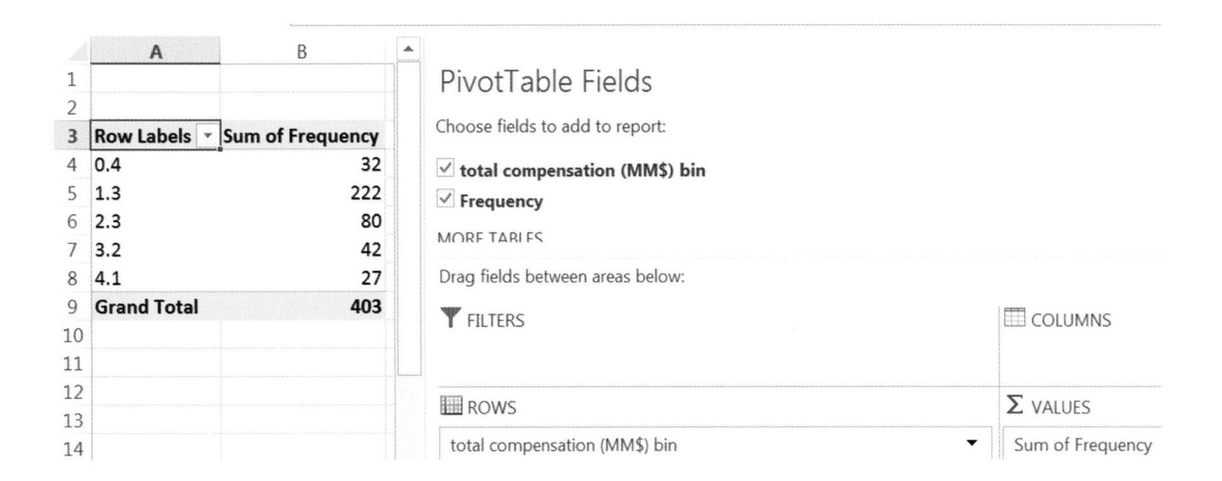

From a cell in the *Sum of Frequency* column, use shortcuts to change *Frequency* to percents:

Alt JT G Tab > Tab dn to %*Grand Total* **Enter**

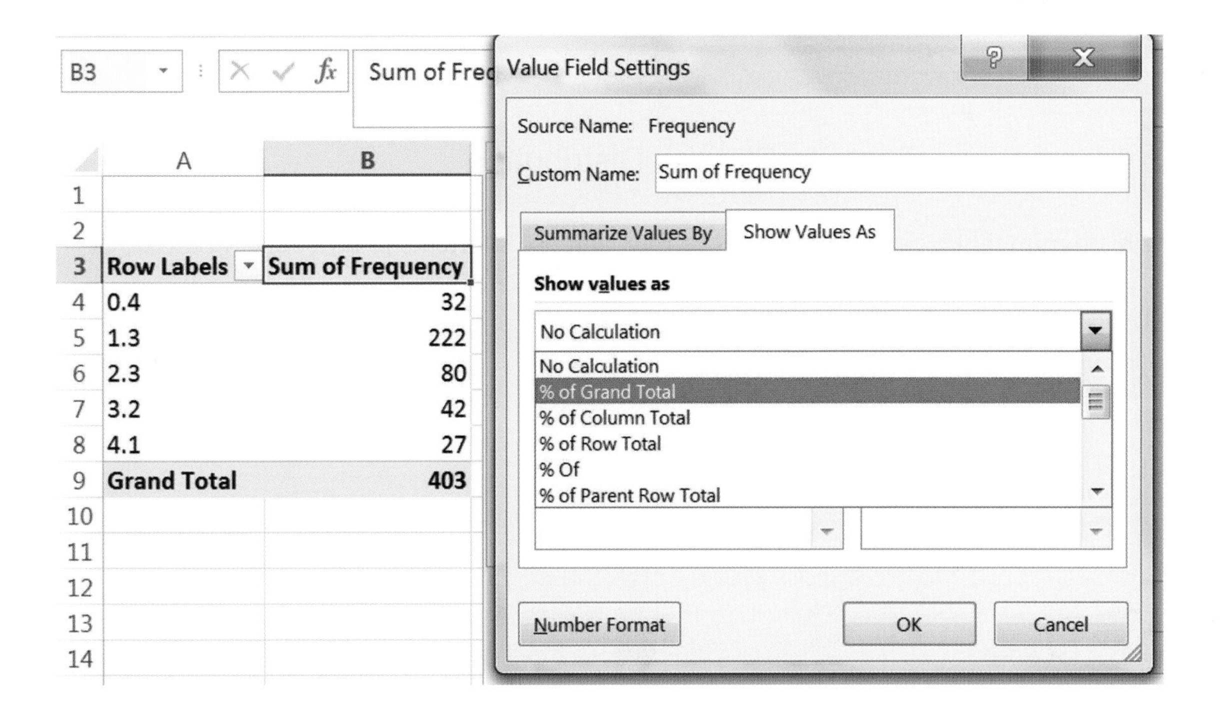

Reduce decimals.

Use shortcuts to make a PivotChart of the distribution:

Alt JT C Enter

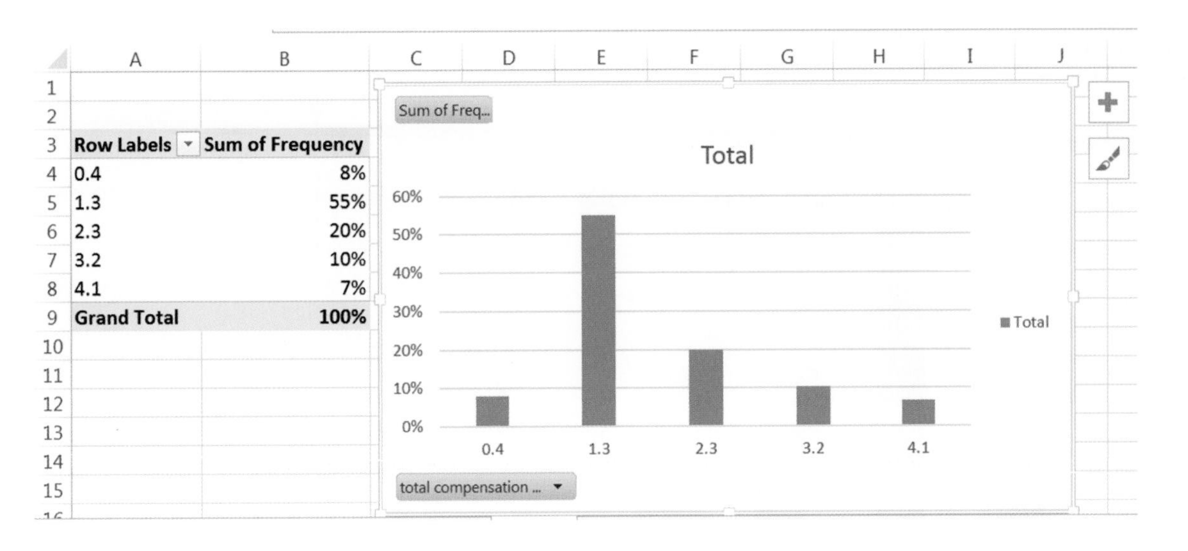

Use shortcuts to choose the ninth layout, which will add axes labels and a title:

Alt JC L Enter

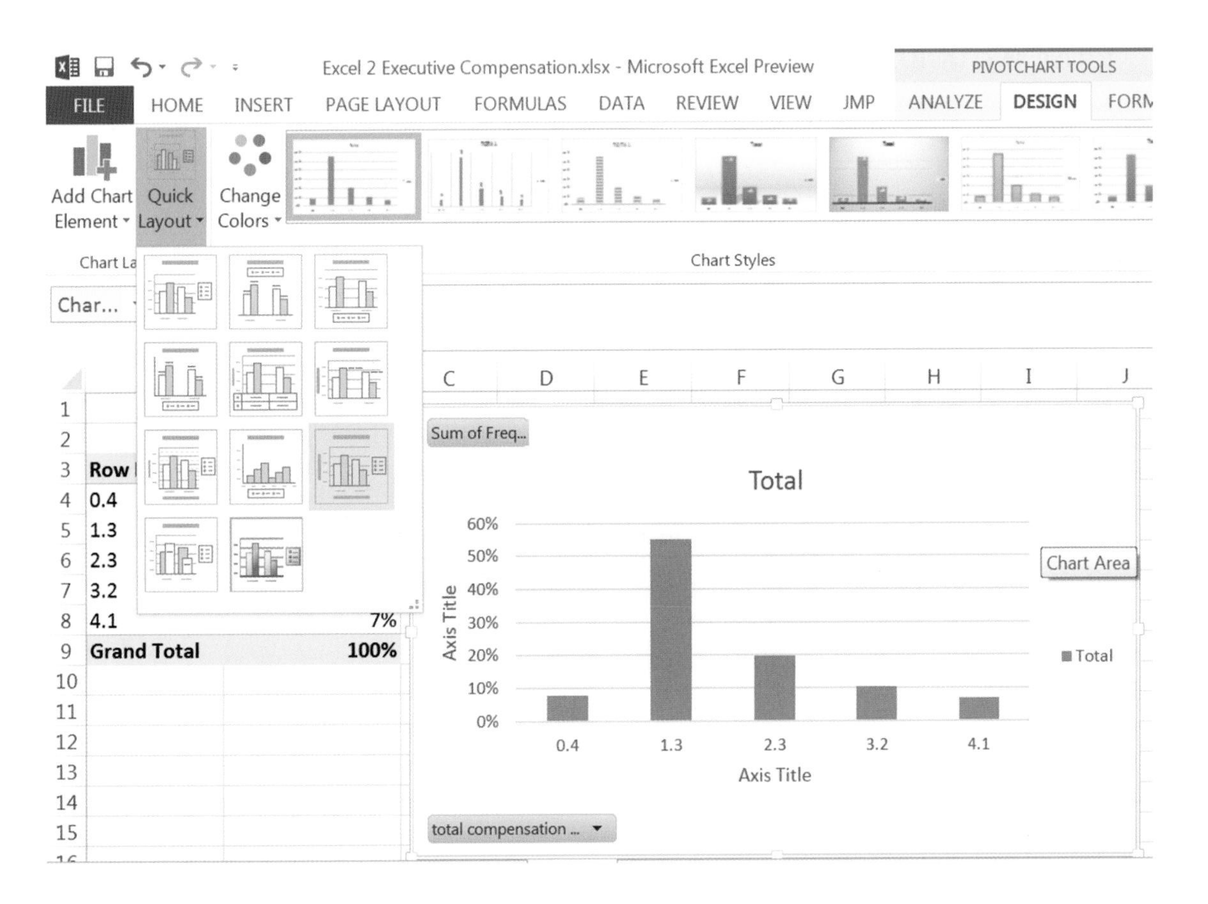

Type in axes labels and title, delete the legend, and use shortcuts to set font size at 12:

Alt H F S 12 Enter

(where shortcuts stand for **H**ome **F**ont **S**ize **12**)

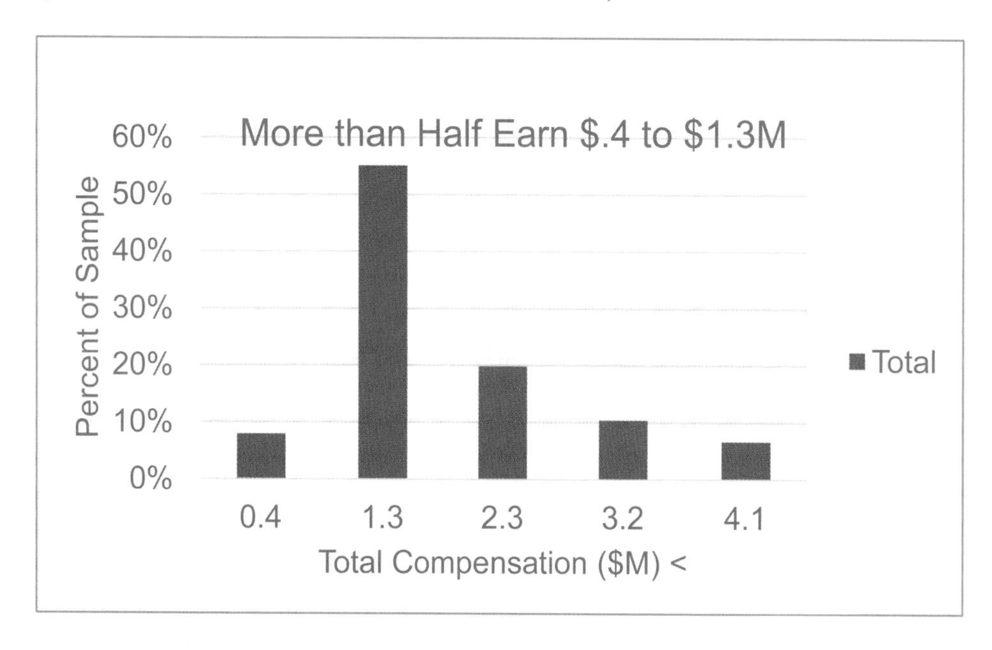

Excel 2.5 Plot a Cumulative Distribution

Return to the data sheet:

CNTL+Page Dn Page Dn

To see the cumulative distribution of *total compensation*, choose **Rank and Percentile, Alt AY3, R dn Enter**

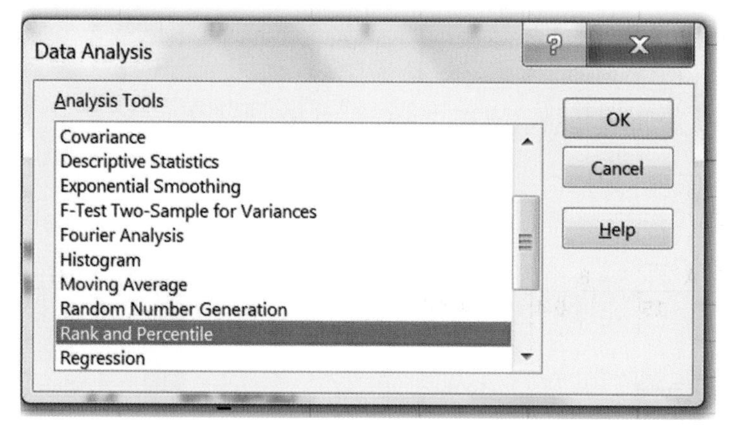

Enter *total compensation* cells, including the label, excluding outliers, Label, **Enter**:

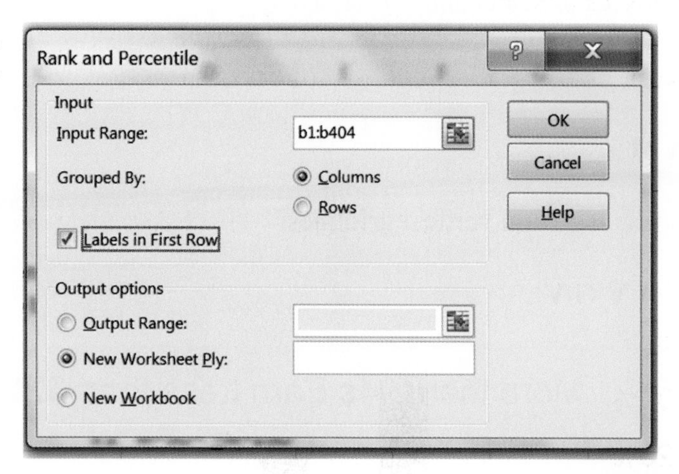

For convenience, select and delete column **C**, **Alt HDC**. (**H** selects the Home menu, **D** selects the Delete menu, and **C** deletes the Column.)

	A	B	C
1	*Point*	*mpensatio*	*Percent*
2	403	4.0	100.00%
3	402	4.0	99.70%
4	401	4.0	99.50%
5	400	4.0	99.20%

Reduce decimals in the *Percent* column **C**.

Select *Total Compensation* in **B** and *Percent* in **C** and then use shortcuts to create a scatterplot of the cumulative distribution, **Alt N D**

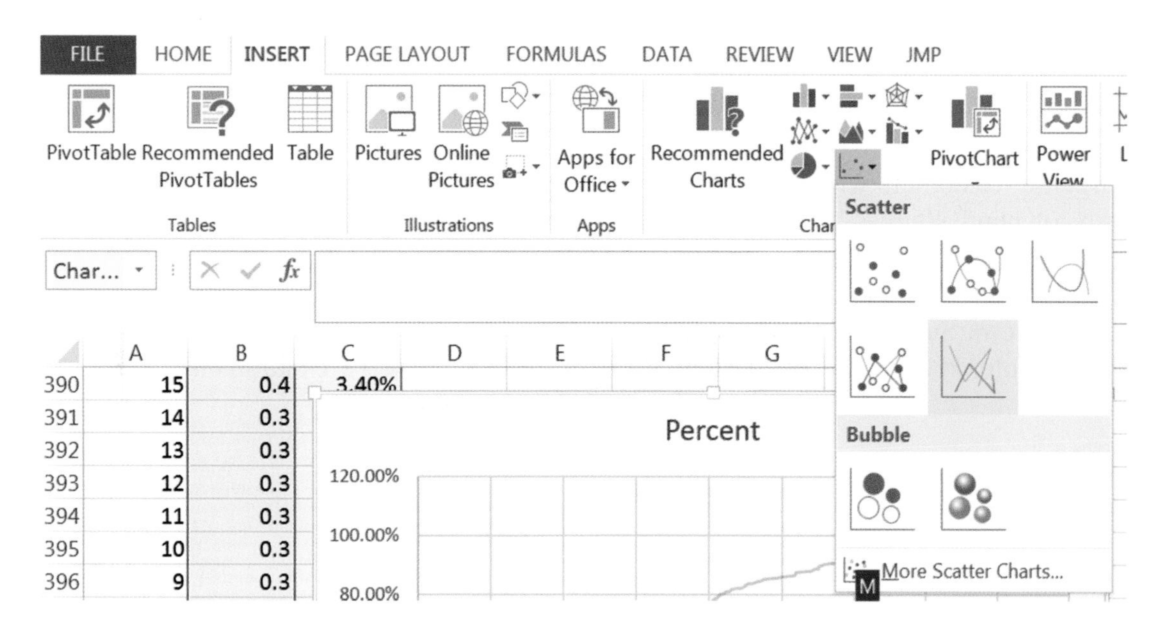

Use shortcuts to choose design layout 1 to add axes labels and title. Delete the legend and adjust font size to 12.

Use shortcuts to select and format axes:

Alt JA E

Alt JA M

Use shortcuts to add vertical gridlines:

Alt JC A V G V

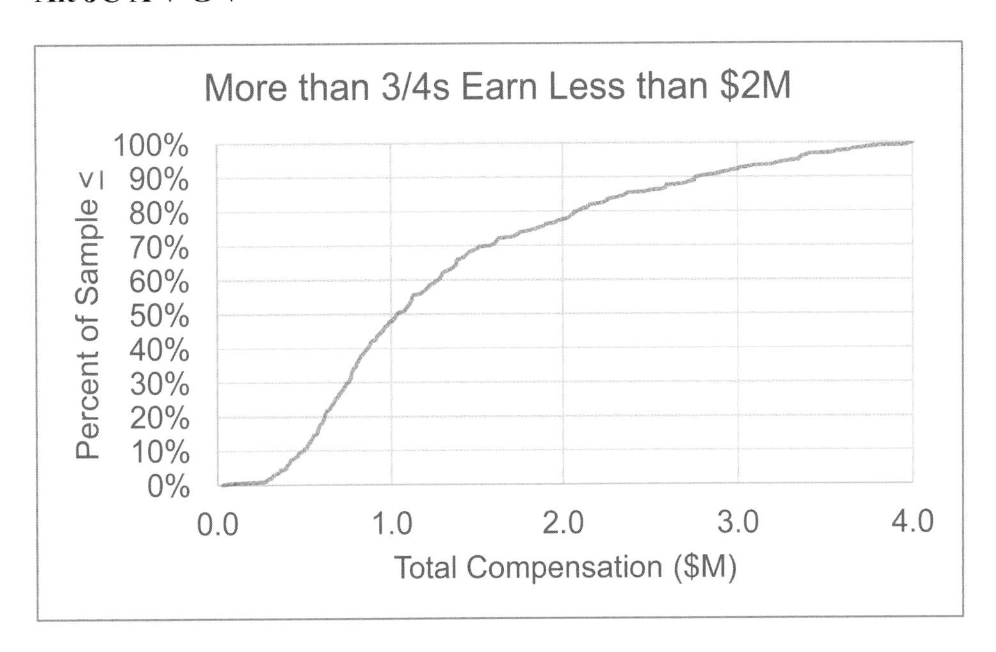

Excel 2.6 Produce a Column Chart of a Nominal Variable

A firm is targeting customers who consult a news source daily. Management wants to compare the popularity of news sources. To facilitate comparisons, we will make a PivotChart from a Gallup Poll of 992 Americans. Data are in **Excel 2.2 News Sources.xls**.

Open **Excel 2.2 News Sources.xls**, select the *News Source* and *% Who Get Daily News Here* data, and insert a column chart **Alt NC**. (where shortcuts activate i**N**sert a **C**olumn chart.)

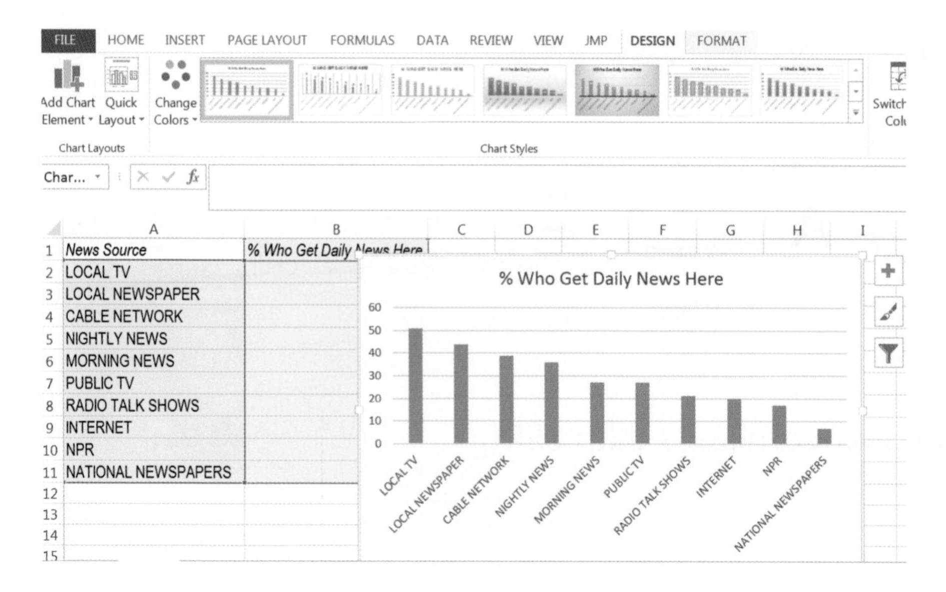

To add vertical margin of error bars, using click inside a column, then use shortcuts

Alt JC A E M

Then enter

Fixed: 3

(where 3 is the approximate margin of error).

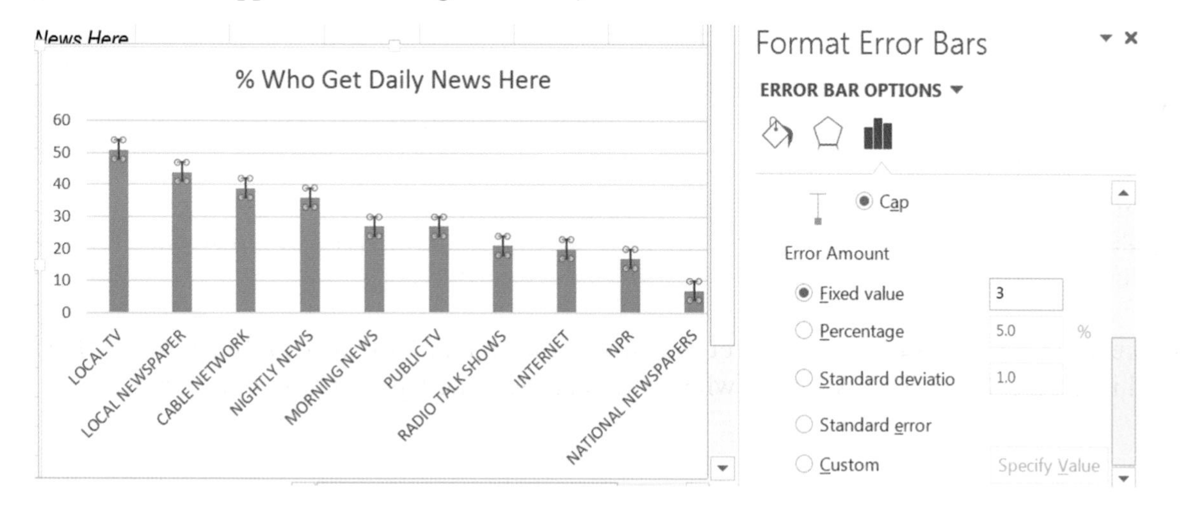

Choose **Design Chart Layout 9**,

Type in a stand alone title and axes titles and add data labels:

Alt JC A D O

Adjust chart height so that all of the *News Source* labels show:

Alt JA H up arrow

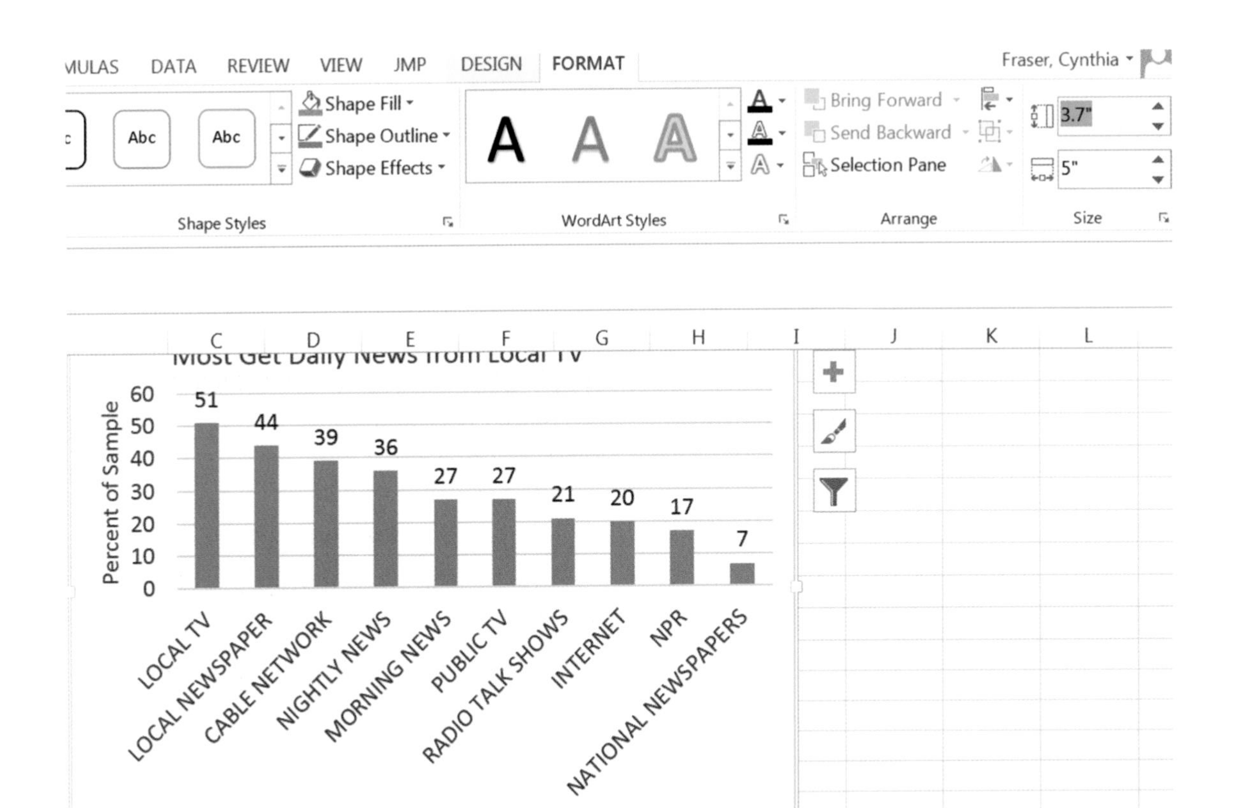

Excel Shortcuts at Your Fingertips

To navigate

to the end of the data
- Cntl+Dn

To select

all of the data
- A1
- Cntl+Shift+right >
- Cntl+Shift+Dn

a column
- Cntl+space bar

a row
- Shift+space bar

To do something with the data

Sort
- Alt A S S
- Tab to Sort by
- Dn to Salary (M$)

Find
- Sample mean from the end of an array
  - Alt M U A
- Standard deviation
  - =STDEV.S(array)
- 25 %
  - =PERCENTILE.INC(array,.25)
- 75 %
  - =PERCENTILE.INC(array,.75)
- Median
  - =MEDIAN(array)
- Skewness
  - =SKEW(array)

Make a histogram
- Alt AY3
- H Enter
- *array* Tab
- *array* Tab
- L Enter

Reduce decimals
- Select data in column A
- Alt H 9

Make a PivotTable
- Alt N V
- Change from counts to percents
  - Alt JT G Tab > Tab to % Grand total
- Make a PivotChart
  - Alt JT C Enter

Find the cumulative distribution
Alt AY3 R down Enter
array Tab
L Enter
Delete a column
Alt H D C
Plot the cumulative distribution
Select values and cumulative percents
Alt N D

Make a column chart
Select categories and values or percents
Alt N C

Alt activates shortcuts menus, linking keyboard letters to Excel menus. Press **Alt,** then release and press letters linked to the menus you want.

Alt Home:

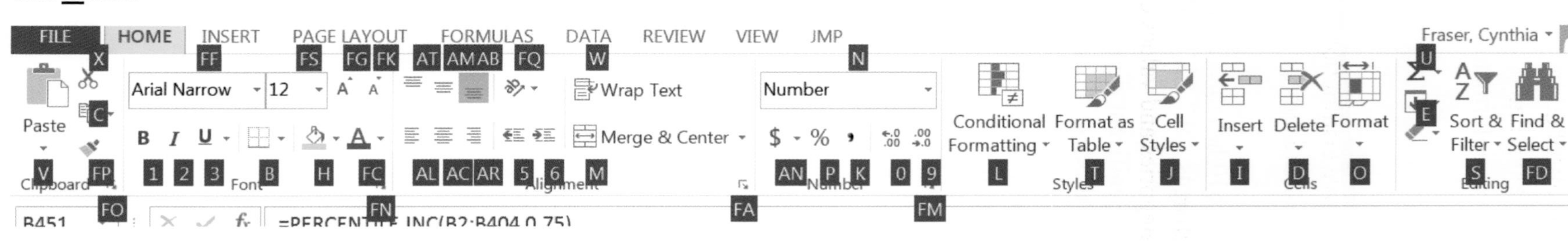

Home menu leys, from left to right, include:

V	paste	FF	Choose a font	FS	Choose a font size	W	Wrap text	9	Reduce decimals	I	Insert
X	cut	1	Bold	FC	Choose font color					D	Delete
C	copy	2	Italicize								
		3	underline								

Other useful menus activated with **Alt** include:

A	Data	N	Insert	W	View

From a chart or plot, **Alt** provides access to chart menus:

JC	Chart design	JA	Chart format

From a PivotTable, **Alt** provides access to PivotTable tool menus, including **JT** for PivotTable ANALYZE (to change how cell values are shown).

From a PivotChart, **Alt** provides access to PivotChart tool menus, including:

JC	PivotChart Design	JA	PivotChart format

Significant Digits Guidelines

The number of significant digits in a number are those which convey information. Significant digits include:

1. All nonzero numbers
2. Zeros between nonzero numbers, and
3. Trailing zeros.

Zeros acting as placeholders aren't counted.

The number 2,061 has four significant digits, while the number 2,610 has three, since the zero is merely a placeholder. The number 0.0920 has three significant digits, "9," "2," and the final, trailing "0." The first two zeros are placeholders that aren't counted.

In rare cases, it is not clear whether zero is a placeholder or a significant digit. The number 40,000 could represent the range 39,500–40,499. In that case, the number of significant digits is one, and the zeros are placeholders. Alternatively, 40,000 could represent the range 39,995–40,004. In this latter case, the number of significant digits is four, since the zeros convey meaning. When in doubt, a number could be written in scientific notation, which is unambiguous. For one significant digit, 40,000 becomes $4 \times E^4$. For four significant digits, 40,000 becomes $4.000 \times E^4$.

Lab 2 Descriptive Statistics

Compensation of 25 Best Paid CEOs

Forbes recently published the compensation packages of the 25 best compensated CEOs in the U.S. These data are in **Lab 2 Compensation of 25 Best Paid CEOs**.xlsx.

I. Describe the compensation of the best paid CEOs.

1. Find the average compensation (M) among the best compensated CEOs:__________
2. Find the standard deviation (SD) of compensation: _______
3. Identify outlier(s) who earn(s) more than 3 SDs above the M: __________________
4. Find average compensation, M, excluding outlier(s): ______
5. Find the standard deviation of compensation, SD, excluding outlier(s): _____
6. Is the distribution of compensation among the best paid CEOs (excluding outlier(s))

   approximately Normal? Y or N Evidence: ______________
7. Make the histogram of compensation for top paid CEOs (excluding outlier(s)).
8. Plot the cumulative distribution of compensation (excluding outlier(s)).
9. What is median compensation among the 25 best paid CEOs? _____
10. What is the Interquartile Range of compensation among the 25 best paid CEOs?

    ____

II. Identify Industries where CEOs are Best Compensated

1. Use a PivotTable to determine the best paid industry: ________
2. What is the best paid industry, excluding outlier(s)? _______
3. In which industries do CEOs earn more than average (excluding outlier(s))?

   ____

Candidate Campaign Contributions

2012 Presidential Candidates' fundraising to date was published in the New York Times in October 2012. These data are in **Lab 2 Candidate Funds**.xlsx.

1. Plot the Candidates' donations by size.

2. What was the modal donation to President Obama? ______

3. What was the modal donation to Romney? _____

4. Which Candidate collected more donations under $200? ______

5. Which Candidate collected more donations of $2,500? ______

Assignment 2-1 Procter & Gamble's Global Advertising

Procter & Gamble spent $5,960,000 on advertising in 51 global markets. This data, from *Advertising Age*, Global Marketing is in **Assignment 2-1 P&G Global Advertising.xls.**

P&G Corporate is reviewing the firm's global advertising strategy, which is the result of decisions made by many brand management teams. Corporate wants to be sure that these many brand level decisions produce an effective allocation when viewed together.

Describe *Procter & Gamble's advertising* spending across the 51 *countries* that make up the global markets.

1. Identify *countries* which are **outliers**:
2. Illustrate advertising levels in countries that are not outliers. Add a "bottom line" chart title.
3. Summarize your analysis by describing *P&G's advertising* in *countries* around the world, excluding outliers.
   Include:
   - one or more measures of central tendency, such as the mean and median,
   - one or more measures of dispersion, such as the standard deviation and range,
   - the similarity of the distribution to a *Normal* distribution

   Be sure to round your answers to two or three significant digits.
4. Considering the entire sample, which advertising strategy describes the P&G strategy better: (i) advertise at a moderate level in many global markets, (ii) advertise heavily to a small number of key markets and spend a little in many other markets.

Assignment 2-2 Best Practices Survey

Firm managers use statistics to advantage. Sometimes when results are lackluster, more significant digits are used, since readers will spend less time digesting results, and results with more significant digits are less likely to be remembered. Sometimes when results are impressive, fewer significant digits are used to motivate readers to digest and remember.

Choose an Annual Report and cite the firm and the year:

1. In the body of the report, what range of significant digits are used to report numerical results? Cite two examples, one with the smallest number of significant digits, one with the largest number of significant digits.
2. In the Financial Exhibits at the end, what range of significant digits are used? Cite two examples, one with the smallest number of significant digits, and one with the largest number of significant digits.
3. Survey the graphics. Cite an example where stand alone title is used to help readers interpret. Cite an example where the title could be more effective, and provide a suggestion for a better title.

Assignment 2-3 Shortcut Challenge

Complete the steps in the first Excel page of Lab 2 (find descriptive statistics, sort to identify and remove outliers, make a PivotTable, make a PivotChart, plot the cumulative distribution) and record your time. If your time is more than 5 min, repeat twice, and then record your best time.

CASE 2-1 VW Backgrounds

Volkswagon management comissioned background music for New Beetle commercials. The advertising message is that the New Beetle is unique. . . "round in a world of squares." To be effective, the background music must support this message.

Thirty customers were asked to write down the first word that came to mind when they listened to the music. The clip is in **Case 2-1 VW background.MP3** and words evoked are contained in **Case 2-1 VW background.xls.** Listen to the clip, then describe market response.

Create a PivotTable of the percent who associate each image with the music and sort rows so that the modal image is first:

1. Create a PivotChart to illustrate the images associated with the background music. (Add a "bottom line" title and round percentages to two significant digits.)
2. What is the modal image created by the VW commercial's background music?
3. Is this music is a good choice for the VW commercial? Explain.

Chapter 3
Hypothesis Tests, Confidence Intervals to Infer Population Characteristics and Differences

Samples are collected and analyzed to estimate population characteristics. Chapter 3 explores the practice of *inference*: how *hypotheses* about what may be true in the population are tested and how population parameters are estimated with *confidence intervals*. Included in this chapter are tests of hypotheses and confidence intervals for:

(i) A population mean from a single sample,
(ii) The difference between means of two populations, or segments from two independent samples, and
(iii) The mean difference within one population between two time periods or two scenarios from two matched or paired samples.

3.1 Sample Means Are Random Variables

The descriptive statistics from each sample of a population are unique. In the example that follows, teams in a New Product Development class each collected a sample from a population to estimate population demand for their concept. Each of the team's statistics are unique, but predictable, since the sample statistics are random variables with a predictable sampling distribution. If many random samples of a given size are drawn from a population, the means from those samples will be similar and their distribution will be Normal and centered at the population mean.

Example 3.1 Thirsty on Campus: Is There Sufficient Demand?

An enterprising New Product Development class has an idea to sell on campus custom-flavored, enriched bottles of water from dispensers which would add customers' desired vitamins and natural flavors to each bottle. To assess profit potential, they need an estimate of demand for bottled water on campus. If demand exceeds the breakeven level of seven bottles per week per customer, the business would generate profit.

The class translated breakeven demand into hypotheses which could be tested using a sample of potential customers. The entrepreneurial class needs to know whether or not demand exceeds seven bottles per consumer per week, because below this level of demand, revenues wouldn't cover expenses. Hypotheses are formulated as *null* and *alternative*. In this case, the null hypothesis states a limiting conclusion about the population mean. This default conclusion is cannot be rejected unless the data indicate that it is highly unlikely.

The null hypothesis is that of insufficient demand, which would lead the class to stop development:

H_0: Campus consumers drink no more than seven bottles of water per week on average:

$$\mu \leq 7$$

C. Fraser, *Business Statistics for Competitive Advantage with Excel 2013: Basics, Model Building, Simulation and Cases*, DOI 10.1007/978-1-4614-7381-7_3, © Springer Science+Business Media New York 2013

Unless sample data indicates sufficient demand, the class will stop development.

In this case, the alternative hypothesis states a conclusion that the population mean exceeds the qualifying condition. The null hypothesis is rejected only with sufficient evidence from a sample that it is unlikely to be true.

In **Thirsty**, the alternate hypothesis supports a conclusion that population demand is sufficient and would lead to a decision to proceed with the new product's development:

H_1: Campus consumers drink more than seven bottles of water per week on average:

$$\mu > 7$$

Given sufficient demand in a sample, the class would reject the null hypothesis and proceed with the project.

Sample statistics are used to determine whether or not the population mean is likely to be less than 7, using the sample mean as the estimate. To test the hypotheses regarding mean demand in the population of customers on campus, each of the 15 student teams in the class independently surveyed a random sample of 30 consumers from the campus. The distribution of means of many "large" ($N \geq 30$) random samples is Normal and centered on the unknown population mean (Fig. 3.1).

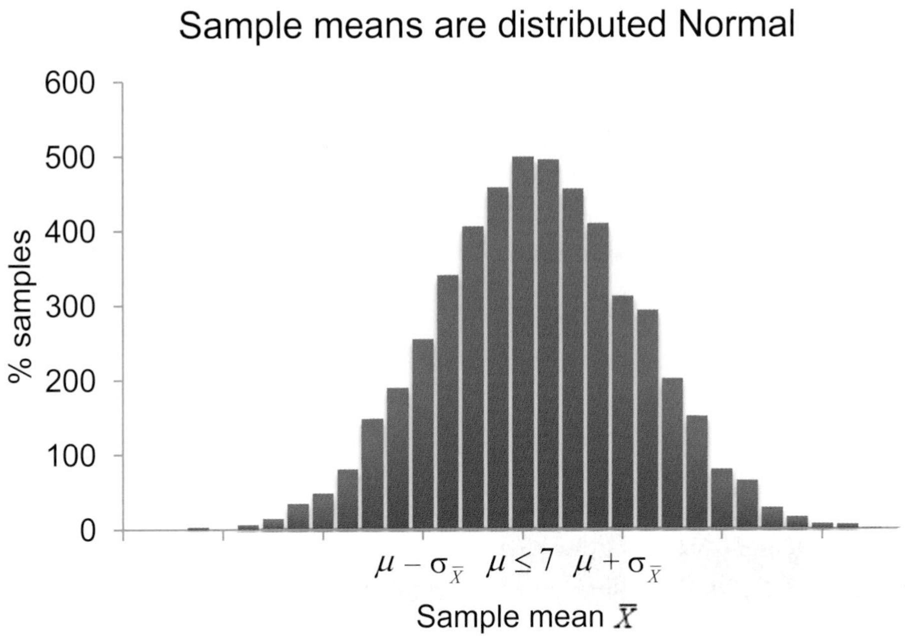

Sample means are distributed Normal

Fig. 3.1 Distribution of sample means under the null hypothesis

On average, across all random samples of the same size N, the average difference between sample means and the population mean is the standard error of sample means:

$$\sigma_{\bar{X}} = \sigma / \sqrt{N}$$

where σ is the standard deviation in the population, and N is the sample size. The standard error is larger when there is more variation in the population and when the sample size is smaller.

With random samples of 30, population mean $\mu = 10.2$ and standard deviation $\sigma = 4.0$, the sampling standard error would be $s_{\bar{X}} = \sigma/\sqrt{30} = 4/5.5 = .7$. From the Empirical Rule introduced in Chap. 2, we would expect 2/3 of the teams' sample means to fall within one standard error of the population mean:

$$\mu - s_{\bar{X}} \le \bar{X} \le \mu + s_{\bar{X}}$$
$$10.2 - .7 \le \bar{X} \le 10.2 + .7$$
$$9.5 \le \bar{X} \le 10.9,$$

and we expect 95 % of the teams' *sample means* to fall within two standard errors of the population mean:

$$\mu - 2s_{\bar{X}} \le \bar{X} \le \mu + 2s_{\bar{X}}$$
$$10.2 - 2(.7) \le \bar{X} \le 10.2 + 2(.7)$$
$$8.8 \le \bar{X} \le 11.6$$

Nearly all of sample means can be expected to fall within three standard errors of the mean, 8.1–12.3.

Each team calculated the sample mean and standard deviation from their sample. Team 1, for example, found that average demand in their sample is 11.2 bottles per week, with standard deviation of 4.5 bottles. Each of team's descriptive statistics from the 15 samples is shown in Fig. 3.2 and Table 3.1.

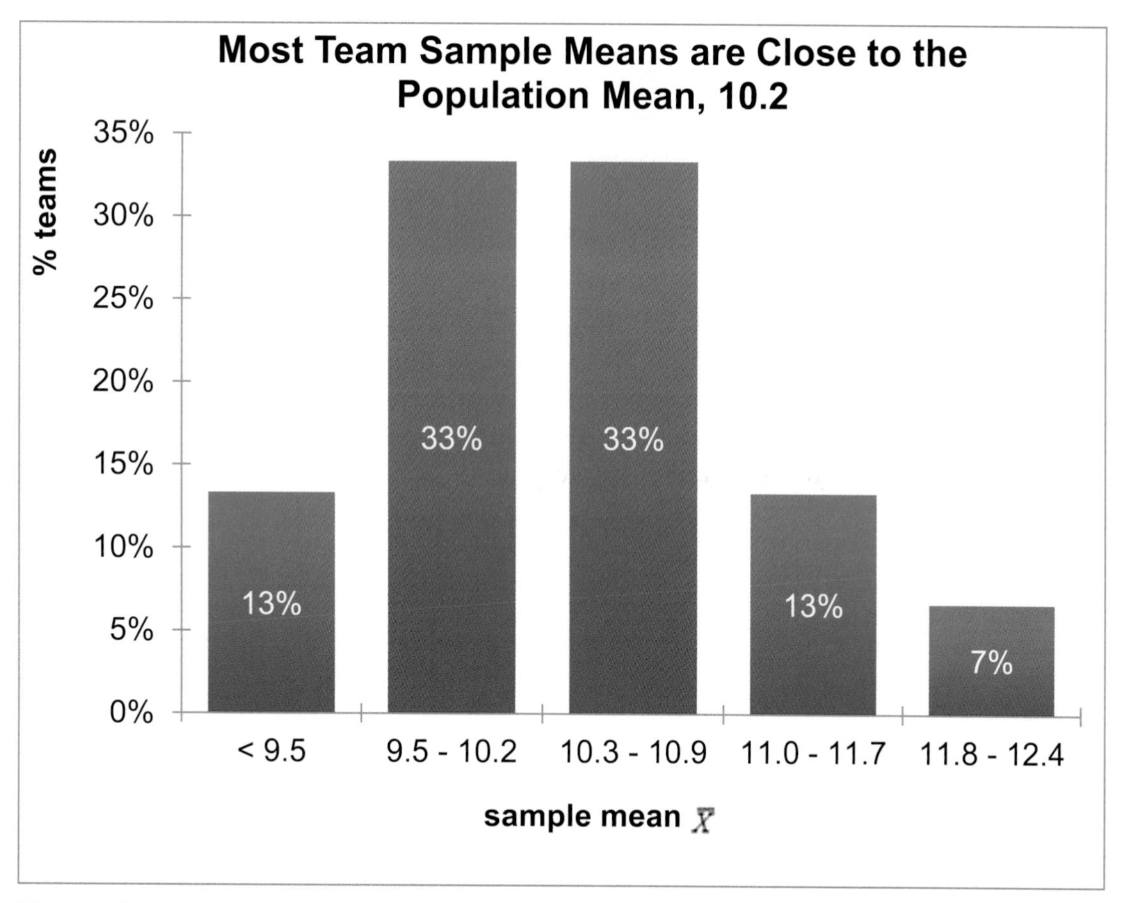

Fig. 3.2 Fifteen teams' samples

Table 3.1 Fifteen teams' samples

Team	**Sample statistics** *Average demand per consumer per week* $\overline{X}$	*Standard deviation* s_i
1	11.2	4.5
2	10.9	4.0
3	10.6	4.3
4	9.5	3.4
5	9.0	3.9
6	10.8	4.6
7	9.6	3.8
8	9.9	4.1
9	9.7	3.7
10	10.7	4.2
11	9.0	3.8
12	9.8	3.6
13	10.5	3.1
14	12.2	4.9
15	11.6	4.2

Sample means across the 15 teams ranged from 9.0 to 12.2 bottles per week per consumer. Each team's sample mean, $\overline{X}$, is close to the true, unknown, population mean, $\mu = 10.2$, and not as close to the hypothetical population mean of 7. Each of the sample standard deviations is close to the true, unknown population standard deviation $\sigma = 4$. In addition, each team's sample statistics are unique.

Since the population standard deviation is almost never known, but estimated from a sample, the standard error is also estimated from a sample, using the estimate of the population standard deviation *s*:

$$s_{\overline{X}} = s / \sqrt{N}$$

When the standard deviation is estimated from a sample (which is nearly always), the distribution of standardized sample means $\overline{X} / s_{\overline{X}}$ is distributed as *Student t*, which is approximately Normal (Fig. 3.3).

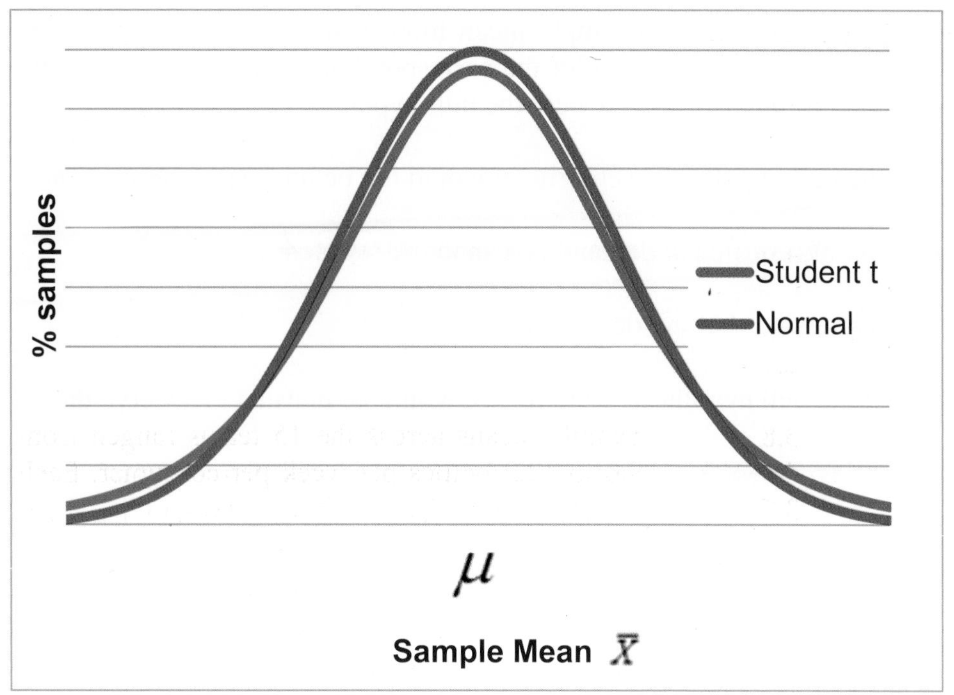

Fig. 3.3 Distribution of sample means

Student t has slightly fatter tails than Normal, since we are estimating the standard deviation. How much fatter the tails are depends on the sample size. *Student t* is a family of distributions indexed by sample size. There is more difference from Normal if a sample size is small. For sample sizes of about 30 or more, there is little difference between Student t and Normal. An estimate of the standard deviation from the sample is close to the true population value if the sample size meets or exceeds 30.

3.2 Infer Whether a Population Mean Exceeds a Target

Each team asks, "How likely is it that we would observe this sample mean, were the population mean seven or less?" From the Empirical Rule, sample means are expected to fall within approximately two standard errors of the population mean 95 % of the time.

Rearranging the Empirical Rule formula, we see that *Student t* counts the standard errors between a sample mean and the population mean:

$$\left(\overline{X} - \mu\right) / s_{\overline{X}} = t_{N-1}$$

A difference between a sample mean and the break-even level of 7 that is more than approximately two standard errors ($t > 2$) is a signal that population demand is unlikely to be 7 or less. In this case, the sample mean would lie to the extreme right in the hypothetical distribution of sample means with center at the hypothetical population mean of 7, where fewer than 5 % of sample means are expected.

In the **Thirsty** example, each team calculated the number of standard errors by which their sample mean exceeded 7. Next, each referred to a table of Student *t* values or used statistical software to find the area under the right distribution tail, called the *p value*. Were true demand less than 7, it would be unusual to observe a sample mean more than t $_{2\alpha=.1;\,29}$ = 1.7 standard errors greater than 7. The larger a *t* value, the smaller the corresponding *p value* will be, and the less likely the sample statistics would be observed were the null hypothesis true:

> *p value* > .05 . . . if the null hypothesis were true, it would not be unusual to observe the data.
>
> The conclusion of insufficient demand H_0 cannot be rejected.
>
> The Team recommends halting development.
>
> *p value* ≤ .05 . . . if the null hypothesis were true, it would be unusual to observe the data.
>
> Reject the null hypothesis.
>
> The Team recommends proceeding with development.

Each team used software to test the hypothesis that demand exceeds 7. Team 8's analyses are illustrated in Fig. 3.4, as an example:

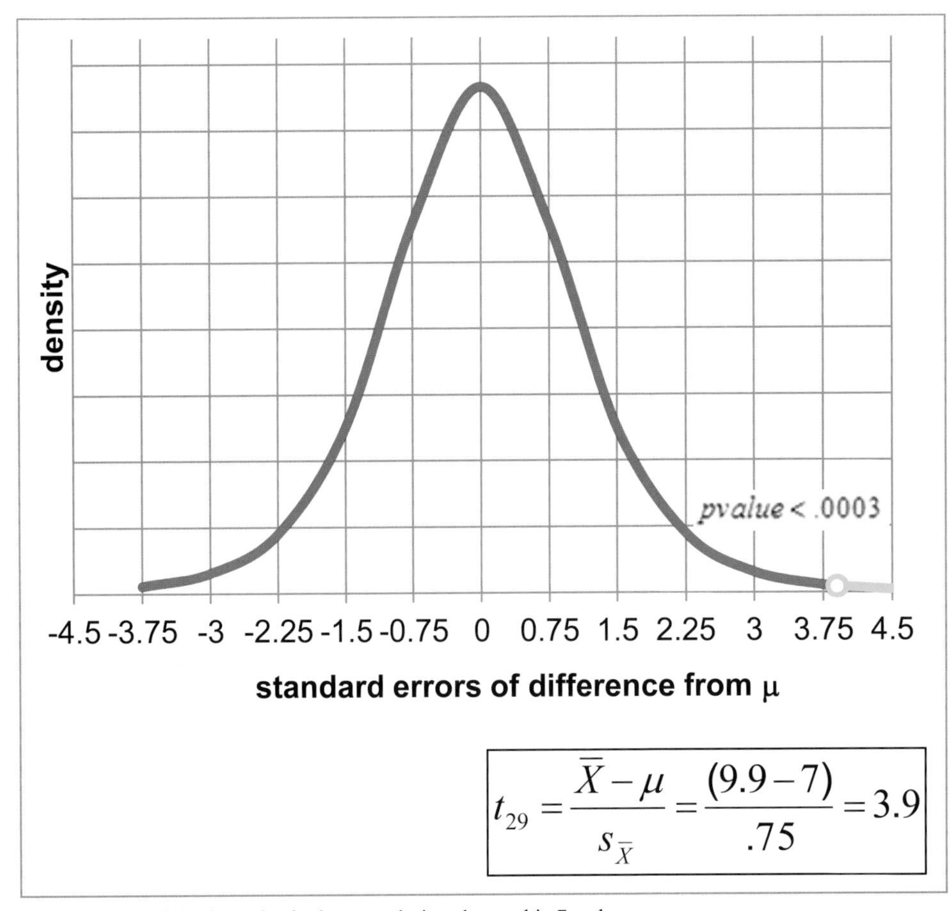

Mean	9.9
sd	4.1
se	.75
t	3.9
p value	.0003

Fig. 3.4 *t test* of the hypothesis that population demand is 7 or less

Reviewing these results Team 8 would conclude:

Demand in our sample of thirty ranged from zero to nineteen bottles per person per week, averaging 9.9 bottles per person per week. With this sample of thirty, the standard error is .75 bottles per week. Our sample mean is 3.9 standard errors greater than breakeven of seven. (The t statistic is 3.9.) Were population demand seven or less, it would be unusual to observe demand of 9.9 in a sample of thirty. The p-value is .0003. We conclude that demand is not seven or less.

In a test of the level of demand for bottles of water, each team used a "one-tail" test. Regardless of how much demand exceeds seven bottles per consumer per week, a team would vote to proceed with development as long as they can be reasonably sure demand exceeds breakeven. They require only that the chance of observing the data be less than 5 %, the *critical p value,* were true demand less than seven. Thus, it is only the area under the right tail that concerns them.

3.3 Critical t Provides a Benchmark

Before statistical software became popular and statistical calculations were done by hand, it was standard practice to conduct a *one tail t test* by finding the *critical t* value for a given sample size which cut off 5 % of the t distribution right tail. *Critical t* values were published in the appendices of texts, indexed by sample size. Comparing a sample *t* statistic with the *critical t* enabled a yes-no test of the null hypothesis. If the sample *t* exceeded the *critical t*, the null hypothesis was rejected.

From the **Thirsty** example, for a sample of 30, the *critical t* value for 29 *df* (= N−1 = 30−1 = 29) is 1.70. Figure 3.5 illustrates *p values* returned by Excel for *t* values with a sample size of 30, or *df* of 29.

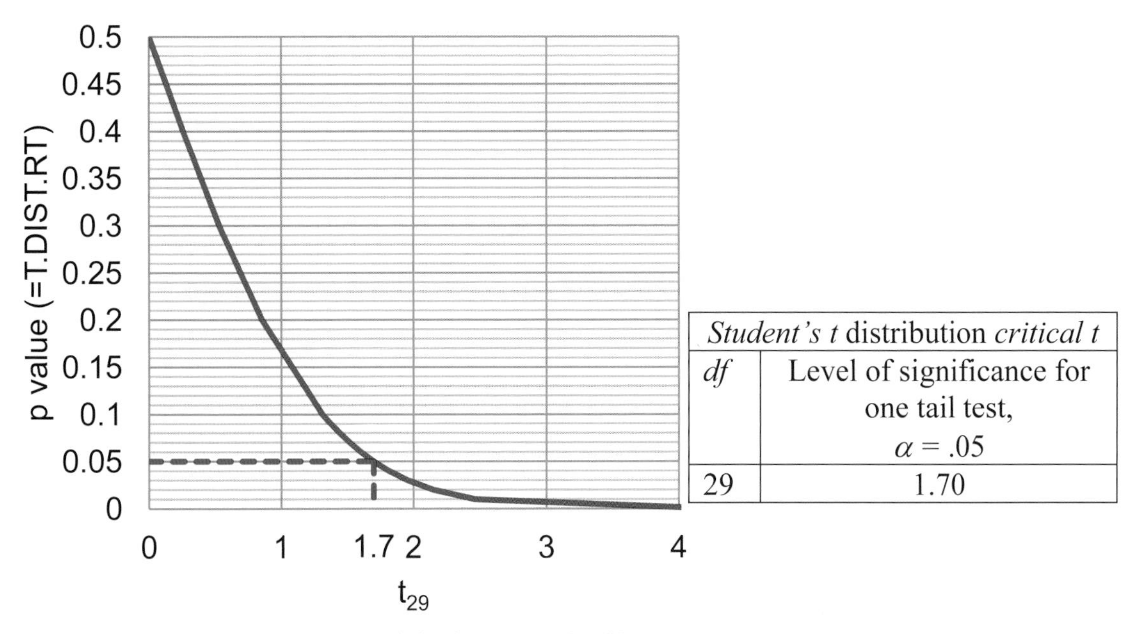

Student's t distribution *critical t*	
df	Level of significance for one tail test, $\alpha = .05$
29	1.70

Fig. 3.5 *p values* returned by Excel for *t* statistics from a sample of 30

Team Eight's sample t of 3.9 exceeds the *critical t*, so the null hypothesis is rejected with 95% (= $1-\alpha$) confidence.

The wide availability of statistical software allows easy determination of the *p value*, providing a more informative estimate of the chance that the sample mean would be observed if the null hypothesis were true. Consequently, it has become standard practice to compare the sample *p value* to the *critical p value* of .05 to test the null hypothesis.

Whether you choose to compare the sample *p value* with the *critical p value* or, alternatively, the sample t to the *critical t value* for a given sample size will lead to the same conclusion. Both comparisons are correct choices.

3.4 Confidence Intervals Estimate the Population Mean

Since the class of entrepreneurs in the **Thirsty** example doesn't know that the population mean is 10.2 bottles per customer per week, each team will estimate this mean using their sample data. Rearranging the formula for a *t test*, we see that each team can use their sample standard error, the Student t value for their sample size and the desired level of confidence to estimate the range that is likely to contain the true population mean:

$$\bar{X} - t_{\alpha/2,N-1} \times s_{\bar{X}} < \mu < \bar{X} + t_{\alpha/2,N-1} \times s_{\bar{X}}$$

where α is the chance that a sample is drawn from one of the sample distribution tails, and $t_{\alpha/2,\,(N-1)}$ is the *critical Student t* value for a chosen level of certainty ($1-\alpha$) and sample size N.

The *confidence level* ($1-\alpha$) allows us to specify the level of certainty that an interval will contain the population mean. Generally, decision makers desire a 95 % level of confidence ($\alpha = .05$), insuring that in 95 out of 100 samples, the interval would contain the population mean. The *critical Student t* value for 95 % confidence with a sample of 30 ($N = 30$) is $t_{\alpha/2,\,(N-1)=29} = 2.05$. In 95 % of random samples of 30 drawn, we expect the sample means to be no further than 2.05 standard errors from the population mean:

$$\bar{X} - 2.05 s_{\bar{X}} \le \mu \le \bar{X} + 2.05 s_{\bar{X}}$$

Each team's sample standard error, margin of error, and 95 % confidence interval from the **Thirsty** example are shown in Table 3.2:

Table 3.2 Confidence intervals from each team's sample

Team i	*Average demand/consumer/week*, $\overline{X}_i$	*Standard deviation* s_i	*Standard error* $s_{\overline{X}}$	*Margin of error* $2.05\, s_{\overline{X}}$	*95 % confidence interval* $\overline{X} \pm 2.05 s_{\overline{X}}$	
1	11.2	4.5	.84	1.7	9.5	12.9
2	10.9	4.0	.74	1.5	9.4	12.4
3	10.6	4.3	.80	1.6	9.0	12.2
4	9.5	3.4	.63	1.3	8.2	10.8
5	9.0	3.9	.72	1.5	7.5	10.5
6	10.8	4.6	.85	1.7	9.1	12.5
7	9.6	3.8	.71	1.5	8.1	11.1
8	9.9	4.1	.75	1.5	8.4	11.4
9	9.7	3.7	.69	1.4	8.3	11.1
10	10.7	4.2	.78	1.6	9.1	12.3
11	9.0	3.8	.71	1.5	7.5	10.5
12	9.8	3.6	.67	1.4	8.4	11.2
13	10.5	3.1	.58	1.2	9.3	11.7
14	12.2	4.9	.91	1.9	10.3	14.1
15	11.6	4.2	.78	1.6	10.0	13.2

In practice, 15 samples would not be collected. A single sample would be selected, just as each individual team did in their market research. Team 8's analysis is shown in Fig. 3.6 as an example:

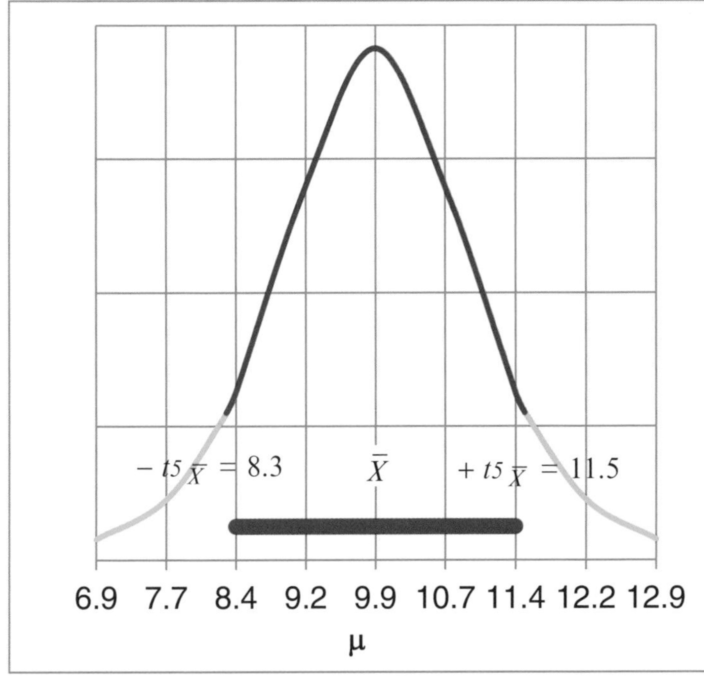

Mean	9.9
Standard error	.75
Critical t	2.1
Margin of error	1.5
95 % lower	8.3
95 % upper	11.5

Fig. 3.6 Confidence interval for bottled water demand μ

Team 8 would conclude:

"Average demand in our sample of thirty is 9.9 bottles per person per week, with a margin of error of 1.5 bottles. It is likely that average campus demand is between 8.3 and 11.5 bottles per person per week."

3.5 Calculate Approximate Confidence Intervals with Mental Math

When the sample size is "large," $N \geq 30$, we can use an approximate $t \simeq 2.0$ to produce approximate confidence intervals with mental math. Using $t \simeq 2$ for an approximate 95 % level of confidence, the 15 student teams each calculated the likely ranges for bottled water demand in the population, shown in Table 3.3.

Table 3.3 Each team's approximate confidence interval

Team_i	Average customer demand/week $\overline{X}_i$	Standard error $s_{\overline{X}}$	Margin of error $2.05\,s_{\overline{X}}$	95 % confidence interval $\overline{X} \pm 2.05 s_{\overline{X}}$		Approximate margin of error $2s_{\overline{X}}$		Approximate 95 % confidence interval $\overline{X} \pm 2s_{\overline{X}}$
1	11.2	.84	1.7	9.5	12.9	1.7	9.5	12.9
2	10.9	.74	1.5	9.4	12.4	1.5	9.4	12.4
3	10.6	.80	1.6	9.0	12.2	1.6	9.0	12.2
4	9.5	.63	1.3	8.2	10.8	1.3	8.2	10.8
5	9.0	.72	1.5	7.5	10.5	1.4	7.6	10.4
6	10.8	.85	1.7	9.1	12.5	1.7	9.1	12.5
7	9.6	.71	1.5	8.1	11.1	1.4	8.2	11.0
8	9.9	.75	1.5	8.4	11.4	1.5	8.4	11.4
9	9.7	.69	1.4	8.3	11.1	1.4	8.3	11.1
10	10.7	.78	1.6	9.1	12.3	1.6	9.1	12.3
11	9.0	.71	1.5	7.5	10.5	1.4	7.6	10.4
12	9.8	.67	1.4	8.4	11.2	1.3	8.5	11.1
13	10.5	.58	1.2	9.3	11.7	1.2	9.3	11.7
14	12.2	.91	1.9	10.3	14.1	1.8	10.4	14.0
15	11.6	.78	1.6	10.0	13.2	1.6	10.0	13.2

With the approximation, Team 8's conclusion remains: expected demand will range from 8.4 to 11.4 bottles per week per customer.

3.6 Margin of Error Is Inversely Proportional to Sample Size

The larger a sample N is, the smaller the *95 % confidence interval* is,

$$\overline{X} - 2s_{\overline{X}} \leq \mu \leq \overline{X} + 2s_{\overline{X}}$$

since the standard error $s_{\bar{X}}$ and *margin of error*, roughly 2 $s_{\bar{X}}$, are inversely proportional to the square root of our size *N*, shown in Fig. 3.7.

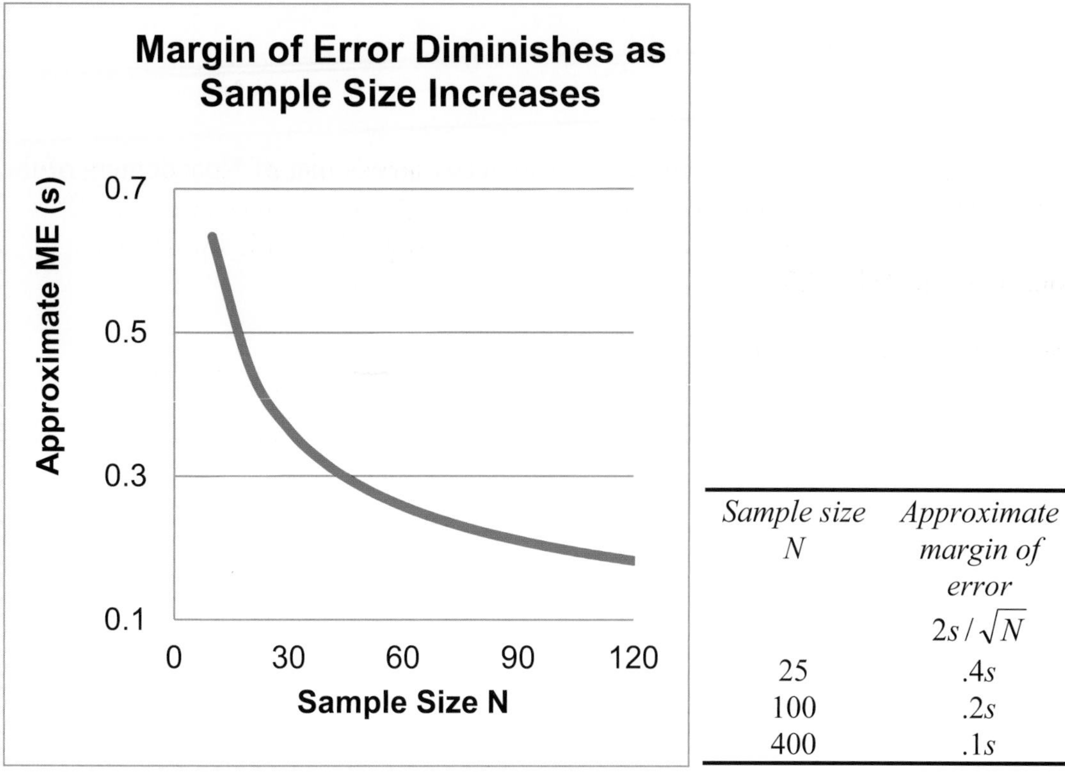

Sample size N	*Approximate margin of error* $2s/\sqrt{N}$
25	.4*s*
100	.2*s*
400	.1*s*

Fig. 3.7 Margin of error, given sample size

To double precision, the sample size must be quadrupled. Gains in precision become increasingly more expensive.

3.7 Determine Whether Two Segments Differ with Student t

Example 3.2 SmartScribe: Is Income a Useful Base for Segmentation?

SmartScribe, manufacturers of a brand of smart pens, would like to identify the demographic segment with the highest demand for its new concept. Smart pens record presentation notes onto a file that can be downloaded. Since the new pens were being sold at a relatively high price, Adopters might have higher incomes. To test this hypothesis, customers at an office supply retail store where sorted into SmartScribe purchasers, which management refers to as The Adopters, and other Nonadopter customers. Random samples from these two segments were drawn and offered a store coupon in exchange for completion of a short survey, which included a measure of annual household income. Fifty-six SmartScribe pen Adopters and 41 Nonadopters completed the survey.

The null hypothesis states the conclusion that the average annual household income of Adopters is not greater than that of Nonadopters.

H_0: Average annual household income of Adopters is equal to or less than that of Nonadopters of the new pen.

$$\mu_{Adopters} \leq \mu_{Nonadopters}$$

or

$$\mu_{Adopters} - \mu_{Nonadopters} \leq 0.$$

Alternatively:

H_1: Average annual household incomes of Adopters exceeds that of Nonadopters of the new pen:

$$\mu_{Adopters} > \mu_{Nonadopters}$$

or

$$\mu_{Adopters} - \mu_{Nonadopters} > 0.$$

If there is no difference in incomes between the two segment samples, or if Adopters earn lower incomes, the null hypothesis cannot be rejected based on the sample evidence.

Average income in the sample of Nonadopters was \$35K, and \$80K, in the sample of Adopters (Fig. 3.8):

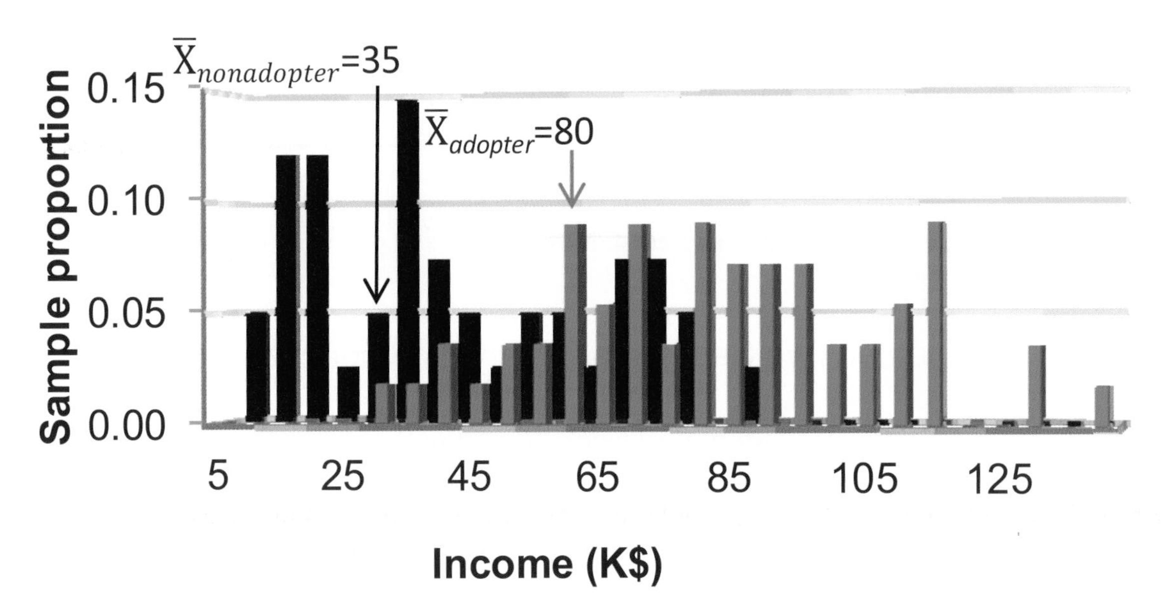

Fig. 3.8 Incomes of samples from two segments

A test of the significance of the difference between the two segments' average annual household incomes is based on the difference between the two sample means, SmartScribe needs to determine whether or not this difference in average incomes,

$$\bar{X}_{Adopters} - \bar{X}_{Nonadopters} = \$80\text{K} - \$35\text{K} = \$45\text{K}$$

is large enough to be significant (Fig. 3.9).

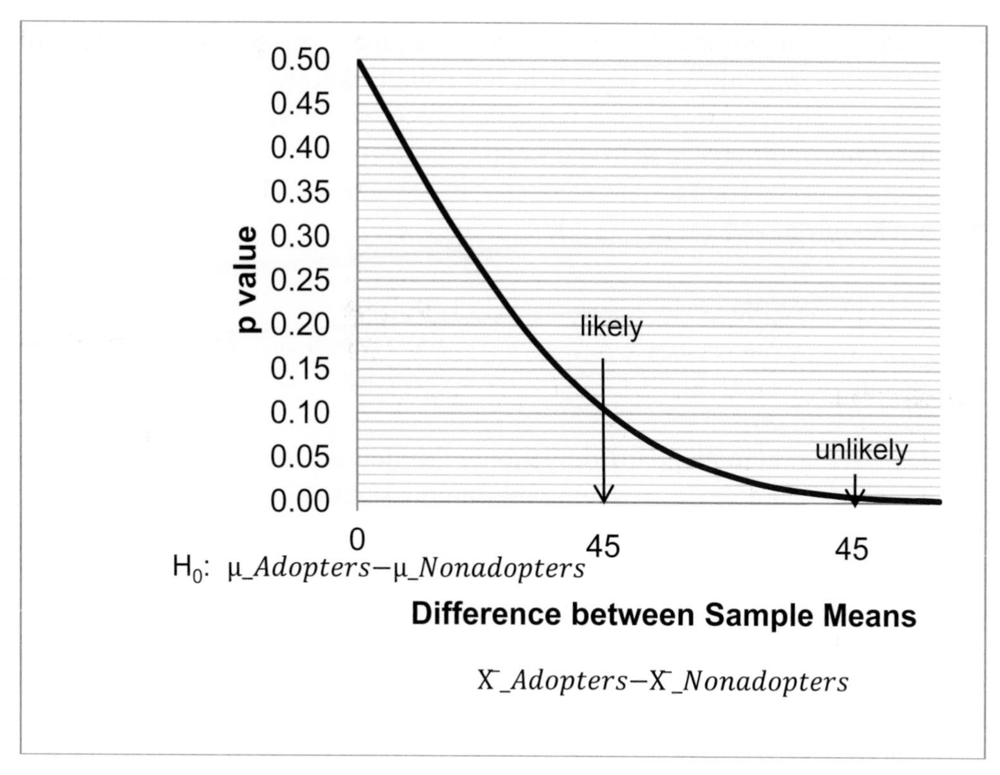

Fig. 3.9 The null hypothesis Adopters earn less or equivalent incomes to Nonadopters

Whether the difference between two sample means is large enough to be significant depends on the amount of dispersion in the two populations, in this case $\sigma_{Adopters}$ and $\sigma_{Nonadopters}$, and the two sample sizes, $n_{Adopters}$ and $n_{Nonadopters}$ in this case. The standard error of the difference between two sample means,

$$s_{\bar{X}_{Adopters}-\bar{X}_{Nonadopters}} = \sqrt{\left(\frac{s_{Adopters}}{\sqrt{n_{Adopters}}}\right)^2 + \left(\frac{s_{Nonadopters}}{\sqrt{n_{Nonadopters}}}\right)^2}$$

$$= \sqrt{s^2_{\bar{X}_{Adopters}} + s^2_{\bar{X}_{Nonadopters}}}$$

captures both the dispersion in the two populations, as well as the two sample sizes.

The standard error of average difference in annual household income (in thousands) is the square root of the two sample standard errors squared, equal to $4.9K in this case:

$$s_{\bar{X}_{Adopters}-\bar{X}_{Nonadopters}} = \sqrt{\left(\frac{25}{\sqrt{56}}\right)^2 + \left(\frac{23}{\sqrt{41}}\right)^2}$$

$$= \sqrt{(3.3)^2 + (3.6)^2}$$

$$= \sqrt{11.2 + 12.9}$$

$$= \sqrt{24.1}$$

$$= 4.9(\$K)$$

This estimate for the standard error of the difference between segment means assumes that the two segment standard deviations may differ. Since it is not usually known whether or not the segment standard deviations are equivalent, this is a conservative assumption.

The number of standard errors of difference between sample means is measured with Student t.

$$t_{90} = (\bar{X}_{Adopters} - \bar{X}_{Nonadopters}) / s_{\bar{X}_{Adopters} - \bar{X}_{Nonadopters}}$$

$$= \$45K / \$4.9K$$

$$= 9.2$$

When the two samples have unique standard deviations, the degrees of freedom for a two segment *t* test depend on both standard deviations and both sample sizes (Fig 3.10):

$$df = \frac{(s_{Adopters}{}^2/N_{Adopters} + s_{Nonadopters}{}^2/N_{Nonadopters})^2}{(s_{Adopters}{}^2/N_{Adopters})^2/(N_{Adopters}-1) + (s_{NOnadopters}{}^2/N_{Nonadopters})^2/(N_{Nonadopters}-1)}$$

$$= \frac{(25^2/56+31^2/41)^2}{(25^2/56)^2/55+(23^2/41)^2/40}$$

$$= 90$$

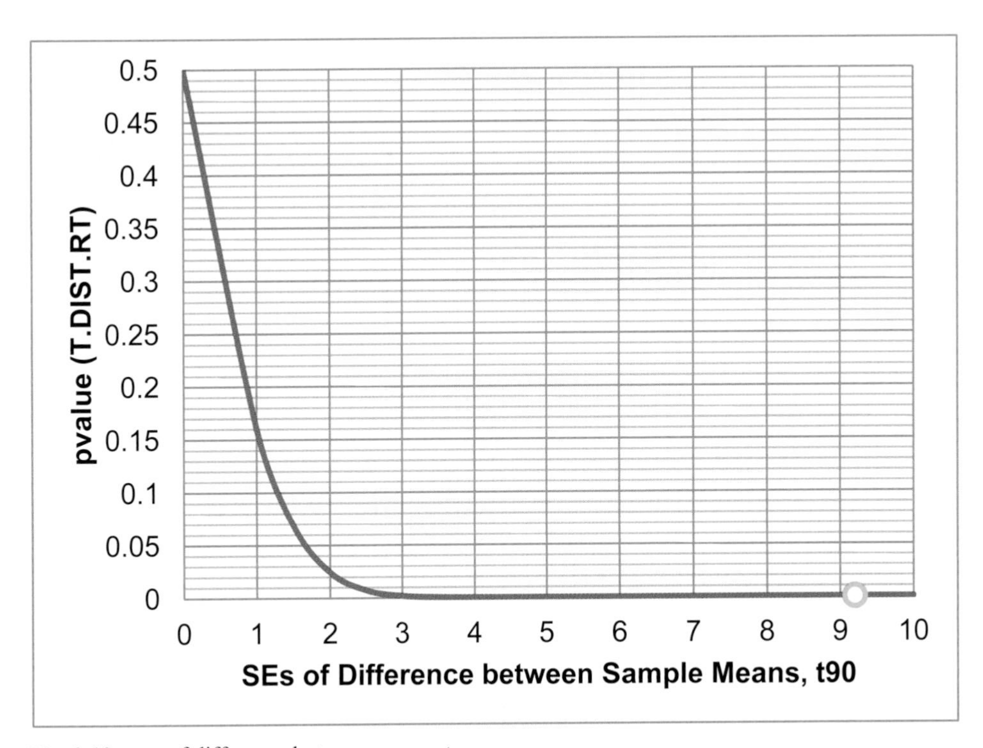

Fig. 3.10 *t* test of difference between segment means

The *p value* for this *t* with 90 degrees of freedom is less than .0001.

From the *t test* of difference between segment incomes, shown in Fig. 3.11, SmartScribe management could conclude:

"In segment samples of 56 Adopters and 41 Nonadopters, the corresponding average segment sample incomes are \$80K and \$35K, a difference of \$45K, more than 9 standard errors. Were there no difference in segment mean incomes in the population, it would be unusual to observe this difference in segment average incomes in the segment samples. Based on sample evidence, we conclude that average incomes of Adopters cannot be less than or equal to the average incomes of Nonadopters. Income is a useful basis for segmentation."

3.8 Estimate the Extent of Difference Between Two Segments

From the sample data, SmartScribe managers estimated the average annual household income difference (in thousands) between Adopters and Nonadopters:

$$\overline{X}_{Adopters} - \overline{X}_{Nonadopters} = \$80K - \$35K = \$45K$$

The 95 % confidence interval around the difference in annual household incomes between Adopters and Nonadopters is made by adding and subtracting the *margin of error*.

The *margin of error* is equal to the two tail *critical t*, with degrees of freedom corresponding to the two sample sizes and α equal to .05 (for 95 % confidence), times the standard error for the difference between sample means:

$$t_{df,\propto/2=.025} \times s_{\overline{X}_{Adopters}-\overline{X}_{Nonadopters}} \cong 2 \times \$4.9K$$

$$\cong \$9.8K$$

The approximate *t,* 2, is used in the example, instead of the *t* which corresponds to a confidence interval for the difference between segments.

The difference between means of the two samples would be no further than \$9.8K from the difference between means in the two populations.

The 95 % confidence interval for the difference between means is \$35K to \$55K:

$$\left(\overline{X}_{Adopters} - \overline{X}_{Nonadopters}\right) - t_{df,\propto/2=.025} \times s_{\overline{X}_{Adopters}-\overline{X}_{Nonadopters}}$$

$$\lesssim \mu_{Adopters} - \mu_{Nonadopters}$$

$$\lesssim \left(\overline{X}_{Adopters} - \overline{X}_{Nonadopters}\right) + t_{df,\propto/2=.025} \times s_{\overline{X}_{Adopters}-\overline{X}_{Nonadopters}}$$

$$\$45K - 2 \times \$4.9K \lesssim \mu_{Adopters} - \mu_{Nonadopters} \lesssim \$45K + 2 \times \$4.9K$$

$$\$45K - \$9.8K \lesssim \mu_{Adopters} - \mu_{Nonadopters} \lesssim \$45K + \$9.8K$$

$$\$35K \lesssim \mu_{Adopters} - \mu_{Nonadopters} \lesssim \$55K$$

Management will conclude that annual household income can be used to differentiate the two market segments, and that Adopters are wealthier than Nonadopters.

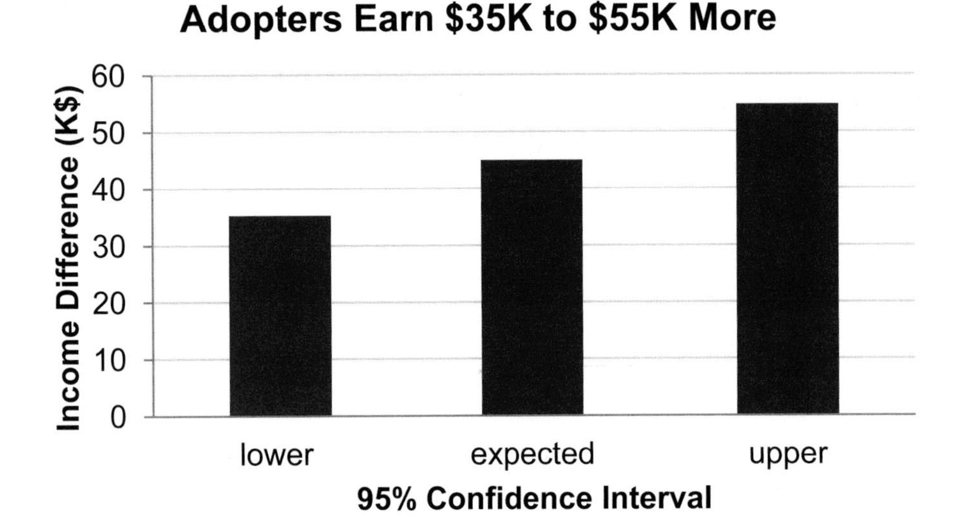

Fig. 3.11 95 % Confidence interval of the difference between segments.

In our samples of 56 Adopters and 41 Nonadopters, the corresponding average difference in income between segment samples is $45K, *and the margin of error of the difference is approximately* $9.8K. *Relative to Nonadopters, we estimate that Adopters earn* $35K *to* $55K *more on average, annually.*

To construct confidence intervals for the difference in means of two samples, we assume that either (i) both segments' characteristics are bell-shaped (distributed approximately Normal) and we've randomly sampled both segments, or (ii) "large" random samples from both segments have been collected.

3.9 Estimate a Population Proportion from a Sample Proportion

Example 3.3 Guinea Pigs

A pharmaceutical company gauges reactions to their products by applying them to animals. An animal rights activist has threatened to start a campaign to boycott the company's products if the animal testing doesn't stop. Concerned managers have hired four public opinion polling organizations to learn whether medical testing on animals is accepted or not.

Four independent pollsters each surveyed 30 Americans and found the proportions shown in Table 3.4 agree that medical testing on animals is morally acceptable:

Table 3.4 Sample approval proportions by poll

Poll	*Sample approval proportion*
1	$P_1 = 16/30 = .53$
2	$P_2 = 19/30 = .63$
3	$P_3 = 17/30 = .57$
4	$P_4 = 21/30 = .70$

If numerous random samples are taken, sample proportions P will be approximately Normally distributed around the unknown population proportion $\pi = .6$, as long as this true proportion is not close to either zero or one.

The standard deviation of the sample proportions *P*, the *standard error of the sample proportion,* measures dispersion of samples of size *N* from the population proportion π:

$$\sigma_{\pi} = \sqrt{\pi \times (1 - \pi)/N}$$

which is estimated with the sample proportion *P*:

$$s_{P} = \sqrt{P \times (1 - P)/N}$$

The four poll organizations would each estimate the proportion of Americans who agree that medical testing on animals is morally acceptable, shown in Table 3.5.

Table 3.5 Confidence interval of approval proportion by poll, $N = 30$

Poll ***i***	*Sample proportion,* P_i	*Standard error,* s_{P_i} ($N = 30$)	*Margin of error for 95 % confidence,* $Z \times s_{P_i} = 1.96 \times s_{P_i}$	*Interval containing the population proportion with 95 % confidence* $P_i \pm Z \times s_{P_i}$
1	.57	.090	.18	.39 to .75
2	.61	.089	.17	.44 to 78
3	.58	.090	.18	.40 to .76
4	.63	.088	.16	.47 to .79

With samples of just 30, margins of error are relatively large and we are uncertain whether a minority or a sizeable majority approves. In practice, polling organizations use much larger samples, which shrink margins of error and corresponding confidence intervals. Had samples of 1,000 been collected instead, the poll results would be as shown in Table 3.6.

Table 3.6 Confidence interval of approval proportion by poll, $N = 1{,}000$

Poll *i*	*Sample proportion,* P_i	*Standard error,* s_{P_i} ($N = 1{,}000$)	*Margin of error for 95 % Confidence,* $Z \times s_{P_i} = 1.96 \times s_{P_i}$	*95% Confidence interval* $P_i \pm Z \times s_{P_i}$
1	.57	.016	.031	.54 to .60
2	.61	.015	.029	.58 to .64
3	.58	.016	.031	.55 to .61
4	.63	.015	.029	.60 to .66

With much larger samples and correspondingly smaller margins of error, it becomes clear that the majority approves of medical testing on animals.

The second polling organization would report:

The majority of a random sample of 1,000 Americans approves of medical testing on animals. 61 % believe medical testing on animals is morally acceptable, with a margin of error of 3 %.

3.10 Conditions for Assuming Approximate Normality

It is appropriate to use the Normal distribution to approximate the distribution of possible sample proportions if sample size is "large" ($N \geq 30$), and both $N \times \mathrm{P} \geq 5$ and $N \times (1-\mathrm{P}) \geq 5$. When the true population proportion is very close to either zero or one, we cannot reasonably assume that the distribution of sample proportions is Normal. A rule of thumb suggests that $\mathrm{P} \times N$ and $(1-\mathrm{P}) \times N$ ought to be at least five in order to use Normal inferences about proportions. For a sample of 30, the sample proportion P would need to be between .17 and .83 to use Normal inferences. For a sample of 1,000, the sample proportion P would need to be between .01 and .99. Drawing larger samples allows more precise inference of population proportions from samples.

3.11 Conservative Confidence Intervals for a Proportion

Polling organizations report the sample proportion and margin of error, rather than a confidence interval. For example, "61 % approve of medical testing on animals. (The margin of error from this poll is 3 % points.)" A 95 % level of confidence is the industry standard. Because the true proportion and its standard deviation are unknown, and because pollsters stake their reputations on valid results, a *conservative* approach, which assumes a true proportion of .5, is used. This conservative approach

$$s_{\mathrm{P}} = \sqrt{.5 \times (1 - .5)/N}$$

yields the largest possible standard error for a given sample size and makes the margin of error ($\mathrm{Z} \times s_{\mathrm{P}}$) a simple function of the square root of the sample size N.

With this conservative approach and samples of $N = 1{,}000$, the pollsters' results are shown in Table 3.7.

Table 3.7 Conservative confidence intervals for approval proportions, $N = 1{,}000$

Poll *i*	*Sample proportion,* P	*Conservative margin of error for 95 % confidence,* $\mathrm{Z} \times s_{\mathrm{P}} = 1.96 \times s_{\mathrm{P}}$	*Conservative 95 % confidence interval* $\mathrm{P} - \mathrm{Z} \times s_{\mathrm{P}} \leq \pi \leq \mathrm{P} + \mathrm{Z} \times s_{\mathrm{P}}$	
1	.57	.031	.54	.60
2	.61	.031	.58	.64
3	.58	.031	.55	.61
4	63	.031	.60	.66

An effective display of proportions or shares is a *pie chart.* The second poll organization used Excel to create this illustration of their survey results, shown in Fig. 3.12:

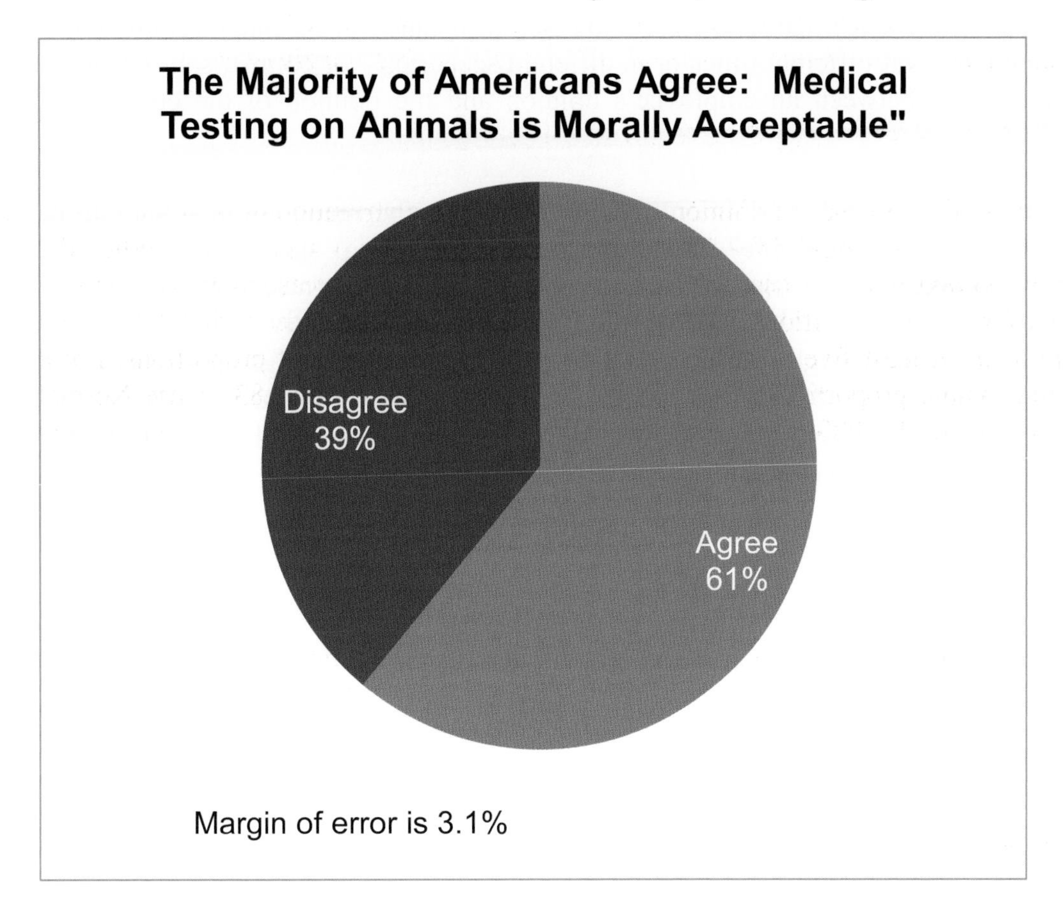

Fig. 3.12 Pie chart of approval percentage

The second polling organization would report:

"*Sixty-one percent of American adults agree that medical testing on animals is morally acceptable. Poll results have a margin of error of 3.1 percentage points. The majority of Americans supports medical testing on animals.*"

Other appropriate applications for confidence intervals to estimate population proportions or shares include:

- Proportion who prefer a new formulation to an old one in a taste test
- Share of retailers who offer a brand
- Market share of a product in a specified market
- Proportion of employees who call in sick when they're well
- Proportion of new hires who will perform exceptionally well on the job

3.12 Assess the Difference Between Alternate Scenarios or Pairs

Sometimes management is concerned with the comparison of means from a single sample taken under varying conditions – at different times or in different scenarios – or comparison of sample pairs, like the difference between an employee's opinion and the opinion of the employee's supervisor:

- Financial management might be interested in comparing the reactions of a sample of investors to "socially desirable" stock portfolios, excluding stocks of firms that manufacture or market weapons, tobacco, or alcohol, versus alternate portfolios which promise similar returns at similar risk levels, but which are not "socially desirable."
- Marketing management might be interested in comparing taste ratings of sodas which contain varying levels of red coloring – do redder sodas taste better to customers?
- Management might be interested in comparing satisfaction ratings following a change which allows employees to work at home.

These examples compare *repeated samples*, where participants have provided multiple responses that can be compared:

- Financial management might also be interested in comparing the risk preferences of husbands and wives.
- Marketing management might want to compare children and parents' preferences for red sodas.
- Management might also be interested in comparing the satisfaction ratings of those employees with their supervisors' satisfaction ratings.

In these examples, interest is in comparing means from *matched pairs*.

In either case of repeated or matched samples, a *t test* can be used to determine whether or not the difference is non-zero. Testing hypotheses that concern a difference between pairs is equivalent to a one sample *t* test. The difference is tested in the same way that a characteristic mean is tested, using a one sample test.

Example 3.4 Are "Socially Desirable" Portfolios Undesirable?

An investment consulting firm's management believes that they have difficulty selling "socially desirable" portfolios because potential investors assume those funds are inferior investments. Socially Desirable funds exclude stocks of firms which manufacture or market weapons, tobacco or alcohol. There may be a perceived sacrifice associated with socially desirable investment which causes investors to avoid portfolios labeled "socially desirable." The null hypothesis is:

H_0: Investors rate "socially desirable" portfolios at least as attractive as equally risky, conventional portfolios promising equivalent returns:

$$\mu_{Socially\ Desirable} - \mu_{Conventional} \geq 0.$$

If investors do not penalize "socially desirable" funds, the null hypothesis cannot be rejected.

The alternative hypothesis is:

> H_1: Investors rate "socially desirable" portfolios as less attractive than other equally risky portfolios promising equivalent returns:

$$\mu_{Socially\ Desirable} - \mu_{Conventional} < 0.$$

Thirty-three investors were asked to evaluate two stock portfolios on a scale of attractiveness (−3 = "Not At All Appealing" to 3 = "Very Appealing"). The two portfolios promised equivalent returns and were equally risky. One contained only "socially desirable" stocks, while the other included stocks from companies which sell tobacco, alcohol and arms. These are shown in Table 3.8.

Table 3.8 Paired ratings of other and socially desirable portfolios

Appeal of conventional portfolio	*Appeal of socially desirable portfolio*	*Difference*	*Appeal of conventional portfolio*	*Appeal of socially desirable portfolio*	*Difference*
−3	1	−4	2	−1	3
−3	2	−5	2	−1	3
−3	3	−6	2	−2	4
−3	3	−6	2	2	0
0	−1	1	2	1	1
0	1	−1	2	2	0
1	−3	4	2	2	0
1	−3	4	2	3	−1
1	−1	2	3	−3	6
1	−1	2	3	−3	6
1	-1	2	3	−3	6
1	1	0	3	−1	4
1	1	0	3	−1	4
1	2	−1	3	−3	6
2	−3	5	3	3	0
2	−3	5	3	3	0
2	−2	4			

From a random sample of 33 investors' ratings of conventional and Socially Desirable portfolios of equivalent risk and return, the average difference is 1.5 points on a 7-point scale of attractiveness.

$$\bar{X}_{dif} = \bar{X}_{SD} - \bar{X}_{C} = -.2 - 1.3 = -1.5$$

With this sample of 33, the standard error of the difference is .6.

$$s_{\overline{X}_{dif}} = \frac{s_{dif}}{\sqrt{N}} = \frac{3.4}{\sqrt{33}} = .6$$

The average difference in attractiveness between the Conventional and the Socially Desirable portfolio is 2.5 standard errors:

$$t_{32} = \frac{\overline{X}_{dif}}{s_{\overline{X}_{dif}}} = -\frac{1.5}{.6} = -2.5$$

The *p value* for $t_{32} = -2.5$, for a sample size of 33, is .009. Were the Socially Desirable portfolio at least as attractive as the Conventional portfolio with equivalent risk and return, it would be unusual to observe such a large sample mean difference in ratings. Based on sample evidence, shown in Fig. 3.13, we conclude that a "socially desirable" label reduces portfolio attractiveness.

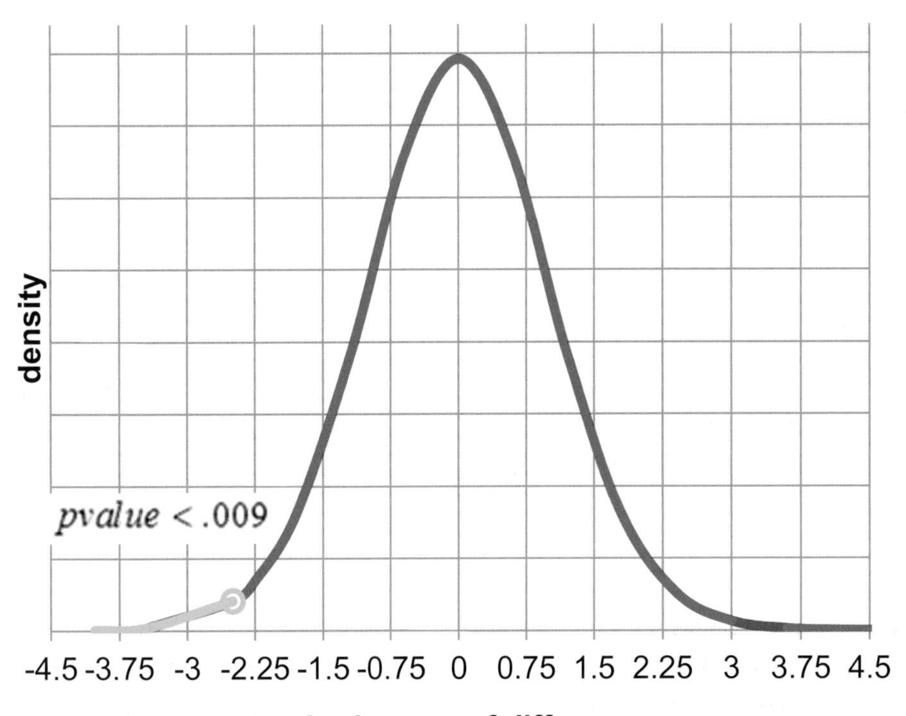

Fig. 3.13 *t test* of differences between paired ratings of socially desirable and conventional portfolios

The 95 % confidence interval for the difference is

$$\overline{X}_{dif} \pm t_{\alpha/2,N-1} s_{\overline{X}_{dif}}$$

$$-1.5 +/- 2.04\ (.6)$$

$$-1.5 +/- 1.2$$

or −2.7 to −.3 on the 7 point scale (Fig. 3.14).

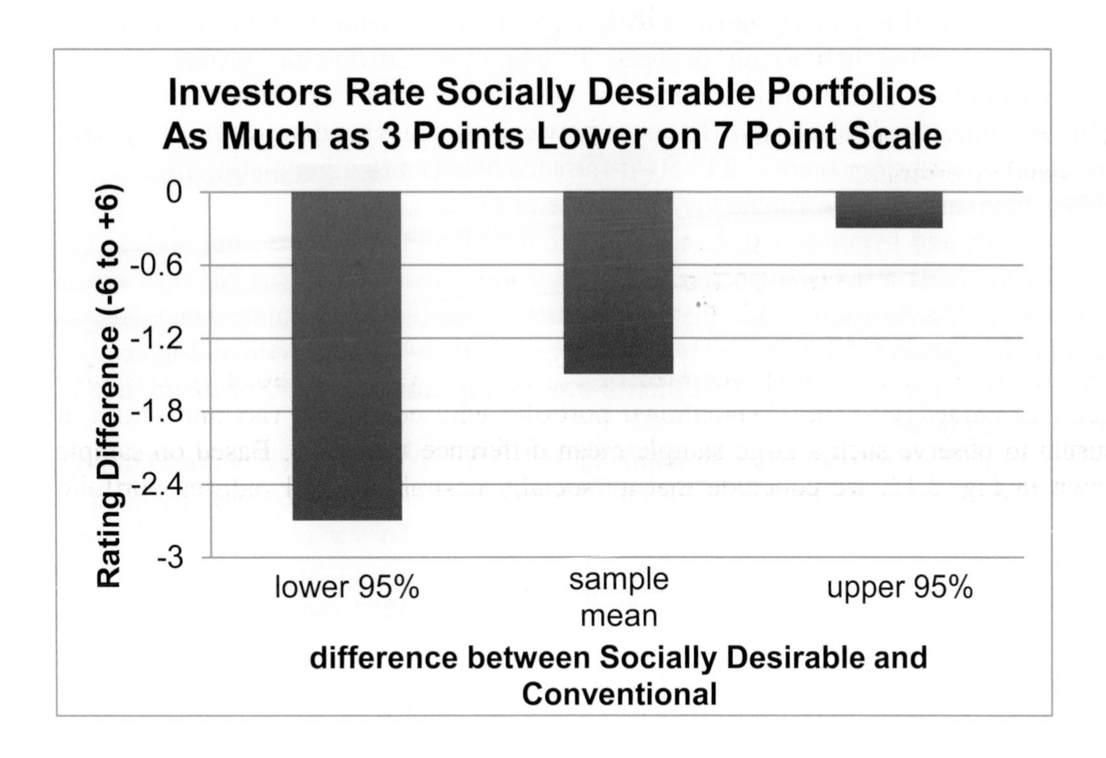

Fig. 3.14 *Confidence interval* of mean difference between paired ratings of socially desirable and conventional portfolios

The investment consultants would conclude:

A "socially desirable" label reduces investors' judged attractiveness ratings. Investors downgrade the attractiveness of "socially desirable" portfolios by about 1–3 points on a 7 point scale, relative to equivalent, but conventional, portfolios.

3.13 Inference from Sample to Population

Managers use sample statistics to infer population characteristics, knowing that inference from a sample is efficient and reliable. Because sample standard errors are approximately *Normally* distributed, we can use the Empirical Rule to build confidence intervals to estimate population means and to test hypotheses about population means with *t tests*. We can determine whether a population mean is likely to equal, be less than, or exceed a target value, and we can estimate the range which is likely to include a population mean.

Our certainty that a population mean will fall within a sample based confidence interval depends on the amount of population variation and on the sample size. To double precision, sample size must be quadrupled, because the margin of error is inversely proportional to the square root of sample size.

Differences are important to managers, since differences drive decision making. If customers differ, segments are targeted in varying degrees. If employee satisfaction differs between alternate work environments, the workplace may be altered. Inference about differences between two populations is similar, and relies on differences between two independent samples. A *t test* can be used to determine whether there is a likely difference between two population means, and with a confidence interval, we can estimate the likely size of difference.

Confidence intervals and hypothesis tests are consistent and complementary, but are used to make different decisions. If a decision maker needs to make a qualitative Yes/No decision, a hypothesis test is used. If a decision maker instead requires a quantitative estimate, such as level of demand, confidence intervals are used. Hypothesis tests tell us whether demand exceeds a critical level or whether segments differ. Confidence intervals quantify demand or magnitude of differences between segments.

Sample statistics are used to estimate population statistics because it is often neither possible nor feasible to identify and measure the entire population. The time and expense involved in identifying and measuring all population elements is prohibitive. To survey the bottled water consumption of each faculty member, student, and staff member on campus would take many hours. An estimate of demand is inferred from a random, representative sample which includes faculty, students, and staff. Though sample estimates will not be exactly the same as population statistics because of sampling error, samples are amazingly efficient if properly drawn and representative of the population.

Excel 3.1 Test the Level of a Population Mean with a One Sample *t test*

Thirsty on Campus. Team 8 wants to know whether the demand for bottled water exceeds a breakeven level of seven bottles per day. To compare the level of demand with to this critical level, we will use a one tail *t test* of *Bottles* purchased per day.

Open **Excel 3.1 Bottled Water Demand.xls.**

Find the sample *mean* and *standard deviation.*

Find the standard error by dividing the sample standard deviation in by the square root of sample size, **30**:

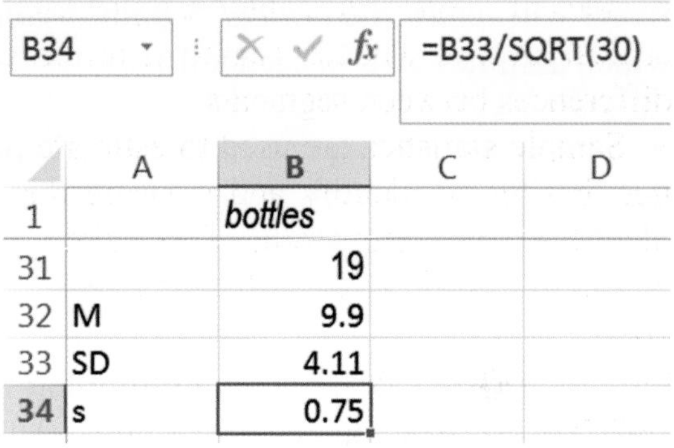

B34 =B33/SQRT(30)

	A	B	C	D
1		*bottles*		
31		19		
32	M	9.9		
33	SD	4.11		
34	s	0.75		

Find the difference between the sample mean and the critical value, **7**, and then divide that difference by the standard error to find *t*.

Find the *p value* for this *t* using the Excel function **TDIST.RT(*t,df*)**. For degrees of freedom, *df*, enter the sample size, minus one, **29** (=30−1).

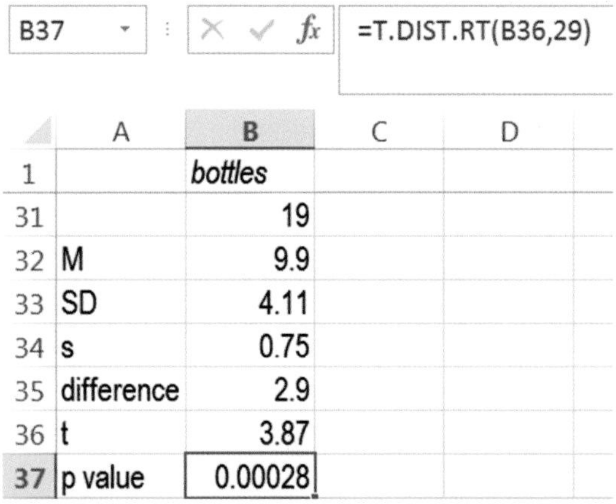

B37 =T.DIST.RT(B36,29)

	A	B	C	D
1		*bottles*		
31		19		
32	M	9.9		
33	SD	4.11		
34	s	0.75		
35	difference	2.9		
36	t	3.87		
37	p value	0.00028		

Excel 3.2 Make a Confidence Interval for a Population Mean

Determine the range which is likely to contain average demand in the population. Construct the 95 % confidence interval for the population mean *Bottles* demanded.

Use the Excel function **T.INV.2T(***probability, df***)** to find the *critical two tail t value* for 95 % confidence.

For *probability*, enter **.05** for a 95 % level of confidence. For *df*, enter the sample size, minus 1, **29**.

(*Note* that Excel requires ∝, and *not* ∝/2. Excel will return the *critical two tail t value* which assumes ∝/2 in each tail.)

B38 =T.INV.2T(0.05,29)

	A	B	C	D
1		*bottles*		
31		19		
32	M	9.9		
33	SD	4.11		
34	s	0.75		
35	difference	2.9		
36	t	3.87		
37	p value	0.00028		
38	critical t	2.05		

Find the *margin of error* by multiplying the *standard error* by the *two tail critical t*.

Add and subtract the *margin of error* from the sample *mean* to find the *95 % upper* and *lower* confidence interval limits:

B41 =B32+B39

	A	B
1		*bottles*
31		19
32	M	9.9
33	SD	4.11
34	s	0.75
35	*difference*	2.9
36	*t*	3.87
37	*p value*	0.00028
38	*critical t*	2.05
39	*me*	1.5
40	*upper 95% ci bound*	8.4
41	*lower 95% ci bound*	11.4

Excel 3.3 Illustrate Confidence Intervals with Column Charts

t-mobile's Service. t-mobile managers have conducted a survey of customers in 32 major metropolitan areas to assess the quality of service along three key areas: coverage, absence of dropped calls, and static. Customers rated t-mobile service along each of these three dimensions using a 5-point scale (1 = poor to 5 = excellent). Management's goal is to be able to offer service that is not perceived as inferior. This goal translates into mean ratings that exceed 3 on the 5-point scale in the national market across all three service dimensions. Make 95 % confidence intervals to estimate the average perceived quality of service.

Open **Excel 3.2 t-mobile.xls**.

95 % Confidence Intervals. Find the sample *mean, standard deviation, standard error, critical t* value, *margin of error*, and *lower and upper 95 %* confidence interval bounds for *coverage, dropped calls, and static ratings.*

C34 =AVERAGE(C2:C33)

	A	B	C	D	E
1	city	service	coverage rating (1=Poor to 5=Excellent)	dropped calls rating (1=Poor to 5=Excellent)	static rating (1=Poor to 5=Excellent)
32	miami	tmobile	2	3	4
33	raleigh	tmobile	1	3	2
34		M	2.25	3.375	2.9375
35		SD	0.98	0.61	0.56
36		s	0.17	0.11	0.10
37		critical t	2.04	2.04	2.04
38		me	0.35	0.22	0.20
39		lower 95%	1.90	3.16	2.73
40		upper 95%	2.60	3.59	3.14

Column chart of confidence intervals. To see the confidence intervals for all three service dimension ratings, first, select and copy row **1**, containing labels, **Cntl+C**, and then paste the copy above the *lower 95 %* confidence interval bounds.

Select row **39**, **Alt HIE**.

(**Alt** activates shortcuts, **H** selects the **H**ome menu, **I** selects **I**nsert menu, and **E** inserts copied cells.)

	A	B	C	D	E
1	city	service	coverage rating (1=Poor to 5=Excellent)	dropped calls rating (1=Poor to 5=Excellent)	static rating (1=Poor to 5=Excellent)
39	city	service	coverage rating (1=Poor to 5=Excellent)	dropped calls rating (1=Poor to 5=Excellent)	static rating (1=Poor to 5=Excellent)
40		lower 95% ci bound	1.90	3.16	2.73
41		upper 95% ci bound	2.60	3.59	3.14

To make a column chart, select the labels and 95 % confidence interval bounds, and then use shortcuts

Alt NC.

(N invokes the **IN**sert menu, and **C** specifies a **C**olumn chart.)

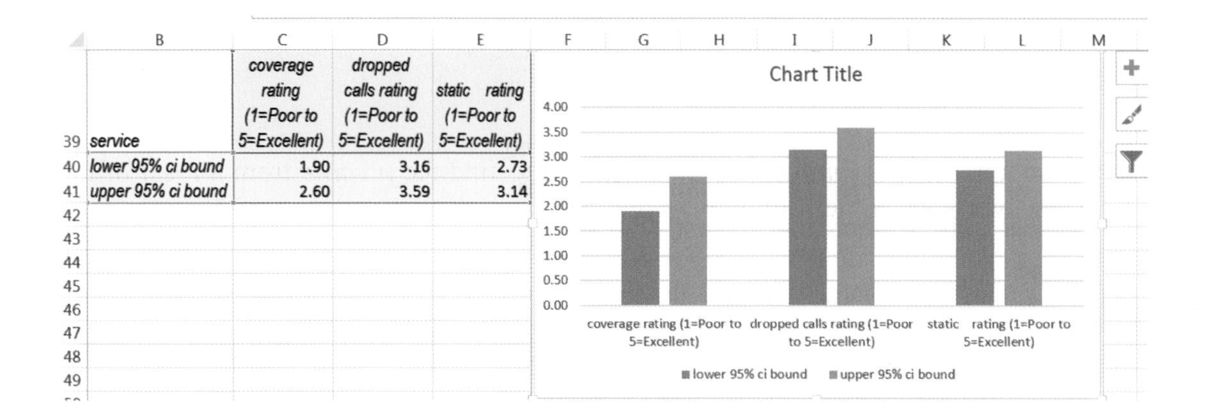

service	coverage rating (1=Poor to 5=Excellent)	dropped calls rating (1=Poor to 5=Excellent)	static rating (1=Poor to 5=Excellent)
lower 95% ci bound	1.90	3.16	2.73
upper 95% ci bound	2.60	3.59	3.14

Choose **Design Layout 9** to add axes labels:

Adjust axes and type in axis label and chart title:

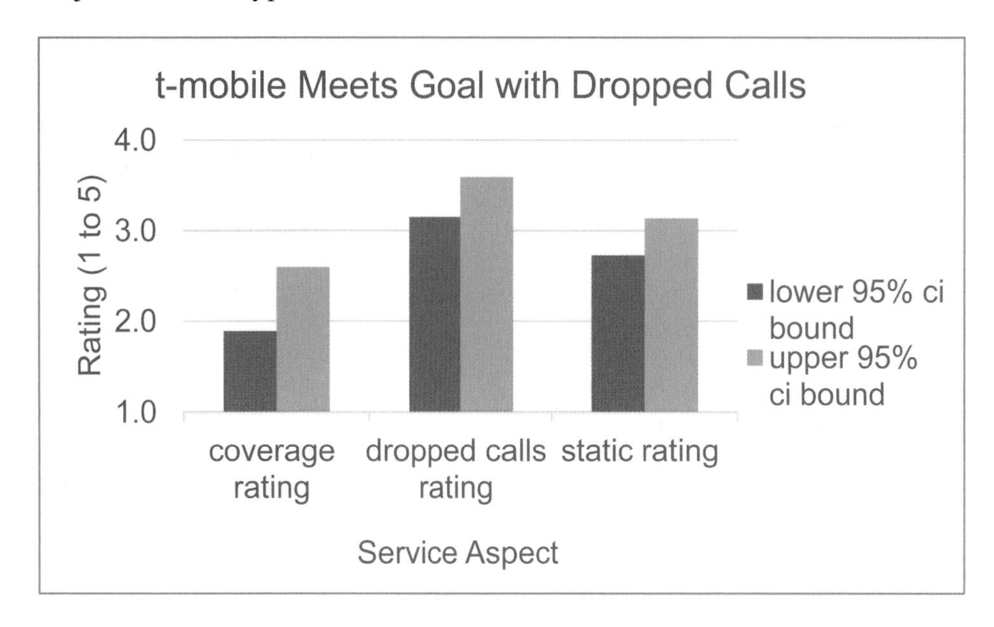

Excel 3.4 Test the Difference Between Two Segment Means with a Two Sample t test

Pampers Preemies. Procter & Gamble management would like to know whether or not household income is a good base for segmentation in the market for their new preemie diaper. Test the hypothesis that average income is greater in the segment likely to try the new diapers than in the segment unlikely to try.

Open **Excel 3.3 Pampers Segment Income.xls**. The first column **A** contains *likely trier income ($K)* and the second column **B** contains *unlikely trier income ($K)*.

Use the Excel function **T.TEST(***array1,array2,tails,type***)** to find the *p value* from a *t test* of the difference between average incomes of the two segments.

For *array1*, enter the sample *likely trier income* values.

For *array2*, enter the sample *unlikely trier income* values.

For *tails*, enter **1** for a *one tail* test, and for *type*, enter **3** to signal a two sample *t test* which allows the standard deviations to differ between segments:

B58 =T.TEST(A2:A57,B2:B57,1,3)

	A	B	C	D
1	*Likely Triers Income*	*Unlikely Triers Income*		
57	156			
58	*p value*	0.000048		

Excel 3.5 Construct a Confidence Interval for the Difference Between Two Segments

Estimate the difference in incomes between the Unlikely and Likely Trier segments.

At the end of the **Excel 3.3 Pampers Segment Income.xls** dataset, find the segment sample means, standard deviations, and standard errors.

A59 =AVERAGE(A2:A57)

	A	B	C
1	*Likely Triers Income*	*Unlikely Triers Income*	
57	156		
58	*p value*	0.000048	
59	80.1	38.5	*M*
60	51.7	48.0	*SD*
61	6.9	7.5	*s*

Find the difference between segment means and the standard error of the difference from the segment sample means and standard errors:

B63 =SQRT(A61^2+B61^2)

	A	B	C
1	*Likely Triers Income*	*Unlikely Triers Income*	
57	156		
58	*p value*	0.000048	
59	80.1	38.5	*M*
60	51.7	48.0	*SD*
61	6.9	7.5	*s*
62	*difference*	41.6	
63	*pooled s*	10.2	

Find the approximate margin of error, which will be twice the standard error.

Make the 95 % confidence interval for the difference by adding and subtracting the margin of error from the mean difference:

B66 =B62+B64

	A	B
1	*Likely Triers Income*	*Unlikely Triers Income*
62	*difference*	41.6
63	*pooled s*	10.2
64	*approximate me*	20.4
65	*lower 95% ci bound*	21.2
66	*upper 95% ci bound*	62.0

Excel 3.6 Illustrate the Difference Between Two Segment Means with Column Chart

Illustrate the difference between average incomes of Likely and Unlikely Triers.

Select the *lower* and *upper 95 %* confidence interval bounds and their labels, and then use short cuts to insert a column chart: **Alt NC**.

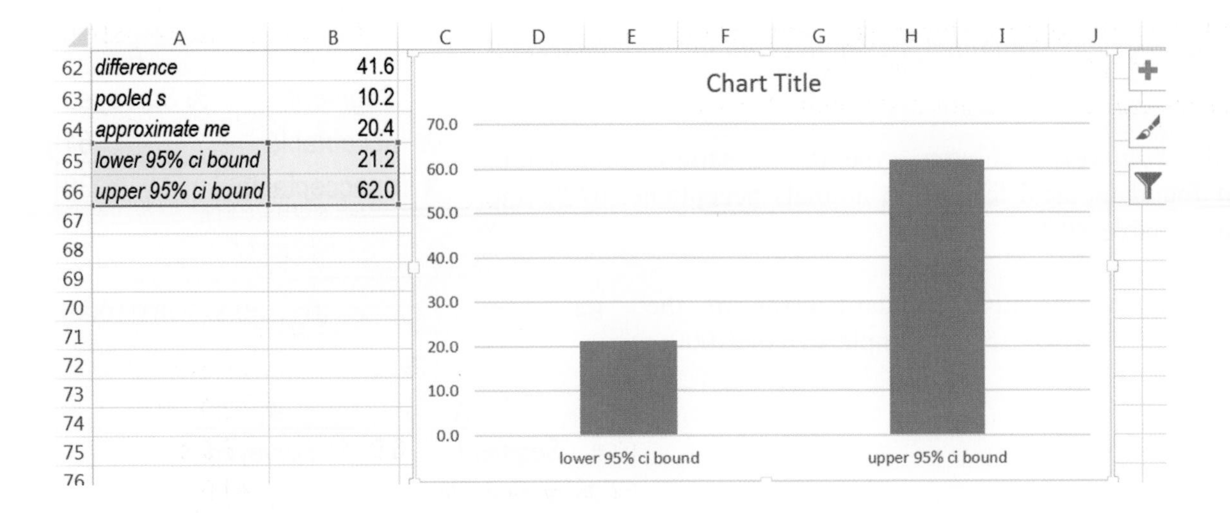

Choose Layout 6, using **Alt JC L**.

Add data labels: **Alt JC A D O**

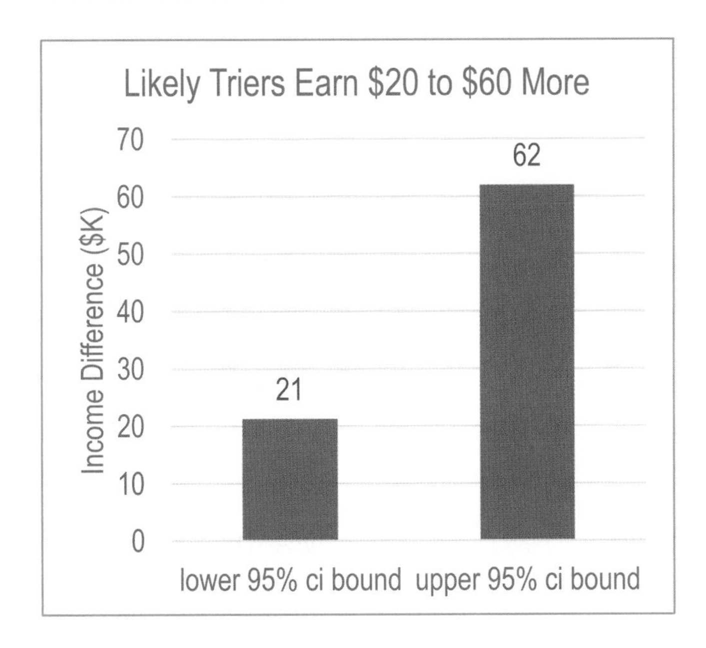

Excel 3.7 Construct a Pie Chart of Shares

Moral Acceptance of Medical Testing on Animals. Construct a pie chart to illustrate how sample ratings of the acceptability of medical testing on animals are split.

Open a new workbook and type in two new columns, *segment* and *%surveyed.*

In the *segment* column, type in *acceptable* and *unacceptable*.

In the *%surveyed* column, type in the sample proportions that found medical testing on animals acceptable, *61 %* and unacceptable *39 %*.

	A	B
1	*Segment*	*% Surveyed*
2	acceptable	61
3	unacceptable	39

Find the conservative standard error of the proportion from P = .5 and sample size of 1,000:

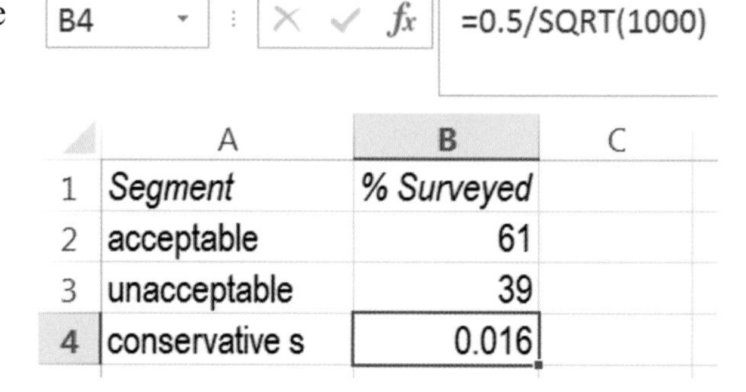

B4 =0.5/SQRT(1000)

	A	B	C
1	*Segment*	*% Surveyed*	
2	acceptable	61	
3	unacceptable	39	
4	conservative s	0.016	

Find the margin of error from the *critical Z* for 95 % confidence (1.96) and the *conservative standard error of the proportion*:

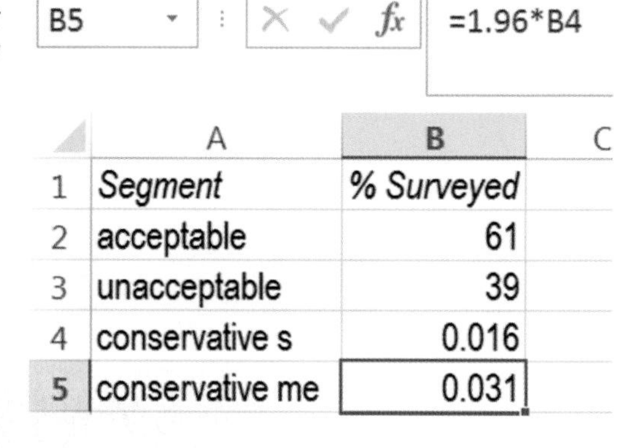

B5 =1.96*B4

	A	B	C
1	*Segment*	*% Surveyed*	
2	acceptable	61	
3	unacceptable	39	
4	conservative s	0.016	
5	conservative me	0.031	

To make a pie chart, select the six label and data cells, and then use shortcuts to insert a pie chart: **Alt NQ**.

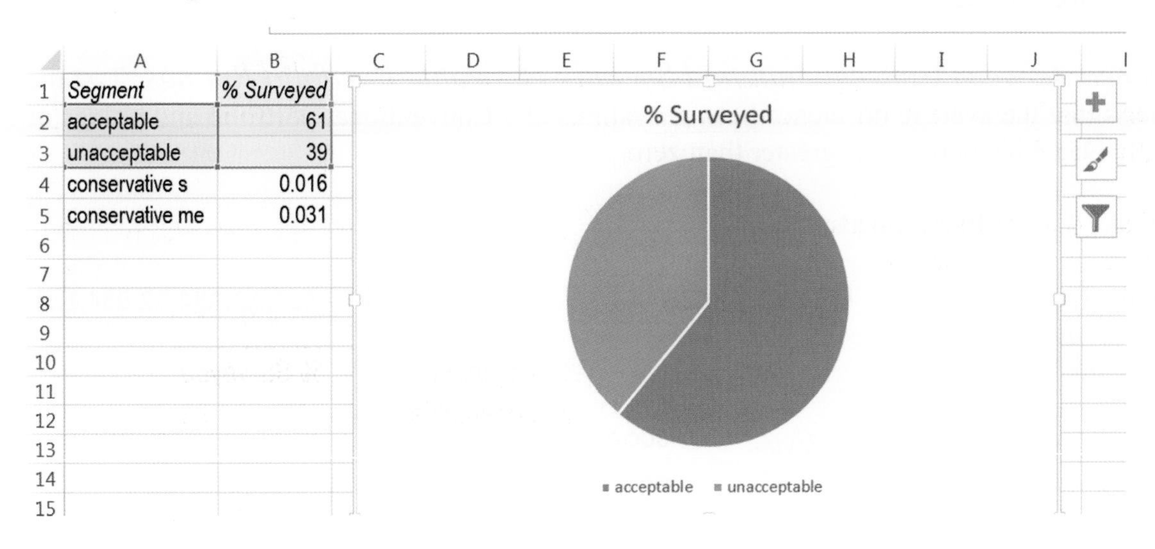

Choose **Design Layout 1**.

To add the margin of error, use shortcuts to insert a text box below the pie: **Alt N ZT X**.

Type in *Margin of error: 3.1 %*:

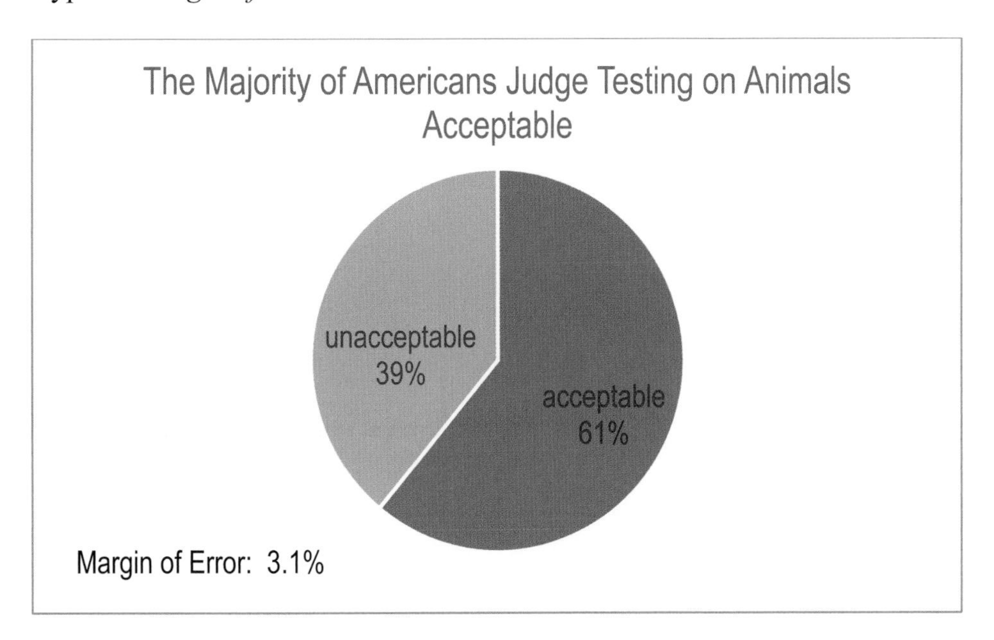

Excel 3.8 Test the Difference in Between Alternate Scenarios or Pairs with a Paired t test

Difference Between Conventional and Socially Desirable Portfolio Ratings. Test the hypothesis that the average difference between ratings of a Conventional portfolio and ratings of a Socially Desirable portfolio is greater than zero.

Open **Excel 3.4 SD Portfolio.xls**.

Use the Excel function **T.TEST (***array1, array2,tails,type***)** to calculate a paired *t test*.

For *array1*, enter the *socially desirable portfolio ratings*. For *array2*, enter the *conventional portfolio ratings*. (It does not matter which array is first.) For *tails***,** enter **1** for a *one tail* test, and for *type*, enter **1** to specify a paired *t test*:

C35 =T.TEST(A2:A34,B2:B34,1,1)

	A	B	C	D
1	socially desirable rating	conventional rating	difference	Socially Desirable difference
33	3	-3	6	
34	3	-3	6	
35		p value	0.0097	

Excel 3.9 Construct a Confidence Interval for the Difference Between Alternate Scenarios or Pairs

To estimate the population difference in investors' ratings of Socially Desirable and Conventional portfolios from sample data, construct a confidence interval of the average rating difference.

Find the mean and standard deviation of the *difference.*

Use the standard deviation of the difference to find the margin of error of the difference with the **CONFIDENCE.T (***alpha, standard deviation, sample size***)**.

For a 95 % confidence interval, enter **.05** for *alpha.*

Enter the *sample size*, **33**. (Do not subtract one—Excel will do this.)

C38 =CONFIDENCE.T(0.05,C37,33)

	A	B	C	D	E
1	socially desirable rating	conventional rating	difference	Socially Desirable difference	
33	3	-3	6		
34	3	-3	6		
35		p value	0.0097		
36		M	-1.45		
37		SD	3.39		
38		me	1.20		

Subtract and add the margin of error from the mean difference to find the *95 % confidence interval bounds* for the difference:

C40 =C36+C38

	A	B	C
1	*socially desirable rating*	*conventional rating*	*difference*
33	3	-3	6
34	3	-3	6
35		*p value*	0.0097
36		M	-1.45
37		SD	3.39
38		*me*	1.20
39		*lower 95% ci bound*	-2.66
40		*upper 95% ci bound*	-0.25

Lab Practice 3 Inference

Cingular's Position in the Cell Phone Service Market

Cingular's managers have conducted a survey of customers in 21 major metropolitan areas to assess the quality of service along three key areas: *coverage, absence of dropped calls*, and *static*. Customers rated Cingular service along each of these three dimensions using a 5-point scale (1 = poor to 5 = excellent). Data are in **Lab Practice 3 cingular.xls**

Management's goal is to be able to offer service that is not perceived as inferior. This goal translates into mean ratings that are greater than 3 on the 5-point scale in the national market across all three service dimensions.

Management can conclude that they have achieved their goal along:

____ *coverage* _____*dropped calls* _____ *static*

Based on this sample, average ratings in all major metropolitan areas are

_________ to _________ for *coverage*,

_________ to _________ for absence of *dropped calls*,

_________ to _________ for *static*, with 95 % confidence.

Value of a Nationals Uniform

The Nationals General Manager is concerned that his club may not be paying competitive salaries. He has asked you to compare Nationals' salaries with salaries of players for the closest team in the National League East, the Phillies. He suspects that the Phillies may win more games because they are attracting better players with higher salary offers. Data are in **Lab Practice 3 Nationals.xls.**

This is a _____tail *t test.*

p value from one tail *t test* of difference in team *salary* means: ______________

The General Manager can conclude that, relative to the Phillies, the Nationals are paid ___Less ___the same.

Extra Value of a Phillies Uniform. If you conclude that the Phillies do earn higher salaries, estimate the average difference at a 95 % level of confidence.

On average, players for the Phillies earn _______ to ________ more than players for the Nationals.

The pooled standard error of the difference in mean salaries is: ____________

Illustrate the 95 % confidence interval for the difference between the two teams' salaries with a column chart.

Confidence in Chinese Imports

Following the recall of a number of products imported from China, the Associated Press-Ipsos Poll asked 1,005 randomly selected adults about the perceived safety of products imported from China. Poll results are below:

"When it comes to the products that you buy that are made in China, how confident are you that those products are safe . . . ?"

Confident	Not Confident	Unsure
%	%	%
42	57	1

Use this data to construct a *conservative 95 % confidence interval* for the *proportion Not Confident* that Chinese imports are safe.

________ to ________ percent are not confident that products made in China are safe.

Illustrate your result with a pie chart which includes the margin of error in a text box. Add a "bottom line" title.

Lab 3 Inference: Dell Smartphone Plans

Managers at Dell are considering a joint venture with a Chinese firm to launch a new smartphone equipped with Microsoft's Windows Phone 8, which offers full Skype compatibility.

I. Estimate the Percent of Smartphone Owners Who Will Replace with Dell

In a concept test of 1,000 smartphone owners, 20 % indicated that they would probably or definitely replace their smartphone with the new Dell concept in the next quarter. Norms from past research suggest that 80 % of those who indicate intent to replace actually will.

Expected Dell smartphone share = 80 % × sample intent proportion: _________

II. Distinguish Likely Dell Smartphone Adopters

Those who indicated that they were likely to switch to the Dell smartphone may be more price conscious than other smartphone owners. In the concept test, participants were asked to rate the importance of several smartphone attributes, including price. These data are in **Lab 3 Inference Dell smartphone.xls**.

1. Do Likely Adopters rate price higher in importance than Unlikely Adopters?

p value: _________ Conclusion: Y or N

2. What is the expected difference in *price importance* between Likely and Unli*kely Adopters?*

   ____

3. Approximate 95 % confidence interval of the difference in *price importance* between Likely and Unlikely Adopters: _______ *to* _______

4. Make a column chart to illustrate your results.

Assignment 3-1The Marriott Difference

There are 51 branded hotels in Washington, DC, owned or managed by Marriott or competitors. The hotel industry in Washington, DC is representative of the hotel industry in cities throughout the U.S. Differences in quality and price distinguish the hotels. Marriott would like to claim that its hotels offer higher average quality lodging than competing hotels and that Marriott's average *starting room price* is equivalent to competitors' average *starting room price.* The dataset **Assignment 3-1 DC Hotels.xls** contains *Guest rating*, a measure of quality, and *starting room price* for Marriott hotels and for competitors' hotels.

1. Can Marriott claim that Marriott hotels are rated higher in quality than competitors' hotels? (Assume a 95 % level of confidence.)
   a. State the null and alternative hypotheses.
   b. State your conclusion in one sentence with words that a technically savvy manager would understand.
   c. State your conclusion in one sentence with words that a manager, not necessarily statistically savvy, would understand.

2. Can Marriott claim that Marriott hotels are priced equivalently to competitors? (Assume a 95 % level of confidence.)
   a. State the null and alternative hypotheses.
   b. State your conclusion in one sentence that a technically savvy manager would understand.
   c. State your conclusion in one sentence with words that a manager, not necessarily statistically savvy, would understand.

Assignment 3-2 Immigration in the U.S.

The FOX News/Opinion Dynamics Poll of (N=) 900 registered voters nationwide, reports public opinion concerning immigrants and proposed immigration legislation:

Join society/ give	Stay separate/ take	Depends (vol.)	Unsure
%	%	%	%
41	36	17	6
Increase	**Decrease**	**No change (vol.)**	**Unsure**
%	%	%	%
24	51	17	8

Use this data to construct ***conservative*** *95 % confidence intervals* for the *proportions* who (i) agree that immigrants joint society/give and (ii) agree that the U.S. should increase the number of legal immigrants.

Briefly summarize the opinions of **all registered voters** using language that American adults would understand.

Illustrate your summary with pie charts embedded in your report.

Be sure to include the margins of error in your pie charts.

Assignment 3-3 McLattes

McDonalds recently sponsored a blind taste test of lattes from Starbucks and their own McCafes. A sample of 30 Starbucks customers tasted both lattes from unmarked cups and provided ratings on a −3 (=worst latte I've ever tasted) to +3 (=best latte I've ever tasted) scale. These data are in **Assignment 3-3 Latte.xls**.

Can McDonalds claim that their lattes taste every bit as good as Starbucks' lattes? (Please use 95 % confidence.)

What evidence allows you to reach this conclusion?

Assignment 3-4 A Barbie Duff in Stuff

Mattel recently sponsored a test of their new Barbie designed by Hillary Duff. The Duff Barbie is dressed in Stuff, Hillary Duff clothing designs, and resembles Hillary Duff. Mattel wanted to know whether or not the Duff Barbie could compete with rival MGA Entertainment's Bratz dolls.

A sample of 30 7-year-old girls attended Barbie parties, played with both dolls, then rated both on a −3 (=Not At All Like Me) to +3 (=Just Like Me) scale. These data are in **Assignment 3-4 Barbie.xls**.

Do the 7-year-olds identify more strongly with the Duff Barbie in Stuff than the Bratz? (Please use 95 % confidence.)

What evidence allows you to reach this conclusion?

CASE 3-1 Yankees v Marlins: The Value of a Yankee Uniform[1]

The Marlins General Manager is disgruntled because two desirable rookies accepted offers from the Yankees instead of the Marlins. He believes that Yankee salaries must be noticeably higher—otherwise, the best players would join the Marlins organization. Is there a difference in salaries between the two teams? If the typical Yankee is better compensated, the General Manager is planning to chat with the Owners about sweetening the Marlins' offers. He suspects that the Owners will argue that the typical Yankee is older and more experienced, justifying some difference in salaries.
Data are in **Case 3-1 Yankees v Marlins Salaries.xls**.
Determine:

- Whether or not Yankees earn more on average than Marlins, and
- Whether or not players for the Yankees are older on average than players for the Marlins.

If you find a difference in either case, construct a *95 % confidence interval* of the expected difference in any season.

Briefly summarize your results using language that the General Manager and Owners would understand, and illustrate with a column chart.

CASE 3-2 Gender Pay

The Human Resources manager of Slam's Club was shocked by the revelations of gender discrimination by WalMart and wants to demonstrate that there is no gender difference in average salaries in his firm. He also wants to know whether levels of responsibility (measured with the Position variable) and experience differ between men and women, since this could explain a difference in salaries.

Case 3-2 GenderPay.xls contains *salaries, positions*, and *experience* of men and women from a random sample of the company records.
Determine:

- Whether or not the sample supports a conclusion that men and women are paid equally,
- Whether average level of *responsibility* differs across genders,
- Whether average *experience* differs across genders.

If you find that the data support the alternate hypothesis that men are paid more, on average, construct a 95 % confidence interval of the expected average difference.

If either average level of *responsibility* or average years of *experience* differs, construct *95 % confidence intervals* of the expected average difference.

Briefly summarize your results using language that a businessperson (who may not remember quantitative analysis) could understand.

---

[1] This example is a hypothetical scenario using actual data.

Illustrate your results with column charts. Choose bottom line titles that help your audience see the results.

Be sure to round your statistics to two or three significant digits.

CASE 3-3 Polaski Vodka: Can a Polish Vodka Stand Up to the Russians?

Seagrams management decided to enter the premium vodka market with a Polish vodka, suspecting that it would be difficult to compete with Stolichnaya, a Russian vodka and the leading premium brand. The product formulation and the package/brand impact on perceived taste were explored with experiments to decide whether the new brand was ready to launch.

The taste. First, Seagrams managers asked, "Could consumers distinguish between Stolichnaya and Seagrams' Polish vodka in a *blind* taste test, where the impact of packaging and brand name were absent?"

Consultants designed an experiment to test the null and alternative hypotheses:

H_0: The taste rating of Seagram's Polish vodka is at least as high as the taste rating of Stolichnaya. The average difference between taste ratings of Stolichnaya and Seagrams' Polish vodka does not exceed zero:

$$\mu_{STOLICHNAYA} - \mu_{POLISH} \leq 0$$

H_1: The taste rating of Seagram's Polish vodka is lower than the taste rating of Stolichnaya. The average difference between taste ratings of Stolichnaya and Seagram's Polish vodka is positive:

$$\mu_{STOLICHNAYA} - \mu_{POLISH} > 0$$

In this first experiment, each participant tasted two unidentified vodka samples and rated the taste of each on a 10-point scale. Between tastes, participants cleansed palates with water. Experimenters flipped a coin to determine which product would be served first: if heads, Seagrams' polish vodka was poured first; if tails, Stolichnaya was poured first. Both samples were poured from plain, clear beakers. The only difference between the two samples was the actual vodka.

These experimental data in **Case 3-3 Polaski Taste.xls** are repeated measures. From each participant, we have two measures whose difference is the difference in taste between the Russian and Polish vodkas.

Test the difference between taste ratings of the two vodkas.

Construct a *95 % confidence interval* of the difference in taste ratings.

Illustrate your results with a PivotChart and interpret your results for management.

The brand & package. Seagrams management proceeded to test the packaging and name, Polaski. The null hypothesis was:

H_0: The taste rating of Polaski vodka poured from a Polaski bottle is at least as high as the taste rating of Polaski vodka poured from a Stolichnaya bottle. The mean difference between taste ratings of Polaski vodka poured from a Stolichnaya bottle and Polaski vodka poured from the Seagrams bottle bearing the Polaski brand name is not exceed zero.

Alternatively, if the leading brand name and distinctive bottle of the Russian vodka affected taste perceptions, the following could be true:

H_1: The mean difference between taste ratings of Polaski vodka poured from Stolichnaya bottle and Polaski vodka poured from the Seagrams bottle bearing the Polaski brand name is positive.

In this second experiment, Polaski samples were presented to participants twice, once poured from a Stolichnaya bottle, and once poured from the Seagrams bottle, bearing the Polaski name. Any minute differences in the actual products were controlled for by using Polaski vodka in both samples. Differences in taste ratings would be attributable to the difference in packaging and brand name.

Thirty new participants again tasted two vodka samples, cleansing their palates with water between tastes. As before, a coin toss decided which bottle the first sample would be poured from: Stolichnaya if heads, Polaski if tails. Each participant rated the taste of the two samples on a 10-point scale.

These data are in **Case 3-3 Polaski Package.xls**.

Test the difference in ratings due to packaging.
Construct a *95 % confidence interval* of the difference in ratings due to the packaging.
Illustrate your results with a PivotChart.
Interpret your results for management.

Chapter 4
Simulation to Infer Future Performance Levels Given Assumptions

Decision makers deal with uncertainty when considering future scenarios. Performance levels depend on multiple influences with uncertain future values. To estimate future performance, managers make assumptions about likely future scenarios and uncertain future values of performance components. To evaluate decision alternatives, the "best" and "worst" *case* outcomes are sometimes compared. Alternatively, *Monte Carlo simulation* can be used to simulate random samples using decision makers' assumptions about performance components, and those random samples can then be combined to produce a distribution of likely future scenarios and outcomes that are less extreme that the "best" and "worst" cases. Inferences from a simulated distribution of outcomes can then be made to inform decision making. "Best" and "worst" case comparisons are contrasted with inferences from Monte Carlo simulation in this chapter.

4.1 Specify Assumptions Concerning Future Performance Drivers

Example 4.1 The **Thirsty** Team 8 partners were concerned that they might either pass up a profitable opportunity or invest in an unprofitable business. Their estimate of average bottles of water demanded per customer per week seemed promising, though they realized that success of the business depended on several factors, each with uncertain values in the future. An estimate of potential revenues from the first year of operation was desired.

Potential revenues depended on uncertain factors: the number of potential customers, the market growth rate, access (restricted or not) to those potential customers, share of customers that the new business could capture, and bottles demanded per customer.

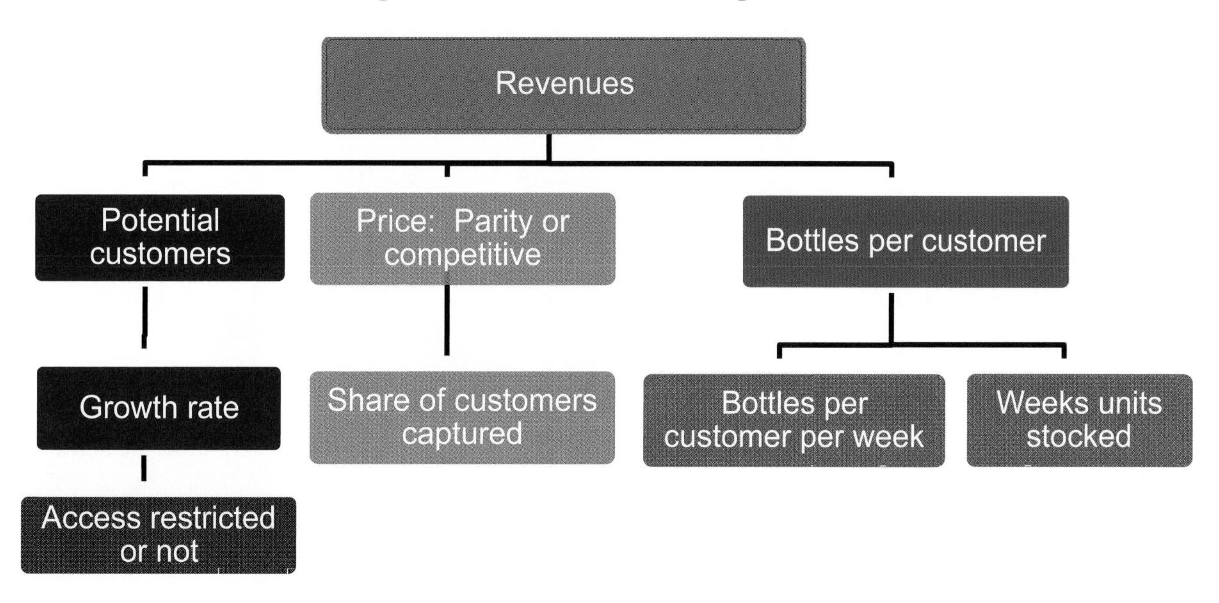

C. Fraser, *Business Statistics for Competitive Advantage with Excel 2013: Basics, Model Building, Simulation and Cases*, DOI 10.1007/978-1-4614-7381-7_4, © Springer Science+Business Media New York 2013

Spreadsheet. The Team created a spreadsheet linking each of the uncertain revenue influences to weekly revenues, given their assumptions about the potential market.

Assumptions concerning possible outcomes for uncertain influences were highlighted in green, with three scenarios considered: the two extremes, "worst case" and "best case", and the expected, "best guess."

The chance of an outcome worse than "the worst" or better than "the best" was assumed to be about 5 %, making the interval from worst case to best case a 95 % confidence interval (Table 4.1).

Table 4.1 Spreadsheet for bottled water revenue

		Assumptions		
		95 % confidence interval		
		worst	*expected*	*best*
(1)	*Potential customers last year (K)*	34.1	34.1	34.1
(2)	*Annual growth %*	2.5 %	3.5 %	4.5 %
(3)	*Potential customers in year 1 (K)* = (1) × (100 % + (2))	35.0	35.3	35.6
(4)	*P(Unrestricted access)*	0 %	67 %	100 %
(5)	*% Customers w access* = (4) × 100 % + (1−(4)) × 80 %	80 %	93 %	100 %
(6)	*Customers accessed (K)* = (3) × (5)	28.0	33.0	35.6
(7)	*Price per Bottle ($) given access* = $1.5−(4) × $.25	$1.50	$1.33	$1.25
(8)	*Share captured at parity price*	10 %	15 %	20 %
(9)	*Share captured at competitive price*	20 %	35 %	50 %
(10)	*Share captured given price* = (100 %−(4)) × (8) + (4) × (9)	10 %	28 %	50 %
(11)	*Customers captured (K)* = (6) × (10)	2.8	9.4	17.8
(12)	*Bottles sold per customer per week*	8	10	12
(13)	*Bottles sold per week (K)* = (11) × (12)	22.4	93.6	213.8
(14)	*Weeks in business*	38	38	38
(15)	*Bottles sold in year 1 (M)* = (13) × (14)/1,000	.85	3.6	8.1
(16)	*Revenues in year 1($M)* = (7) × (13)	$1.28	$4.74	$10.2

Potential customers in year 1. Potential customers include faculty, staff, and students on campus, currently 34.1K.

University admissions had been growing between 3 % and 4 % in recent years, so future growth between 2.5 % and 4.5 % is anticipated with 95 % confidence. Hiring of faculty and staff is expected to grow at similar rates to accommodate the student population.

The potential market in the first year of business is:

Potential customers (K) = *Potential customers last year (K)* × (100 % + *annual growth%*)

= 34.1K × (100 % + *annual growth%*)

= 34.1K × 102.5 % = 35.0K in the worst case,
= 34.1K × 104.5 % = 35.6K in the best case,
= 34.1K × 103.5 % = 35.3K in the expected case.

Access. If the new business is successful in gaining approval to place units in dorms, 100 % of the potential market would have access. Without this approval, restricted access for vending units would reach an about 80 % of the potential market.

Customers accessed (K) = *%Accessed* × *Potential customers (K)*

= 80 % × 35.0K = 28.0K in the worst case,
= 100 % × 35.6K = 35.6K in the best case,

The Team assumed that chance of unrestricted access, *P*(*Unrestricted Access*), was about 67 %:

Customers accessed (K) = 67 % × (100 % × 35.6K)

+ (100 %–67 %) × (80 % × 35.0K) in the expected case.

Price. Bottled water on campus sells for $1.50 from vending units and in campus eateries. If access is unrestricted, the Team assumes that the volume of business to be great enough to enable volume discounts on plastic bottles and natural flavorings. In this case, a lower price of $1.25 could be charged, which would be assumed to stimulate trial and repeat sales.

Share. With restricted access and a parity price, the Team assumes that the business could capture at least 10 % of the market, and possibly as much as 20 %. With unrestricted access and the lower price, they assume that at least 20 % of the market would be captured, and that 50 % share would be possible.

Customers captured (K) = *Share captured* × *Customers accessed (K)*

= 10 % × 28.0K = 2.8K in the worst case,
= 50 % × 35.6K = 17.8K in the best case,

= (67 % × 35 % + (100 %–67 %) × 15 %) × 33.0K = 9.4K expected.

From their market research, the Team estimates that the average number of bottles of water demanded per customer per week falls within the range of 8–12, with 95 % confidence, and an average of 10 bottles per customer per week is expected.

Given this level of demand per customer, weekly sales would be

Bottles sold per week (K) = Bottles per customer per week × Customers captured (K)

= 8 × 2.8K = 22.4K in the worst case,
= 12 × 17.8K = 213.8K in the best case,
= 10 × 9.4K = 93.6K expected.

The Team assumes that the business will operate during the 38 weeks in which classes are in session. Therefore, volume in the first year, in millions (M), would be:

Bottles sold (M) = 38 × *Bottles sold per week (K)*/1,000

= 38 × 22.4K/1,000 = .85M in the worst case,
= 38 × 213.8K/1,000 = 8.1M in the best case,
= 38 × 93.6K/1,000 = 3.6M expected.

At those potential volumes, with the two alternative prices, revenue in the first year would be:

Potential revenue ($M) = Price × Bottles sold (M),

= $1.50 × .85M = $1.3M in the worst case,
= $1.25 × 8.1M = $10.2M in the best case,
= $1.33 × 3.6M = $4.7M expected.

4.2 Compare Best and Worst Case Performance Outcomes

Best versus *Worst.* If worst case outcomes occurred (slower growth, restricted access, parity price at $1.50, 10 % share, low demand per customer), revenue would be just $1.3M in the first year, making the investment unattractive. However, if best case outcomes occurred (faster growth, unrestricted access, competitive price at $1.25, 50 % share, high demand per customer), revenue would be $10.2M, making the investment extremely attractive.

These extreme outcomes differ widely. How likely are these two extremes?

Based on the Team's assumptions, the chance of the worst case outcome is equal to the joint probability assumed for the four uncertain influences:

P(the worst case outcome) = P(annual growth ≤ 2.5%) × P(access restricted)

× *P(Share ≤ 10 %)*

× *P(Demand ≤ 8 bottles per customer per week)*

The chance that annual market growth would be as low as 2.5 %, *P(annual growth* $\leq$ *2.5 %)*, the low end of the 95 % confidence interval, is 2.5 %.

The chance that share would be as low as 10 %, *P(Share* $\leq$ *10 %)*, the low end of the 95 % confidence interval, is 2.5 %.

The chance that demand would be as low as 8 bottles per customer per week, *P(demand* $\leq$ *8)*, the low end of the 95 % confidence interval, is 2.5 %.

Therefore, considering the chance of each of these unfortunate outcomes, the chance the revenue could be as low as $1.3M is:

P(the worst case outcome) = 2.5 % × 33 % × 2.5 % × 2.5 % = .00052 % = .0000052, or one in 200,000 (= 1/.0000052), making the worst case extremely unlikely. The Team could be 95 % certain that, given their assumptions, the worst case would not occur.

Based on the Team's assumptions, the chance that the best case outcome would occur is equal to the joint probability of four fortunate circumstances:

P(best case outcome) = *P(annual growth* $\geq$ *4.5 %)* × *P(access unrestricted)*

× *P(Share* $\geq$ *50 %)*

× *P(Demand* $\geq$ *12 bottles per customer per week)*

The chance that annual market growth would be as high as 4.5 %, *P(annual growth* $\geq$ *4.5 %)*, the high end of the 95 % confidence interval, is 2.5 %.

The chance that share would be as high as 50 %, *P(Share* $\geq$ *50 %)*, the high end of the 95 % confidence interval, is 2.5 %.

The chance that demand would be as high as 12 bottles per customer per week, *P(demand* $\geq$ 12*)*, the high end of the 95 % confidence interval, is 2.5 %.

Therefore, considering the chance of each of these fortunate outcomes, the chance the revenue could be as high as $10.2M is:

P(best case outcome) = 2.5 % × 67 % × 2.5 % × 2.5 % = .0011 % = .000011, or one in 100,000 (= 1/.000011), making the best case extremely unlikely, as well. The Team could be 95 % certain that the best case outcome would also not occur.

Both the worst case and the best case outcomes were clearly not likely enough to warrant consideration. What range of revenues actually was likely?

To quantify the risks and produce a range of likely revenues that could actually occur, the Team decided to use Monte Carlo simulation. They could then incorporate the uncertainty, given their assumptions, into their forecast. Results would show the distribution of possible outcomes and their likelihoods under the Team's assumptions, and they would be able to determine a 95 % confidence interval for possible outcomes.

4.3 Spread and Shape Assumptions Influence Possible Outcomes

Spread and Shape Assumptions. The Team updated their revenue spreadsheet, specifying the spread and shape for each of the uncertain influences:

		Assumptions		
		Expected	SD = 95 % CI/4 or range	Distribution
(1)	*Potential customers last year (K)*	34.1		
(2)	*Annual growth %*	3.5 %	2.5–4.5 %	Uniform
(3)	*Potential customers in year 1 (K)* = (1) × (100 % + (2))	35.3		
(4)	*P(Unrestricted access)*	67 %		Binomial
(5)	*% Customers w access* = (4) × 100 % + (1−(4)) × 8 0 %	93 %		
(6)	*Customers accessed (K)* = (3) × (5)	33.0		
(7)	*Price per Bottle ($) given access* = $1.5−(4) × $.25	$1.33		
(8)	*Share captured at parity price*	15 %	2.5 %	Normal
(9)	*Share captured at competitive price*	35 %	7.5 %	Normal
(10)	*Share captured given price = (100 %−(4)) × (8) + (4) × (9)*	28 %		
(11)	*Customers captured (K)* = (6) × (10)	9.4		
(12)	*Bottles sold per customer per week*	10	1	Normal
(13)	*Bottles sold per week (K)* = (11) × (12)	93.6		
(14)	*Weeks in business*	38		
(15)	*Bottles sold in year 1 (M)* = (13) × (14)/1,000	3.6		
(16)	*Revenues in year 1 ($M)* = (7) × (15)	$4.7		

4.4 Monte Carlo Simulation of the Distribution of Performance Outcomes

The distribution of performance outcomes, *revenues in year 1*, in the **Thirsty** case, depend on the distributions of performance influences. With assumptions for center, spread, and shape of each influence now specified in their spreadsheet, the Team drew simulated samples for each. Formulas in their spreadsheet then combined the simulated samples to produce the distribution of possible revenues in year 1.

Growth possibilities. Because recent annual growth had been close to zero, the Team assumed a uniform distribution for growth in the next year. A uniform distribution of possible growth values would give equal chances to all levels within the range of possibility, 2.5–4.5 %.

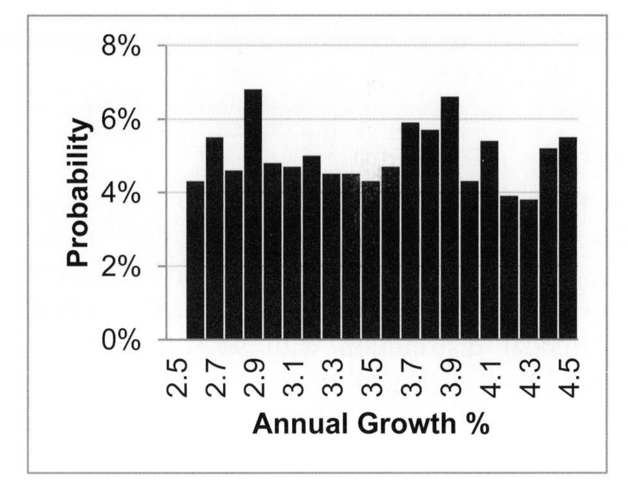

Fig. 4.1 Simulated sample of possible annual growth values

A random sample of 1,000 simulated possible growth values was drawn (Fig. 4.1).

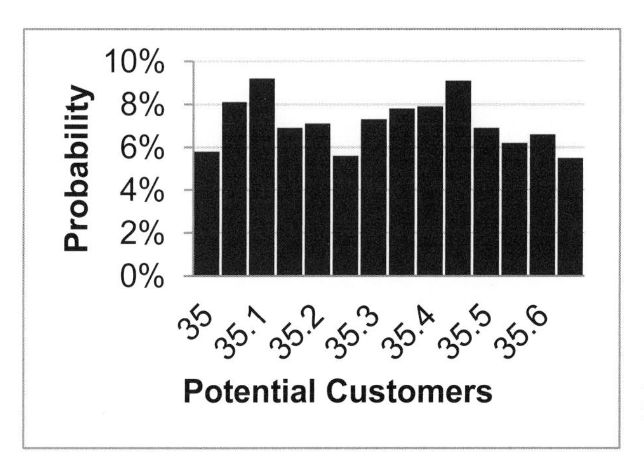

Fig. 4.2 Simulated sample of possible potential customers

With the uniform distribution of potential growth rates, the distribution of possible values for potential customers is uniform (Fig. 4.2).

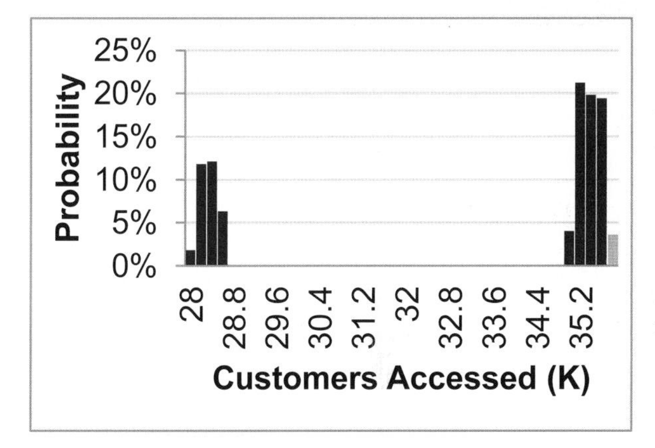

Fig. 4.3 Simulated sample of customers accessed

Access. A random sample of 1,000 possible unrestricted access outcomes was drawn, with the probability for a favorable outcome set at 67 %.

With the random sample of growth rates, possible outcomes for customers accessed is bimodal. The 95 % confidence interval is 28.0–35.6K (Fig. 4.3).

Share. With restricted access, the parity price of $1.50 would be charged. With unrestricted access, the competitive price of $1.25 could be charged, which the Team assumes would yield a higher share of customers captured.

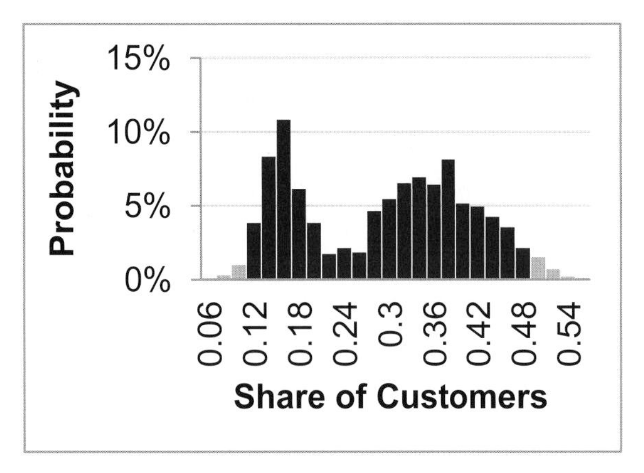

Fig. 4.4 Simulated distribution of possible shares

A random sample of 1,000 possible shares at each price was drawn, and the share corresponding to the each randomly selected *access* outcome and price was chosen, yielding the bimodal distribution with 95 % confidence interval 12–48 % (Fig. 4.4).

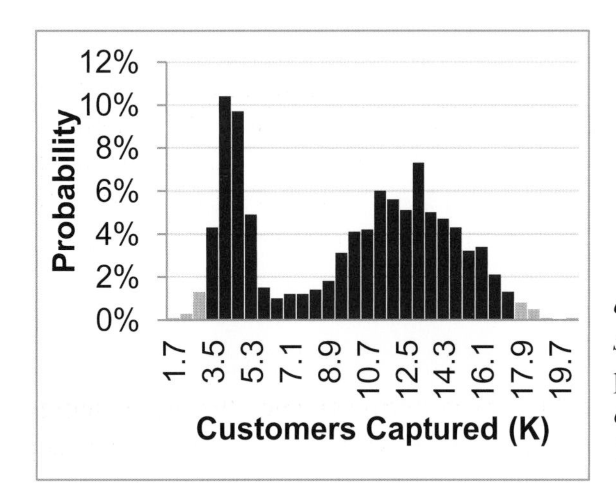

Fig. 4.5 Simulated sample of possible annual growth values

Combining the random sample of *customers accessed* with the random sample of possible *shares of customers*, given access outcome and price, yields the random sample of *customers captured* with 95 % confidence interval 3.5–17.3K (Fig. 4.5).

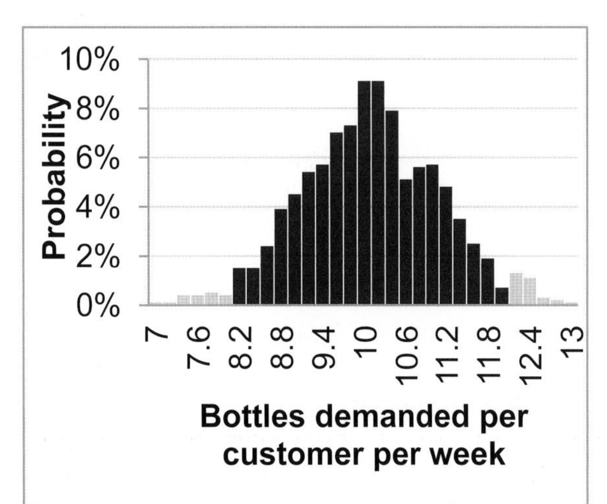

Fig. 4.6 Simulated sample of market shares

Demand per customer. A random sample of 1,000 Normally distributed *bottles per customer per week* was drawn, producing a 95 % confidence interval 8.1–12.0 bottles per customer per week (Fig. 4.6):

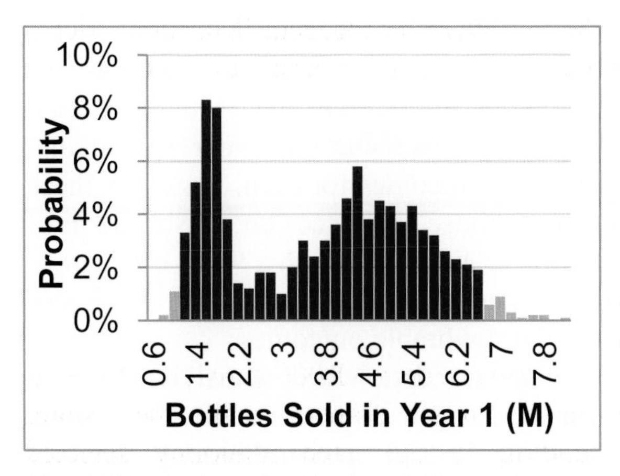

Fig. 4.7 Simulated sample of bottles sold in year 1

Bottles sold in year 1. Combining the random sample of *customers captured* with the random sample of *bottles per customer per week* and 38 weeks in the operating year provides the distribution of possible volumes in *bottles sold* in year 1 with 95 % confidence interval 1.2–6.6M (Fig. 4.7).

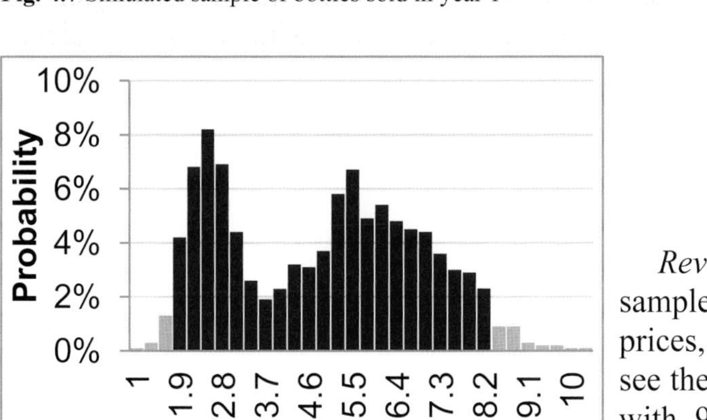

Fig. 4.8 Simulated sample of revenues in year 1

Revenues in year 1. Combining the random sample of *bottles sold in year 1* with the sample of prices, given *access* outcomes, the Team could see the distribution of possible *revenues in year 1*, with 95 % confidence interval $1.9–9.2M (Fig. 4.8).

This range in likely *revenues*, $7.3M (= $9.2M−$1.9M) is narrower and, therefore, more informative, than the comparison of the extremely unlikely best and worst case outcomes, whose wider range was $8.9M (= $10.2M−$1.3M).

If their assumptions were correct, the Team could expect business of at least $1.9M. Expected revenues in year 1 are $4.7M, given the Team's assumptions, and the median of possible revenues is $4.9M, given assumptions: There was a 50 % chance that the business would produce revenues of at least $4.9M in year 1.

4.5 Monte Carlo Simulation Reveals Possible Outcomes Given Assumptions

Decisions concerning investments or allocation of resources depend on inference of likely future performance outcomes. Those performance outcomes hinge on the values that multiple uncertain influences will take on in the future. Monte Carlo simulation offers a view of the possibilities, given the assumptions we make about each of those uncertain influences.

It is naïve and misleading to focus on the "best" and "worst" case scenarios. Multiple influences are always at play. The chance that all influences will take on the least favorable value, producing the "worst" case outcome is virtually zero. The chance that all influences will assume the most favorable value is also virtually zero. While these two extreme outcomes provide a range of possibility, it is an exaggerated range. Attractive decision alternatives may appear to be unattractive in a "worst" case. Unattractive decision alternatives may similarly appear to be attractive in a "best" case. It is much more productive and realistic to link performance drivers together in a spreadsheet, specify assumptions about the center, spread, and shape of each influence, then simulate a distribution of likely outcomes for each. Together, these reflections of our assumptions enable us to see the results of those influences on a distribution of future performance outcomes, with corresponding descriptive statistics. With a 95 % confidence interval for outcomes, based on assumptions, decision makers are much more likely to choose favorable investments or resource allocations and to avoid unfavorable outcomes.

Monte Carlo simulation is a powerful means to generate data when actual data is not available. . . either because it has not yet occurred, or because it is unaccessible. Simulation offers the additional advantage of allowing us to see how our multiple assumptions will together combine to produce possible outcomes. Decision making hinges on assumptions, and simulation provides a reflection of those assumptions.

Excel 4.1 Set Up a Spreadsheet to Link Simulated Performance Components

Use Team 8's assumptions to link revenue influences together in a spreadsheet.

Potential customers. Potential customers are the product of the existing market, 34.1K faculty, staff and students on campus, and the annual growth rate, which has been 3–4 % in recent years.

Team 8 assumes that growth in the next year could be slightly lower or higher than in recent years, 2.5–4.5 %.

Enter *market growth 2.5–4.5 %* and generate a random sample of 1,000 possible market growth values using Excel's Random Number Generation:

Alt AY3 R Enter

In the Random Number Generation menu, request

Number of Variables 1
Number of Random Numbers 1,000
Distribution Uniform
Between .025 **and** .045
Output Range B2:B1001

By entering:

1 Tab 1,000 Tab U Tab Tab Tab .025 Tab .045 Tab Tab Tab B2:B1001

Enter

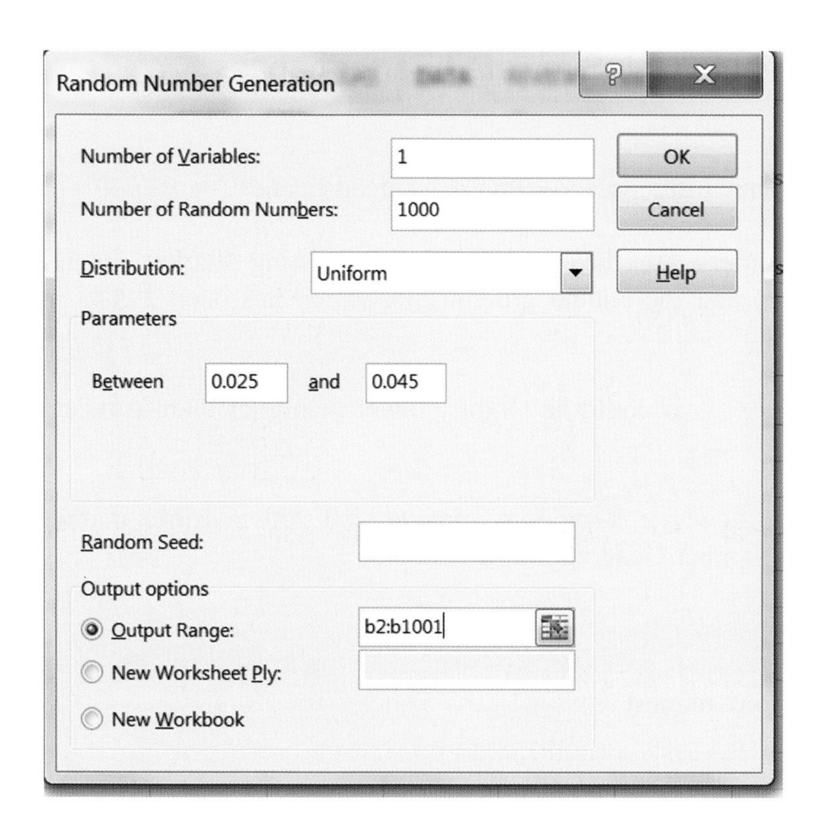

This will add the Uniformly distributed random sample of 1,000 annual market growth rates between 2.5 % and 4.5 % below the label:

The last five growth rates in the simulated sample are shown here.

(Your random sample will differ from the sample drawn here.)

	A	B
1		*market growth 2.5 to 4.5%*
997		0.033
998		0.028
999		0.038
1000		0.043
1001		0.039

Make a random sample of *customers in year 1* from existing customers, 34.1K, and each of the simulated market growth rates:

$= 34.1 \times (1.00 + \textit{market growth})$

=34.1*(1 + B2)

for row 2.

Select this new cell and double click lower left corner to fill in the new column.

(Your sample of *customers(K)* will be slightly different from this sample.)

C2 =34.1*(1+B2)

	A	B	C
1		market growth 2.5 to 4.5%	customers (K) in year
2		0.027	35.0
3		0.026	35.0
4		0.029	35.1
5		0.043	35.6
6		0.036	35.3

Find the 95 % prediction interval for *customers (K) in year 1* using the PERCENTILE.INC(*array,k*).

Enter .975 for *k* to find the upper 95 % prediction interval bound.
Enter .025 for *k* to find the lower 95 % prediction interval bound.

C1003 =PERCENTILE.INC(C2:C1001,0.025)

	A	B	C	D	E
1		market growth 2.5 to 4.5%	customers (K)		
1001		0.033	35.2		
1002		97.50%	35.62		
1003		2.50%	34.98		

Excel 4.2 View a Simulated Sample with a Histogram

To see the simulated sample of potential customers, make a histogram.

To identify the range of values that will be shown in the histogram, use functions **min(***array***)** and **max**(*array*):

=MIN(C2:C1001)

B	C
market growth 2.5 to 4.5%	*customers (K)*
97.50%	35.62
2.50%	34.98
max	35.63
min	34.95

To see a continuous distribution, we need approximately 30 histogram *bins*.

Divide the range (= max−min) by 30 to find *bin width*:

=(C1004-C1005)/30

B	C
market growth 2.5 to 4.5%	*customers (K)*
97.50%	35.62
2.50%	34.98
max	35.63
min	34.95
bin width	0.023

Add bins, making the first bin the *minimum* value, the second bin equal to the first bin plus *one* bin width (locking the bin width column reference).

Select the first two bins, the grab the lower right corner and drag to fill more bins.

Stop after you've reached the *maximum*, (35.63, with this sample).

C1010 | =C1009+C$1006

	A	B	C
1		*market growth 2.5 to 4.5%*	*customers (K)*
1005		min	34.95
1006		bin width	0.023
1007			
1008			*customers (K) bins*
1009			34.95
1010			34.98
1011			35.00
1012			35.02

Create a histogram of *customers(K)*, using the *customers (K) bins.*

Fill bins within the 95 % confidence interval a unique shade:

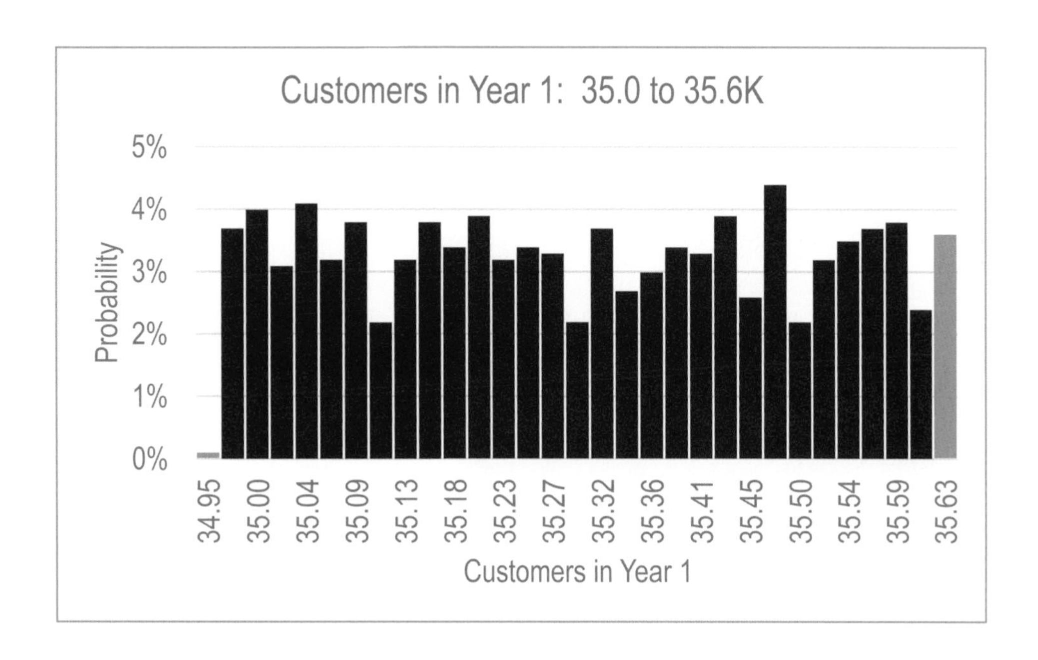

Access. The Team assumes that there is a 67 % chance that they will be able to place vending units in any campus location, including dorms.

Insert a new column D, **Alt NC**, with label *Access*, and generate a random sample of *access* outcomes using Excel Random Number Generation.

In the Random Number Generation menu, request

Number of Variables 1
Number of Random Numbers 1,000
Distribution Binomial
p Value .67
Number of trials 1
Output Range D2:D1001

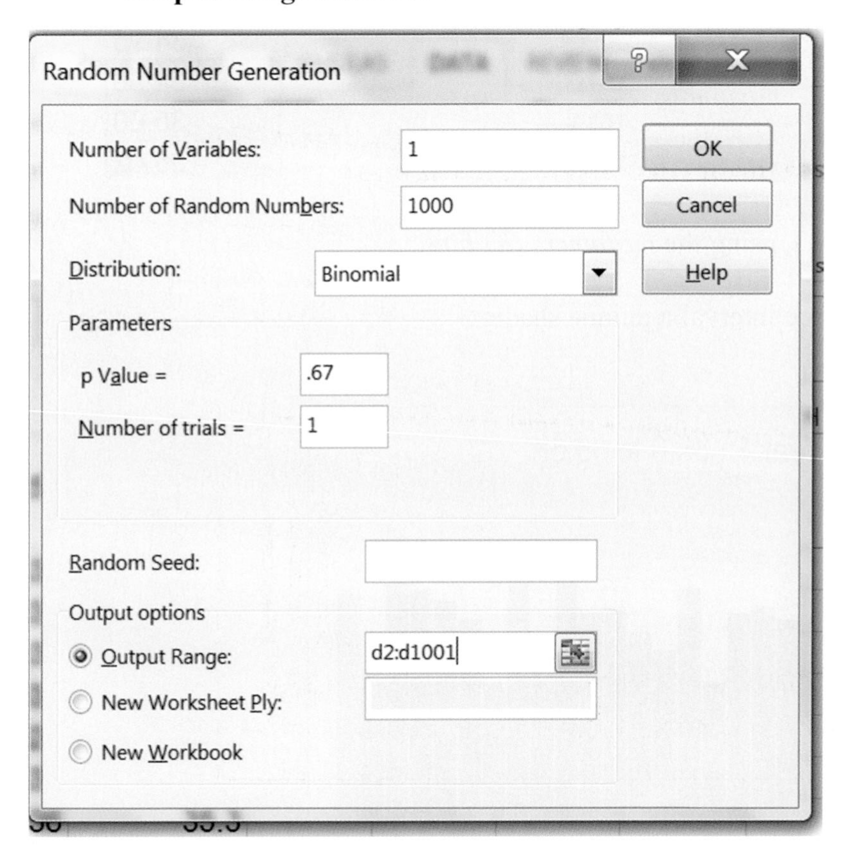

If access is restricted, dorms would be off limits, and the Team assumes that only 80 % of potential customers could be reached.

Find the sample of *customers (K) given access* from the product of

(80 % + *access* × 20 %) × *customers*:

=(.8 + .2 * D2) * C2

in row 2.

Fill in the column.

(Your random sample will differ from the sample drawn here.)

E2 fx =(0.8+0.2*D2)*C2

	B	C	D	E
1	*market growth 2.5 to 4.5%*	*customers (K)*	*Access*	*Customers (K) w Access*
2	0.032	35.2	1	35.2
3	0.031	35.2	1	35.2
4	0.039	35.4	0	28.3
5	0.033	35.2	1	35.2
6	0.025	35.0	1	35.0

Find the *95 % prediction interval*, *maximum* and *minimum customers (K) given access*, create histogram bins, and produce a histogram of *customers (K) given access:*

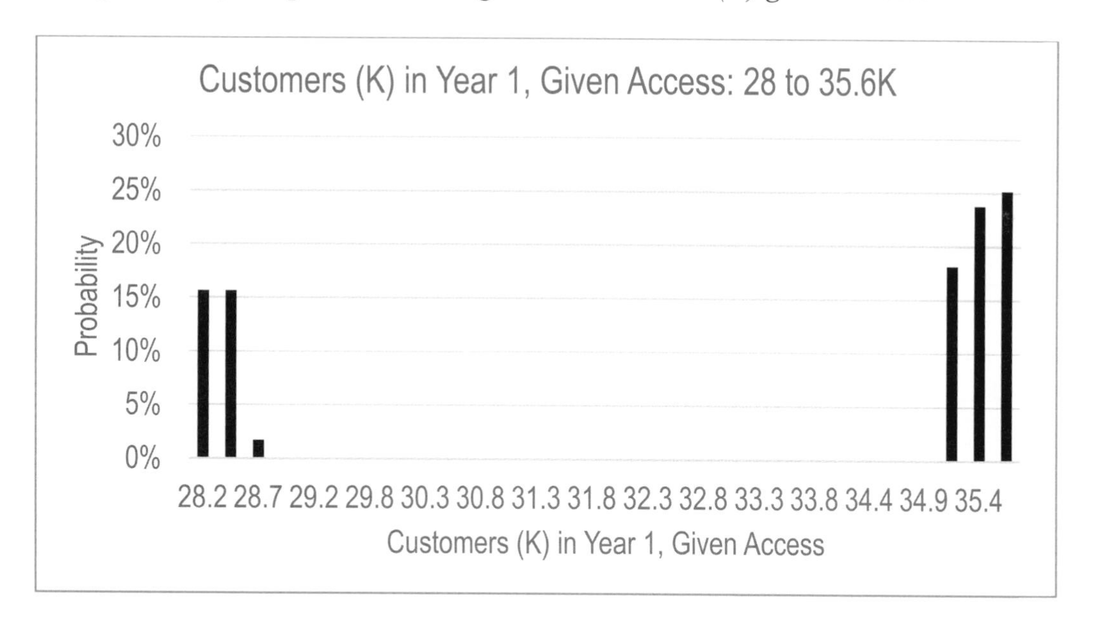

Price. The Team assumes that full access would translate to higher volume and discounted unit costs, enabling a lower competitive price of $1.25 per bottle to be charged, instead of the parity price of $1.50.

Find *price ($)* from each *access* outcome

= 1.50−.25 × *access*

= 1.5−.25 * D2

for row 2.

Fill in the column.

F2 | f_x =1.5-0.25*D2

	D	E	F
1	Access	Customers (K) w Access	Price given access
2	1	35.2	1.25
3	1	35.2	1.25
4	0	28.3	1.5
5	1	35.2	1.25
6	1	35.0	1.25
7	1	35.4	1.25
8	1	35.3	1.25

Find the Standard Deviation, Given Assumptions. At the lower price, $1.25, with unrestricted access, the Team assumes that at least 20 %, and as much as 50 %, of potential customers will be captured, with the most likely share equal to 35 %, and share possibilities distributed Normal.

It is assumed that there is only a 2.5 % chance that share could be less than 20 %, and only a 2.5 % chance that share could be greater than 50 %. Therefore, there is a 95 % chance that share will fall within the range 20–50 %, making the standard deviation approximately equal to one quarter of this range, by the Empirical rule:

SD = (50 %−20 %)/4 = 7.5 %

Insert a new with label *share given access* and use Excel Random Number Generation to simulate a sample of 1,000 Normally distributed values.

In the Random Number Generation menu, request

Number of Variables 1
Number of Random Numbers 1,000
Distribution Normal

Mean .35
Standard deviation .075
Output Range G2:G1001

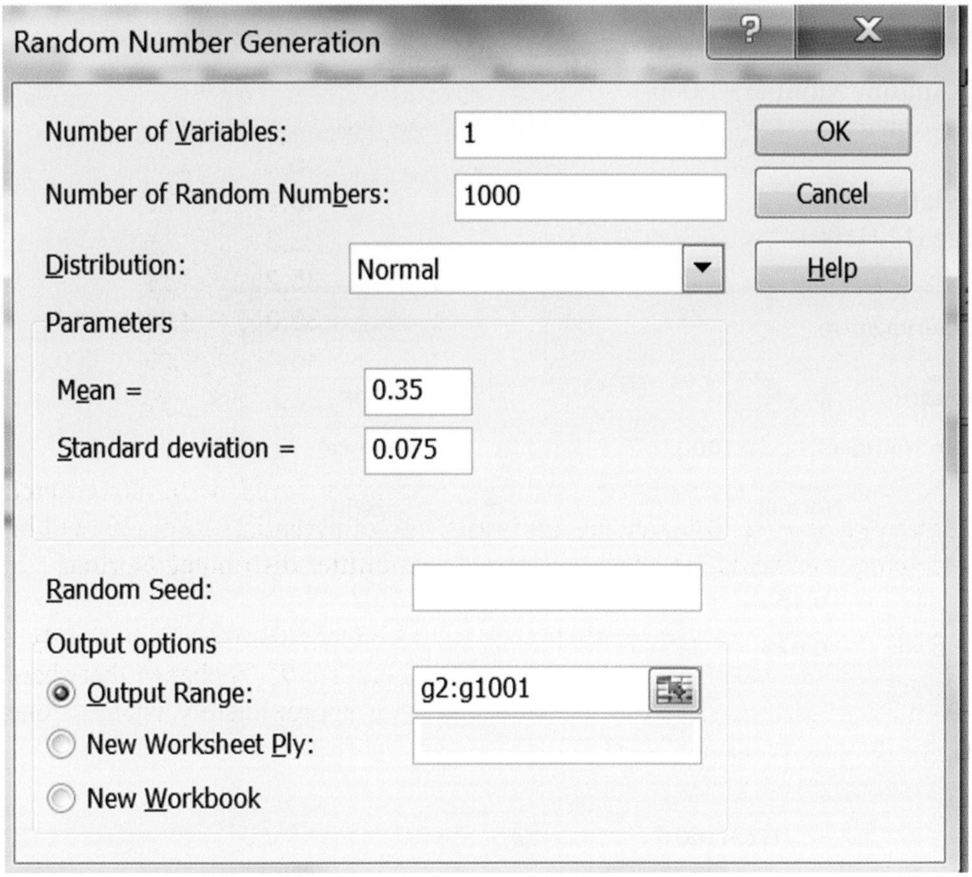

Random Number Generation

Number of Variables: 1
Number of Random Numbers: 1000
Distribution: Normal
Parameters
Mean = 0.35
Standard deviation = 0.075
Random Seed:
Output options
Output Range: g2:g1001
New Worksheet Ply:
New Workbook
OK
Cancel
Help

With restricted access and the higher, parity price, $1.50, the Team assumes that at least 10 %, and as much as 20 %, of customers with access could be captured, with the most likely share in the middle at 15 %.

It is assumed that there is only a 2.5 % chance that *share given restricted access, higher price*, would be less than 10 %, and that there is only a 2.5 % chance that share could be greater than 20 %. Therefore, from the Empirical Rule, it is assumed that the share standard deviation is one quarter of this range:

SD = (20 %−10 %)/4 = 2.5 %

Insert a new column labeled *share given restricted access* and simulate a Normally distributed sample of 1,000 values.

In the Random Number Generation menu, request

Number of Variables 1
Number of Random Numbers 1,000
Distribution Normal
Mean .15
Standard deviation .025
Output Range H2:H1001

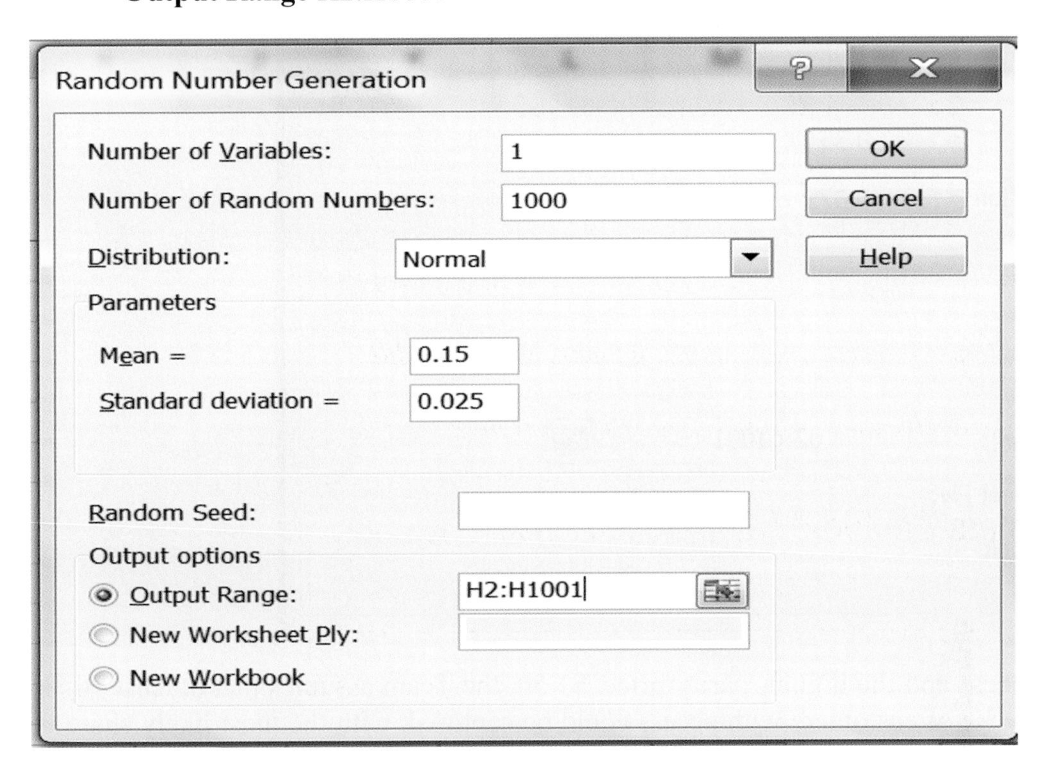

Insert a new column labeled *customers captured*, with values from simulated samples for *access*, *customers given access*, *share given access*, *lower price* and *share given restricted access*, *higher price*:

= (*access* × *share given access*
+ (1−*access*) × *share given restricted access*) × *customers given access*

There is a small chance that a simulated share could be slightly less than zero. In such a case, ask Excel to substitute zero by using **MAX(***share*,0):

= (D2 * **MAX**(G2,0) + (1−D2) * **MAX**(H2,0)) * E2

for row 2.

Fill the column.

(Your random sample will differ from the sample drawn here.)

I2 =(D2*MAX(G2,0)+(1-D2)*MAX(H2,0))*E2

	D	E	F	G	H	I
1	*Access*	*Customers (K) w Access*	*Price given access*	*share given access*	*share given restricted access*	*customers captured (K)*
2	1	35.2	1.25	0.304	0.169	10.7
3	1	35.2	1.25	0.214	0.165	7.5
4	0	28.3	1.5	0.384	0.173	4.9
5	1	35.2	1.25	0.339	0.158	11.9
6	1	35.0	1.25	0.273	0.169	9.6
7	1	35.4	1.25	0.312	0.161	11.0

Find the *95 % prediction interval, maximum, minimum,* and *bin width.*

Create bins and a histogram of *customers captured (K) given access:*

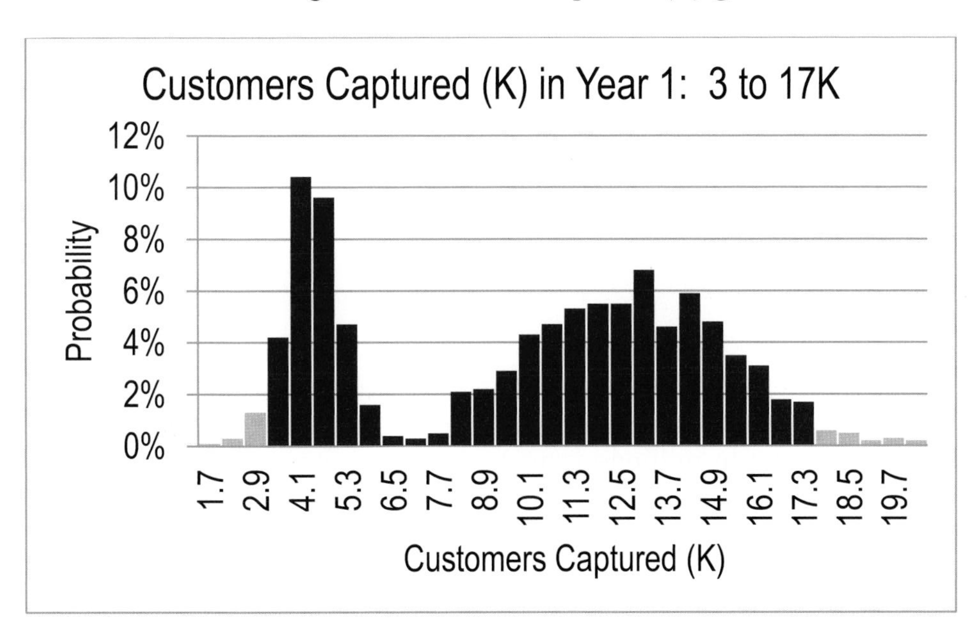

Demand per customer. From earlier market research, the Team believes that average bottles demanded per customer per week falls within the range 8–12, with 95 % confidence. The Team assumes that demand is Normally distributed, with standard deviation equal to one quarter of the 95 % confidence interval:
SD = (12−8)/4 = 1

Insert a new column and generate a sample of 1,000 Normally distributed levels for *bottles per customer per week.*

In the Random Number Generation menu, request

Number of Variables 1
Number of Random Numbers 1,000
Distribution Normal
Mean 10
Standard deviation 1
Output Range J2:J1001

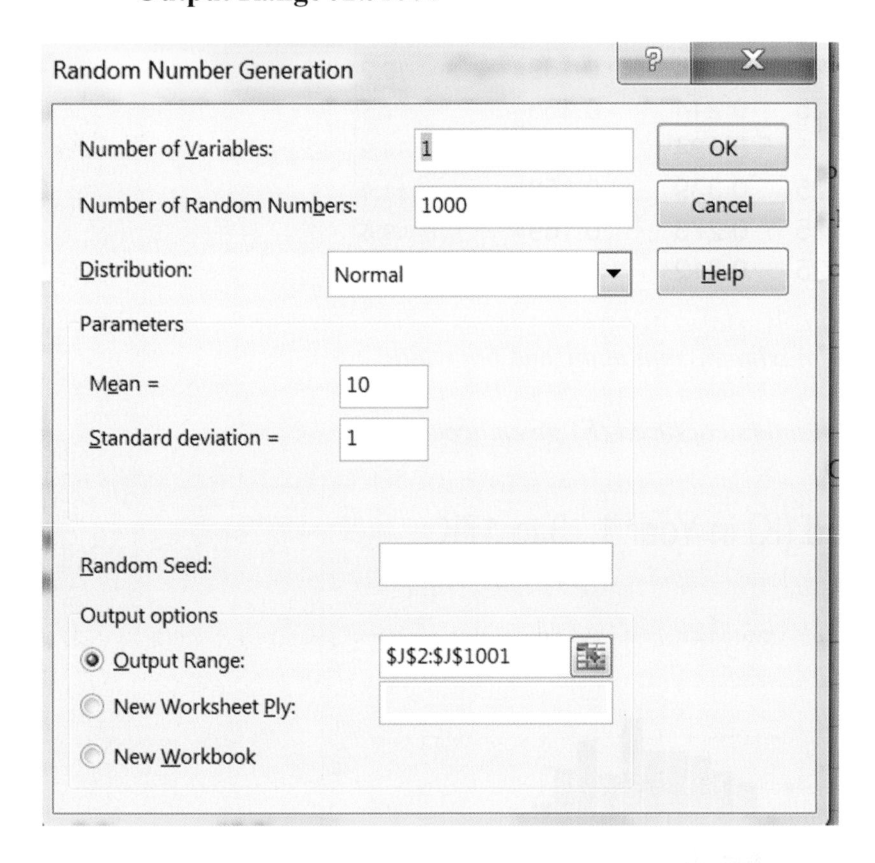

Insert a new column *bottles sold (M)* in millions from *customers captured (K) given access* and *bottles per customer per week*, assuming that the vending units will be stocked during the 38 weeks in which classes are in session:

= 38 weeks × *bottles per customer per week*

× *customers captured (K) given access*/1,000

= 38 * J2 * I2/1,000

in row 2.

Fill in the column.

(Your random sample will differ from the sample drawn here.)

K2 | fx =38*I2*J2/1000

	I	J	K
1	customers captured (K)	bottles per customer per week	bottles sold (M)
2	10.7	10.2	4.2
3	7.5	10.9	3.1
4	4.9	9.9	1.8
5	11.9	9.5	4.3
6	9.6	10.0	3.6
7	11.0	10.3	4.3

Find the expected sales volume, *95 % confidence interval*, *minimum* and *maximum.*

Create bins and a histogram for possible volumes of *bottles sold in year 1:*

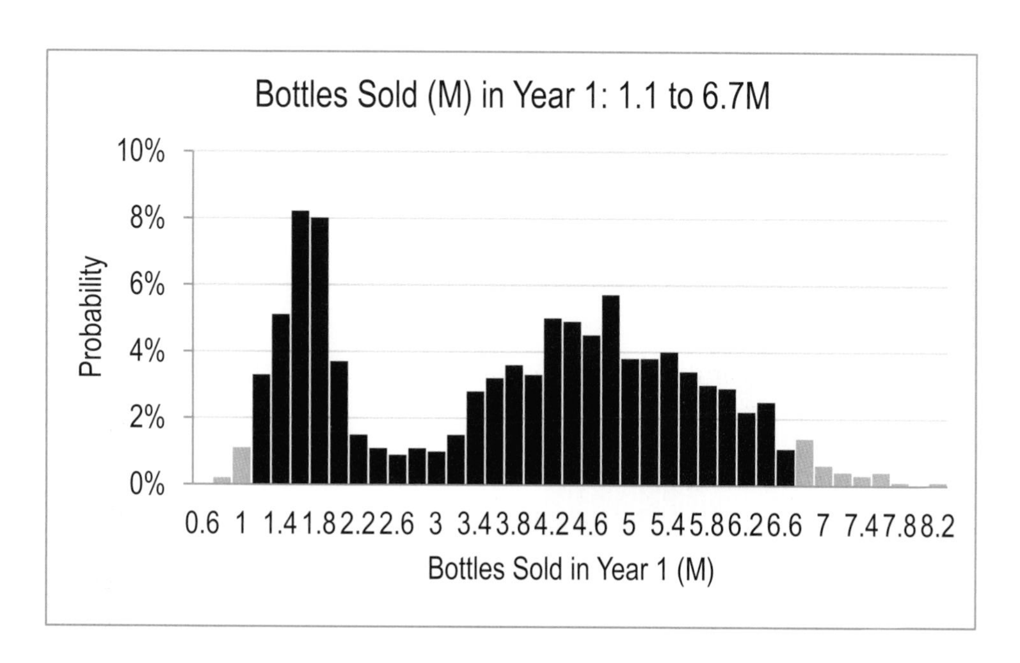

Insert a new column *revenues ($M)*, from the simulated samples of *price ($) given access* and *bottles sold (M)*:

= *price ($) given access* × *bottles sold (M)*

= F2 * K2

for row 2.

Fill in the column.

(Your random sample will differ from the sample drawn here.)

L2 =F2*K2

	F	G	H	I	J	K	L
1	*Price given access*	*share given access*	*share given restricted access*	*customers captured (K)*	*bottles per customer per week*	*bottles sold (M)*	*revenues ($M)*
2	1.25	0.304	0.169	10.7	10.2	4.2	5.2
3	1.25	0.214	0.165	7.5	10.9	3.1	3.9
4	1.5	0.384	0.173	4.9	9.9	1.8	2.8
5	1.25	0.339	0.158	11.9	9.5	4.3	5.4
6	1.25	0.273	0.169	9.6	10.0	3.6	4.5

Find the *95 % prediction interval*, *minimum*, *maximum*, and *bin width*.

Create bins and a histogram of *revenues ($M)* in year 1:

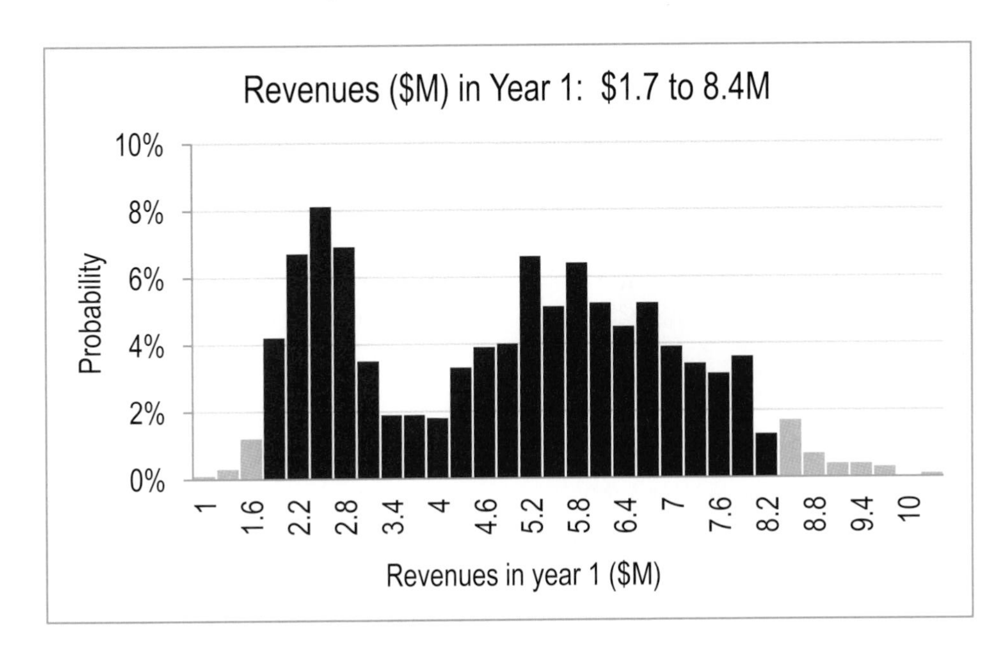

Lab 4 Inference: Dell Smartphone Plans

Managers at Dell are considering a joint venture with a Chinese firm to launch a new smartphone equipped with Microsoft's Windows Phone 8, which offers full Skype compatibility.

I. *Find the Distribution of Potential Dell Smartphone Shares*

Managers agree that 16 % is a reasonable estimate of the new smartphone's share of the smartphone market. Apple and Samsung control 53 % of the market, with the rest of the market split by several smartphone brands, including HTC and Palm.

Use Excel to create a sample of 1,000 *Dell Smartphone share* proportions from a Normal distribution with mean equal to the sample proportion expected to purchase.

1. To choose a standard deviation for your simulation:

    a. Use subjective judgment to estimate the lowest possible *Dell Smartphone share* (which would be too high only 2.5 % of the time): _____

    b. Use subjective judgment to estimate the highest possible *Dell Smartphone share* (which would be too low only 2.5 % of the time): _____

    c. Estimate the standard deviation of *Dell Smartphone share* by dividing the likely range, a. to b., by 4, following the Empirical Rule: ___

2. Create a histogram to illustrate the distribution of potential *Dell Smartphone shares.*

II. *Simulate Dell Shipments to Assess Potential, Given Assumptions*

Managers want to estimate potential *shipments of Dell smartphones* in the third quarter of 2012 with 95 % confidence. *Shipments of Dell smartphones* would be driven by *world smartphone shipments*, *quarterly growth in world shipments*, and *Dell smartphone share.*

World smartphone shipments in the third quarter of 2011 were 600K.

The world smartphone market has been growing 2.5–3.5 % in the recent quarters relative to same quarters the year before. Management assumes that quarterly growth over past year could be as low as 2.0 %, and possibly as high as 4.0 %, given growth in developing markets.

1. Use Excel to draw a random sample of 1,000 Uniformly distributed *quarterly growth rates.*

    Find the distribution of possible *world smartphone shipments* from third quarter *world smartphone shipments* and the sample of simulated *quarterly growth rates.*

    Make a histogram of *world smartphone shipments* in the third quarter of 2012.

2. Use your samples of simulated *Dell smartphone share* and *World smartphone shipments* to find the distribution of possible *Dell smartphone shipments.*

   Make a histogram of *Dell smartphone shipments* in the third quarter of 2012.

3. What is the 95 % confidence interval for *Dell smartphone shipments* in the third quarter of 2012?

   __________ to ____________

4. Several conservative managers worry that *Dell smartphone share* could be less than 10 % (the lower bound of their 95 % confidence interval) and quarterly *growth in world shipments of smartphones* could slow to 2 %, producing a forecast of *Dell smartphone shipments* of just 61.2K in the third quarter of 2012.

   What is the chance that this unfortunate "worst case" could occur? _________

CASE 4-1 American Girl in Starbucks

Mattel and Warner Brothers are considering a partnership with Starbucks to promote their new American Girl movie. Starbucks previously backed Lionsgate's "Akeelah and the Bee," which earned $19 million. In exchange for $8 million, Starbucks would install signage and stickers in 6,800 of its stores, print American Girl branded cup sleeves, and sell the picture's soundtrack for a 3 month period. Materials for the movie would also appear on the company's website during that period. Starbucks claims 44 million customers in the 6,800 stores.

In a pretest of the promotion during one week in one Starbucks store, 184 of the 924, **or 20 %** of Fast Card customers served that week agreed that they had heard of the movie when surveyed by phone the following week.

Mattel managers believe that roughly 35 % of those who are aware of the movie will buy tickets.

Managers assumed that there is only a 2.5 % that the percent buying tickets would be less than 5 %, and there is only a 2.5 % change that the percent buying tickets will be more than 9 %.

The distribution of *tickets purchased* depends on the number of guests that ticket purchasers bring. Ninety-five percent of movie-goers are expected to bring 1–3 family members or friends.

Mattel would earn $1 royalty from each ticket.

1. Illustrate the distributions of *percent buying tickets* and *tickets purchased*, given assumptions.

2. Illustrate the distribution royalties, given assumptions.

3. A conservative manager advises that in the "worst case," royalties could be as low as $4.4M, in which case Mattel would lose $3.6M (*= payment to Starbucks − royalties).*

   What is the chance that this "worst case" could occur, given assumptions?

CASE 4-2 Can Whole Foods Hold On?

Organic food revenue has been increasing at annual rates of 15–20 %, and there are no signs of it slowing down.

Organic food sales were roughly $33B in 2010, and had grown to **$39.2B** in 2011.

Whole Foods opened a record 21 stores during its fiscal year 2007 and accounted for almost half of all U.S. organic food revenue in 2007. In 2011, WFM Revenue grew to **$10.1 B** from **311** stores.

By Matt Thalman, The Motley Fool
Nov. 29, 2011

*Grocery stores, convenience stores, and even full service gas stations are on every corner nowadays. With the market so saturated, how can anyone make money? Well, on paper **Whole Foods Market** looks like it's found a way. The company's business approach has allowed it to expand rapidly, adding 34 new stores in the past two years, and it plans to add at least 52 more stores in the next two years. Same-store sales growth was 8.4% in 2011, up from 6.5% the year before.*

WFM *share* of organic food sales in a given year depends on *organic food sales*, the number of *WFM stores*, and *sales per WFM store:*

$$WFM\ share_t = WFM\ stores_t \times sales\ per\ WFM\ store_t / organic\ food\ sales_t$$

In a given year, *organic food sales* depend on the *organic food growth rate*, 15–20 % in recent years:

$$Organic\ food\ sales_t = (1 + organic\ food\ growth\ rate)^{(t-2011)} \times organic\ food\ sales_{2011}$$

WFM stores depend on the *WFM store growth rate*, 4–7 % in recent years:

$$WFM\ stores_t = (1 + WFM\ store\ growth\ \mathrm{rate})^{(t-2011)}\ WFM\ stores_{2011}$$

Sales per WFM store depend on *same store sales growth*, 6–9 % in recent years:

$$Sales\ per\ WFM\ store_t = (1 + same\ store\ sales\ growth)^{(t-2011)} \times sales\ per\ WFM\ store_{2011}$$

Can WFM hold on?

Forecast WFM's market share of the organic food sales market in 2020.

1. Present histograms of $organic\ food\ sales_{2020}$, $WFM\ stores_{2020}$, and $sales\ per\ WFM\ store_{2020}$ and $WFM\ share_{2020}$, based on your assumptions.

2. What is the 95 % confidence interval for $WFM\ share_{2020}$?

3. A conservative manager forecasts that, in the "worst case," $WFM\ share_{2020}$ could be as low as 12 %. Based on your assumptions, how likely is such a "worst case?"

Chapter 5
Quantifying the Influence of Performance Drivers and Forecasting: Regression

Regression analysis is a powerful tool for quantifying the influence of continuous, *independent, drivers* X on a continuous *dependent, performance* variable Y. Often we are interested in both explaining how an independent decision variable X drives a dependent performance variable Y and also in predicting performance Y to compare the impact of alternate decision variable X values. X is also called a *predictor,* since from X we can predict Y. Regression allows us to do both: quantify the nature and extent of influence of a performance driver and predict performance or response Y from knowledge of the driver X.

With regression analysis, we can statistically address these questions:

- Is variation in a dependent, performance, response variable Y influenced by variation in an independent variable X?

If X is a driver of Y, with regression, we can answer these questions:

- What percent of variation in performance Y can be accounted for with variation in driver X?
- If driver X changes by one unit, what range of response can we expect in performance Y?
- At a specified level of the driver X, what range of performance levels Y are expected?

In this chapter, simple linear regression with one independent variable is introduced, and we explore ways to address each of these questions linking a continuous driver, which may be a decision variable, to a continuous performance variable. We also explore the link between correlation and simple linear regression, since the two are closely related.

5.1 The Simple Linear Regression Equation Describes the Line Relating a Decision Variable to Performance

Regression produces an equation for the line which best relates changes or differences in a continuous, dependent performance variable Y to changes or differences in a continuous, independent driver X. This line comes closest to each of the points in a scatterplot of Y and X:

$$\hat{y} = b_0 + b_1 \times X$$

Where $\hat{y}$ is the expected value of the dependent performance, or response, variable, called "*y hat*",

X is the value of an independent variable, decision variable, or driver,

b_0 is the *intercept* estimate, which is the expected value of y when X is zero,

b_1 is the estimated *slope* of the regression line, which indicates the expected change in performance $\hat{y}$ in response to a unit change from the driver's average $\overline{X}$.

C. Fraser, *Business Statistics for Competitive Advantage with Excel 2013: Basics, Model Building, Simulation and Cases*,
DOI 10.1007/978-1-4614-7381-7_5, © Springer Science+Business Media New York 2013

Example 5.1 HitFlix Movie Rentals

An owner of a movie rental vending business is planning to add a new vending units and needs to decide how many titles to offer in each. The plan is to design the new units to stock and offer 100 titles, but the owner thinks a larger number of titles might generate more revenue. Titles offered may drive revenues. The null and alternate hypotheses which the owner would like to test are:

> H_0: Titles offered X has no effect on movie rental revenues y.
> H_1: Titles offered X drives movie rental kiosk revenues y.

Scatterplots of titles offered, X, and annual revenues, y, for a random sample of 52 vending units from the chain are shown in Fig. 5.1 from Excel:

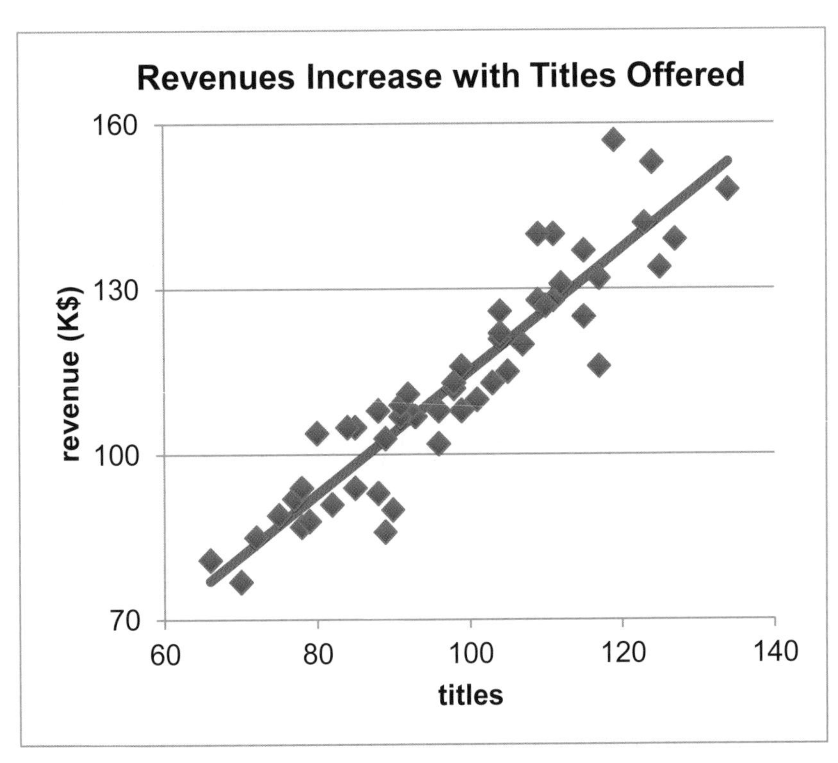

Fig. 5.1 Vending unit revenues by titles

The scatterplot indicates that revenues may be a linear function of titles offered. For each additional title, average annual revenues are higher by about \$1.1K or \$1,100.

Expected revenues $\hat{y}$ increase at a constant rate of \$1,100 with each increase of one title offered.

Because variation in revenues y is related linearly to variation in titles X, the linear regression line is a good summary of the data:

$$\widehat{revenues}(\$K) = 3.4(\$K) + 1.1(\$K\ per\ title) \times titles$$

In this example, the intercept estimate b_0 is 3.4($K). Were a vending unit to offer zero titles (which isn't likely), expected revenue would be $3,400. The estimated slope b_1 is 1.1($K per title), indicating an average increase in revenue of $1,100 is associated with an increase of one additional title.

5.2 Test and Infer the Slope

Because the true impact β_1 of a driver X on performance y is unknown, this *slope*, or *coefficient*, is estimated from a sample. This estimate b_1 and its sample standard error s_{b_1} are also used to test the hypothesis that variation in X drives variation in y:

H_0: Variation in the independent variable X does not drive variation in the dependent variable y.

or

H_0: The regression slope is zero: $\beta_1 = 0$.

Alternatively,

H_1: Variation in the independent variable X drives variation in the dependent variable y.

or

H_1: The regression slope is not zero: $\beta_1 \neq 0$.

In many instances, from experience or logic, we know the likely direction of influence. In those instances, the alternate hypothesis requires a *one tail* test:

H_1: The independent variable X positively influences the dependent variable y.

or

H_1: The regression slope is greater than zero: $\beta_1 > 0$.

This one sided alternate hypothesis describes an upward slope. A similar alternate hypothesis could be used when logic or experience suggests a downward slope. In the **Movie Rentals** example, if revenue did not depend on titles offered, the scatterplot would resemble a spherical cloud and the regression line would be flat at the dependent variable mean $\overline{Y}$, as in Fig. 5.2.

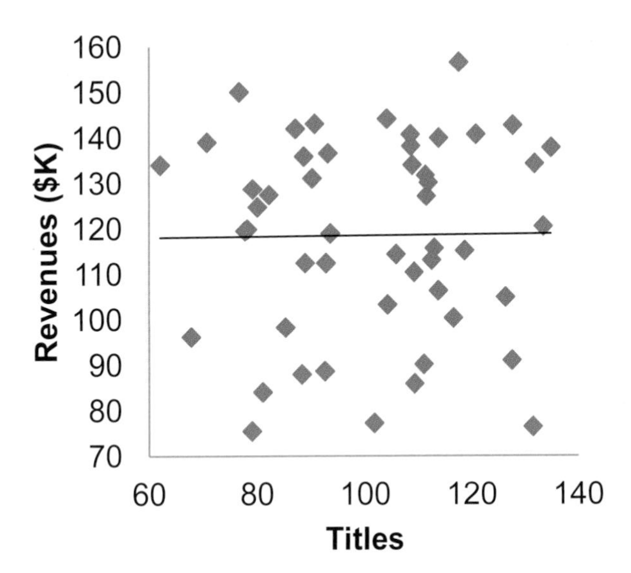

Fig. 5.2 *X* does not drive *y*, and the regression line slope is flat ($b_1 = 0$)

Sample slopes b_1 are Normally distributed around the population slope β_1, which is zero, under the null hypothesis. Whether the sample slope is consistent with the null hypothesis depends on its distance from zero and dispersion of the sample slope distribution. Figure 5.3 illustrates two possibilities. The sample slope to the left is "close" to the hypothetical population slope of zero. The sample slope on the right is "far" from the hypothetical population slope, providing evidence that the population slope is unlikely to be zero.

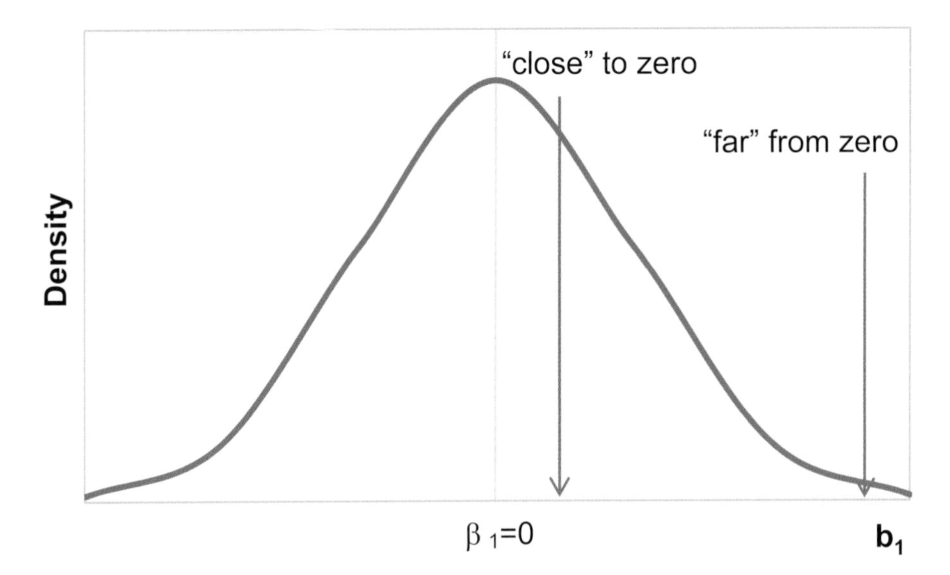

Fig. 5.3 Distribution of sample slope under the null hypothesis

To judge whether a sample slope is close or far from zero, the *standard error of the slope* is needed. The slope standard error depends on unexplained variation, the sample size, and population dispersion, and is equal to .064 in the **Movie Rentals** sample:

$$s_{b_1} = \sqrt{\frac{\Sigma(y-\hat{y})^2/(N-2)}{\Sigma(X-\bar{X})^2}} = .064$$

To form a conclusion about the significance of the slope, calculate the number of standard errors which separate the slope estimate b_1 from zero.

$$t_{N-2} = b_1/s_{b_1} = 1.12/.064 = 17.4$$

In **Movie Rentals**, the slope estimate is more than 17 standard errors from zero.

From both experience and logic, the movie rental business owner had a good idea that titles offered has a positive impact on revenues, so a *one tail* test is appropriate, corresponding to the alternate hypothesis is that the slope is positive. Figure 5.4 illustrates *p values* for *t* statistics for a sample of 52. A *t* values of 17.4 is far from zero, with corresponding *p value* < .0001.

There is a very small chance that we would observe the sample data were titles offered not driving revenues. From our sample evidence, we reject the null hypothesis of a flat slope and accept the alternate hypothesis of a positive slope. Sample evidence suggests that titles offered has a positive impact on revenues.

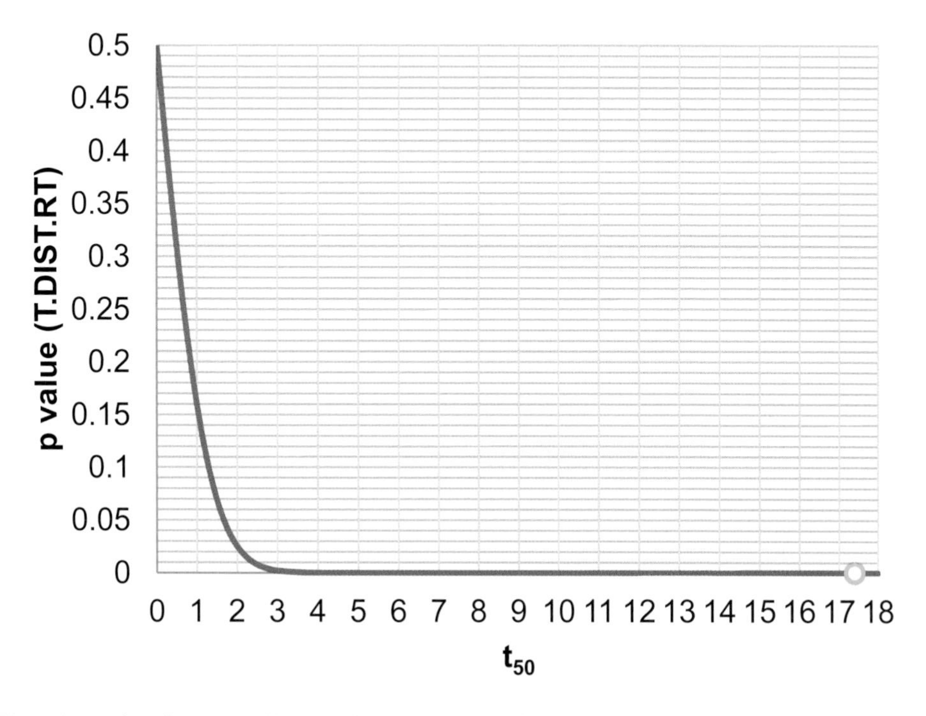

Fig. 5.4 *p values* for a one tail *t* test with 50 degrees of freedom

Excel does these calculations for us. The slope and intercept estimates are labeled *Coefficients* in Excel, shown in Table 5.1, on the left. To the right of the coefficient estimates are their standard errors, *t* statistics, and *p values,* as well as 95 % confidence intervals for the population intercept and slope.

Table 5.1 Coefficient estimates, standard errors and *t tests*

	Coefficients	*Standard error*	*t stat*	*p value*	*Lower 95 %*	*Upper 95 %*
Intercept	3.43	6.382	.5	.5931	−9.39	16.25
Titles	1.12	.064	17.4	.0000	.99	1.24

Excel assumes that we have no prior information concerning the direction of driver influence, and so Excel provides a *two tail p value.* Divide the Excel *p value* by 2 to find the *one tail p value.*

There is a 95 % chance that the true population slope will fall within $t_{.05,(N-2)}$ standard errors of the slope estimate:

$$b_1 - t_{.05,50} \times s_{b_1} < \beta_1 < b_1 + t_{.05,50} \times s_{b_1}$$

$$1.12 - 2.01 \times (.064) < \beta_1 < 1.12 + 2.01 \times .064$$

$$.99 < \beta_1 < 1.24$$

The impact of one additional title on vending unit revenue is .99 ($K) to 1.24 ($K), or $990 to $1,240.

5.3 The Regression Standard Error Reflects Model Precision

Using the regression formula, we can predict the expected revenue $\hat{y}$ for a vending unit offering a given number of titles *X*. Table 5.2 contains predictions for five vending units offering different numbers of titles:

Table 5.2 Expected revenue

b_0	$+ b_1 \times$ *Titles*	$= Rev\hat{e}nue(\$K)$
3.4	+ 1.1 × 70	= 80
3.4	+ 1.1 × 80	= 91
3.4	+ 1.1 × 90	= 102
3.4	+ 1.1 × 110	= 124

The differences between expected and actual revenue are the *residuals* or errors. Errors from these four vending units are shown in Table 5.3 and Fig. 5.5.

Table 5.3 Residuals from the regression line

Titles X	*Actual revenue($K)* y	*Expected revenue ($K)* $\hat{y}$	*Residual ($K)* $e = y - \hat{y}$
80	87	91	−4
100	102	113	−11
110	140	124	16
120	142	135	7

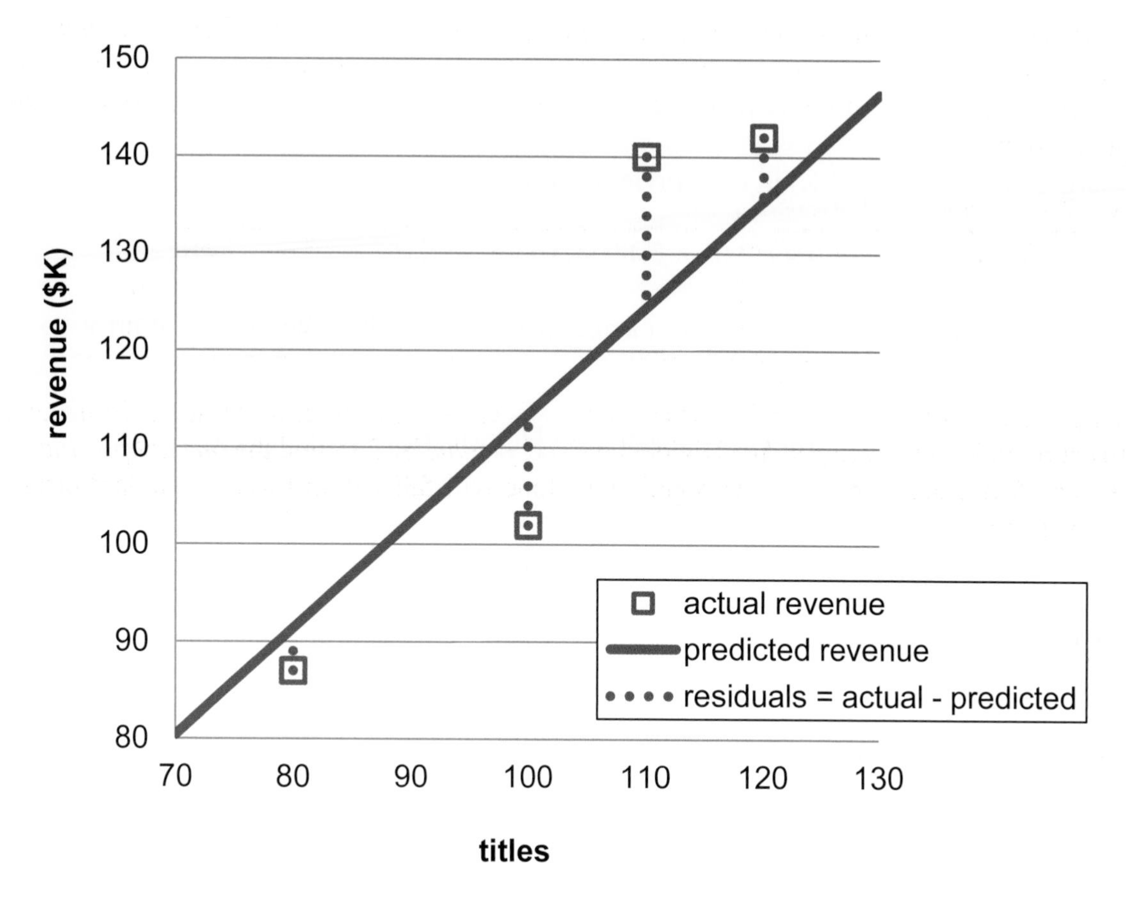

Fig. 5.5 Four residuals from the regression line

The *Sum of Squared Errors* in a regression,

$$SSE = \sum e_i^2 = \sum(y_i - \hat{y})^2 = \sum(y_i - b_0 - b_1 \times X_i)^2$$

$$= 2{,}800$$

is the portion of total variation in the dependent variable, *SST,* which remains unexplained after accounting for the impact of variation in *X*. In the **Movie Rentals** sample, *SSE* is 2,800. The *Least Squares* regression line is the line with the smallest SSE of all possible lines relating *X* to *Y*.

The regression *standard error,* equal to the square root of *Mean Square Error*, reflects the precision of the regression equation.

$$s_{\hat{y}} = \sqrt{SSE/(N-2)} = \sqrt{MSE}$$

In the **Movie Rentals** regression, the standard error is 7.44($K):

$$s_{\hat{y}} = \sqrt{2{,}800/50} \quad = \sqrt{55} \quad = 7.44(\$K)$$

The forecast *margin of error* is approximately twice the standard error:

$$me = t_{.05,residual\ df} \times s_{\hat{y}}$$

where the residual degrees of freedom are N−2 for a regression model with one driver (and the intercept).

The margin of error is 15.0($K) in the **Movie Rentals** regression:

$$me = 2.01 \times 7.44(\$K) = 15.0(\$K)$$

We expect forecasts to be no further from actual performance than the margin of error 95 % of the time.

5.4 Prediction Intervals Estimate Average Population Response

95 % prediction intervals

$$\hat{y} \pm me$$

for vending units offering various numbers of titles are shown in Table 5.4 and Fig. 5.6.

Table 5.4 Individual 95 % prediction intervals

Titles	*Expected revenue ($K)* $\hat{y}$	*Standard error* $s_{\hat{y}}$	*Margin of error me* $t_{.05,50} \times s_{\hat{y}}$	*95 % prediction interval* $\hat{y} \pm me$	
70	82	7.4	15	67	96
100	115	7.4	15	100	130
130	149	7.4	15	134	163

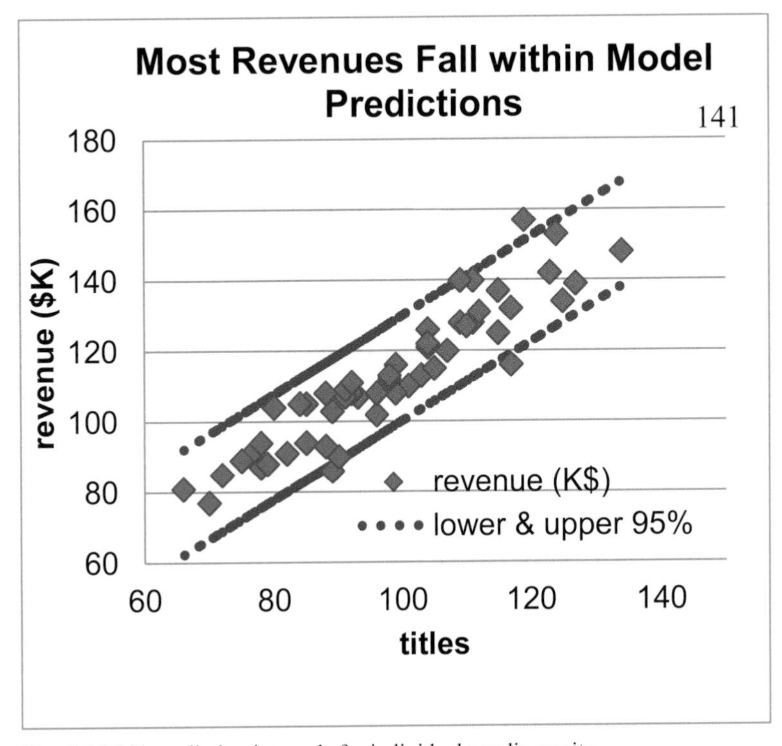

Fig. 5.6 95 % prediction intervals for individual vending units

5.5 *Rsquare* Summarizes Strength of the Hypothesized Linear Relationship and *F* Tests its Significance

ANOVA, an acronym for *Analysis of Variance,* focuses on explained and unexplained variation. The difference, SST - SSE, called the *Regression Sum of Squares, SSR,* or *Model Sum of Squares,* is the portion of total variation in y influenced by variation in X (Fig. 5.7).

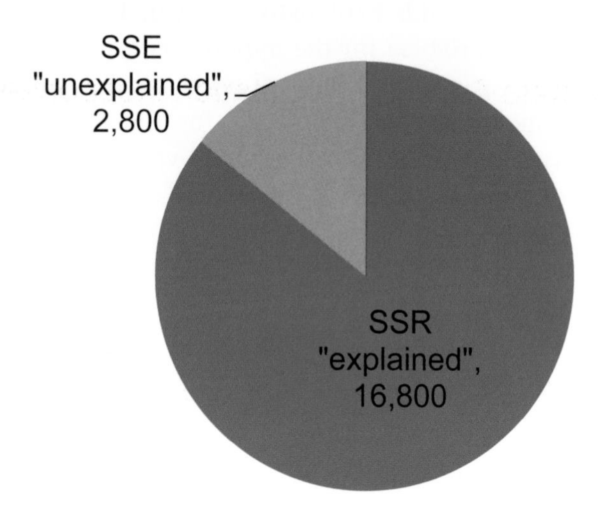

Fig. 5.7 ANOVA showing explained variation, SSR, and unexplained variation, SSE

RSquare is the ratio of explained to unexplained variation:

$$RSquare = SSR/SST$$

RSquare reflects the power of the driver X in explaining variation in performance y. At the extremes, *RSquare* is zero if no variation is explained and one, if all of the variation is explained. In the **Movie Rentals** example, *RSquare* is .86, or 86 %.

$$RSquare = 16{,}800/19{,}600 = .86$$

Variation in titles offered X accounts for 86 % of the variation in revenues y. Other factors account for the remaining 14 %.

A test of the null hypothesis that the independent variable does not influence the dependent variable in the population is equivalent to a test that RSquare is equal to zero, and no variation in the dependent variable is explained by variation in the independent variable:

H_0: Variation in X does not drive variation in y,

or H_0: $RSquare = 0$

Versus

H_1: Variation in X does drive variation in y,

or H_1: $RSquare > 0$

Adding independent variables to a model adds explanatory power. Explained variation, *SSR,* is divided by the number of independent variables for the hypothesis test on variation explained per independent variable. Unexplained variation, *SSE,* is divided by the sample size, less the number of variables in the model (including the intercept), for the relevant comparison of variation explained per independent variable, *MSR,* to unexplained variation for a model of given size and sample size, *MSE*. This ratio of mean squares is distributed as an *F,* and the particular *F* distribution is indexed by model size and sample size. The numerator degrees of freedom is the number of predictors and the denominator degrees of freedom is the sample size, less the number of variables in the model (including the intercept).

$$F_{regression\ df,residual\ df} = \frac{SSR/regression\ df}{SSE/residual\ df} = \frac{MSR}{MSE}$$

The *F* statistic can also be determined with *RSquare*:

$$F_{regression\ df,residual\ df} = \frac{RSquare/regression\ df}{(1 - RSquare)/residual\ df}$$

In **Movie Rentals,** a relatively large proportion of variation is explained by just one driver, making the *F* statistic large:

$$F_{1,50} = \frac{16{,}800/1}{2{,}800/50} = \frac{16{,}800}{55} \cong 300$$

or

$$= \frac{.86/1}{(1-.86)/50} = \frac{.86}{.0028} \cong 300$$

F distributions are skewed with minimum value of zero. *p values* for *F* statistics with 1 and 50 degrees of freedom are shown in Fig. 5.8. In the **Movie Rentals** regression, the $F_{1,50}$ equal to 300 has a *p value* < .0001, providing evidence that the sample data would not be observed if *RSquare* were zero.

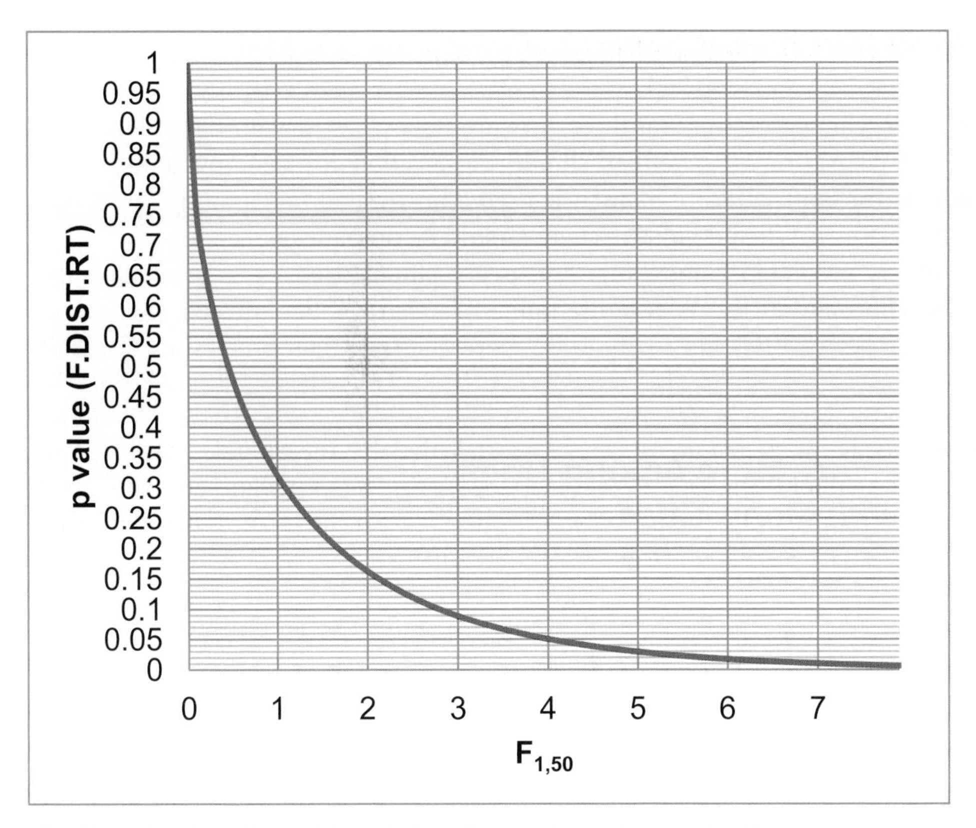

Fig. 5.8 *p values* from *F* tests with one independent variable and sample size 52

In Excel, regression model fit and ANOVA statistics appear in two tables. *RSquare* and the *standard error* appear in SUMMARY OUTPUT, which is followed by the ANOVA table with Regression, Residual, and Total *SS* (*SSR, SSE,* and *SST*), Regression and Residual MS (*MSR* and *MSE*), *F*, and *significance F* (*p value* for the *F* test). The SUMMARY OUTPUT and ANOVA tables from Excel for the **Movie Rental** regression are shown in Table 5.5.

Table 5.5 Model summary of fit and ANOVA table

SUMMARY OUTPUT					
Regression Statistics					
RSquare	.86				
Standard Error	7.44				
Observations	52				
	df	*SS*	*MS*	*F*	*Significance F*
ANOVA					
Regression	1	16,800	16,800	303	.0000
Residual	50	2,800	55		
Total	51	19,500			

5.6 Analyze Residuals to Learn Whether Assumptions are Met

We assume when we use linear regression that the errors are uncorrelated with the independent variable. Explanation and prediction of revenues should be as good for vending units with a limited number of titles, as for units offering many titles. To confirm that this assumption is met, look at a plot of the residuals by predicted values. There should be no pattern.

A plot of the residuals by predicted values, Fig. 5.9, is not pattern free. The residuals show more variation for units with many titles. Within the range of existing titles offered, predictions for units with a limited number of titles are likely to be more accurate than predictions for units offering many titles.

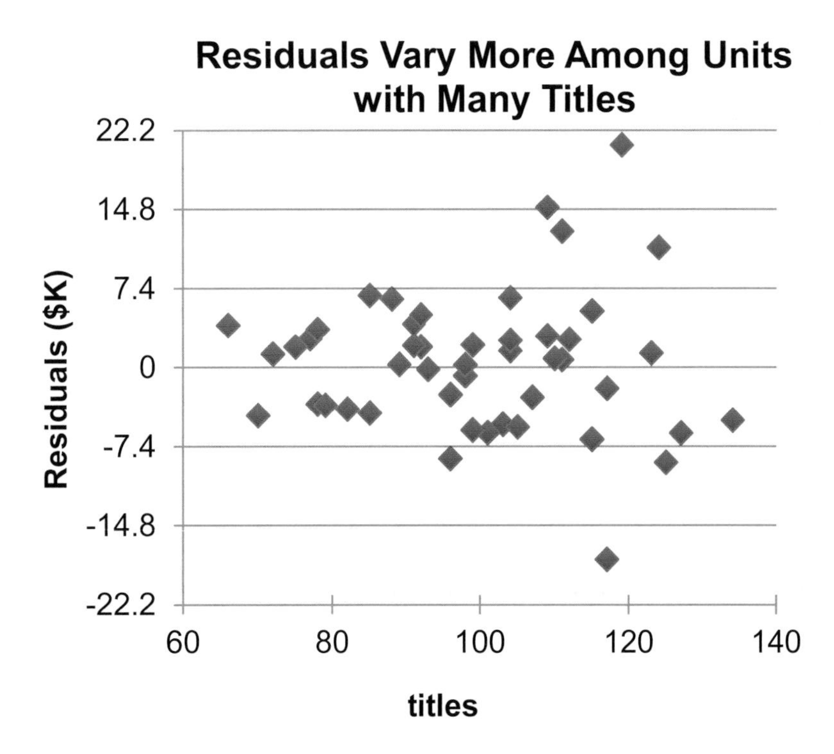

Fig. 5.9 Residuals by titles

This situation, in which residual variation is nonconstant, is termed *heteroskedasticity.* A remedy may be rescaling either the dependent variable, the independent variable, or both, perhaps to natural logarithms.

Linear regression assumes that the residuals are *Normally* distributed. The distribution of residuals, shown in Fig. 5.10, is bell-shaped and has skewness of .1. Roughly 95 % of predictions are within the margin of error, $15.0K, of actual revenues.

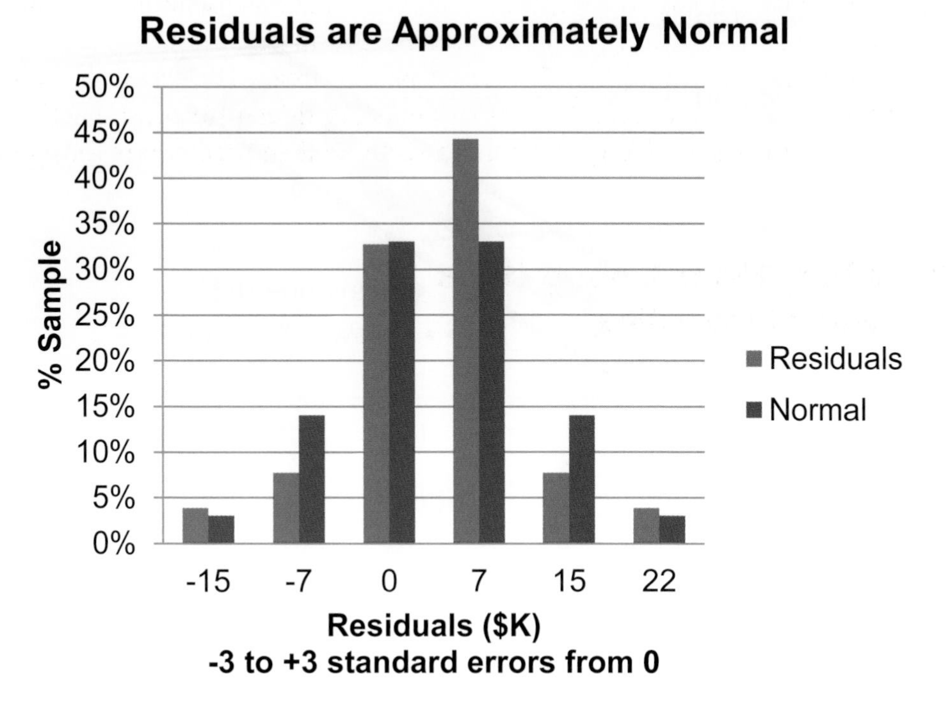

Fig. 5.10 Approximately normal residuals

5.7 Explanation and Prediction Create a Complete Picture

From the regression analysis, the **HitFlix Movie Rental** owner can

- Conclude that titles offered drives revenues,
- Estimate the extent that titles offered drives revenues, and
- Predict revenues at alternate levels of titles offered.

In the presentation of results to management, the owner would conclude:

"Sample evidence suggests that the number of titles offered drives vending unit revenues.

Variation in titles offered accounts for 86 % of the variation in revenues among a random sample of 52 vending units.

With knowledge of titles offered, revenue can be estimated with a margin of error of $15K.

For additional each title offered, a revenue increase of $990 to $1,240 can be expected.

Comparing expected revenue from vending units offering 100 and 130 titles, the 30 additional titles are expected to generate $34K more revenue.

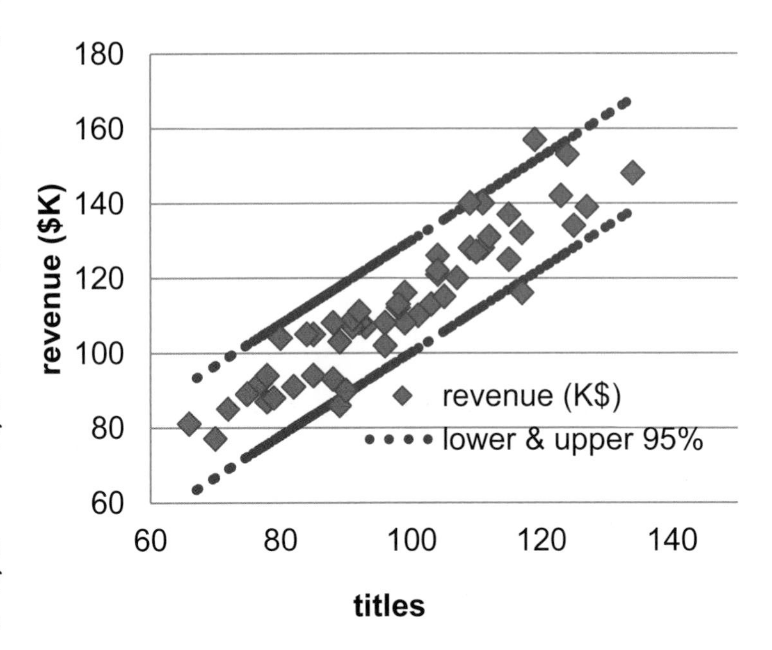

$$\hat{revenue}(\$K) = 3.4(\$K)^a + 1.1(\$K)^a \times titles$$

RSquare: .86[a]

[a]*Significant at .0001*

Titles offered	*Expected revenue*
100	$115,000
130	$149,000

The HitFlix owner presented results of his regression analysis by illustrating the regression line with 95 % confidence prediction intervals on top of the actual data. This demonstrates how well the model fits the data.

5.8 Present Regression Results in Concise Format

The regression equation is included, in standard format, with the dependent variable on the left, RSquare below the equation, and significance levels of the model and parameter estimates indicated with superscripts:

$$\hat{y} = b_0^a + b_1^a \times X$$

RSquare=___[a]

[a]*Significant at* ___,

where the variable names and units are specified.

Significance is reported at two levels, .05 and .01.

p values greater than .01, but less than or equal to .05 are reported as significant at .05.

p values less than or equal to .01 are reported as significant at .01 (Table 5.6).

Table 5.6 Two levels of significance reported

	Reported as significant at
$.01 < p\ value \leq .05$	.05
$p\ value \leq .01$	.01

For the general business audience, the verbal description with graphical illustration conveys all of the important information. The three additional lines provide the information that statistically savvy readers will want in order to assess how well the model fits and which parameter estimates are significant.

5.9 Assumptions We Make When We Use Linear Regression

Often a group of independent variables together jointly influence a dependent variable. If we attempt to explain or predict a dependent variable with an independent variable, but omit a third (or fourth) important influence, results will be misleading. If just one from the group is included in a regression, it may seem to be responsible for the joint impact of the group. It will seem that the independent variable chosen is more important than it actually is. Chapters 10 and 11 introduce diagnosis of *multicollinearity,* the situation in which predictors are correlated and jointly influence a dependent variable.

Linear regression of time series data assumes that the unexplained portion of a model, the residuals, are stable over time, and that predictions do not get better or worse with time. Patterns uncovered in the data are stable over time. Chapter 11 introduces diagnosis of and remedies for *autocorrelated* errors which break this assumption and vary with time.

Linear regression assumes that the dependent variable, which is often a performance variable, is related linearly to the independent variable, often a decision variable. In reality, few relationships are linear. More often, performance increases or decreases in response to increases in a decision variable, but at a diminishing rate. The dependent variable is often limited. Revenues, for example, are never negative and are limited (probably at some very high number) by the number of customers in a market. In these cases, linear regression doesn't fit the data perfectly. Extrapolation beyond the range of values within a sample can be risky if we assume constant response when response is actually diminishing or increasing. Though often not perfect reflections of reality, linear relationships can be useful approximations. In Chap. 13, we will explore simple remedies to improve linear models of nonlinear relationships by simply rescaling to square roots, logarithms or squares.

5.10 Correlation Reflects Linear Association

A correlation coefficient ρ_{xy} is a simple measure of the strength of the linear relationship between two continuous variables, X and y. The sample estimate of the population correlation coefficient ρ_{xy} is calculated by summing the product of differences from the sample means $\overline{X}$ and $\overline{Y}$, standardized by the standard deviations s_x and s_y:

$$r_{xy} = \frac{1}{(N-1)} \sum \frac{(x_i - \overline{X})}{s_x} \frac{(y_i - \overline{Y})}{s_y},$$

where x_i is the value of X for the i'th sample element, and y_i is the value of Y for the i'th sample element.

When x and y move together, they are positively correlated. When they move in opposite directions, they are negatively correlated.

Example 5.2 HitFlix Movie Rentals

Table 5.7 contains titles offered x and revenues y from a sample of eight vending units:

Table 5.7 Titles stocked and revenues ($K) for eight vending units

Vending unit	*Titles offered* x	*Revenues ($K)* y	*Vending unit*	*Titles offered* x	*Revenues ($K)* y
1	110	75	5	150	115
2	110	80	6	160	135
3	120	85	7	170	140
4	130	105	8	170	145
			Sample mean	**140**	**$110**

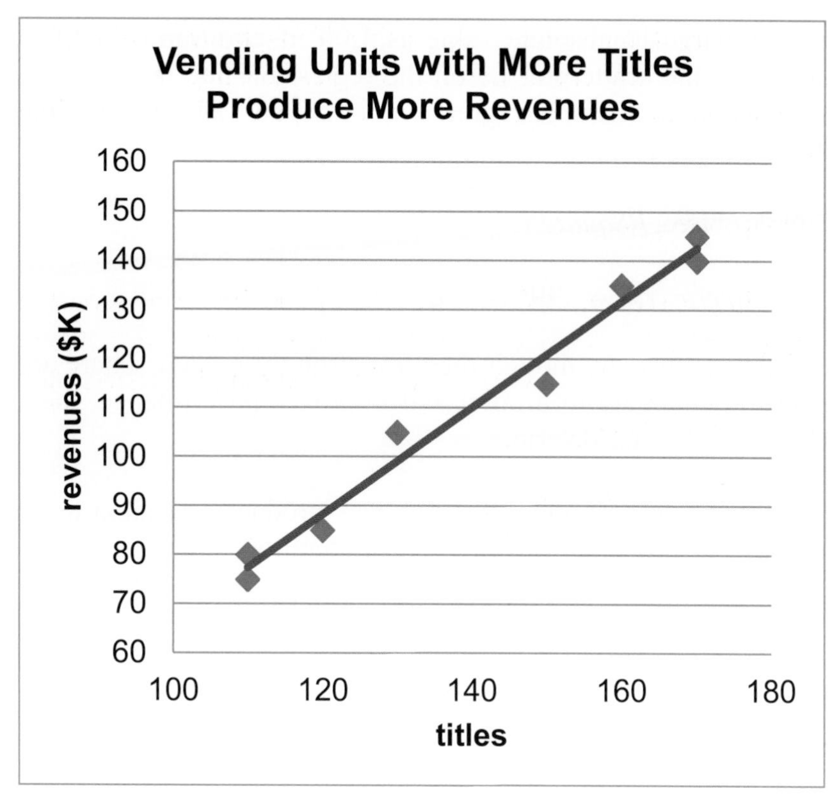

Fig. 5.11 Vending unit revenues ($K) by titles stocked

A scatterplot in Fig. 5.11 reveals that units which stock more titles also have greater revenues.

Differences from the sample means and their products are shown in Table 5.8.

Table 5.8 Differences from sample means and crossproducts

	Titles stocked			*Revenues ($K)*			
Unit	x_i	$\overline{X}$	$x_i - \overline{X}$	y_i	$\overline{Y}$	$y_i - \overline{Y}$	$(x_i - \overline{X}) \times (y_i - \overline{Y})$
1	110	140	−30	75	110	−35	1050
2	110	140	−30	80	110	−30	900
3	120	140	−20	85	110	−25	500
4	130	140	−10	105	110	−5	50
5	150	140	10	115	110	5	50
6	160	140	20	135	110	25	500
7	170	140	30	140	110	30	900
8	170	140	30	145	110	35	1050

The sample standard deviations are $s_x = 25.6$ titles and $s_y = 28.2$ ($K).

The correlation coefficient is:

$$r_{xy} = \frac{1}{(8-1)}\left[\frac{1050 + 900 + 500 + 50 + 50 + 500 + 900 + 1050}{(25.6)(28.2)}\right]$$

$$= \frac{1}{7}\left[\frac{5000}{722}\right]$$

$$= .990$$

A correlation coefficient can be as large in absolute value as 1.00, if two variables were perfectly correlated. All of the points in the scatterplot lie on the regression line in that case. *RSquare*, which is the squared correlation in a simple regression, would be 1.00, whether the correlation coefficient were −1.00 or +1.00.

In the **HitFlix Movie Rentals** example above, *RSquare* is

$$RSquare = r_{xy}^2 = .99^2 = .98$$

In some cases, *x* and *y* are not related *linearly*, though they *are* strongly related. There are situations, for example, where more is better up to a point and improves performance, then, *saturation* occurs and, beyond this point, response deteriorates.

- Without enough advertising, customers will be not aware of a new product. Spending more increases awareness and improves performance. Beyond some saturation point, customers grow weary of the advertising, decide that the company must be desperate to advertise so much, and switch to another brand, reducing performance.
- A factory with too few employees *x* to man all of the assembly positions would benefit from hiring. Adding employees increases productivity *y* up to a point. Beyond some point, too many employees would crowd the facility and interfere with each other, reducing performance.

5.11 Correlation Coefficients are Key Components of Regression Slopes

Correlation coefficients are closely related to regression slopes. From the correlation between *x* and *y*, as well as their sample standard deviations s_x and s_y, the regression slope estimate can be calculated:

$$b_1 = r_{xy}\frac{s_y}{s_x}$$

Similarly, from the regression slope estimate and sample standard deviations s_x and s_{xy}, the correlation coefficient can be calculated:

$$r_{xy} = b_1\frac{s_x}{s_y}$$

The *t* test of hypothesis that a slope is zero

$$H_0: \beta_1 = 0$$

Versus

$$H_1: \beta_1 \neq 0$$

is equivalent to the t test used to test the hypothesis that a correlation is zero:

$$H_0: \rho_{xy} = 0$$

Versus

$$H_1: \rho_{xy} \neq 0:$$

$$t_{N-2} = \frac{b_1}{s_{b_1}} = \sqrt{N-2}\frac{r_{xy}}{\sqrt{1-r_{xy}^2}}$$

In the **Movie Rentals** example, with the correlation coefficient, $r_{titles,revenues} = .99$, and the sample standard errors, $s_{titles} = 26.5$ and $s_{revenues} = 28.2$, the regression slope estimate can be calculated,

$$b_{titles} = .99\frac{28.2}{26.5} = 1.09$$

as well as the t value to test the hypothesis that the slope is zero or, equivalently, that the correlation is zero:

$$t_6 = \sqrt{6}\frac{.990}{\sqrt{1-.990^2}} = 16.8$$

with p *value* < .01.

Based on sample evidence, there is little chance that titles stocked and vending unit revenues are uncorrelated.

Corresponding simple regression results are shown in Table 5.9.

Table 5.9 Regression of revenue by titles

SUMMARY OUTPUT

Regression Statistics	
R Square	.98
Standard Error	4.38
Observations	8

ANOVA

	Df	*SS*	*MS*	*F*	*Significance F*	
Regression	1	5,435	5,435	283.0	3E-06	
Residual	6	115	19			
Total	7	5,550				
	Coefficients	*Standard Error*	*t Stat*	*p value*	*Lower 95 %*	*Upper 95 %*
Intercept	−42.17	9.177	−4.6	.0037	−64.63	−19.72
Titles Stocked	1.09	.065	16.8	.0000	.93	1.25

Example 5.3 Pampers

Procter & Gamble hoped that targeted customers who value fit in a preemie diaper would use price as a quality of fit cue and prefer a higher-priced diaper. Ideally, fit importance would be negatively correlated with price responsiveness. In the concept test of the new preemie diaper using a sample of 97 preemie mothers, price responsiveness was measured as the difference between trial intentions at competitive and premium prices, each measured on a 5-point scale (1="Definitely Will Not Try" to 5="Definitely Will Try"). Fit importance was measured on a 9-point scale (1="Unimportant" to 9="Very Important"). The correlation between price responsiveness and fit importance from Excel are shown in Fig. 5.12:

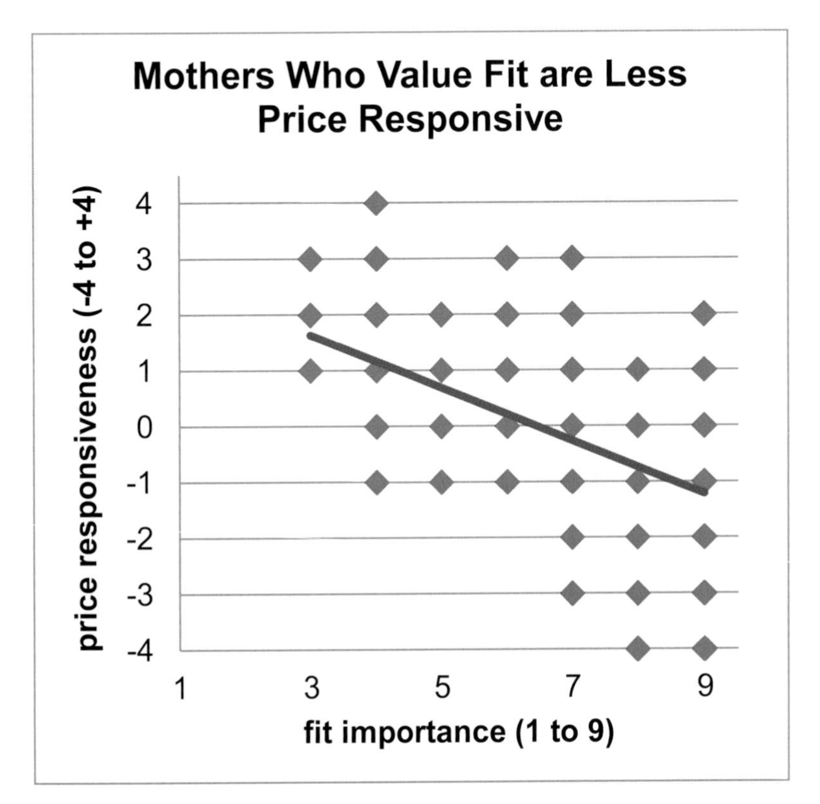

Correlation	-.45
t	-4.9
p value	< .0001

Fig. 5.12 Correlation between price responsiveness and fit importance

The correlation between price responsiveness y and fit importance x is moderately large and negative:

$$r_{xy} = -.45$$

The lower the importance of fit to a preemie mom, the greater her responsiveness to a price reduction.

Regression analysis from Excel, shown in Table 5.10, quantifies this negative, linear relationship:

Table 5.10 Regression of price responsiveness by fit importance

SUMMARY OUTPUT							
Regression Statistics							
R Square	.20						
Standard Error	1.66						
Observations	97						
ANOVA							
	df	*SS*	*MS*	*F*	*Significance F*		
Regression	1	67	67	24.1	.00001		
Residual	95	262	2.8				
Total	96	329					
	Coefficients	*Standard Error*	*t Stat*	*p value*	*Lower 95 %*	*Upper 95 %*	
Intercept	3.06	.65	4.7	.0000	1.77	4.35	
Fit Importance	−.47	.10	−4.9	.0000	−.67	−.28	

From results of correlation and regression analysis, Procter & Gamble management concluded:

"Price responsiveness is negatively correlated with fit importance of diapers to preemie mothers. Variation in fit importance accounts for 20% of the variation in price responsiveness. Though not a large influence on price responsiveness, fit importance does drive responsiveness, along with other factors. A difference between "Moderately Important" and "Important", which is a two-point difference on the 9-point importance scale, reduces price responsiveness by about one (.5 to 1.3) scale point on a 9-point responsiveness scale.

It is likely that preemie mothers seeking a high quality diaper with superior fit find claims of superior fit at a lower price unbelievable. A higher price supports the higher quality, superior fit image."

5.12 Correlation Complements Regression

The correlation coefficient summarizes direction and strength of linear association between two continuous variables. Because it is a standardized measure, taking on values between −1 and +1, it is readily interpretable. Unlike regression analysis, it is not necessary to designate a dependent and an independent variable to summarize association with correlation analysis. Later, in the context of multiple regression analysis, the correlations between independent variables will be an important focus in our diagnosis of multicollinearity, introduced in Chaps. 8 and 9.

Correlation analysis should be supplemented with visual inspection of data. It would be possible to overlook strong, nonlinear associations with small correlations. Inspection of a scatterplot will reveal whether or not association between two variables is linear.

Correlation is closely related to simple linear regression analysis:

- The squared correlation coefficient is *RSquare*, our measure of percent of variation in a dependent variable accounted for by an independent variable.

- The regression slope estimate is a product of the correlation coefficient and the ratio of the sample standard deviation of the dependent variable to sample standard deviation of the independent variable.
  - Slope estimates from simple linear regression are unstandardized correlation coefficients.
  - Correlation coefficients are standardized simple linear regression slope estimates.

5.13 Linear Regression is Doubly Useful

Linear regression handles two modeling jobs, quantification of a driver's influence and forecasting. Regression models quantify the direction and nature of influence of a driver on a response or performance variable. Regression models also enable forecasts and to compare decision alternatives. This latter use of regression to answer "what if" questions, *sensitivity analysis*, is an important tool for decision making.

Excel 5.1 Build a Simple Linear Regression Model

Impact of Titles Offered on HitFlix Movie Rental Revenues. Use regression analysis to explore the linear influence of differences in *titles* offered on *revenue ($K)* differences across a random sample of 52 movie rental vending units.

Open **Excel 5.1 HitFlix Movie Rental Revenues.xls.**

Use shortcuts to run regression:

Alt AY3, R, down, down, Enter

For **Input Y Range**, enter label and observations on the dependent variable, *revenues ($K),* and for **Input X Range,** enter label observations on the independent variable, *titles.* Specify **Labels,** and **Residuals:**

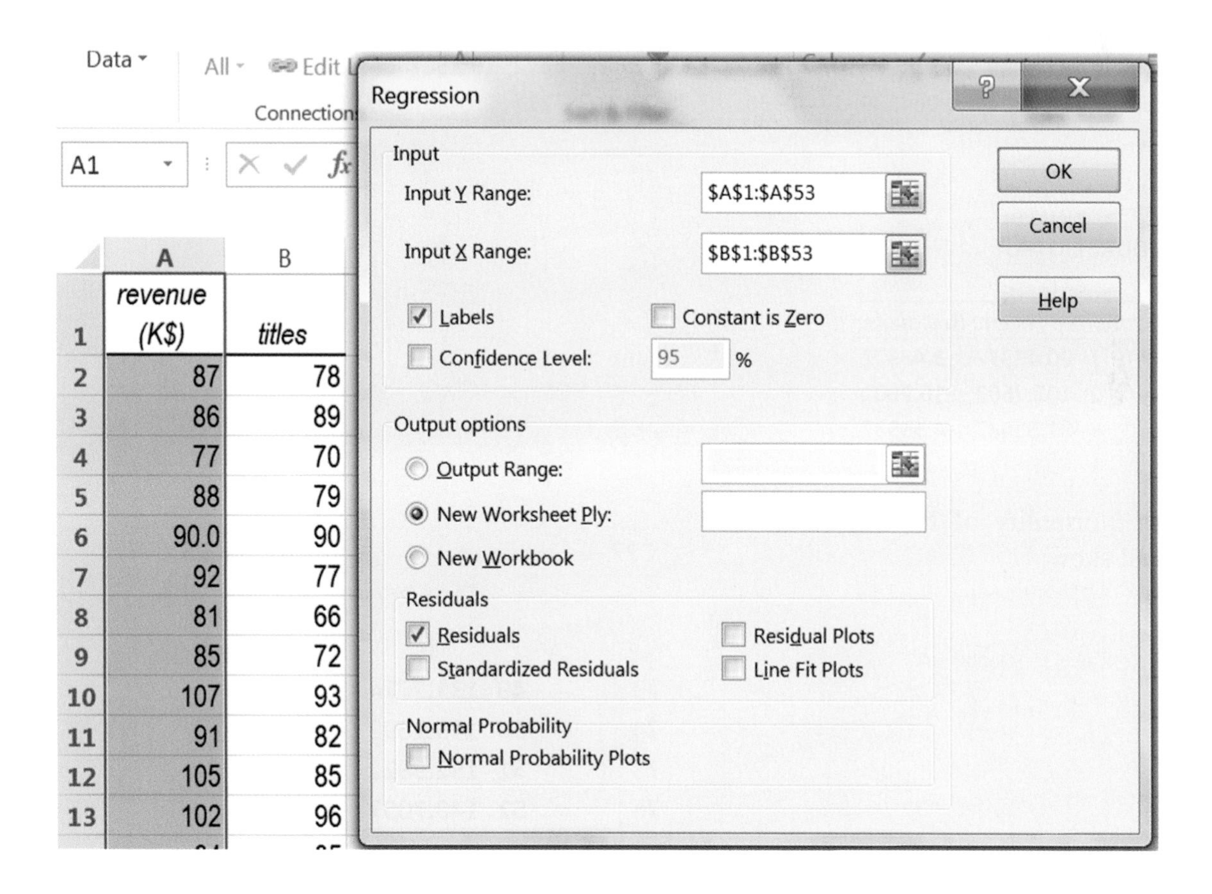

	A	B	C	D	E	F	G	H	I
1	SUMMARY OUTPUT								
2									
3	*Regression Statistics*								
4	Multiple R	0.926426							
5	R Square	0.858266							
6	Adjusted F	0.855431							
7	Standard I	7.440716							
8	Observatic	52							
9									
10	ANOVA								
11		*df*	*SS*	*MS*	*F*	*gnificance F*			
12	Regressior	1	16762.77	16762.77	302.7724	7.4E-23			
13	Residual	50	2768.213	55.36425					
14	Total	51	19530.98						
15									
16		*Coefficients*	*andard Erro*	*t Stat*	*P-value*	*Lower 95%*	*Upper 95%*	*ower 95.0%*	*pper 95.0%*
17	Intercept	3.432283	6.381939	0.537812	0.593093	-9.38622	16.25078	-9.38622	16.25078
18	titles	1.116044	0.064139	17.40036	7.4E-23	0.987217	1.244871	0.987217	1.244871
19									
20									
21									
22	RESIDUAL OUTPUT								
23									
24	*Observatio*	*rted revenu*	*Residuals*						
25	1	90.48372	-3.48372						
26	2	102.7602	-16.7602						
27	3	81.55537	-4.55537						

To assess Normality of the residuals, find the residual skew:

C77 =SKEW(C25:C76)

	A	B	C	D
73	49	131.7774	-6.77736	
74	50	145.1699	-6.16989	
75	51	152.9822	-4.9822	
76	52	140.7057	1.29429	
77			0.100238	

Excel 5.2 Construct Prediction Intervals

To see 95 % prediction intervals for a vending unit offering a specific number of *titles*, return to the data sheet, **CNTL+Page Dn,** select and copy *Revenues* and *Titles,* return to the regression sheet, **CNTL+Page Up,** and paste next to residuals:

	A	B	C	D	E
24	)bservatio	rted revenu	Residuals	*revenue (K$)*	*titles*
25	1	90.48372	-3.48372	87	78
26	2	102.7602	-16.7602	86	89
27	3	81.55537	-4.55537	77	70
28	4	91.59977	-3.59977	88	79
29	5	103.8763	-13.8763	90.0	90

Use the regression equation to find *predicted revenues ($K),* locking cell references to the coefficients with **fn 4,**

$= \boldsymbol{b_0} + \boldsymbol{b_{titles}} \times \boldsymbol{titles}$

Which will be the following in row 25:

=$B$17 + $B$18*E25

(The intercept will always appear in **B17.)**

Fill in the column:

F76 =$B$17+$B$18*E76

	A	B	C	D	E	F
24	)bservatio	rted revenu	Residuals	*revenue (K$)*	*titles*	*predicted revenues (K$)*
25	1	90.48372	-3.48372	87	78	90.5
26	2	102.7602	-16.7602	86	89	102.8
27	3	81.55537	-4.55537	77	70	81.6
28	4	91.59977	-3.59977	88	79	91.6
29	5	103.8763	-13.8763	90.0	90	103.9

To find the *margin of error* needed to produce *95 % prediction intervals*, first find the *critical t.*

In the cell next to the *standard error,* **C7,** find the *critical t* value which corresponds to a 95 % confidence level and the *residual degrees of freedom* with the Excel function **T.INV.2T(***probability, df***).**

For *probability*, enter .05 for a 95 % level of confidence, and for *df* enter the residual degrees of freedom from **B13** of the regression sheet.

(The *residual dfs* always appear in cell **B13).**

Multiply the *standard error* (which always appears in **B7**) by the *critical t* to find the *me*:

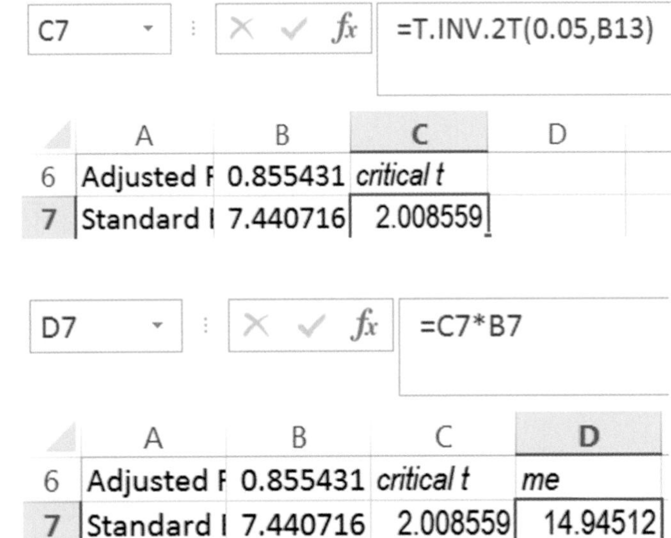

Find the *lower 95 %* and *upper 95 %* prediction interval bounds by adding and subtracting the margin of error to and from *predicted revenues,* locking the *margin of error* cell reference with **fn 4**:

Fill in the columns.

G76 =F76-$D$7

	D	E	F	G	H
6	me				
7	14.94512				
8					
9					
10					
11	*MS*	*F*	*ignificance F*		
12	16762.77	302.7724	7.396E-23		
13	55.36425				
14					
15					
16	*t Stat*	*P-value*	*Lower 95%*	*Upper 95%*	*ower 95.0%*
17	0.537812	0.593093	-9.386218	16.25078	-9.38622
18	17.40036	7.4E-23	0.9872168	1.244871	0.987217
19					
20					
21					
22					
23					
24	*revenue (K$)*	*titles*	*predicted revenues (K$)*	*lower 95% pi bound*	*upper 95% pi bound*
25	87	78	90.5	75.5	105.4
26	86	89	102.8	87.8	117.7

To make a scatterplot of *the 95 % prediction intervals* with actual *revenues*, first move *predicted revenues* to the right of the *95 % prediction intervals.*

Select and cut *predicted revenues*, **Cntl+X**,

and then, from cell **I24**, "drop them in here,"

	F	G	H	I
24	*predicted revenues (K$)*	*lower 95% pi bound*	*upper 95% pi bound*	
25	90.5	75.5	105.4	
26	102.8	87.8	117.7	
27	81.6	66.6	96.5	
28	91.6	76.7	106.5	

Alt H I E:

	F	G	H
24	*lower 95% pi bound*	*upper 95% pi bound*	*predicted revenues (K$)*
25	75.5	105.4	90.5
26	87.8	117.7	102.8
27	66.6	96.5	81.6
28	76.7	106.5	91.6

(**Cntl+X** cuts selected cells. **Alt H I E** selects the **H**ome menu and **I**nsert function and inserts cut or copied c**E**lls to the left of the selected column or cell.)

Then use **Cntl+X** and **Alt H I E** to move *titles* to the left of *revenues*.

Select *titles*, actual *Revenues,* and *95 % lower* and *upper* prediction intervals and make a scatterplot:

Alt N D

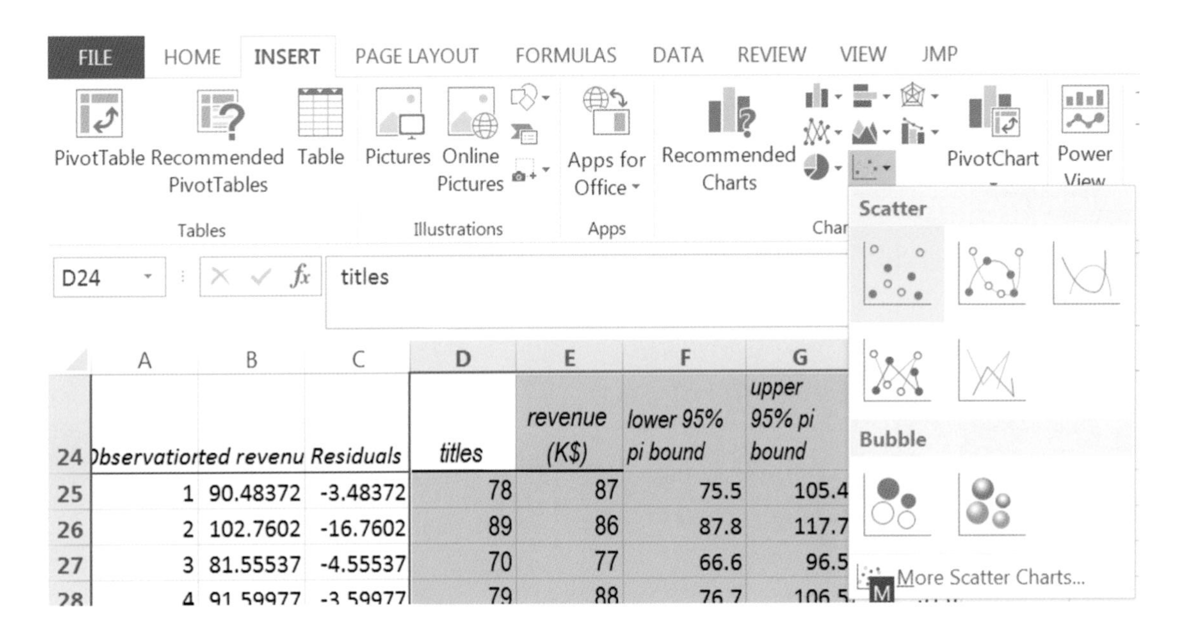

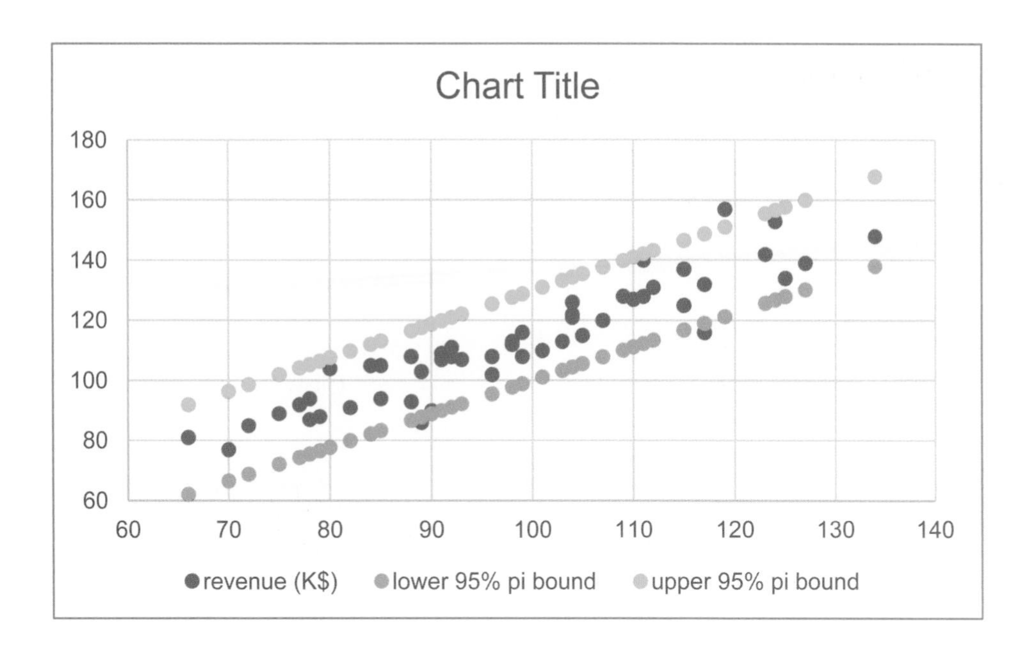

Change the *lower* and *upper 95 % pi bounds* to lines:

Alt JC C Tab dn dn dn dn (to Customer), **Tab Tab** (to *Revenue*), **Tab dn** to Scatter,

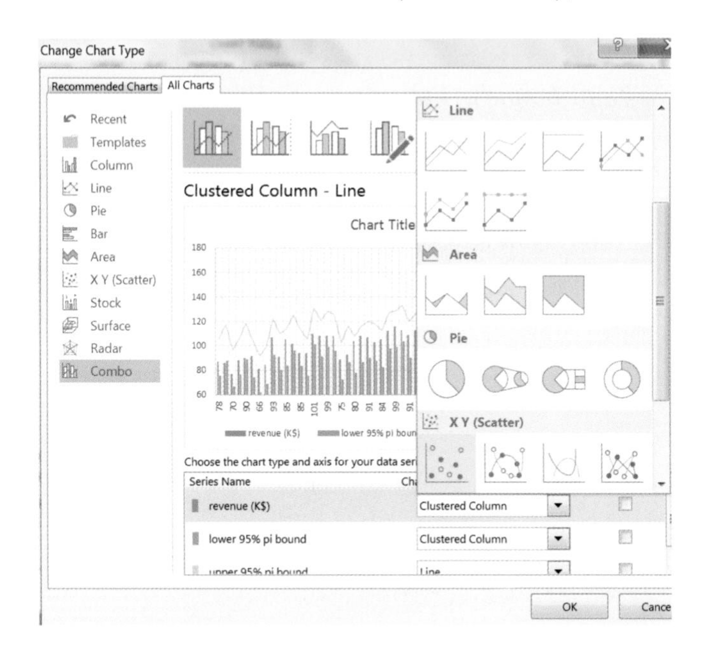

Enter, Tab spacebar, Tab Tab Tab Tab Tab Tab, dn (to *lower 95 % pi bound*), **Tab, dn** to Scatter with Straight Line,

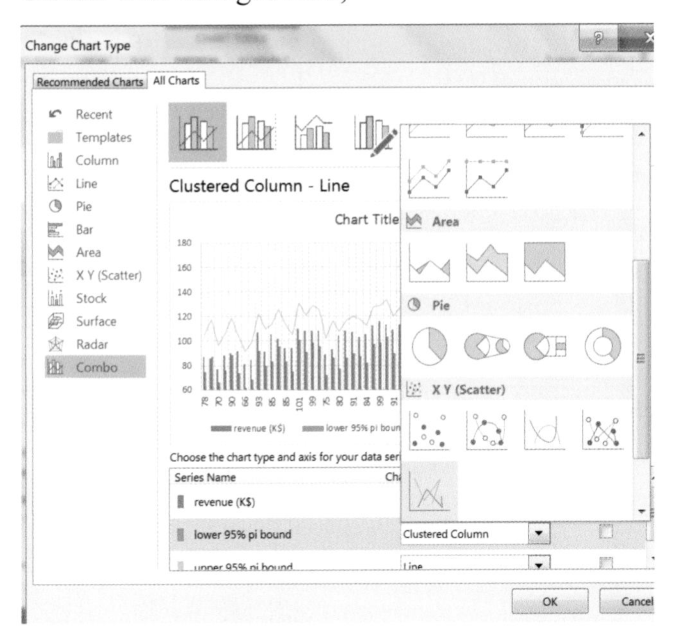

Enter, Tab spacebar,

Tab Tab Tab Tab Tab Tab, dn, (to *upper 95 % pi bound*), **Tab, dn** to Scatter with Straight Line,

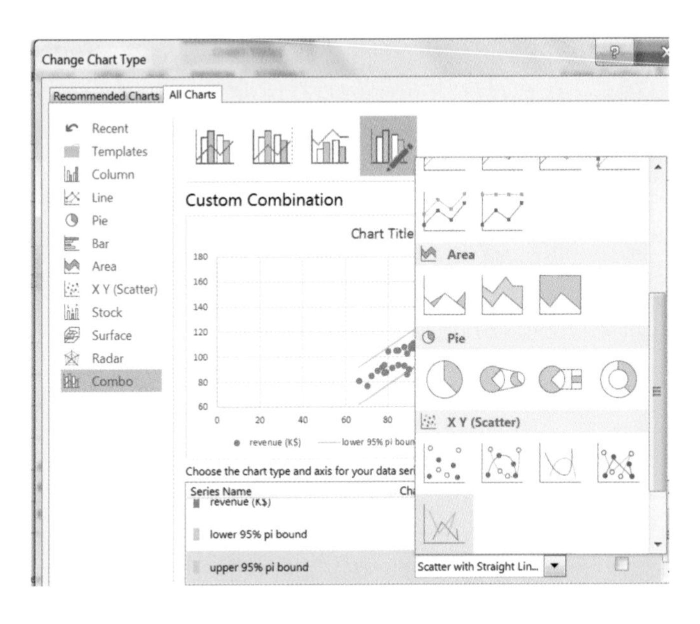

Enter, Tab spacebar,Tab (to OK), **Enter.**

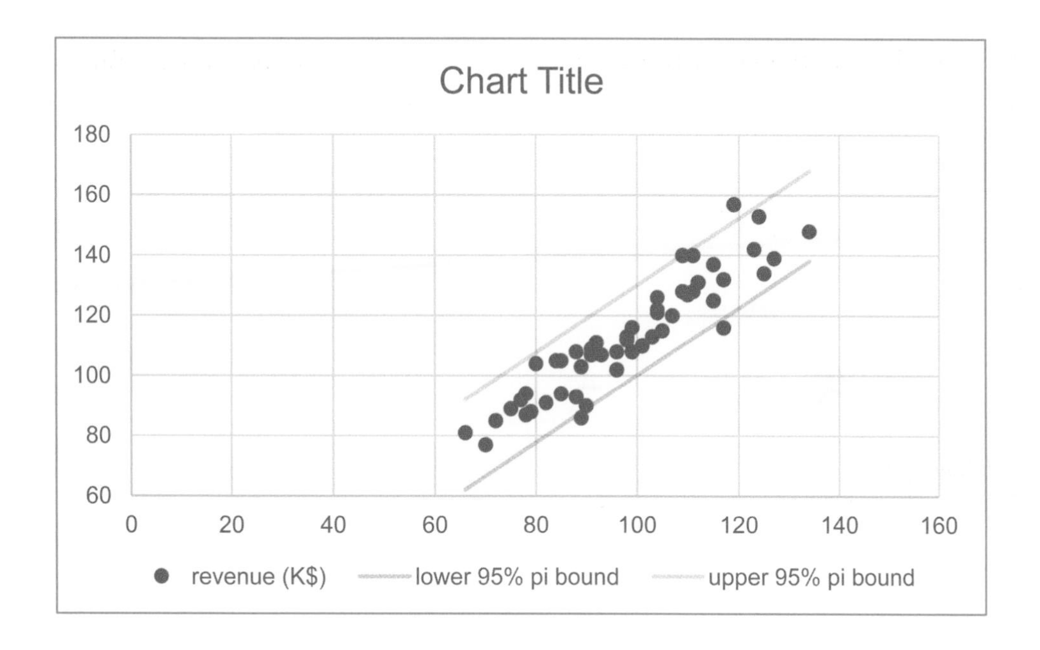

Adjust the axes to make better use of white space, using

Alt JA E

to select an axis,

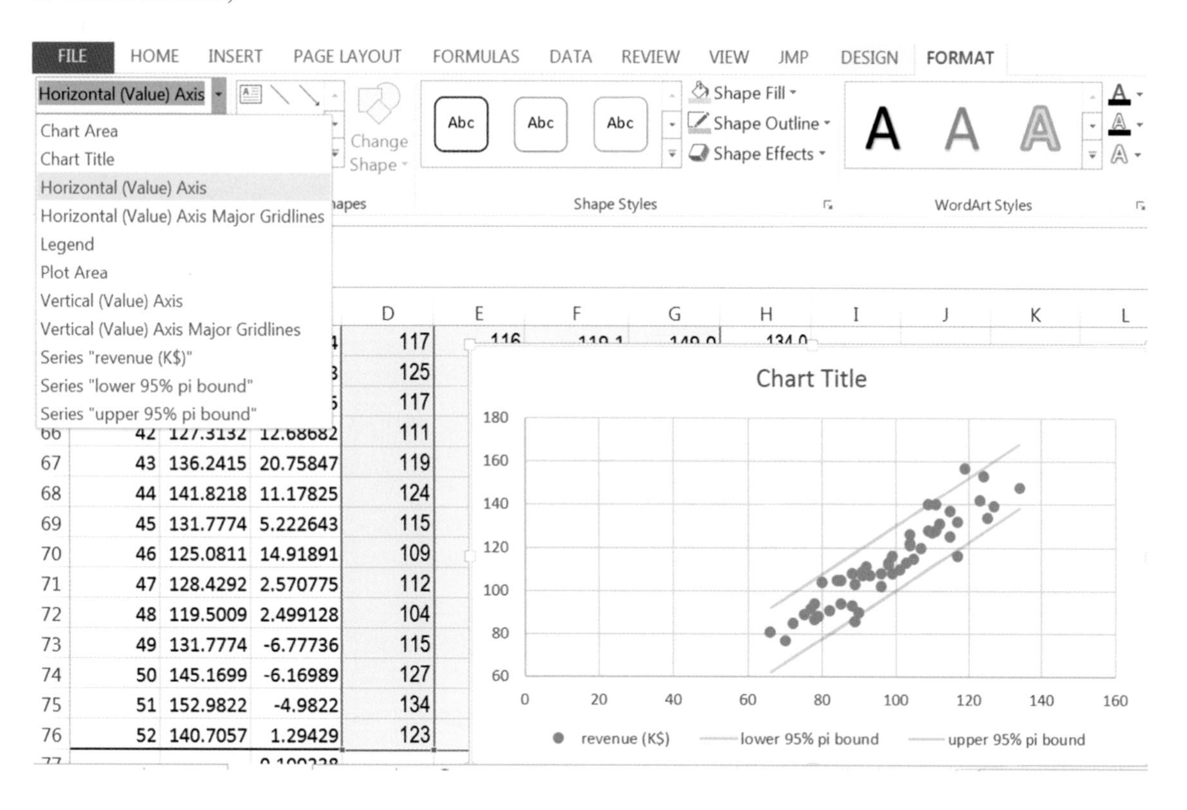

and then

Alt JA M

to format.

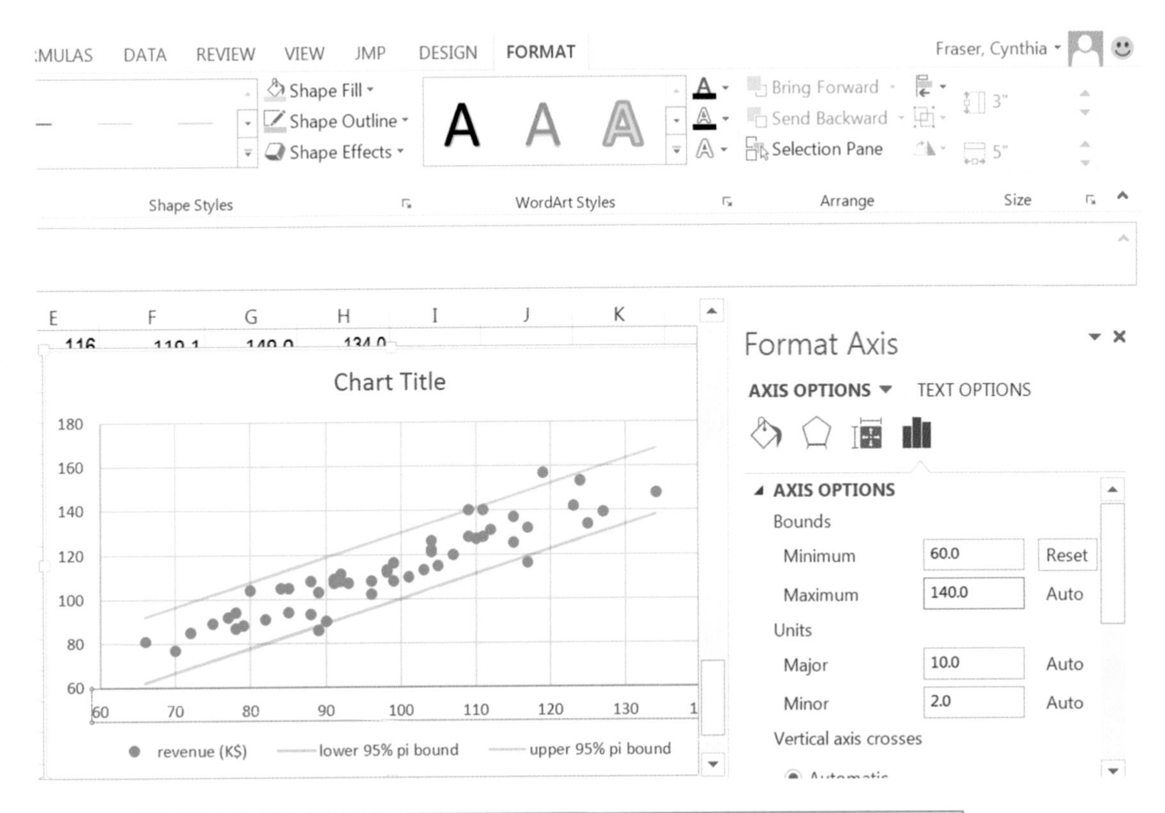

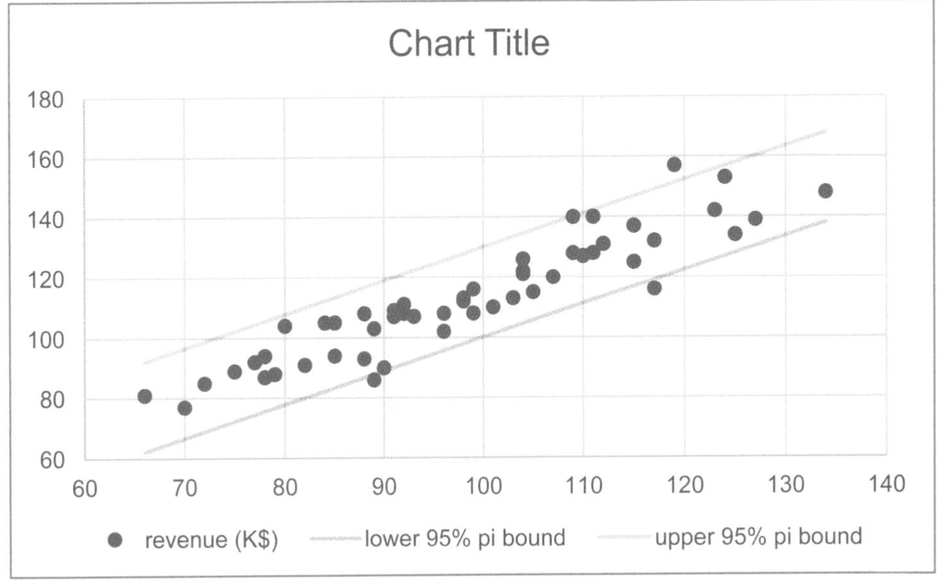

Change the *upper 95 % pi bound* to the color of the *lower 95 % pi bound* using

Alt JA E, dn to *upper 95 % pi bound*

Alt JA SO to desired color.

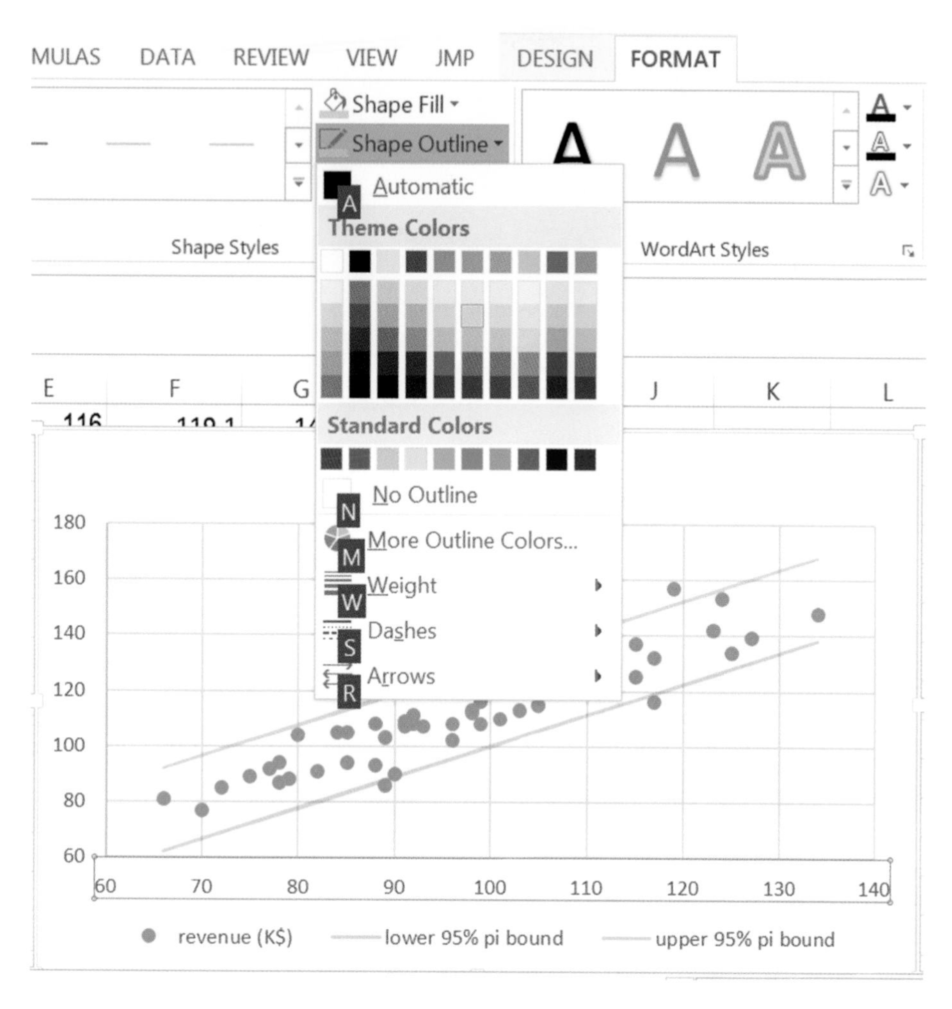

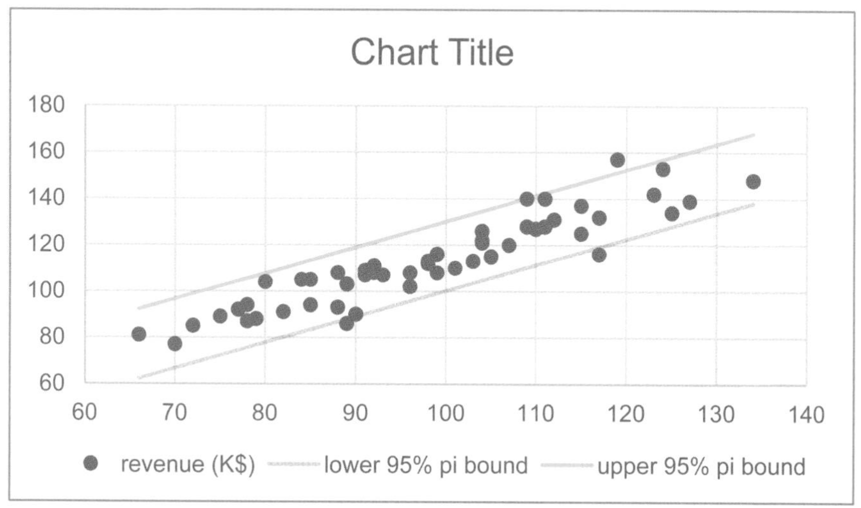

Choose the first Layout from

Alt JA L,

and add axis titles and a chart title that conveys the primary regression result.

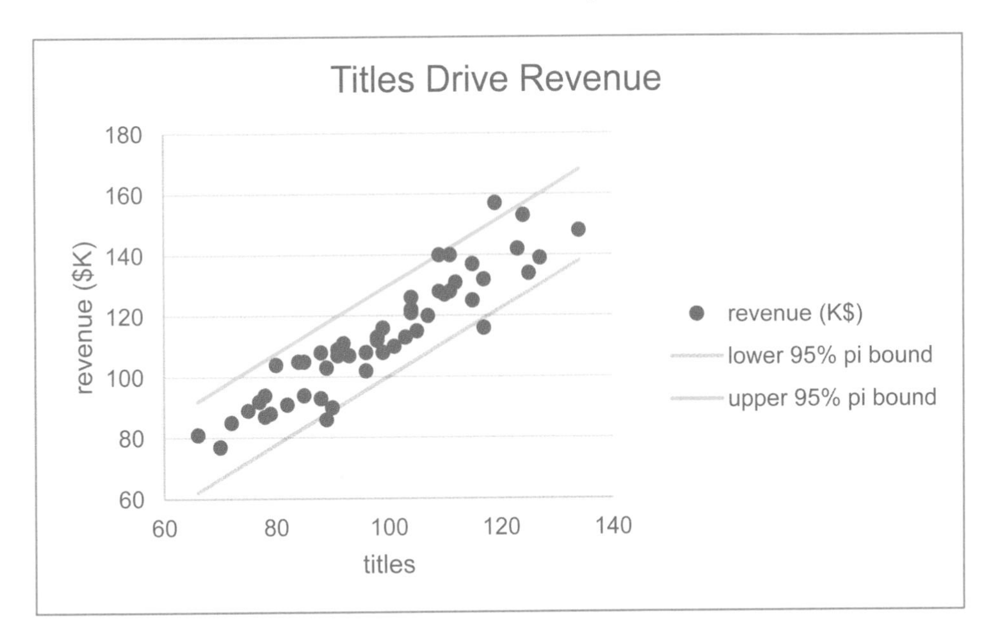

Excel 5.3 Find Correlations Between Variable Pairs

Management would like to know whether there is an association between the perceived importance of diaper fit and price responsiveness among preemie mothers.

Fit importance ratings and *price responsiveness* from a concept test sample of 97 preemie mothers are in **Excel 5.2 Pampers Price Responsiveness.xls.**

At the bottom of the dataset, use the Excel function **CORREL(***array1,array2***)** to find the correlation between *fit importance* rating and *price responsiveness.*

B99 =CORREL(A2:A98,B2:B98)

	A	B	C	D
1	fit importance	price responsiveness		
96	4	4		
97	3	1		
98	3	2		
99	correlation	-0.45		

Lab Practice 5 Oil Price Forecast

Rolls-Royce is concerned that demand for their jet engines may decline if the price of oil rises. Management believes that the growing Indian and Chinese economies, declining US crude oil production and World supply, and rising US crude oil imports may be driving oil prices.

Lab Practice 5 Oil Price Forecast.xls contains 19 years of annual data on

spot price of Oklahoma crude oil_t,
$China\ GDP(\$B)_{t-4}$,
$India\ GDP(\$B)_{t-4}$,
$US\ production(K\ barrels/day)_{t-4}$,
$World\ supply(M\ barrels/day)_{t-4}$, and
$US\ imports(K\ barrels/day)_{t-4}$.

Build a simple regression model to estimate the impact of **ONE** of these hypothetical drivers on crude oil *spot price*. Round all of your numerical answers to two or three significant digits.

1. Present your regression equation in standard format:

2. What percent of variation in *spot prices* can be accounted for by variation in the driver that you chose? ____________

3. How close to actual *spot prices* could you expect a forecast to be 95 % of the time? __

4. What range in *spot prices* could management be 95 % certain to expect in 2016, given
   an increase in *China GDP* to $3,200 billion in 2012,
   an increase in *India GDP* to $2,300 billion in 2012,
   constant *US production* to 8,500 thousand barrels/day in 2012,
   an increase in *World supply* to 90 million barrels/day in 2012,
   OR
   a decrease in *US imports* of 2,500 thousand barrels/day in 2012?
   (Choose the **ONE** driver that you chose for your model.) __________ to __________

5. Make a scatterplot of *95 % individual prediction intervals* and *actual spot prices* by *year* and attach. Insert a chart title that summarizes your conclusion.

Lab 5 Simple Regression Dell Slimmer PDA

Dell is considering the introduction of an ultraslim PDA which would fit in a shirt pocket, come in an array of colors and be sold in WalMarts. Dell withdrew its Axim PDA after share fell to 3 %. Developers want to be sure that the new PDA will offer the features most desired by the target segments of young, lower income high school students and service workers. Managers believe from past research that there are three PDA lifestyle segments.

- ***Younger Players***. The **youngest** segment, **high school** students, who are fashion conscious and technically savvy. Some PDAs in this segment are provided by **higher income** parents. PDAs are primarily used to text message, play music and video games. Penetration in this segment is low.
- ***Older Players***. **High school graduates** employed in service jobs. These users are the least technically savvy. PDAs are a luxury used to play music and video games. Penetration in this segment is the lowest.
- ***Professionals and Soon to Be***. **College students and college graduates**. This segment is technically savvy and uses PDA software in classes or on the job. PC connectivity is important, though text messaging and music are also important. This market is saturated and most purchases are upgrades.

Palm and HP cater to the *Professionals* and *Soon to Be* segments.
Dell is targeting *Younger* and *Older Players*, hoping to avoid competition.
The new PDA would be ultra slim and also fit in a shirt pocket (unlike the withdrawn Axim).

Data from a concept test of 14–34 year olds in **Lab 5 Dell Slimmer.xls** include

- Measures of the importance of *thinness and ability to fit in a shirt pocket*, on a 1- to 9-point scale (1=*unimportant…9=extremely important)*
- *Key demogra*phics: *age,* household *income* (in thousands), and years of *education*

1. **Importance of thinness.** Use a *t test* to determine whether "*thinness*" is an important attribute to customers like those surveyed. An attribute is considered important if the average customer rating is **greater than 5** on the 9-point scale.

   A ___one tail ____two tail *t test* is required.

   The null hypothesis is: ____

   The alternate hypothesis is: ____________________________________

   Management can conclude that 14–34 years olds
   rate *thinness* important (**at least 5** on a 9 point scale): ___Y ___N

2. Construct a *95 % confidence interval* for the average *importance* of "*thinness*" **in the population** and illustrate your result with a clustered column chart.

Margin of error: ______

Average importance of "*thinness:*" _____ to _____ on a 9-point scale.

3. Demographics that drive *thinness importance.* Use simple regression to identify demographics which drive the *importance of "thinness"* and the variation in *thinness importance* explained by variation in each demographic.

Demographic	*p value*	Drives *thinness importance*	% var in *thinness importance* explained
Age		Y or N	
Education		Y or N	
Income		Y or N	

4. Find the expected difference in *thinness importance* associated with each demographic difference in the sample. (If a potential driver is not significant, leave its row blank.)

Demographic	Expected difference in *thinness importance* due to demographic difference
Age (years)	
Education (years)	
Income ($k)	

5. Illustrate *one* of the significant driver's influence with a scatterplot showing **population** average response to driver differences by adding the line of fit with *95 % prediction intervals*.

CASE 5-1 GenderPay (B)

The Human Resources manager of Slam's Club was shocked by the recent revelations of gender discrimination by Wal-Mart ("How Corporate America is Betraying Women," Fortune, January 10, 2005), but believes that the employee salaries in his company reflect levels of responsibility (and not gender). You have been asked to analyze this hypothetical link between level of responsibility and salary.

Case 5-1 GenderPay.xls contains employee *salaries* and levels of *responsibility* from a random sample of employees.

1. Determine whether or not *responsibility* drives *salaries*.

   If level of *responsibility* drives *salaries*, determine
   a. The percent of variation in *salaries* which can be accounted for by variation in level of *responsibility*
   b. The margin of error in forecasts of *salaries* from level of *responsibility*
   c. With 95 % certainty, how much *expected salary* **in the population** changes with each additional *responsibility* level

2. The Human Resources manager noticed that many employees are working at *responsibility* **level 5**. Determine how much payroll might be reduced, on average, if a **level 5** employee were replaced with a new **level 1** employee with similar experience.

3. Present the model that you built, including
   a. The regression equation in the standard format
   b. A scatterplot of *salaries* by level of *responsibility* with *95 % prediction intervals*
   c. A chart title that helps your audience see your conclusion

Be sure to round your results to two or three significant digits.

CASE 5-2 GM Revenue Forecast[1]

General Motors Management would estimate the percent of customers who will return to again choose a GM car. GM's award-winning customer Loyalty has been widely publicized, though in 2009, for the first time in 9 years, Toyota and Honda overtook GM, claiming the annual Automotive Loyalty Awards.

Toyota and Honda Win Top Honors in 14th Annual Event

SOUTHFIELD, Mich. (January 12, 2010)—Toyota and Honda took top honors in R. L. Polk & Co.'s 14th Annual Automotive Loyalty Awards, which were presented this evening at the 2010 Automotive News World Congress in Detroit.

Edging out General Motors for the first time in 9 years, Toyota ranked number 1 in Overall Loyalty to Manufacturer, indicative of the manufacturer's ability to retain previous customers.

Honda also was a big winner, taking top honors in the Overall Loyalty to Make category.

"Maintaining a solid loyal customer base is not easy, but it is essential to survive in today's competitive environment," said Stephen Polk, chairman, president and CEO of R. L. Polk & Co. "Tonight's winners are all excellent examples of what customer retention can do for your brand and your bottom line."

About the Polk Automotive Loyalty Awards.

The Polk Automotive Loyalty Awards recognize manufacturers for superior owner loyalty performance. Loyalty is determined when a household that owns a new vehicle returns to market and purchases or leases another new vehicle of the same model or make. For a complete list of current and past Polk Automotive Loyalty Award winners, please visit http://usa.polk.com/Company/Loyalty/.

Table
Polk Automotive Loyalty Award Winners – 2009 Model Year

Categories	Winners	Loyalty %
Overall Awards		
Overall Loyalty to Manufacturer	Toyota	58.60%
Overall Loyalty to Make	Honda	54.86%

Case 5-2 General Motors Revenue.xls contains 5 years of quarterly data, including:

quarter,
revenues, revenues
revenues q-4, lagged revenues from four quarters ago,

[1]This case is a hypothetical scenario using actual data.

Build a simple regression model to estimate the impact of past year *revenues* on current *revenues*. Round your numerical results to two or three significant digits.

1. Present your regression equation in standard format.
2. What percent of variation in *revenues* can be accounted for by past *revenues*?
3. How close to actual *revenues* could you expect a forecast to be 95 % of the time?
4. What range in percents of this quarter's GM *revenues* could management be 95 % certain will repeat next year?
5. Present a scatterplot of *95 % individual prediction intervals* with *actual revenues* by *quarter*. Insert a chart title that summarizes your conclusion.

Assignment 5-1 Impact of Defense Spending on Economic Growth

Some experts have suggested that the U.S. economy thrives when the Nation is involved in global conflict. **Assignment 5-1 Defense.xls** contains quarterly *GDP* and past quarter *Defense* spending in billion dollars.

Create a scatterplot and calculate the correlation coefficient to see whether or not *GDP* and *defense spending* are related linearly.

Fit a simple linear regression to estimate the impact on quarter *GDP* of changes in past quarter *defense spending*.

Analyze the residuals.
Are they

- Pattern-free?
- Approximately *Normally* distributed?

Summarize your results, in a **single-spaced** report, **12 pt font**, with **one embedded figure** and your **regression equation in standard format**.
Round to two or three significant digits.
Choose a chart title which summarizes your conclusions.
Use language that policy-makers could easily understand, whether or not they have recently taken statistics.
Include in your report:

- Whether or not past quarter *defense spending* is correlated with *GDP*
- The percent of variation in *GDP* that can be explained by variation in past quarter *defense spending*
- The margin of error in forecasts of *GDP* from past quarter *defense spending*,
- The expected range of possible impacts on *GDP* of a **$1 billion increase** in past quarter *defense spending*

In a technical footnote, include your conclusions from your residual analysis.

Chapter 6
Naïve Forecasting with Regression

In some circumstances, managers want to estimate the *trend*, or stable level of growth in performance, in order to produce a longer term forecast. Regression allows estimation of trend, using the time period as the independent variable. Forecasting based on trend is *naïve,* because the focus is not on understanding why performance varies across time periods, but only on forecasting what future performance would look like if the future performance resembled past performance.

Naïve forecasting based on trend is quick and easy and provides a view of the future, under the status quo. In Chap. 11, more sophisticated forecasting, which includes explanation of variation across time periods, will be introduced.

6.1 Use Naïve Forecasts Estimate Trend

Example 6.1 Concha y Toro.

Concha y Toro vineyards, a Chilean global multinational, is the largest producer of wines in Latin America. The firm produces and exports wine to 135 markets in Europe, North, Central and South American, Asia, and Africa. Concha y Toro has been successful in developing wine varieties that appeal to the unique tastes of the various global segments.

In 2010, the Board of Directors was reviewing performance. While the firm's wines had earned multiple distinctions, export revenues had grown only 4 % since 2009, which was a startling slow down, relative to revenue growth as high 43 % in recent years. Some conservative managers were convinced that revenue growth over the next 5 years could be as little as 1 %. The Board needed a forecast of 2015 export revenues.

Revenues depended on both volume and prices. The modeling team elected to estimate the trends in both volume and prices in the two largest export regions, Europe and the U.S. In the U.S. market, volume had grown consistently, but at a much slower rate than volume growth in European, as Fig. 6.1 illustrates. While volume growth in Europe was celebrated, there was concern that growth in the U.S. had fallen short of that desired. Wine consumption in the U.S. was growing faster than in other global regions, presenting an opportunity that could not be ignored. In European markets, prices had been stable, in spite of quality improvements, as Fig. 6.1 illustrates. Price increases in the U.S. markets were positive responses to quality improvements, though stable prices in Europe were concerning.

In order to inform future discussions, regressions were used to estimate the annual trends in both sales volume and prices for both major export regions. Trends were estimated with the following models:

$$sales\ vol\hat{u}me(ML)_{i,t} = a_{0_i} + a_{1_i} \times t$$

$$pr\hat{i}ce\ (\$/L)_{i,t} = b_{0_i} + b_{1_i} \times t,$$

C. Fraser, *Business Statistics for Competitive Advantage with Excel 2013: Basics, Model Building, Simulation and Cases*,
DOI 10.1007/978-1-4614-7381-7_6, © Springer Science+Business Media New York 2013

where i is the i'th global region, and t is the year.

The slope estimates, a_{1_i} and b_{1_i}, will provide estimates of average annual change in sales volume and price.

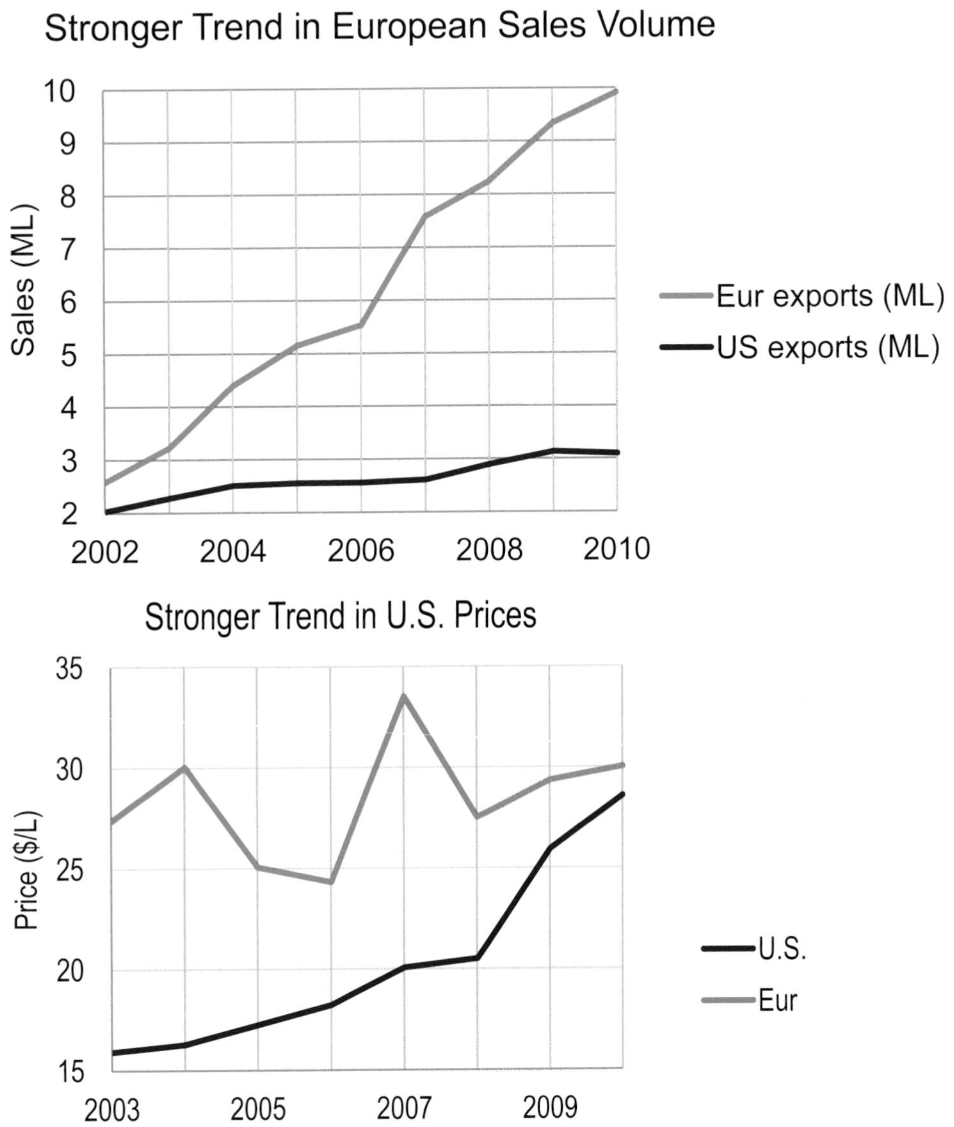

Fig. 6.1 Growth in sales volumes and prices across global regions

Regression results, shown in Tables 6.1 and 6.2, suggest that sales volumes are increasing annually in all both regions, since both slope estimates (coefficients for year t) are significant, providing evidence that the neither population slope is less than or equal to zero.

Table 6.1 Sales trend (ML) by region

Regression statistics: U.S. sales (ML)	
R square	.93
Standard error	.103
Observations	9

ANOVA

	df	*SS*	*MS*	*F*	*Significance F*
Regression	1	.974	.974	91.6	.00003
Residual	7	.074	.011		
Total	8	1.047			

	Coefficients	*Standard error*	*t stat*	*p value*	*Lower 95 %*	*Upper 95 %*
Intercept	−253	27	−9.5	.00003	−32	−19
Year	.13	.013	9.6	.00003	.10	.16

Regression statistics: Eur sales (ML)	
R square	.99
Standard error	.337
Observations	9

ANOVA

	df	*SS*	*MS*	*F*	*Significance F*
Regression	1	55.66	55.66	490.4	.0000001
Residual	7	.79	.11		
Total	8	56.45			

	Coefficients	*Standard error*	*t stat*	*p value*	*Lower 95 %*	*Upper 95 %*
Intercept	−1926	87	−22.1	.0000001	−2132	−1720
Year	.963	.043	22.1	.0000001	.860	1.07

Expected annual growth in volume has been much stronger in Europe (averaging .96 ML per year) than in the U.S. segment (averaging .13 ML per year).

$$\widehat{sales}(ML)_{U.S.,t} = -253^{a} + .13^{a} \times t$$

RSquare: $.93^{a}$

$$\widehat{sales}(ML)_{Eur,t} = -1{,}926^{a} + .96^{a} \times t$$

RSquare: $.99^{a}$

[a]Significant at .01 or better.

Table 6.2 Price trend ($/L) by region

Regression statistics: U.S. price ($/L)	
R square	.88
Standard error	1.74
Observations	8

ANOVA

	df	SS	MS	F	Significance F
Regression	1	131	131	43.5	.0006
Residual	6	18	3.0		
Total	7	149			

	Coefficients	Standard error	t stat	p value	Lower 95%	Upper 95%
Intercept	−3528	538	−6.6	.0006	−4845	−2211
Year	1.8	.27	6.6	.0006	1.1	2.4

Regression statistics: Eur price ($/L)	
R square	.097
Standard error	3.05
Observations	8

ANOVA

	df	SS	MS	F	Significance F
Regression	1	6.0	6.0	.6	.45
Residual	6	56	9.3		
Total	7	62			

	Coefficients	Standard error	t stat	p value	Lower 95%	Upper 95%
Intercept	−729	945	−.8	.47	−3042	1584
Year	.38	.47	.8	.45	−.78	1.53

The annual increase in prices, on the other hand, is significant, providing evidence that the population slope is greater than zero, only in the U.S. markets (with average annual of increase of $1.77/L).

$$\hat{price}(\$/L)_{U.S.,t} = -3{,}530^{a} + 1.77^{a} \times t$$

$$RSquare: .88^{a}$$

The average annual price change in Europe has been negligible:

$$\hat{price}(\$/L)_{Eur,t} = -729 + .39 \times t$$

$$RSquare: .097$$

[a]Significant at .01 or better.

Adding and subtracting the margins of error to predictions using these regression equations, the modeling team illustrated forecasts with 95 % prediction intervals for sales volume and prices, shown in Fig. 6.2.

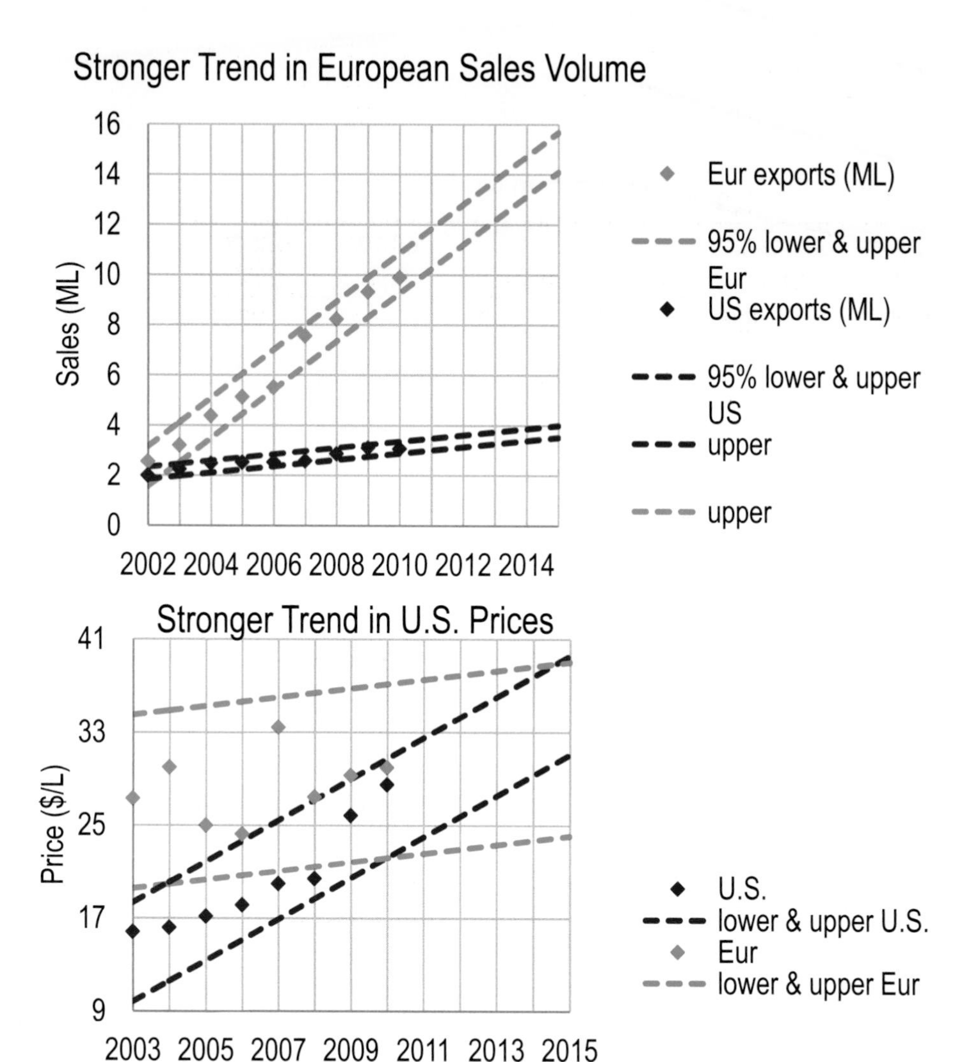

Fig. 6.2 Sales volume and price trends by global region

Comparing the forecasts, the modeling team noticed that sales volumes were remarkably consistent in the U.S., producing forecasts with a relatively narrow margin of error. Sales volume in the U.S. was growing consistently, but at a low rate. The positive trend in price was stronger in U.S., though the trend in sales volume was weaker.

The modeling team used the export market forecasts to produce a consolidated forecast of export revenues in the two major export markets:

$$major\ export\ market\ rev\hat{e}nues\ (\$M)_t = \sum pr\hat{\imath}ce\ (\$/L)_{i,t} \times sa\hat{l}es\ (L)_{i,t}$$

These consolidated major export market revenue forecasts are shown in Fig. 6.3.

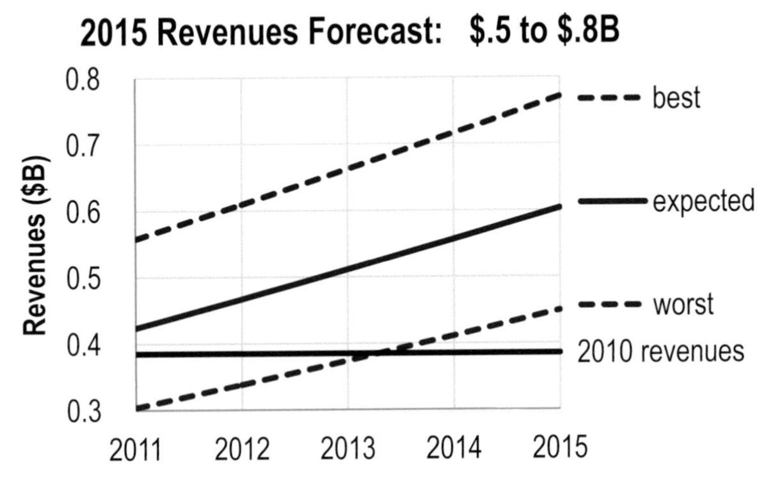

Fig. 6.3 Consolidated revenue forecast

The consolidated forecasts aggregated the two global segments. If volumes and prices were at the upper 95 % confidence interval bound, consolidated export revenues in the two major segments could double 2010 revenues by 2015. In this optimistic case, revenue growth could be as high as 15 % per year.

If, on the other hand, volumes and prices were at the lower 95 % confidence interval bound, consolidated major export market revenues would exceed 2010 revenues only slightly by 2014, and would be less in interim years, 2011–2013. Annual revenue growth would be just 3 % per year. In the expected case, major export segment revenues would grow by 9 %, annually, to $.6B by 2015.

The range of possibilities, from worst case to best case, was large. In 2015, for example, the worst case forecast was $.5B, and the best case forecast was $.8B, a range of $.3B. However, neither of these extreme outcomes was likely.

The chance that sales volume or price would be at the lower 95 % interval boundary in either of the two global regions is 2.5 %. The joint probability that worst case outcomes for both sales volume and price would occur is less than one tenth of one percent, .06 % (= 2.5 % × 2.5 %). The joint probability that worst case sales volumes and prices would occur in both global regions is smaller still, .00004 % (= 2.5 % × 2.5 % × 2.5 % × 2.5 %), or about one in three million (= 1/.000004):

$$p(worst\ case) = \prod p(worst\ sales\ volume)_i\, p(worst\ price)_i$$

$$= .025^4 = .0000004$$

The best case outcome is equally unlikely, making the interval from worst to best a 99.99992 % prediction interval.

6.2 Use Naïve Forecasts to Set Assumptions with Monte Carlo

In order to be useful to management, a 95 % prediction interval was needed for consolidated revenues. Using expected 2015 sales volumes and prices and the corresponding standard errors from the naïve models, the modeling team used Monte Carlo simulation to find the 95 % prediction interval for consolidated major export market revenues in 2015 (Figs. 6.4 and 6.5).

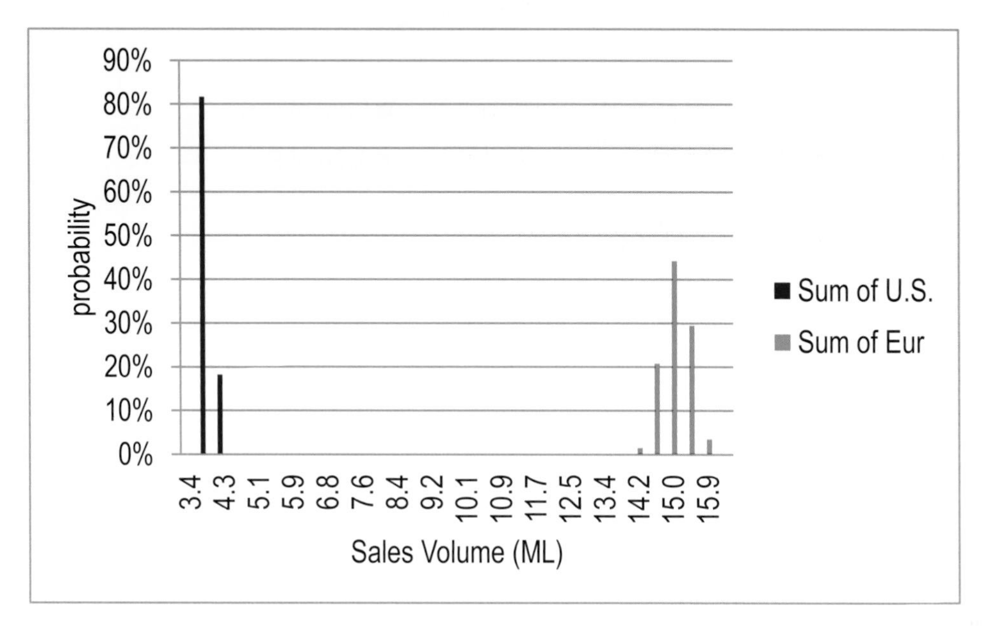

Fig. 6.4 Simulated samples of 2015 sales volumes

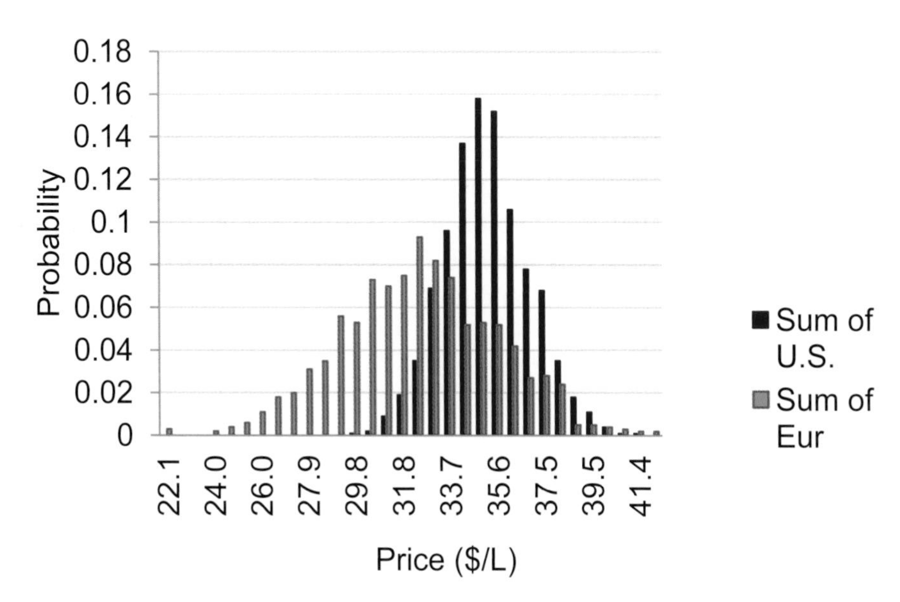

Fig. 6.5 Simulated samples of 2015 prices

Combining the simulated samples of possible 2015 sales volumes and prices for the two global segments yields a sample of possible 2015 revenues in each segment (Fig. 6.6):

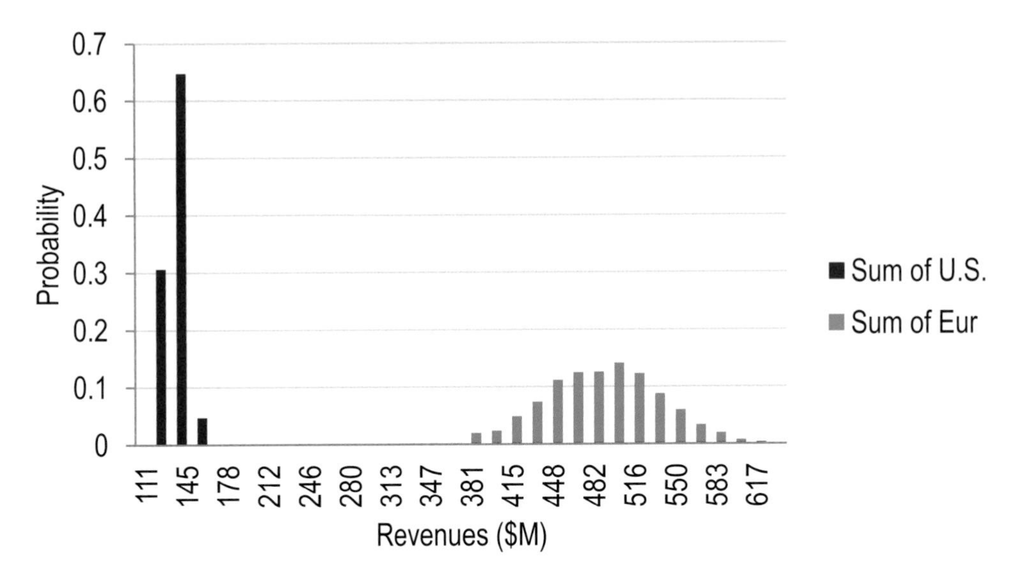

Fig. 6.6 Simulated samples of 2015 revenues

2015 revenues would be higher in Old World markets in Europe than in New World markets in the U.S. Based on the assumption that sales volume and price trends would continue, revenues from exports to Europe would continue to grow at about 10 % annually, to at least $380M. Based on the same assumptions, revenues from exports to the U.S. would grow at just 8 % annually, to levels less than $150M.

Consolidated revenue possibilities in 2015 under the assumption of continued sales volume and price trends were estimated by combining possibilities for the two global segments, revealing the 95 % prediction interval $.5B to $7B, a range of $.2B.

Compared with the unlikely best and worst case forecasts, which differed by $.3B, management now had a much better picture of 2015 possibilities, under the assumption of continuing sales volume and price trends. Revenues in the two major export markets would grow annually at 6–15 % if current trends continued. Relative to annual growth in those markets of 17 % over the five preceding years, the forecast was just the news that management needed to motivate pursuit of a new strategy for improved growth and performance (Fig. 6.7).

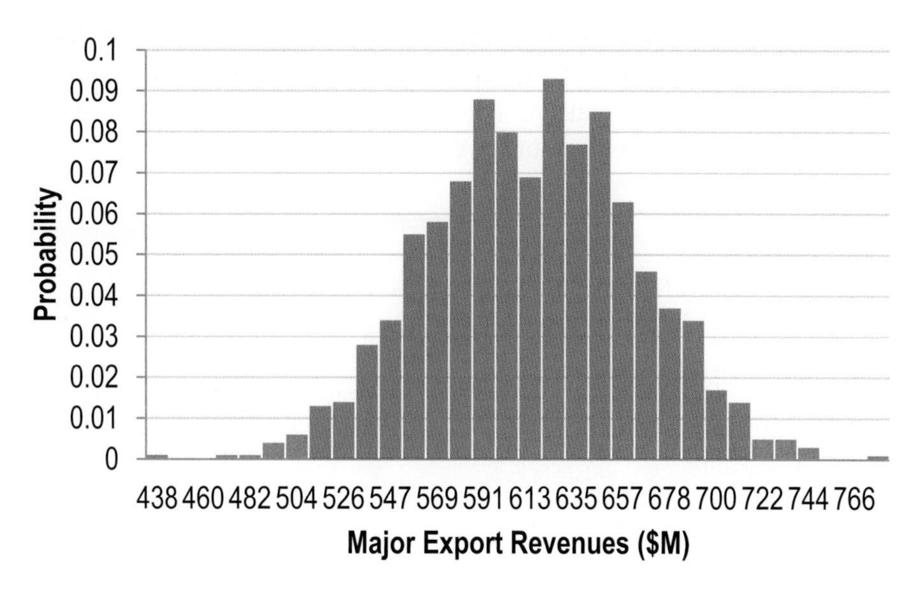

Fig. 6.7 2015 Revenue possibilities assuming the status quo

6.3 Naïve Forecasts Offer Quick and Simple Forecasts of Trend

Regression offers two complementary capabilities to managers – the ability to identify performance drivers and the ability to forecast performance. In some cases, managers are primarily interested in forecasting performance, and in such a case, a naïve model to estimate trend is a quick and easy solution.

Naïve models assume that the future will mimic the past and that previous trends are stable and will continue. Since the driver in a naïve model, the time period, varies systematically by an increase of one unit in each time period, the regression slope quantifies the trend in performance. Naïve models are effective, since most performance variables, as well as their drivers, such as GDP and population, tend to move together across time periods.

The obvious sacrifice made when a naïve model is used to forecast is explanation. Performance may be increasing at a consistent rate, but managers may not be certain which of several drivers are responsible. To explain performance variation, models that include potential drivers are built. More sophisticated time series models are introduced in Chap. 11. When explanation and the identification of drivers is desired, one must move beyond naïve models.

Naïve models, estimating systematic trend and the amount of systematic variation in performance (through the standard error) provide one option for quantifying Monte Carlo analysis to see future outcomes, based on the status quo. While managers may elect to work toward enhancing future performance, relative to the past, using trend estimates with Monte Carlo provides a benchmark, a picture of what is expected to happen in the future, if nothing changes.

Naïve models may not be valid for forecasting, especially if conditions within the business or the economy are changing. In Chap. 11, a means to test validity is introduced, where a model is built from all but the two most recent observations. If those two recent points fit within the 95 % prediction interval, the model is valid for forecasting, and can be recalibrated from the full time series.

Autocorrelation, which is the presence of pattern in the residuals, is likely with time series data. Business performance is vulnerable to economic cycles and seasonality. Chapter 11 introduces a test to screen for positive autocorrelation. When positive autocorrelation is present, additional variables can be added to a model to account for cycles or seasonality.

Excel 6.1 Estimate Trend with a Naïve Model

Concha y Toro executives desire a forecast of 2015 revenues from domestic and export markets. Begin by building naïve models of sales volume and price trends for domestic and export markets of Concha y Toro wines. Data in **concha y toro revenues.xlsx** contain sales and prices by year for wines sold in domestic and export markets.

Run regressions with *domestic sales (ML)* in Input Y Range and *year* in Input X Range.

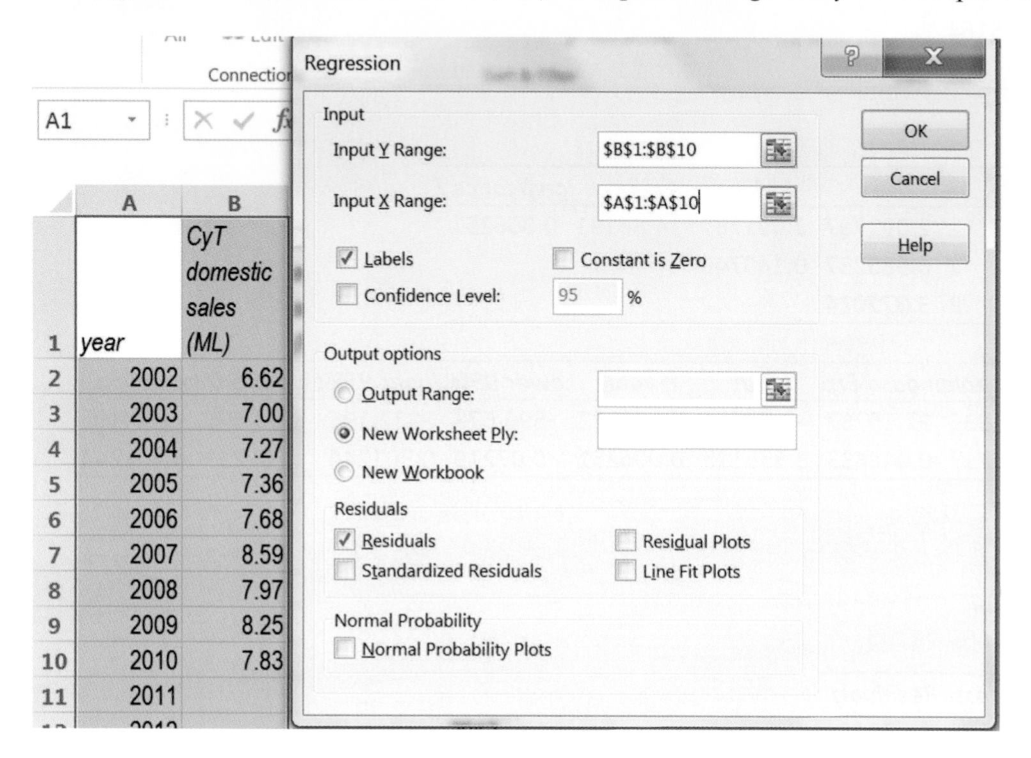

Excel output provides a model of trend in domestic wine sales.

	A	B	C	D	E	F	G	H	I
1	SUMMARY OUTPUT								
2									
3	*Regression Statistics*								
4	Multiple R	0.824505							
5	R Square	0.679808							
6	Adjusted F	0.634067							
7	Standard I	0.375164							
8	Observati(	9							
9									
10	ANOVA								
11		*df*	*SS*	*MS*	*F*	*gnificance F*			
12	Regressior	1	2.091787	2.091787	14.86191	0.006251			
13	Residual	7	0.985237	0.140748					
14	Total	8	3.077024						
15									
16		*Coefficients*	*andard Err*	*t Stat*	*P-value*	*Lower 95%*	*Upper 95%*	*ower 95.0%*	*pper 95.0%*
17	Intercept	-366.936	97.15767	-3.77671	0.006922	-596.678	-137.195	-596.678	-137.195
18	year	0.186717	0.048433	3.855115	0.006251	0.07219	0.301244	0.07219	0.301244
19									
20									
21									
22	RESIDUAL OUTPUT								
23									
24	*)bservation*	*T domesti*	*Residuals*						
25	1	6.870578	-0.25258						

The trend in domestic sales is positive. Domestic sales have been increasing by an average of .19MLs per year:

$$domes\hat{t}ic\ sales\ (ML)_t = -370^a + .19^a \times t$$

[a]Significant at .01 or better.

Repeat this process to build naïve models for *export sales (ML), domestic prices ($/L),* and *export prices ($/L).*

	A	B	C	D	E	F	G	H	I
1	SUMMARY OUTPUT								
2									
3	*Regression Statistics*								
4	Multiple R	0.991655							
5	R Square	0.98338							
6	Adjusted F	0.981006							
7	Standard I	0.592667							
8	Observati	9							
9									
10	ANOVA								
11		*df*	*SS*	*MS*	*F*	*gnificance F*			
12	Regressior	1	145.483	145.483	414.1811	1.73E-07			
13	Residual	7	2.458782	0.351255					
14	Total	8	147.9417						
15									
16		*Coefficients*	*andard Err*	*t Stat*	*P-value*	*Lower 95%*	*Upper 95%*	*ower 95.0%*	*pper 95.0%*
17	Intercept	-3112.66	153.4852	-20.2798	1.78E-07	-3475.59	-2749.72	-3475.59	-2749.72
18	year	1.55715	0.076513	20.35144	1.73E-07	1.376225	1.738075	1.376225	1.738075
19									
20									
21									
22	RESIDUAL OUTPUT								
23									
24	*bservatior*	*yT export*	*Residuals*						
25	1	4.759289	0.700711						
26	2	6.316439	-0.00244						

The trend in export sales is positive. Export sales have been increasing by an average of 1.6MLs per year:

$$\widehat{export\ sales}\ (ML)_t = -3110^a + 1.6^a \times t$$

[a]Significant at .01 or better.

	A	B	C	D	E	F	G	H	I
1	SUMMARY OUTPUT								
2									
3	*Regression Statistics*								
4	Multiple R	0.790516							
5	R Square	0.624916							
6	Adjusted F	0.562402							
7	Standard I	3.31682							
8	Observati(	8							
9									
10	ANOVA								
11		*df*	*SS*	*MS*	*F*	*'gnificance F*			
12	Regressior	1	109.9736	109.9736	9.996425	0.019523			
13	Residual	6	66.00777	11.0013					
14	Total	7	175.9814						
15									
16		*Coefficients*	*andard Err(*	*t Stat*	*P-value*	*Lower 95%*	*Upper 95%*	*ower 95.0%*	*pper 95.0%*
17	Intercept	-3226.15	1026.92	-3.14158	0.020028	-5738.93	-713.368	-5738.93	-713.368
18	year	1.618153	0.511796	3.161712	0.019523	0.365832	2.870474	0.365832	2.870474
19									
20									
21									
22	RESIDUAL OUTPUT								
23									
24	*)bservatior*	*T domesti(*	*Residuals*						
25	1	15.00988	-3.729						
26	2	16.62803	-1.3535						

The trend in domestic prices is positive. Domestic prices have been increasing by an average of 1.6 ($/L) per year:

$$domêstic\ price\ (\$/L)_t = -3200^a + 1.6^a \times t$$

[a]Significant at .01 or better.

	A	B	C	D	E	F	G	H	I
1	SUMMARY OUTPUT								
2									
3	*Regression Statistics*								
4	Multiple R	0.576916							
5	R Square	0.332832							
6	Adjusted F	0.221637							
7	Standard I	2.994282							
8	Observati(	8							
9									
10	ANOVA								
11		*df*	*SS*	*MS*	*F*	*'gnificance F*			
12	Regressior	1	26.83651	26.83651	2.993235	0.134337			
13	Residual	6	53.79433	8.965722					
14	Total	7	80.63084						
15									
16		*Coefficients*	*andard Err*	*t Stat*	*P-value*	*Lower 95%*	*Upper 95%*	*ower 95.0%*	*pper 95.0%*
17	Intercept	-1579.37	927.0591	-1.70363	0.139339	-3847.8	689.0636	-3847.8	689.0636
18	year	0.799353	0.462028	1.730097	0.134337	-0.33119	1.929894	-0.33119	1.929894
19									
20									
21									
22	RESIDUAL OUTPUT								
23									
24	*)bservatior*	*:yT export*	*Residuals*						
25	1	21.73487	0.865826						

The trend in export prices is not significant. Export prices have been flat.

Excel 6.2 Produce Naïve Forecasts

Forecast *domestic sales* through 2015 below regression output.

Move to the data page
(Cntl+Page Dn),

select and copy data in the *year* and *domestic sales (ML)* columns,

move back to the regression output page (**Cntl+Page Up**), and

paste next to *residuals*:

24	)bservatio	/T domesti	Residuals	year	CyT domestic sales (ML)
25	1	6.870578	-0.25258	2002	6.62
26	2	7.057294	-0.05429	2003	7.00
27	3	7.244011	0.022989	2004	7.27
28	4	7.430728	-0.07073	2005	7.36
29	5	7.617444	0.058556	2006	7.68
30	6	7.804161	0.783839	2007	8.59
31	7	7.990878	-0.01988	2008	7.97
32	8	8.177594	0.070406	2009	8.25
33	9	8.364311	-0.53831	2010	7.83

Find *predicted* domestic sales with the regression equation.

(Lock the cell references to the intercept and slope with **fn 4**.)

(Leave two columns free for lower and upper 95 % prediction intervals, which you will want to plot with *domestic sales*.)

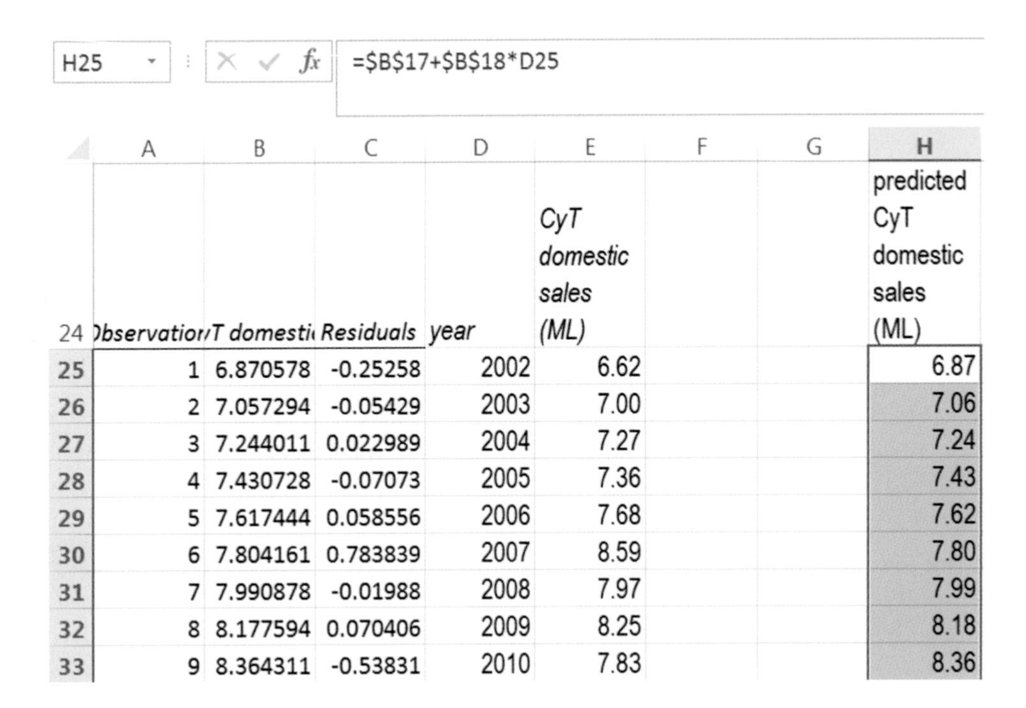

H25 =$B$17+$B$18*D25

	A	B	C	D	E	F	G	H
24	)bservatio	/T domesti	Residuals	year	CyT domestic sales (ML)			predicted CyT domestic sales (ML)
25	1	6.870578	-0.25258	2002	6.62			6.87
26	2	7.057294	-0.05429	2003	7.00			7.06
27	3	7.244011	0.022989	2004	7.27			7.24
28	4	7.430728	-0.07073	2005	7.36			7.43
29	5	7.617444	0.058556	2006	7.68			7.62
30	6	7.804161	0.783839	2007	8.59			7.80
31	7	7.990878	-0.01988	2008	7.97			7.99
32	8	8.177594	0.070406	2009	8.25			8.18
33	9	8.364311	-0.53831	2010	7.83			8.36

In cell **C7**, next to the *standard error*, which is always in cell **B7,** find the *critical t* for the *residual dfs*, which is always in cell **B13,** using the function **TINV**(.05,**B13**):

C7 =TINV(0.05,B13)

	A	B	C	D
3	*Regression Statistics*			
4	Multiple R	0.824505		
5	R Square	0.679808		
6	Adjusted F	0.634067	*critical t*	
7	Standard I	0.375164	2.36	

Find the *margin of error* by multiplying the *standard error* by the *critical t*:

D7 =B7*C7

	A	B	C	D
3	*Regression Statistics*			
4	Multiple R	0.824505		
5	R Square	0.679808		
6	Adjusted F	0.634067	*critical t*	*me*
7	Standard I	0.375164	2.36	0.89

Find the *lower* and *upper* 95 % prediction interval bounds by subtracting the *margin of error* from *predicted values*, locking the cell reference to the *margin of error* with **fn 4**:

F25 =H25-$D$7

	D	E	F	G	H
24	*year*	*CyT domestic sales (ML)*	*lower 95% pi bound*	*upper 95% pi bound*	predicted CyT domestic sales (ML)
25	2002	6.62	5.98	7.76	6.87
26	2003	7.00	6.17	7.94	7.06
27	2004	7.27	6.36	8.13	7.24
28	2005	7.36	6.54	8.32	7.43
29	2006	7.68	6.73	8.50	7.62

Repeat this process to forecast *export sales.*

Reuse the formula for *predicted* (and *lower* and *upper*).

After pasting a copy of *year* and *export sales,* move to the *domestic sales* regression page **(Cntl+Page Up),** copy the three columns for *lower, upper* and *predicted,* move back to the *export sales* regression page **(Cntl+Page Down)**, and paste.

Excel will apply your formulas to this *export sales* data.

(Note that the three columns will be identical until you've created the *margin of error*.)

	D	E	F	G	H
24	year	*CyT export sales (ML)*	*lower 95% pi bound*	*upper 95% pi bound*	predicted CyT export sales (ML)
25	2002	5.46	4.76	4.76	4.76
26	2003	6.31	6.32	6.32	6.32
27	2004	7.37	7.87	7.87	7.87
28	2005	9.50	9.43	9.43	9.43
29	2006	10.55	10.99	10.99	10.99

Reuse formulas for *critical t* and the *margin of error*. Move to the *domestic sales* regression page (**Cntl+Page Up**), copy the two labels and formulas, move back to the *export sales* page, and paste.

Excel will reuse the formulas and the forecast will now be complete:

	A	B	C	D
3	*Regression Statistics*			
4	Multiple R	0.991655		
5	R Square	0.98338		
6	Adjusted F	0.981006	*critical t*	*me*
7	Standard I	0.592667	2.36	1.40

	D	E	F	G	H
24	year	*CyT export sales (ML)*	*lower 95% pi bound*	*upper 95% pi bound*	predicted CyT export sales (ML)
25	2002	5.46	3.36	6.16	4.76
26	2003	6.31	4.92	7.72	6.32
27	2004	7.37	6.47	9.28	7.87
28	2005	9.50	8.03	10.83	9.43
29	2006	10.55	9.59	12.39	10.99

Repeat this process to forecast *domestic prices* and *export prices,* reusing formulas for *predicted, lower, upper, critical t* and the *margin of error*:

	D	E	F	G	H
24	year	CyT domestic price ($/L)	lower 95% pi bound	upper 95% pi bound	predicted CyT domestic price (ML)
25	2003	11.3	6.89	23.13	15.01
26	2004	15.3	8.51	24.74	16.63
27	2005	24.2	10.13	26.36	18.25
28	2006	20.1	11.75	27.98	19.86
29	2007	21.5	13.37	29.60	21.48

	D	E	F	G	H
24	year	CyT export price ($/L)	lower 95% pi bound	upper 95% pi bound	predicted CyT export price (ML)
25	2003	22.6	14.41	29.06	21.73
26	2004	24.4	15.21	29.86	22.53
27	2005	19.7	16.01	30.66	23.33
28	2006	22.5	16.81	31.46	24.13
29	2007	29.2	17.61	32.26	24.93

Excel 6.3 Plot Naïve Model Fits and Forecasts

To plot the fit and forecast for *domestic sales* and *export sales*, move to the export sales regression sheet,

copy the *export sales* column, including blank cells corresponding to years through 2015, and the *lower* and *upper* columns,

move to the domestic sales regression sheet, and

paste between *upper* and *predicted* columns:

From the cell containing the *predicted* label (where you want *export sales* to go), “drop it in here:”

Alt H I E R

Remove formulas to preserve the *lower* and *upper* values:

Alt H V S U

	D	E	F	G	H	I	J
24	*year*	*CyT domestic sales (ML)*	*lower 95% pi bound*	*upper 95% pi bound*	*CyT export sales (ML)*	*lower 95% pi bound*	*upper 95% pi bound*
25	2002	6.62	5.98	7.76	5.46	3.36	6.16
26	2003	7.00	6.17	7.94	6.31	4.92	7.72
27	2004	7.27	6.36	8.13	7.37	6.47	9.28
28	2005	7.36	6.54	8.32	9.50	8.03	10.83
29	2006	7.68	6.73	8.50	10.55	9.59	12.39

Select *year, CyT domestic sales (ML), lower* and *upper, CyT export sales (ML), lower* and *upper* data, and insert a scatterplot:

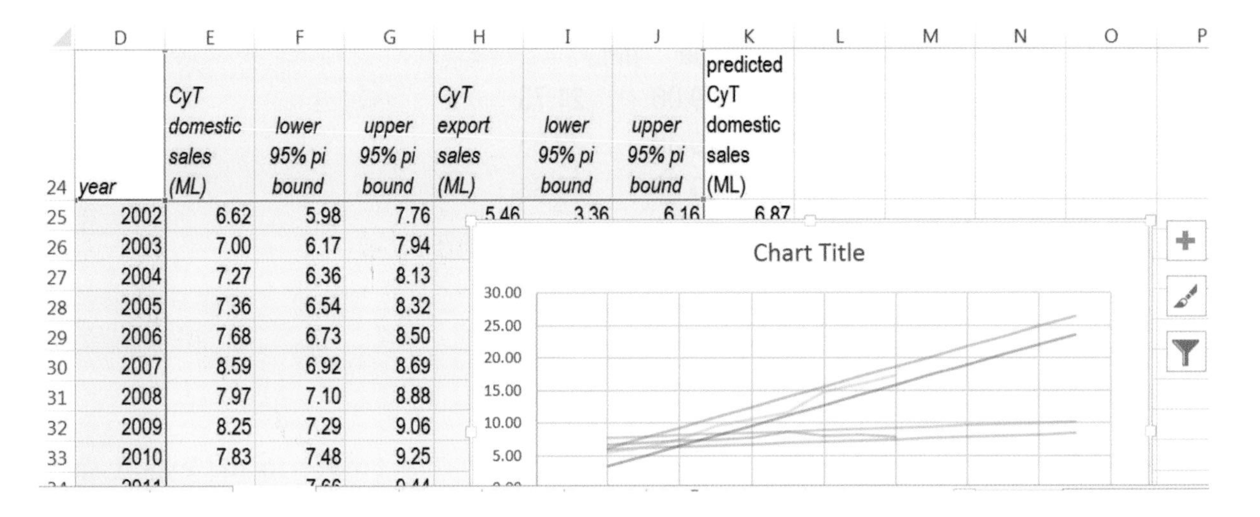

	D	E	F	G	H	I	J	K
24	*year*	*CyT domestic sales (ML)*	*lower 95% pi bound*	*upper 95% pi bound*	*CyT export sales (ML)*	*lower 95% pi bound*	*upper 95% pi bound*	predicted CyT domestic sales (ML)
25	2002	6.62	5.98	7.76	5.46	3.36	6.16	6.87
26	2003	7.00	6.17	7.94				
27	2004	7.27	6.36	8.13				
28	2005	7.36	6.54	8.32				
29	2006	7.68	6.73	8.50				
30	2007	8.59	6.92	8.69				
31	2008	7.97	7.10	8.88				
32	2009	8.25	7.29	9.06				
33	2010	7.83	7.48	9.25				

Reformat axes, add axes labels, change the *domestic sales* and *export sales* series to markers, and recolor *lower* series to match *upper* series:

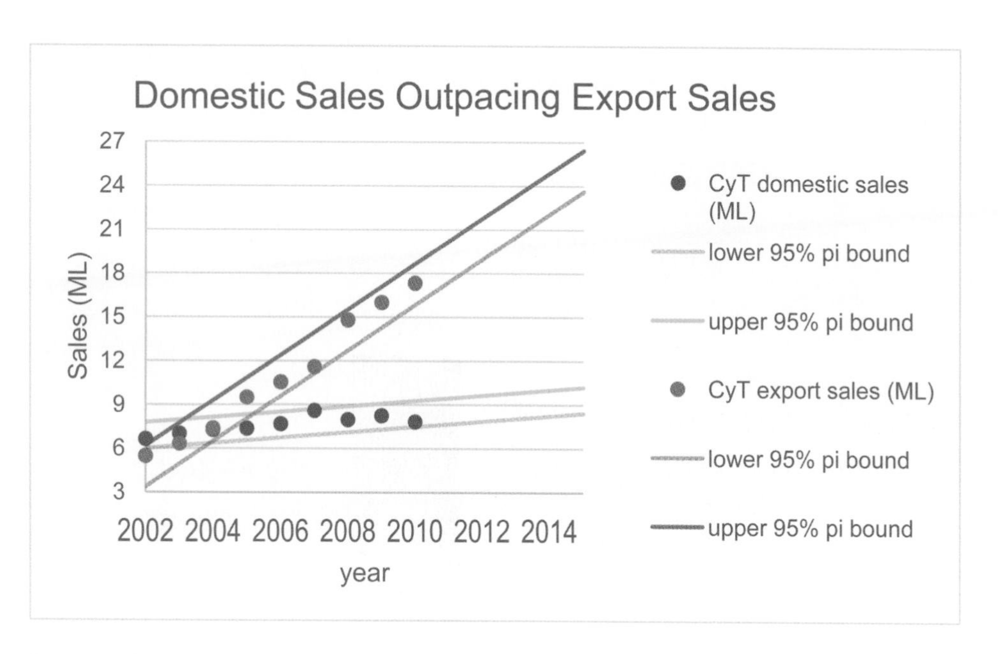

Repeat the process to plot the fits and forecasts for *domestic* and *export prices*:

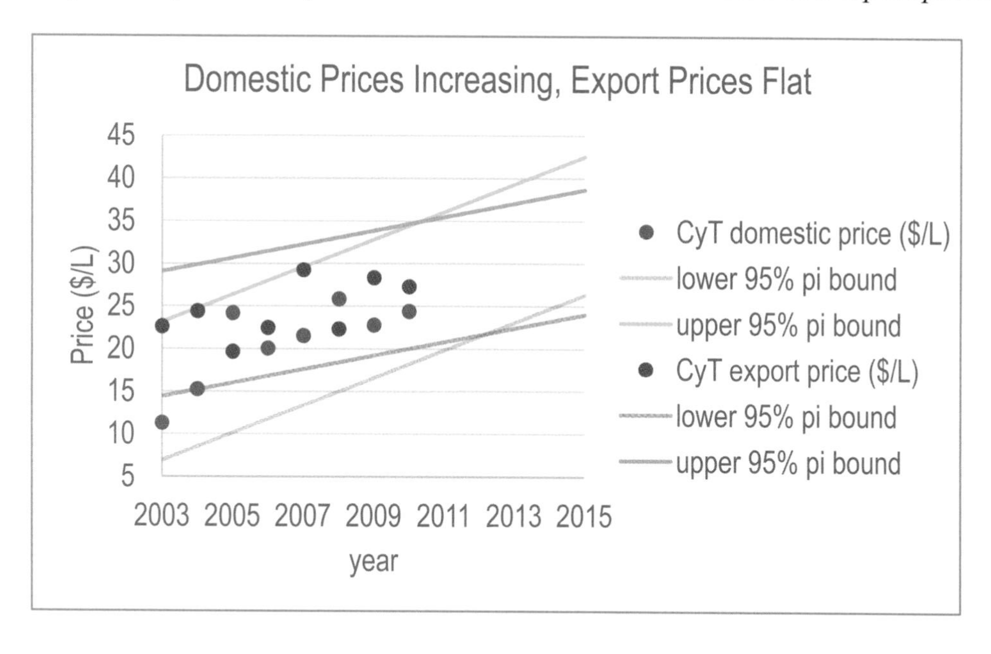

Lab 6 Forecast Concha y Toro Sales in Developing Export Markets

With forecasts for market segments in Europe and the U.S., executives are convinced that forecasts for the remaining export segments would provide invaluable information. Build naïve forecasts for export segments in Asia and Latin America.

Time series of annual sales volumes and prices in these two export segments are in **Concha y Toro developing segments.xlsx.**

1. If annual growth in export sales volume is stable, what volume will Concha y Toro export to Asia in 2015?

   ___________ to _____________

2. If annual growth in export sales volume is stable, what volume will Concha y Toro export to Latin America in 2015?

   ___________ to _____________

3. Plot your fits and forecasts of sales volumes in export markets in Asia and Latin America.

4. If annual growth in prices is stable, what average price will Concha y Toro wines bring in export markets in Asia in 2015?

   ___________ to _____________

5. If annual growth in prices is stable, what average price will Concha y Toro wines bring in export markets in Latin America in 2015?

   ___________ to _____________

6. Plot your fits and forecasts of average prices in export markets in Asia and Latin America.

7. What is the worst case for 2015 revenues in consolidated export markets in Asia and Latin America?

   ____

8. What is the probability that revenues in consolidated export markets in Asia and Latin America will be as low as the worst case in 2015?

   __________ OR 1 in _____________

Use expected 2015 sales volume and the standard error from the sales volume regression to simulate a sample of 1,000 hypothetical 2015 sales volumes for exports to Asia.

Use expected 2015 average price and the standard error from the price regression to simulate a sample of 1,000 hypothetical 2015 average prices for exports to Asia.

From the samples of hypothetical 2015 sales volumes and 2015 average prices, find 1,000 hypothetical 2015 revenues from exports to Asia.

Simulate samples for 2015 sales volume and 2015 average price for exports to Latin America, and use those to find a sample of hypothetical 2015 revenues from exports to Latin America.

Use the two samples of hypothetical 2015 revenues from exports to Asia and Latin America to find a sample of hypothetical consolidated 2015 revenues from exports to developing segments.

9. What is the 95 % prediction interval for consolidated 2015 revenues from exports to developing segments?
10. Make a column chart to illustrate 2015 revenue possibilities, given the assumption of stable growth in sales volume and prices in export markets in Asia and Latin America.

CASE 6-1 Can Arcos Dorados Hold On?

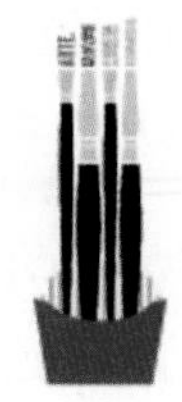

Arcos Dorados ("golden arches," in Spanish), headquartered in Argentina, is the largest franchiser of McDonalds restaurants in the World. Arcos Dorados operates restaurants in Brazil, the Caribbean, North and Central Latin America (NOLAD), and South America (SLAD). (Arcos Dorados listed its shares on the New York Stock Exchange in April, 2011, making its CEO, Colombian Woods Staton, a billionaire.)

With increasing wealth in Latin America, more families are electing to dine out, and fast food revenues, as well as McDonalds' revenues, have been growing at a steady clip. Arcos Dorados claimed the largest share of the fast food business in Latin America in 2009, 12.4 %. Though maintaining the largest share, Arcos Dorados' share slipped to 10.4 % in 2010. Mr. Staton is concerned that the business may be losing ground.

Arcos Dorados.xlsx contains annual revenues per restaurant and number of restaurants in the four global Latin business segments, Brazil, Caribbean, NOLAD, and SLAD, for years 2007 through 2011. Data from earlier years is not publicly available, since Arcos Dorados issued shares recently; therefore, "valid" models of revenues per restaurant and number of restaurants cannot be built. However, naïve models can be built, to forecast future performance based on historical trends.

1. In which global segments is the trend in revenue per restaurant increasing?

   ___ Brazil ___ Caribbean ___ NOLAD ___ SLAD

2. Illustrate your fits and forecasts for revenue per restaurant in the four global segments on a single graph and embed below:

3. In which global segments is the trend in number of restaurants increasing?

   ___ Brazil ___ Caribbean ___ NOLAD ___ SLAD

4. Illustrate your first and forecasts for revenue per restaurant in the four global segments in a single graph and embed below:

5. Forecast 2016 consolidated revenues, reporting the 95 % prediction interval:

Fast Food Market and Growth in Latin America. The fast food industry in Latin America and the Caribbean was valued at $29.0B in 2010. The industry has been growing at an average annual rate of 9.6 %, though industry experts have suggested that growth over the next 6 years could be as high as 11.4 %, though there is only a 2.5 % chance that growth would exceed 11.4 %. Management agrees that the chance of growth less than 9.6 % could occur with only a 2.5 % chance. Average annual growth of 10.5 % is expected over the next 6 years.

6. Determine the 95 % prediction interval for Arcos Dorados' market share in 2016:

7. Illustrate the distribution of possible Arcos Dorados' market shares in 2016 with a column chart and embed, below:

8. What is the chance that Arcos Dorados' 2016 market share will exceed the 2010 level of 10.4 %?

9. Some Arcos Dorados executives worry that 2016 market share could be as low as 7.7 %, if revenues per restaurant and restaurant expansion slow, and if the fast food industry in Latin America and the Caribbean grows as fast as 11.4 % each year. What is the probability that 2016 market share could be as low as 7.7 %, based on management's assumptions?

Chapter 7
Marketing Segmentation with Descriptive Statistics, Inference, Hypothesis Tests and Regression

CASE 7-1 Segmentation of the Market for Preemie Diapers

Deb Henretta is about to commit substantial resources to launch *Pampers Preemies*. The following article from the *Wall Street Journal* describes Procter & Gamble's involvement in the preemie diaper market:

New York, N.Y.

P&G Targets the 'Very Pre-Term' Market
Wall Street Journal

THE TARGET MARKET for Procter & Gamble Co.'s newest diaper is small. Very small.

Of the nearly half a million infants born prematurely in the U.S. each year, roughly one in eight are deemed "very pre-term," and usually weigh between 500 and 1,500 g (1–3 lb). Their skin is tissue-paper-thin, so any sharp edge or sticky surface can damage it, increasing the chance of infection. Their muscles are weak, and unlike full-term newborns, excessive handling can add more stress that in turn could endanger their health.

Tiny as they are, the number of premature infants is increasing – partly because of improved neonatal care: From 1985 to 2000, infant mortality rates for premature babies fell 45 %, says the National Center for Health Statistics. Increasingly, such babies are being born to older or more affluent women, often users of fertility drugs, which have stimulated multiple births.

It's a testament to the competitiveness of the $19 billion global diaper market that a behemoth like Procter & Gamble, a $40 billion consumer-products company, now is focusing on a niche that brought in slightly more than $1 million last year; just 1.6 % of all births are very pre-term. But P&G sees birth as a "change point," at which consumers are more likely to try new brands and products. Introducing the brand in hospitals at an important time for parents could bring more Pampers customers, the company reasons.

P&G's Pampers, which is gaining ground on rival Kimberly-Clark, but still trails its Huggies brand, has made diapers for premature infants for years. (P&G introduced its first diaper for "pre-emies" in 1973; Kimberly-Clark in 1988), but neither group had come up with anything that worked well for the very smallest of these preemies.

The company that currently dominates the very-premature market is Children's Medical Ventures, Norwell, Mass., which typically sells about four million diapers a year for about 27 cents each. The unit of Respironics Inc., Murrysville, Pa., has been making its "WeePee" product for more than a decade. But the company, which also makes incubator covers, feeding tubes and extra small bathtubs for preemies, hadn't developed certain features common in mass-market diapers, such as softer fabric coverings.

C. Fraser, *Business Statistics for Competitive Advantage with Excel 2013: Basics, Model Building, Simulation and Cases*,
DOI 10.1007/978-1-4614-7381-7_7, © Springer Science+Business Media New York 2013

By contrast, P&G's preemie diapers, which it started distributing to hospitals in August, sell for about 36 cents each; about 4 cents more than P&G's conventional diapers. P&G's "Preemie Swaddler" fits in the palm of an adult's hand and has no adhesives or hard corners. It closes with mild velcro-like strips and is made of breathable fabric, not plastic. It has an extra layer of fabric close to the infant's skin to avoid irritation.

Children's Medical Ventures is coming out with another size of the WeePee, and plans to introduce velcro-like closures, a development the company says was in the works before P&G came out with a rival diaper. The new diapers won't cost any more, Children's Medical Ventures says.

P&G says the new diaper is the natural extension of its Baby Stages initiative, which took effect in February 2002 when P&G revamped its Pampers brand in the U.S. to cater to various stages of a baby's development. Working with very small preemies helps the company better understand infant development and become "more attuned to new products they might need," says Deb Henretta, president of P&G's global baby-care division.

But the marketing director for Children's Medical Ventures believes the increasing affluence of preemie parents is a greater inducement for big companies to enter the market. In the past, the typical mother of a preemie was poorer, often a teenager, but today more preemie "parents tend to be older, well-educated, and have money for things like fertility treatments," says Cathy Bush, marketing director for Children's Medical Ventures.

The competition may raise the bar for the quality of diapers for these smallest of preemies. P&G says the parents of premature babies are demanding better products. "They have much higher expectations than they did years ago," Ms. Henretta says.

Neonatal nurses have all sorts of opinions about the relative merits of Preemie Swaddlers and WeePees. Pat Hiniker, a nurse at the Carilion Roanoke Community Hospital in Virginia, says the new Pampers diaper, while absorbent, is too bulky for small infants. Allison Brooks of Alta Bates Hospital in Berkeley, Calif., says P&G's better absorbency made the babies less fidgety when they needed to be changed. "That sounds small, but you don't want them wasting their energy on squirming around," she says. "They need all their energy to grow."

In any case, if health professionals have their way, the very-premature market will shrink, or at least stop growing. The March of Dimes recently launched a $75 million ad campaign aimed at stemming the rise of premature births. P&G is donating 50,000 diapers to the nonprofit organization.

Before resources are dedicated, Deb wants to confirm that preemie parents are attracted to the *Pampers Preemies* concept of superior comfort and fit. She commissioned a concept test to assess consumers' intentions to try the product. There is evidence that motivations, preferences and motivations of preemie mothers may differ and differences may be linked to lifestyle or demographics. In such a case, promotional material ought to target particular segments.

If the concept is attractive to at least one segment, would commercialization produce revenues sufficient to justify the investment? Deb requires a forecast of future revenues in order to make sound decisions.

The Market for Preemie Diapers

The market for preemie diapers is unusual in that the first diapers that a preemie baby wears are chosen by the hospital. Procter & Gamble is banking on positive experiences with *Pampers Preemies* in the hospital and consumer brand loyalty once baby goes home. If parents see *Pampers Preemies* in the hospital, are satisfied with their performance, and find them widely available at the right price, parents may adopt the *Pampers Preemies* brand after their infant comes home. Satisfaction and brand loyalty to *Pampers Preemies* could then lead to choice of other Pampers products as their baby grows. If the concept test indicates that consumers' intentions to try are high, then the results will be included in promotional materials and selling efforts to hospital buyers.

Preemie Parent Segments

Based on focus group interviews and market research, Deb's team has learned that there are five broad segments of preterm parents:

- Younger (15–19), unemployed mothers who live with their parents. These young mothers are inexperienced and their pregnancies are unplanned. They tend to differ widely in their attitudes and preferences, and so a further breakdown is necessary:
    - ***Younger, Single, Limited Means***. The means of these young mothers are limited, and they are highly responsive to low prices and price promotions. Due to lack of knowledge about prenatal care and lack of access to healthcare, the preterm birth rate is relatively high in this segment, and minorities make up a disproportionately large proportion of this segment.
    - ***Younger, Single, with Means***. These young mothers have their parents' resources at their disposal and want the best diapers. They are inexperienced consumers and could be attracted by a premium diaper. Brand name appears to be very important to these young women, and they believe that better mothers rely on name brands seen on television. This segment has access to healthcare and the majority of mothers in this segment are ethnically white.
- ***Young*** (20–35) mothers tend to be married and have adequate resources. Their pregnancies tend to be planned and this segment is virtually indistinguishable from the larger segment of disposable diaper users for full-term babies. This group has the fewest preterm births.
- ***Later in Life Moms*** (35–39) and ***Latest*** (40+) mothers tend to be wealthier, more highly educated professionals with higher incomes. A large proportion has no other children and has undergone fertility treatment. Multiple preemie births are more likely in this segment. Some of these mothers are single parents. This group is particularly concerned about functional diaper features and wants the best diaper their dollars can buy. They are willing to pay for a premium diaper perceived as the highest quality, offering superior fit and comfort. This segment is predominantly ethnically white.

The Concept Test

A market research agency has conducted a concept test of *Pampers Preemies* to gauge interest among consumers in a variety of potential target markets. Ninety mothers with preemies who had been born at two local hospitals were asked to fill out a survey about purchase intensions after trying the product on their babies. These mothers were members of the three largest ethnic segments in the U.S.: Black, Hispanic, and Caucasian. If that data supports the launch, Deb will need to know which functional feature(s) to stress in advertising and the type of mother and family to feature in the ads. Therefore, questions regarding attribute importance and demographic information were also collected in the survey.

Data from the concept test is contained in **Case 7-1 Pampers Concept test.xls**. Below is an overview of the questions asked in the survey, the manner in which they were coded, and the variable names contained in the dataset (which are in italics).

Trial Likelihood

Participants were asked, "How likely would you be to try Pampers Preemies if they were available in the store where you normally buy diapers and were sold at a price of $X.XX per diaper?"
The interviewer flipped a coin and inserted the "premium" price of $0.36, for heads, and the "value" price of $0.27, for tails.
Responses were coded as follows:

Definitely Would Not Try = .05
Probably Would Not Try = .25
Maybe Would Try = .5
Probably Would Try = .75
Definitely Would Try = .95

Attribute Importance

Participants were asked, "How important is each of the following attributes to you when choosing a diaper?" for the attributes:

"brand name" (*brand importance*),
"comfort/fit" (*fit importance*),
"keeps baby dry/doesn't leak" (*staysdry importance*) and
"natural composition" (*natural importance*),

Responses were given on a 9 point scale, where "1" = "Not Important at All" and "9" = "Extremely Important."

Demographic Information

Consumers were asked to report their age (*age*), annual household income (*income*), family size including the new baby (*family size*), the number of other children in the home (*other children*), and their ethnicity (*ethnicity*).

Data Recoding

A new variable, *likely trier*, was created from the intention to try question. "Likely triers" were identified using a "Top two box rule" (i.e., those who indicated that they "Probably" or "Definitely" would try the product). Therefore, for *intent* ≥ .75, *likely trier* = 1; otherwise *likely trier* = 0.

Information Needed

Deb's team needs an estimate of revenue potential, plus additional information on target segments and attribute importances.

I. Revenue Potential

Deb's team has devised a method to estimate potential revenues, based on demographics. Their logic is explained below.

The potential market for Pampers Preemies depends on two key demographic factors.

Births in a year are a product of *women 15–44* who could have babies and the *birthrate* among those women. The number of women of childbearing age has been increasing. Medical advances and changing demographics, including immigration, have led to an increasing birthrate among women of childbearing age. Despite the increase in women of childbearing age, the birthrate has been slowly declining since 2007.

The number of very preterm births in a year is the product of number of births and the chance that a newborn will be very preterm, the very preterm birthrate. Advances in infertility treatments have led to more births by *older, predominantly white*, high risk mothers. Immigration has led to more births by the *youngest, predominantly Hispanic* mothers, many with little information about prenatal care and lack of access to adequate healthcare. The *very preterm* percentage of births is expected to increase in future years. The number of very preterm babies is thought to vary by *ethnicity*.

Preterm Diaper Market. *Very preterm diaper sales volume* is the product of the average number of days a very preterm baby remains very preterm, approximately 30, the average number of diapers used per day, approximately 9, and the sum of *very preterm babies* in each segment *i:*

$$\begin{aligned} \textit{Very preterm diaper sales volume}_t &= \textit{30 days per very preterm baby} \\ &\times \textit{9 diapers per very preterm baby per day} \\ &\times \textstyle\sum_i \textit{very perterm babies}_{i,t} \end{aligned} \tag{7.1}$$

Procter & Gamble's Preemie Business. From past experience, Procter & Gamble managers have learned that 75 % of the proportion of *Likely Triers,* the *trial rate*, become loyal customers in the first year. Managers expect the *trial rate* to depend on price *j, premium* or *value:*

$$\textit{P\&G share of diaper purchases}_j = .75 \times \textit{trial rate}_j \tag{7.2}$$

Procter & Gamble consolidated *very preterm diaper sales volume* would depend on *very preterm diaper sales volume* and *P&G share of diaper purchases* at price *j*.

$$\begin{aligned} \textit{P\&G very preterm diaper sales volume}_{j,t} &= \textit{P\&G share of diaper purchases}_j \\ &\times \textit{very preterm diaper sales volume}_t \end{aligned} \tag{7.3}$$

Consolidated *revenue($)* would then be the product of *price j*, and *P&G very preterm diaper sales volume* at *price j:*

$$Revenue(\$)_{j,t} = price(\$)_j \ x \ P\&G \ very \ preterm \ diaper \ sales \ volume_{j,t} \quad (7.4)$$

To effectively evaluate revenue potential, managers believe the uncertainties in the number of *very preterm births* by ethnic segment and *trial rate* by price must be incorporated in analyses.

Case 7-1 Very Preterm Births.xlsx contains time series of very preterm births for the three largest ethnic segments.

1. Build a naïve model of *very preterm births* to forecast expected *very preterm births* in 2013 and to find the standard error for each of the ethnic segments.
2. Use Monte Carlo simulation to find a distribution of possible *very preterm births* in 2013 for each ethnic segment.

Illustrate possible *very preterm births* in 2013, showing the distributions for each of the ethnic segments side by side in a column chart.

3. From the distribution of possible *very preterm births* in 2013 in each segment, find consolidated *very preterm diaper sales volume* in 2013, using the equation in (7.1) in your spreadsheet.

Illustrate the distribution of possible *very preterm diaper sales volume* in 2013 with a column chart.

4. From the concept test sample, find the *trial rate* for the two alternate prices, *premium* and *value*, and the *conservative standard error* for these two proportions
5. Use Monte Carlo simulation to find a distribution of possible *trial rates* for the two alternate prices, *premium* and *value.*
6. From the simulated samples of possible *trail rates* for *premium* and *value* prices, find the distribution of possible *P&G shares of diaper purchases*, using the Eq. 7.2 in your spreadsheet.

Il*lustrate the distribution of* possible *P&G shares of diaper purchases* at the *premium* price, side by side, with the distribution of possible *P&G shares of diaper purchases* at the *value* price with a column chart.

7. From the sample of possible consolidated *very preterm diaper sales volume* in 2013 (Eq. 7.1) and the distribution of possible *P&G shares of diaper purchases* (Eq. 7.2) for the two alternative prices, find the distribution of possible *P&G very preterm diaper sales volumes* in 2013 for *premium* and *value* prices, using Eq. 7.3 in your spreadsheet.
8. From the distributions of possible *P&G preterm diaper sales volumes* in 2013 (Eq. 7.3) at the two alternative prices, find the distribution of possible consolidated *revenues* in 2013 for *premium* and *value* prices, using Eq. 7.4 in your spreadsheet.

Illustrate possible 2013 *revenues* with a column chart, showing the two distributions (at *premium* and at *value* prices) side by side.

Find the 95 % prediction intervals for possible *2013 revenues* for *premium* and for *value* prices.

II. Additional Information Needed

1. Demographic differences between Likely and Unlikely Triers and identification of lifestyle segments most likely to try
(a) Test suspected population differences between Likely and Unlikely Triers using a two sample *t test* along each of the demographics, *Age*, *Income*, *Family size*, and *Number of other Children*
(b) For each signi*ficant demographic difference betwee*n Li*kely and Unlikely Trier s*egments, estimate the extent of difference between Likely and Unlikely Trier segments in the population.
The **Likely v Unlikely** worksheet in **Case 7-1 pampers concept test.xls** has been sorted by trier segment for these tests.

Illustrate significant differences with a column chart.

(c) From differences in (a), identify the lifestyle segments which you believe will be most attracted to the concept (*younger detached, younger committed, young, later in life moms, latest in life moms*).
2. Identification of attributes likely to be considered important by Likely Triers

The worksheet page, **Likely Triers Only,** of **Case 7-1 pampers concept test.xls** contains importance ratings from the segment of *Likely Triers* only.
Determine which attributes are likely to be considered important to the segment of *Likely Triers* from sample ratings of *brand importance*, *fit importance*, *staysdry importance*, and *natural importance.*
To qualify as an important attribute, the average importance rating for that attribute by the *Likely Trier* segment would exceed 7 on a 9-point scale.

Illustrate your results with a column chart of the 95 % lower and upper confidence interval bounds.

Team Assignment

To prepare for the case discussion, your Team should estimate revenue potential and find the additional information needed by Deb Henretta, listed above.

I. PowerPoint Presentation to Management. Each Team is responsible for the presentation to management of *revenue* forecasts at the two alternate prices and information regarding attribute importances and target segments.

To facilitate your presentation, construct no more than eight PowerPoint slides that illustrate your key results, using the guidelines from class.

- Slide 1 introducing your team
- Up to six slides presenting your results
- One concluding slide with Take Aways

Use graphs, rather than tables. Round to no more than three significant digits. Use fonts no smaller than 24 pt, including text in graphs. Adjust axes.

II. Memo to Management. Each Team is also responsible for creating a single page, single-spaced memo, using 12 pt font, presenting your analysis to P&G management.

Include one embedded figure which illustrates a key result. Include additional pages with exhibits containing graphics which are referred to in your memo. Exhibits should contain only graphs, and only graphs that are referred to in the memo. Round to no more than three significant digits.

7.1 Use PowerPoints to Present Statistical Results for Competitive Advantage

PowerPoint presentations are a powerful tool that can greatly enhance your presentation of the results of your analysis. They are your powerful sidekick. Tonto to your Lone Ranger. PowerPoints help your audience member key points and statistics and make available graphics to illustrate and enhance the story you are telling.

The key to effective use of PowerPoints for presenting your results for competitive advantage is to be sure that they are not competing with you. PowerPoints with too much text draw audience attention away from you. Cliff Atkinson, in his 2008 book, beyond bullet points, (Microsoft Press) explains clearly how audience members process information during PowerPoint presentations and why you should move beyond bullet points in the design of your PowerPoints. Much of the material that follows reflects Mr. Atkinson's wisdom, and his book is a recommended investment.

7.1.1 Audience Brains Are Designed to Process and Remember Information

Our brains are ingeniously designed to filter and process large amounts of information, selecting the most relevant to be stored in long term memory. Only a small portion of incoming information gains admission into working memory, and only some portion of information processing in working memory survives and is stored in long term memory (Fig. 7.1).

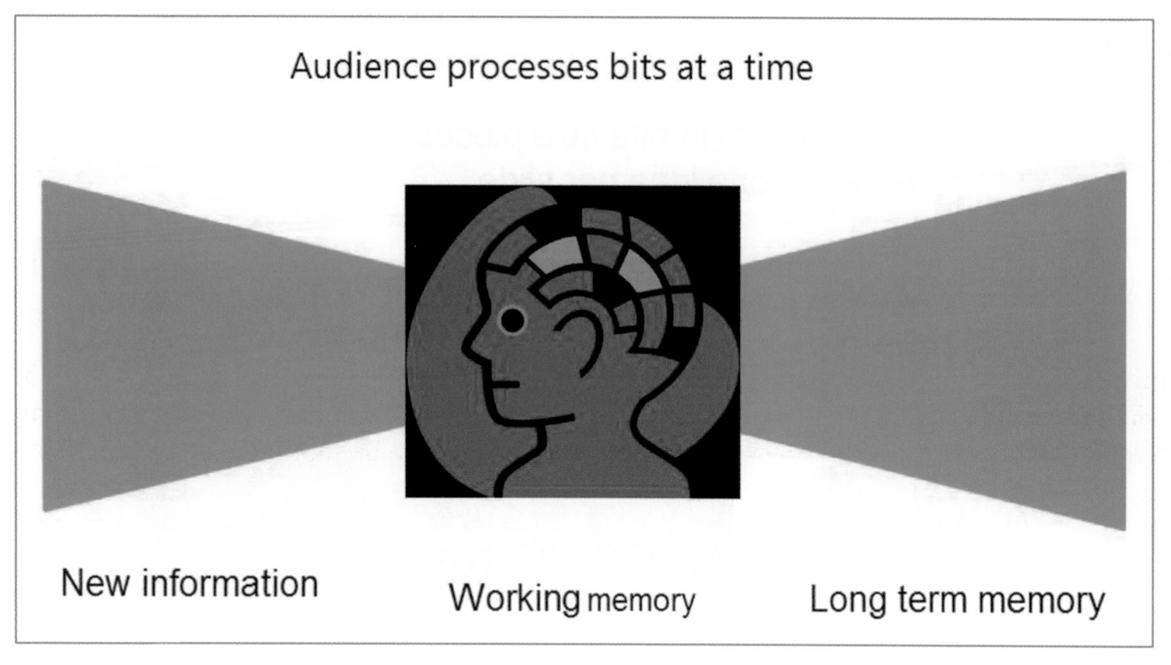

Fig. 7.1 Information processing in a given moment

The goal is to help your audience filter information, direct their attention to your key results and interpretation, so that your message will be remembered.

7.1.2 Limit Text to a Single Complete Sentence per Slide

Since brains process only a few select bits of information in a given moment, increase the chance that the key points in your presentation become those select elements. If your slides are loaded with text, the critical point has a small chance of being processed in working memory (Fig. 7.2).

Fig. 7.2 Limited processing in working memory

Work to design your slides so that each slide presents a single idea. Use only one complete sentence per slide (Fig. 7.3).

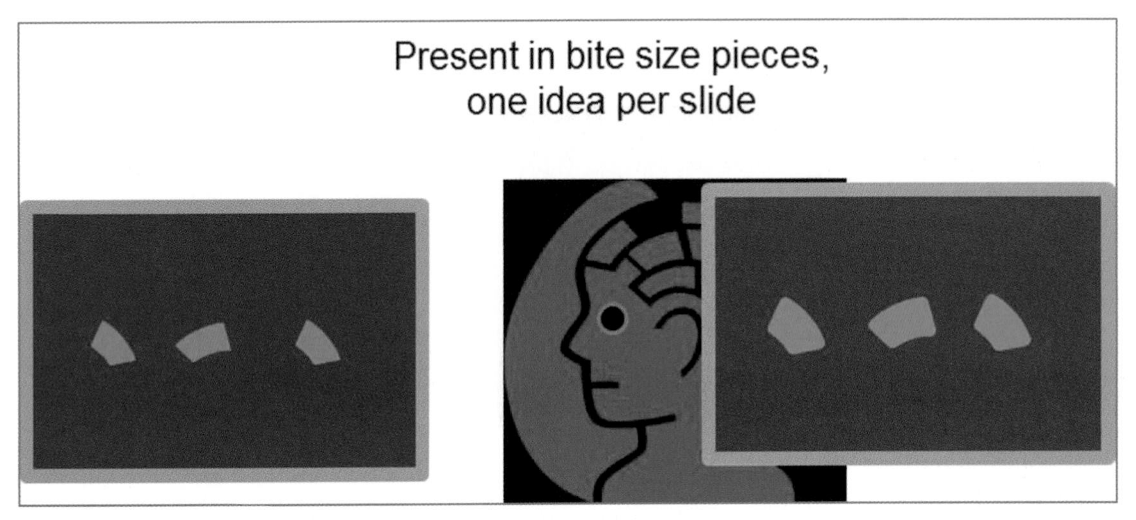

Fig. 7.3 Present one idea at a time

7.1.3 Pause to Avoid Competing with Your Slides

Brains process a single channel at a time. Attention is directed toward either visuals or audio in any given moment (Fig. 7.4).

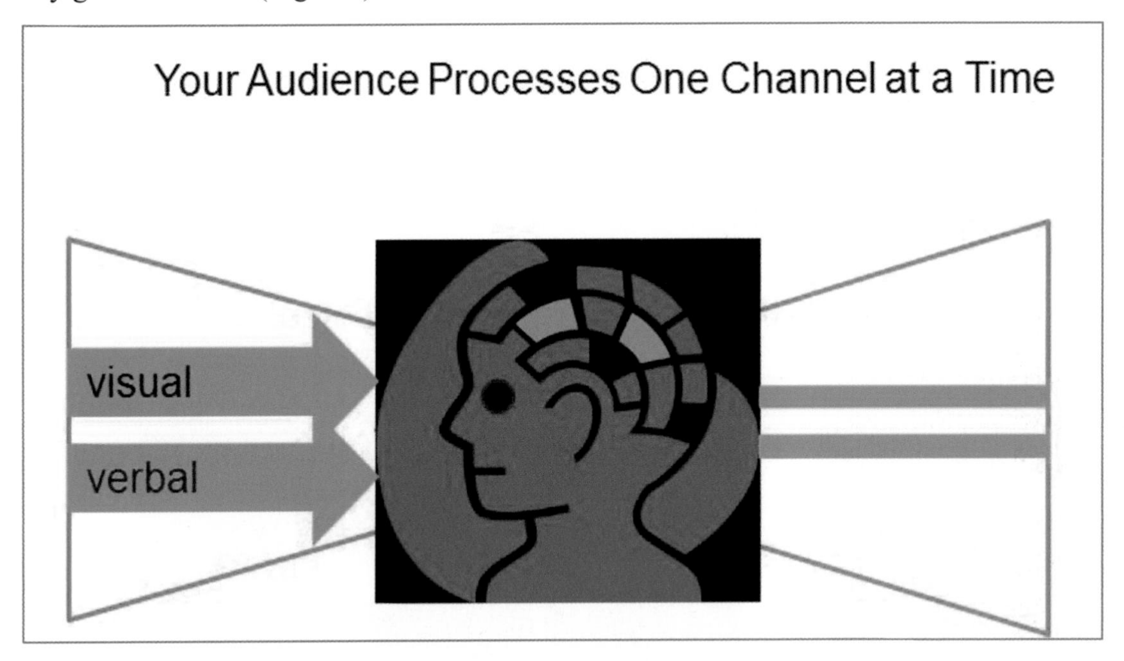

Fig. 7.4 Processing of a single channel in a given moment

Your PowerPoints should complement the story that you are delivering. Your PowerPoints should not compete with you for attention. You are the star in the focus of attention. Your PowerPoints should play a supporting role. Pause to allow time for the audience to process that single idea, and then elaborate and explain. This will avoid competition between your slides and you.

7.1.4 Illustrate Results with Graphs Instead of Tables

Tables are effective elements in reports which convey a lot of information for readers to refer to and ponder. Tables are not processed in seconds, which is the time available to process each of your PowerPoint slides (Fig. 7.5).

Fig. 7.5 Information overload from tables

Synthesize the results in your tables into graphs. Graphs organize your results and illustrate key take aways. Well designed graphs can be processed in seconds, allowing audience attention to flow from a slide back to you, the speaker and, ideally, the focus of attention (Fig. 7.6).

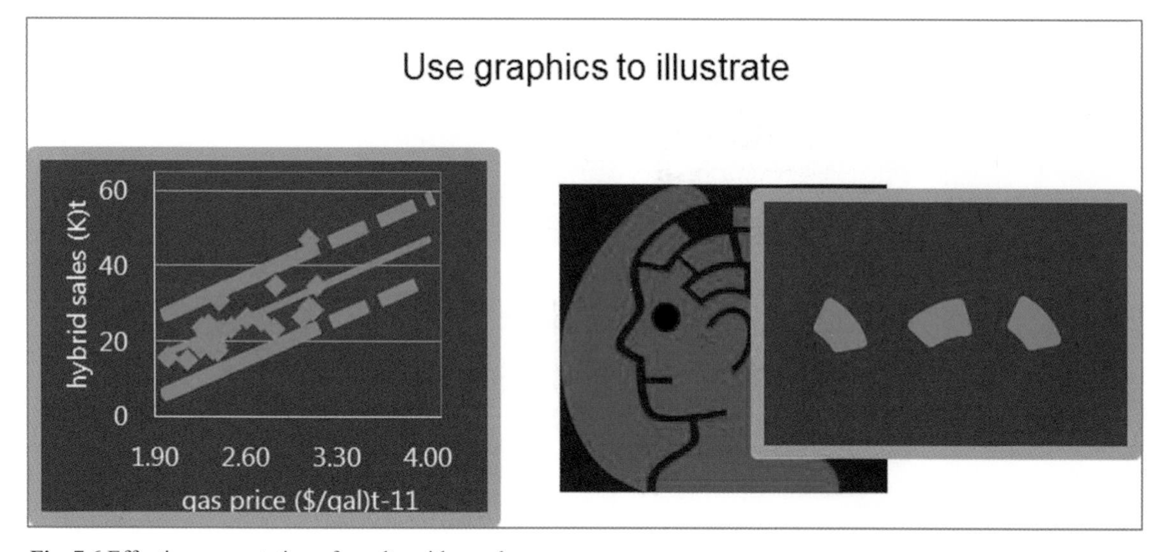

Fig. 7.6 Effective presentation of results with graphs

An effective slide contains a single complete sentence, the *headline*, and a graph to illustrate.

7.1.5 Start PowerPoint Design in Slide Sorter

Insure that your PowerPoints are organized effectively. Build your deck by beginning in Slide Sorter view. Choose the main points that you want the audience to remember (Fig. 7.7).

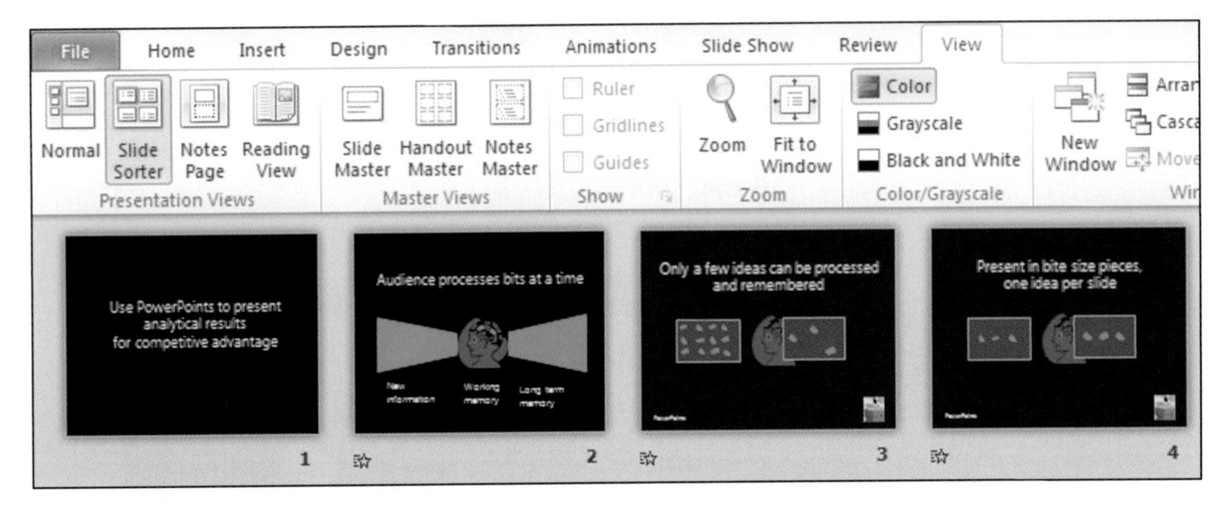

Fig. 7.7 Slide sorter view

Next, add slides with supporting information the main point slides (Fig. 7.8).

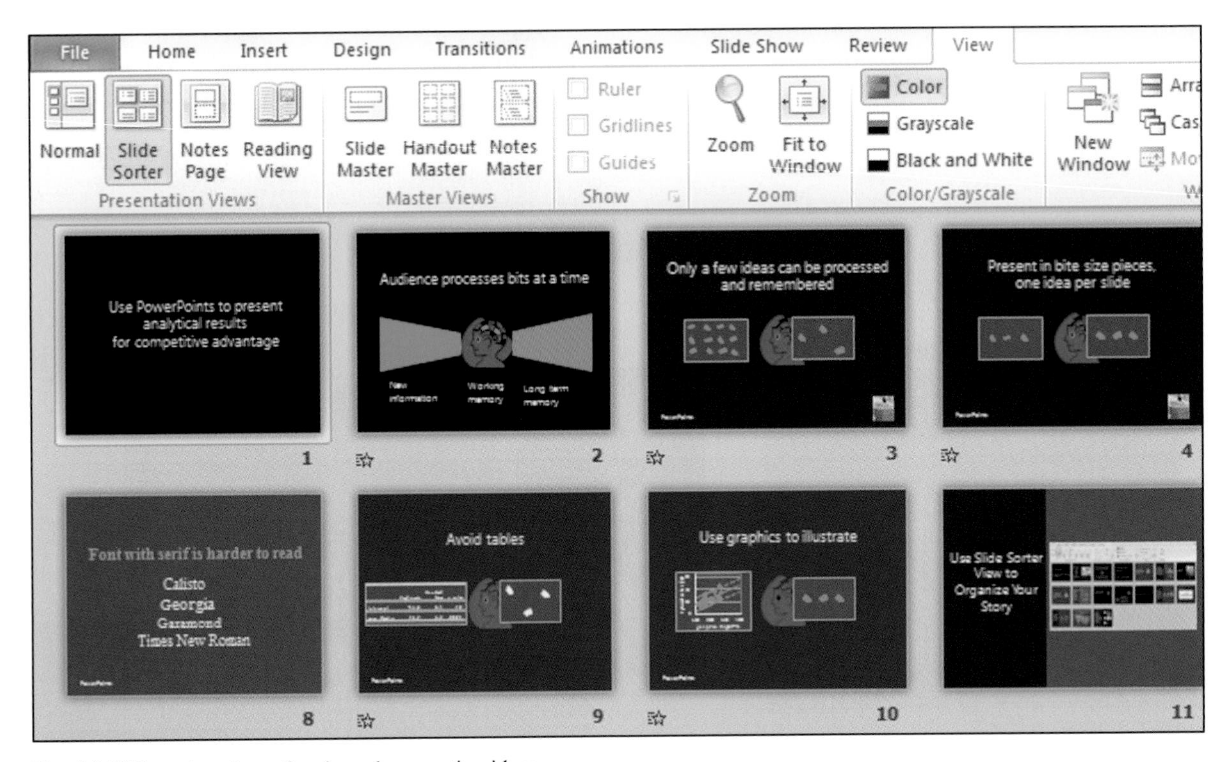

Fig. 7.8 Slide sorter view of main and supporting ideas

7.1.6 Put Supporting Text in Slide Notes

Presenters sometimes worry that they will forget the story. For insurance, they include all of the text to be delivered in their slides. You can guess the consequence. Audience members attempt to read and process all of the text in the slides. To do this, they must ignore the presenter. In the few seconds that a slide appears, there is much too little time to read and process all of the text. As a result, the audience processes only remnants of the story. Audience members are frustrated, because at the end of the presentation, they have incomplete information that doesn't make sense.

In addition to supporting your presentation, your PowerPoints deliver an impression. Slides filled with text deliver the impression that the presenter lacks confidence. When audience members fail to process slides laden with text and tables, the natural conclusion is that the speaker is ineffective. "She spoke for fifteen minutes, but I can't remember what she said. Made no sense."

Audience members can reach a second, unfortunate conclusion in cases where a presenter has simply converted report pages into slides. Slides converted from reports are crammed with text and tables, and too often look like report pages, with white backgrounds, black text. This sort of unimaginative PowerPoint deck delivers the impression that the speaker is lazy.

In contrast, slides with a single, complete sentence headline and graph deliver the impression that the presenter is confident. After easily processing the slides and then focusing on explanation and elaboration delivered by the presenter, audience members understand and remember the story.

If you present a single idea in each slide, you will remember what you want to say to explain the idea and add elaboration. The audience will focus on your presentation, since you will provide the missing links.

You can have the best of both worlds. You can include your explanation and elaboration of the main points in the slide Notes. The Notes are not seen during your presentation, but they are available later. Provide handouts at the end of your presentation from the Notes view of your slides (Fig. 7.9).

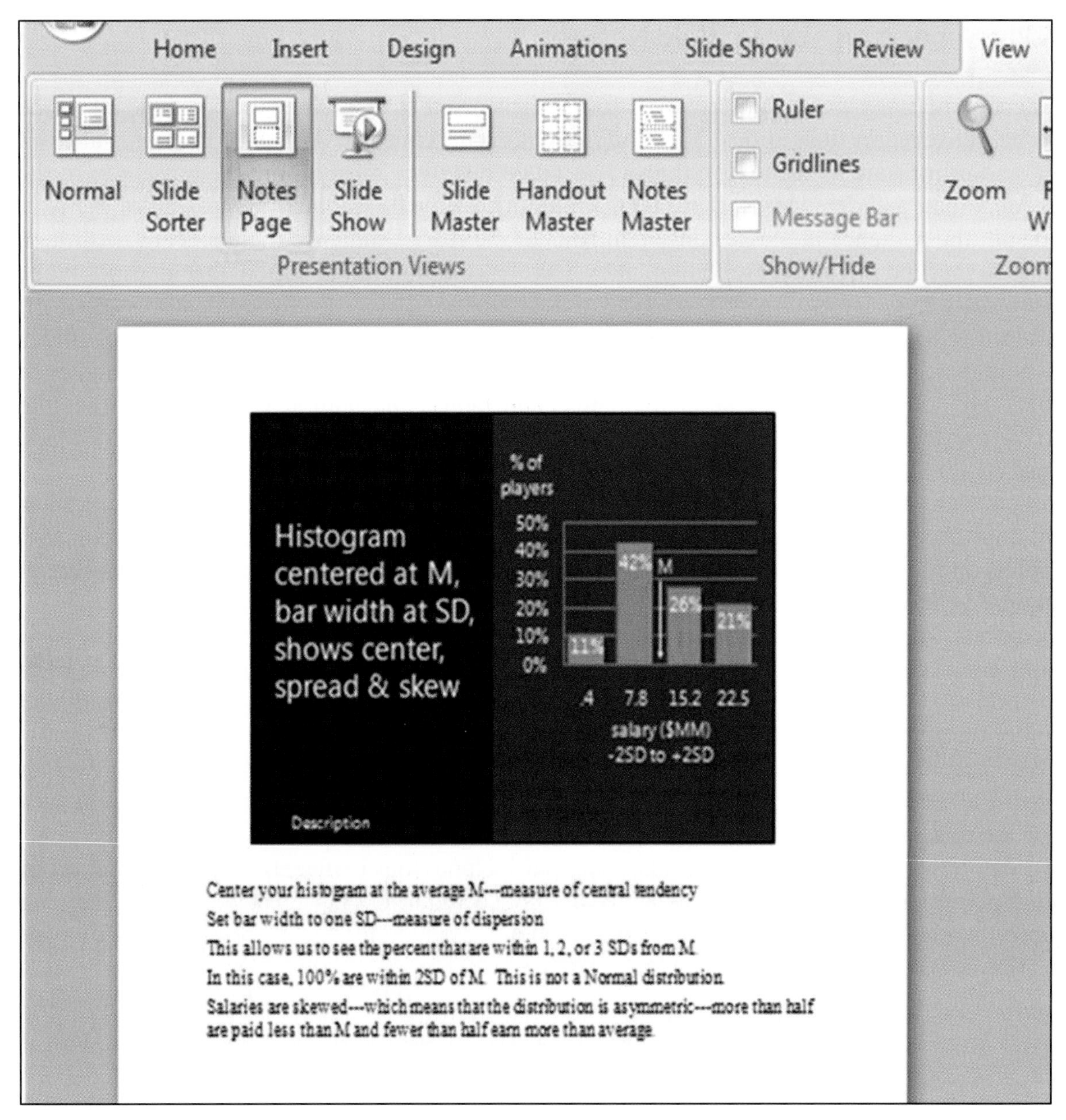

Fig. 7.9 Notes view

7.1.7 Choose a Slide Design that Reduces Distraction

Design your slides so that elements are minimally distracting. You want the audience to be able to quickly and easily see the idea in each slide and then focus on you for explanation and elaboration.

7.1.8 Use a Font that Can Be Easily Read

Use *at least 24 pt* font so that audience members can easily read your headline, numbers, and labels. If you include numbers in your graphs, in the axes or as data labels, they must be easily read. (Be sure to round your numbers to two or three significant digits, also.) Axes labels, and other text must also be easily read. Any font smaller than 24 pt will challenge easy reading.

Choose a *sans serif* font (Ariel, Lucida, or Garamond). *Sans serif* fonts, without "feet" are easier to read in PowerPoints. *San serif* characters, without extra lines, are clearer in slides. (The opposite is true for reports, where the *serif* enhances reading ease.) If you have any doubts about readability, test your slides in a room similar in size and shape to the presentation location.

7.1.9 Choose Complementary Colors and Limit the Number

In cases where the slides will be presented in a darkened room, the background should be darker than the title and key words. Choose a medium or darker background, with complementary, contrasting, lighter text color. PowerPoints in this setting are more like television, movies, and internet media, and less like books or reports, and should feature darker backgrounds like those you see in movie credits.

When presentations are in well lit rooms, backgrounds can be lighter than title and key words. In a light setting, PowerPoints resemble text pages, with lighter backgrounds and darker text colors.

If we see more than five colors on a slide (including text), our brains overload and we have difficulty processing the message and remembering it. Limit the number of distinct colors in each slide.

7.2 Write Memos that Encourage Your Audience to Read and Use Results

Memos are the standard for communication in business. They are short and concise, which encourages the intended audience to read them right away. Memos which present statistical analysis to decision makers:

- Feature the bottom line in the subject line,
- Quantify how the bottom line result influences decisions,
- Are ideally confined to one single spaced page,
- Include an attractive, embedded graphic which illustrates the key result.

Many novice analysts copy and paste pages of output. The output is for consumption by analysts, whose job it is to condense and translate output into general business language for decision makers. Decision makers need to be able to easily find the bottom line results without referring to a statistics textbook to interpret results. It is our job to explain in easily understood language how the bottom line result influences decisions. For the quantitative members of the audience, key statistics are included.

On the following page is an example of a memo which might have been written by the quantitative analysis team at Procter & Gamble to present a key result of a concept test of Pampers Preemies to brand management.

Notice that:

- The subject line contains the bottom line result,
- Results are illustrated,
- Results are described in general business English.

Description of the concept test and results are condensed and translated. Brand management learns from reading the memo what was done, who was involved, what results were, and what implications are for decision making.

MEMO

Re: Importance of Fit Drives Trial Intention
To: Pampers Preemies Management
From: Procter & Gamble Quantitative Analysis Team
Date: October 2007

Summary

Results of a concept test of the Pampers Preemies suggest that the *Importance of fit* drives trial intentions, supporting the expected market salience of superior diaper fit.

Sample and method

The Concept Test Sample. The Preemies concept with premium price was described to a convenience sample of 60 preemie mothers in three hospitals in Cincinnati. Demographics of this sample mirror national demographics of preemie mothers and are representative of all preemie mothers.

Data scales

Concept Test Measures. The mothers indicated intent to purchase on a five-point scale (.05 = "Definitely Won't Try" . . . .95 = "Definitely Will Try") and rated the importances of diaper attributes, including fit, brand, capability to protect from insults, and natural composition on balanced 9-point scales (1 = "Unimportant" . . .9 = "Very Important").

Results

Fit is more important to Likely Triers

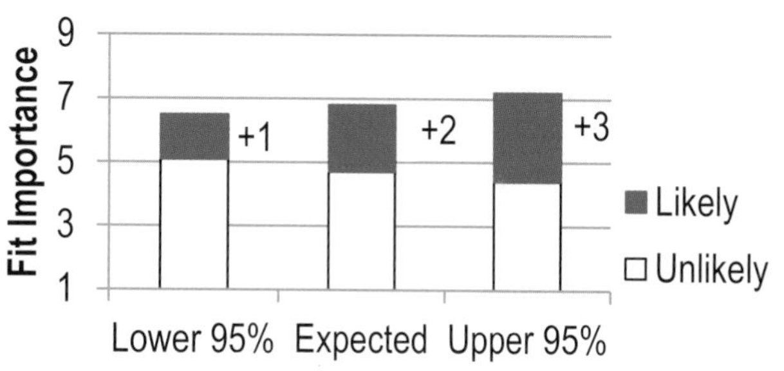

Differences in fit importance account for 6 % of the differences in trial intention. Comparing mothers who rate fit moderately important (5 on the 9-point scale) with those who rate fit very important (9), the difference in intention is expected to be .4, which translates into the difference between mothers who "might try" and "definitely will try."

Conclusion

Likely and Unlikely Trier segments differ, suggesting a focus on fit

Likely Triers value diaper fit more than Unlikely Triers. This key difference coincides with a differential advantage of Pampers Preemies, suggesting a focus for promotional material.

Additional considerations

Other attributes, as well as demographics, potentially drive trial intent

Other attributes, including brand, composition, capability to keep baby dry, and price, probably also affect intent. Demographics are likely to affect diaper attitudes, as well as intent to try Pampers Preemies.

Chapter 8
Finance Application: Portfolio Analysis with a Market Index as a Leading Indicator in Simple Linear Regression

Simple linear regression of stock rates of return with a Market index provides an estimate of *beta*, a measure of risk, which is central to finance investment theory.

Investors are interested in both the mean and the variability in stock price growth rates. Preferred stocks have higher expected growth – expected *rates of return* – shown by larger percentage price increases over time. Preferred stocks also show predictable growth – low variation – which makes them less risky to own. A portfolio of stocks is assembled to diversify risk, and we can use our estimates of portfolio *beta* to estimate risk.

8.1 Rates of Return Reflect Expected Growth of Stock Prices

Example 8.1 Wal-Mart and Apple Returns

Figure 8.1 contains plots of share prices of two well known companies, Wal-Mart and Apple, over a 60 month period, July 2007 to July 2012.

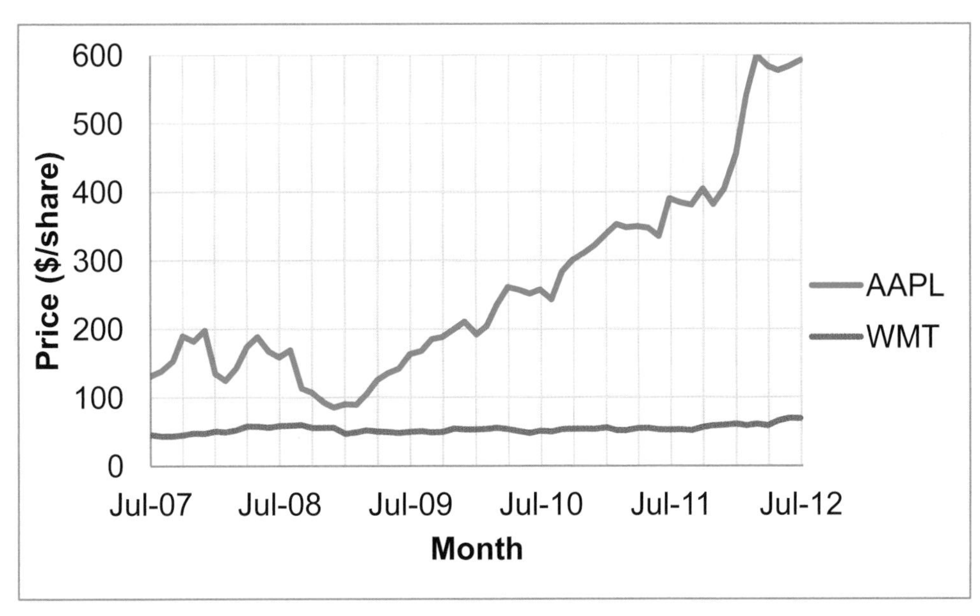

Fig. 8.1 Monthly share prices of Wal-Mart and Apple, July 2007 to July 2012

It is important to note that although prices in some months were statistical outliers, those unusual months were not excluded. A potential investor would be misled were unusually high or low prices ignored. Extreme values are expected and included, since they influence conclusions

C. Fraser, *Business Statistics for Competitive Advantage with Excel 2013: Basics, Model Building, Simulation and Cases*, DOI 10.1007/978-1-4614-7381-7_8, © Springer Science+Business Media New York 2013

about the appeal of each potential investment. The larger the number of unusual months, the greater the dispersion in a stock price, and the riskier the investment.

To find the growth rate in each of the stock investments, calculate the monthly percent change in price, or *rate of return, RR*:

$$RR_{stock,t} = \frac{(price_{stock,t} - price_{stock,t-1})}{price_{stock,t-1}}$$

where t is month.

Investors seek stocks with higher average rates of return and lower standard deviations. They would prefer to invest in stocks that exhibit higher expected, average growth and less volatility or risk. The standard deviation in the rate of return captures risk. If a stock price shows little variability, it is a less risky investment.

Figure 8.2 illustrates monthly rates of return in Wal-Mart and Apple stocks and a Market index, the S&P 500, over the 5 year period:

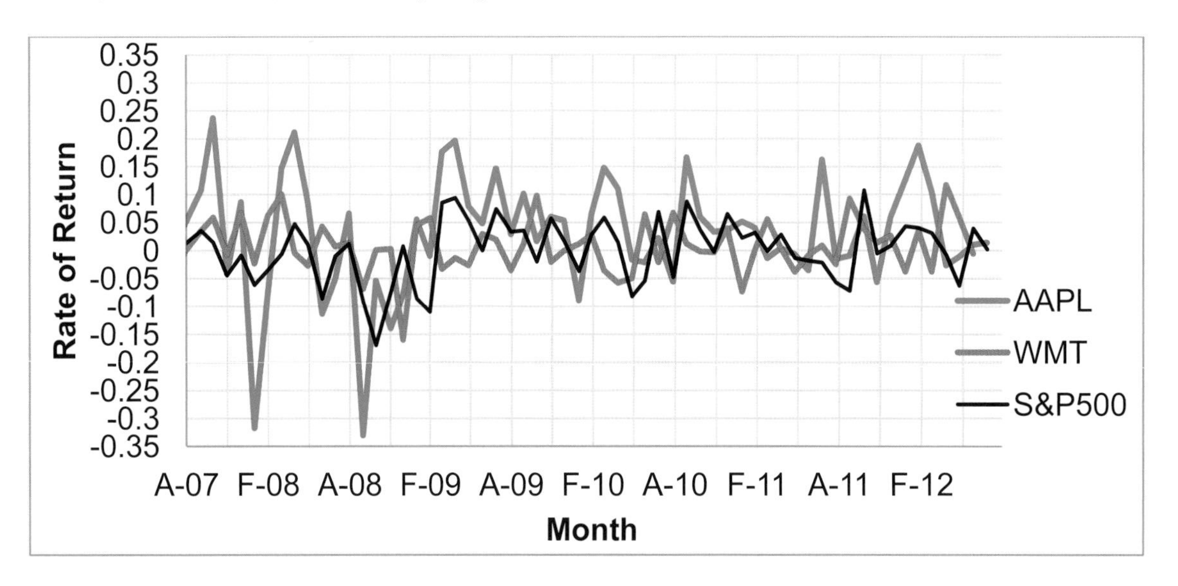

Fig. 8.2 Monthly rates of return of Wal-Mart and Apple, August 2007 to July 2012

Table 8.1 Monthly rates of return of Wal-Mart and Apple Stock, August 2007 to July 2012

Monthly rate of return					
WAL-MART		*APPLE*		*S&P 500*	
M	.0080	*M*	.031	*M*	.00048
SD	.048	*SD*	.107	*SD*	.055
Minimum	−.16	*Minimum*	−.33	*Minimum*	−.17
Maximum	.12	*Maximum*	.24	*Maximum*	.11

From Table 8.1, notice that Apple's mean monthly rate of return of 3.1 % exceeds the Market mean monthly rate of return of .048 %, though Apple stock prices are more volatile: the standard deviation in monthly rates of return is .107, compared with the Market's standard deviation of .055. The greater expected return from Apple comes at the cost of added risk. Over this 5 year period, Wal-Mart stock had a monthly rate of return, .80 %, higher than the Market, but lower than Apple stock, and its prices are less volatile than either the Market or Apple stock, with a standard deviation in monthly rates of return of .048.

We would report to a potential investor:

- *Over the 60 months examined, Apple offers a greater expected monthly rate of return of 3.1 %, relative to the S&P500 Index of the Market, with expected monthly return of .048 %, but at higher risk with standard deviation in return .11* versus *.055 for the Market.*
- *In this 5 year period, Wal-Mart offered an expected rate of return higher than the Market, .80 %, with lower risk, reflected in the lower standard deviation in return .048.*
- *Over the 5 year period, Apple stock delivered a higher rate of return than Wal-Mart stock, though Wal-Mart stock was less volatile.*

8.2 Investors Trade Off Risk and Return

Investors seek stocks which offer higher expected rates of return *RR* and lower risk. Relative to a Market index, such as the S&P 500, which is a composite of 500 individual stocks, many individual stocks offer higher expected returns, but at greater risk. Market indices are weighted averages of individual stocks. Like other weighted averages, a Market index has an expected rate of return in the middle of the expected returns of the individual stocks making up the index. An investor attempts to choose stocks with higher than average expected returns and lower risk.

8.3 Beta Measures Risk

A Market index reflects the state of the economy. When a time series of an individual stock's rates of return is regressed against a Market index, the simple linear regression slope β indicates the expected percent change in a stock's rate of return in response to a percent change in the Market rate of return. β is estimated with b using a sample of stock prices:

$$\hat{R}R_{stock,t} = a + b \times RR_{Market,t}$$

Where $RR_{stocki,t}$ is the estimated rate of return of a stock i in month t, and

$RR_{Market,t}$ is the rate of return of a Market index in month t.

In this specific case, the simple linear regression slope estimate b is called *beta.* Beta captures Market specific risk. If, in response to a percent change in the Market rate of return, the expected change in a stock's rate of return b is greater than one, the stock is more volatile, and exaggerates Market movements. A 1 % increase in the Market value is associated with an expected change in the stock's price of more than 1 % change. Conversely, if the expected change in a stock's rate of return b is less than one, the stock dampens Market fluctuations and is less risky. A 1 % change in the Market's value is associated with an expected change in the stock's price of less than 1 %. Beta reflects the amount of risk a stock contributes to a well diversified portfolio.

Recall from Chap. 5 that the sample correlation coefficient between two variables r_{xy} is closely related to the simple regression slope estimate b:

$$b = r_{X,y} \times \frac{s_y}{s_X}$$

In a model of an individual stock's rate of return against a Market index, the estimate of beta b is directly related to the sample correlation between the individual stock's rate of return and the Market rate of return:

$$\hat{beta}_{stock_i} = b_{stock_i} = r_{stock_i,Market} \times \frac{s_{stock_i}}{s_{Market}}$$

The estimate of beta is a direct function of the sample correlation between an individual stock's rate of return and the Market rate of return, as well as Market sample variance.

Stocks with rates of return that are more strongly correlated with the Market rate of return and those with larger standard deviations have larger betas.

Notice in Fig. 8.2 that Wal-Mart stock has a smaller variance than Apple stock or the Market. Wal-Mart is a less risky investment. Notice also that both stocks tend to move with the Market, though Apple moves more and Wal-Mart moves less.

It would not be surprising to find that Apple stock returns are riskier than Wal-Mart returns, since iPhones and iPads are relatively expensive, luxury items. In boom cycles, companies that sell luxuries do more business. Wal-Mart sells many necessities, including food, personal care products, cleaning products, and clothing. The demand for these products is affected less by economic swings, making Wal-Mart stock relatively less correlated with Market swings, and, hence, less risky.

Table 8.2 contains sample correlation coefficients, standard deviations, and betas for both of the stocks using 5 years of monthly data.

Table 8.2 Correlations, standard deviations, covariances and betas for August 2007 to July 2012

	Correlation with the market $r_{stock,Market}$	*SD*	*Beta* b_{stock}
SP500 RR		.055	
Wal-Mart RR	.35	.048	.31[a,b]
Apple RR	.64	.11	1.23[a]

[a]Significant at .01
[b]Significantly less than 1.0 at a 95 % confidence level

Correlations between each of the stocks' returns and the Market are positive, indicating that they do move with the Market.

Apple returns are more strongly correlated with the Market index returns than Wal-Mart returns. Apple returns are also more volatile. Because Apple rates of return are both more strongly correlated with the Market and more volatile than Wal-Mart returns, Apple stock will have a larger beta than Wal-Mart stock.

Betas b_{stocki} are shown in the last column of Table 8.3. A percent increase in the Market produces:

- Less than 1 % expected increase in the Wal-Mart stock price, and
- A 1 % expected increase in the Apple stock price,

Beta estimates are shown in Table 8.3 and Fig. 8.3.

Table 8.3 Estimates of betas

Wal-Mart

SUMMARY OUTPUT						
Regression Statistics						
R Square	.124					
Standard Error	.046					
Observations	60					
ANOVA	*df*	*SS*	*MS*	*F*	*Significance F*	
Regression	1	.017	.0170	8.2	.006	
Residual	58	.120	.0021			
Total	59	.137				
	Coefficients	*Standard Error*	*t Stat*	*p value*	*Lower 95 %*	*Upper 95 %*
Intercept	.008	.006	1.3	.185	−.004	.020
S&P RR	.307	.107	2.9	.006	.093	.522

Apple

SUMMARY OUTPUT						
Regression Statistics						
R Square	.405					
Standard Error	.083					
Observations	60					
ANOVA	*df*	*SS*	*MS*	*F*	*Significance F*	
Regression	1	.272	.272	39.4	.0000	
Residual	58	.400	.007			
Total	59	.672				
	Coefficients	*Standard Error*	*t Stat*	*p value*	*Lower 95 %*	*Upper 95 %*
Intercept	.031	.011	2.9	.0058	.009	.052
S&P RR	1.226	.195	6.3	.0000	.835	1.617

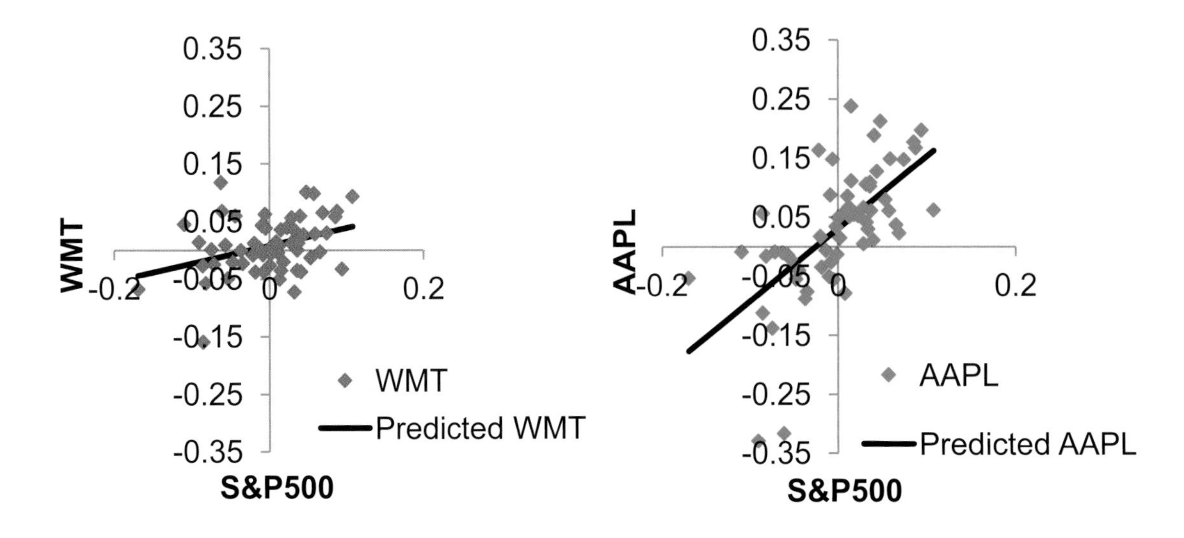

$\hat{R}R_{Wal-Mart_t} = .0079 + .31^a \times S\&P500_t$
RSquare: $.12^a$

$\hat{R}R_{Apple_t} = \mathbf{.031}^a + \mathbf{1.23}^a \times S\&P\mathbf{500}_t$
RSquare: $.40^a$

[a]*Significant at* .01

Fig. 8.3 Response of Wal-Mart and Apple stocks to the Market

Relative to Wal-Mart, Apple rates of return have both a higher correlation with the Market and a larger standard deviation, components of specific risk, producing the larger beta.

Comparing betas, a potential investor would conclude:

"Wal-Mart stock, with an estimated beta less than one ($b_{Wal\text{-}Mart} = .31$), is a low risk investment. Wal-Mart returns dampen Market swings. With a percent increase in the Market, we expect to see an average increase of .31% in Wal-Mart's price.

Apple stock, with an estimated beta of one ($b_{Apple}=1.23$) is riskier than Wal-Mart, and mirrors Market movement. With a percent increase in the Market, we expect to see an average increase of about one percent, 1.23 %, in Apple's price."

8.4 A Portfolio Expected Return, Risk and Beta Are Weighted Averages of Individual Stocks

An investor is really interested in the expected return and risk of her portfolio of stocks. These are weighted averages of the expected returns and betas of the individual stocks in a portfolio:

$$E(RR_P) = \sum_i w_i \times E(RR_i)$$

$$b_P = \sum_i w_i \times b_i$$

Where $E(RR_P)$ is the expected portfolio rate of return,
w_i is the percent of investment in the i'th stock,
$E(RR_i)$ is the expected rate of return of the i'th stock,
b_P is the portfolio beta estimate,
b_i is the beta estimate of the ith stock,

Example 8.2 Three Alternate Portfolios

An Investment Manager has been asked to suggest a portfolio of two stocks from three being considered by a client: Wal-Mart, Google and IBM. The prospective investor wanted to include computer stock in his portfolio and had heard that IBM was a desirable "Blue Chip." She suspected that holding both Google and IBM stocks might be risky, were the computer industry to falter.

To confidently advise her client, the Investment Manager compared three portfolios of two equally weighted stocks from the three requested options. Individual stock weights in each portfolio equal one half. Table 8.4 contains the expected portfolio rates of return and betas for the three possible combinations:

Table 8.4 Expected portfolio returns and beta estimates

Portfolio	*Expected portfolio return* $\sum E(RR_i)/2$	$E(RR_P)$	*Portfolio beta estimate* $\sum b_i/2$	b_P
Wal-Mart + Google	(.008 + .007)/2	.008	(.31 + 1.08)/2	.70
Wal-Mart + IBM	(.008 + .011)/2	.010	(.31 + .67)/2	.49
Google + IBM	(.007 + .011)/2	.009	(1.08 + .67)/2	.88

Alternatively, she could find expected portfolio returns and betas with software, and this would be the practical way to compare more than a few portfolios. Figure 8.4 shows expected (mean) rates of return and regression beta estimates for the three portfolios from Excel:

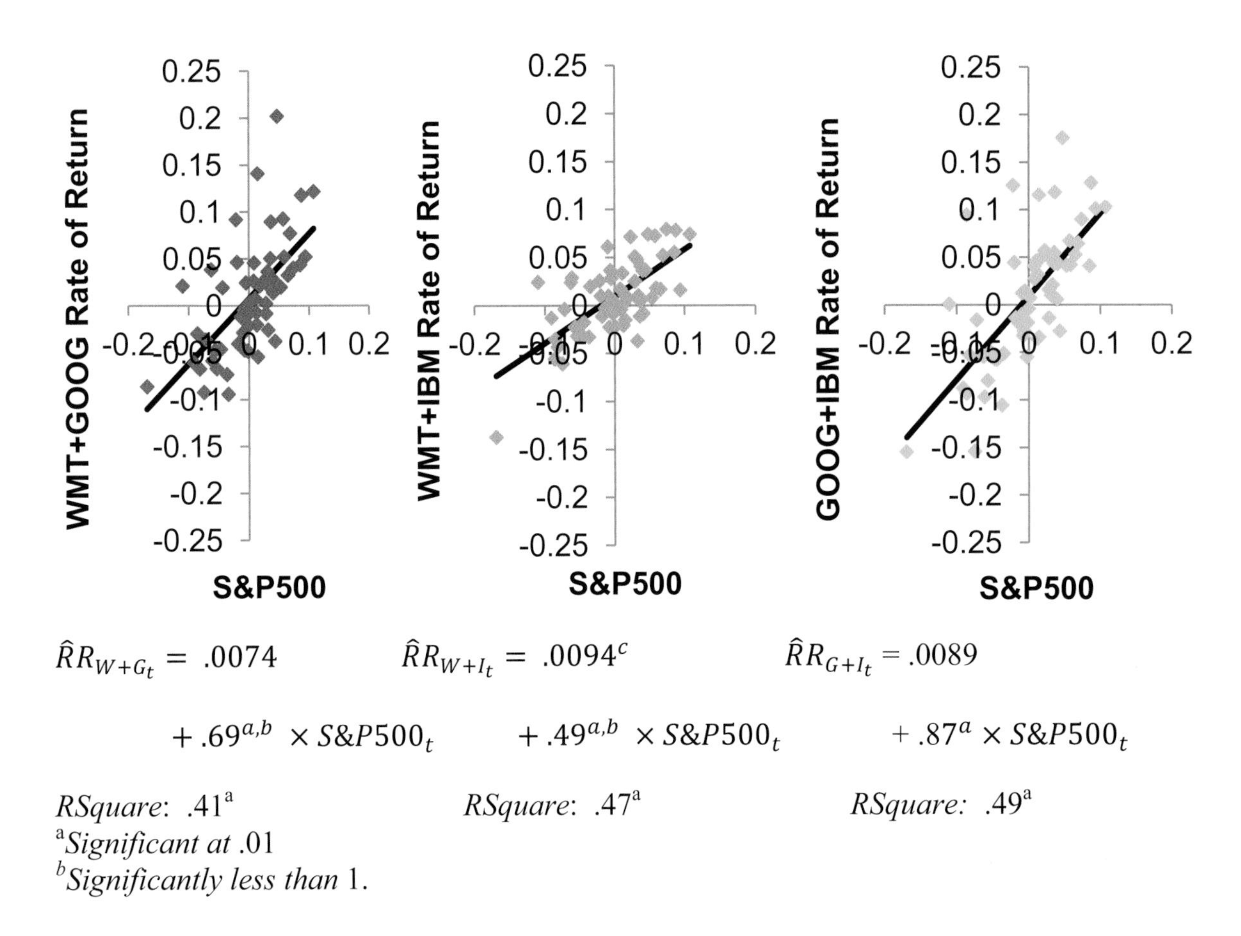

Fig. 8.4 Beta estimates of three alternate portfolios

8.5 Better Portfolios Define the Efficient Frontier

In the comparison of alternative portfolios, the Investment Manager wanted to identify alternatives which promised greater expected return without greater risk – or, alternatively, those which reduced risk without reducing return. Better portfolios, which promise the highest return for a given level of risk, define the *Efficient Frontier.* To see the Efficient Frontier, she made a scatterplot of portfolio expected rates of return by portfolio risk. Those relatively efficient portfolios lie in the upper left.

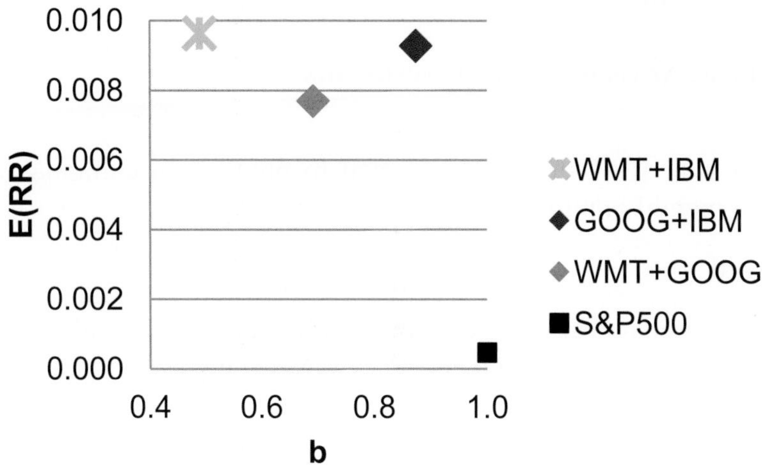

Fig. 8.5 Relatively efficient portfolios offer greater expected return and lower risk

Comparing portfolios in Fig. 8.5, the Investment Manager found that all three portfolios outperform the Market. The Wal-Mart + IBM portfolio dominates the other combinations with Google, offering higher expected rate of return and lower risk:

$$E(RR)_{WMT+IBM} = .010 > E(RR)_{IBM+GOOG} = .0093 > E(RR)_{WMT+GOOG} = .0077$$

$$b_{WMT+IBM} = .49 < b_{WMT+GOOG} = .69 < b_{IBM+GOOG} = .87$$

The second best choice would depend on the prospective investor's risk preference. The combination of both computer stocks, IBM and Google, offers a higher return that Wal-Mart with Google, but is also riskier.

The Investment Manager presented results of her analysis with recommendations in this memo to her client:

MEMO

Re: Recommended Portfolio is Diversified
To: Ms. Rich N. Vest
From: Christine Kasper, Investment Advisor, Stellar Investments
Date: July 2012

The portfolio combining Wal-Mart with IBM stocks is expected to outperform combinations with Google, promising an expected monthly return of 1.0 %.

Alternate portfolios were compared

Portfolios containing two from the candidate set of three stocks, Wal-Mart, IBM, and Google have been compared to assess their expected returns and risk levels. Assessments were based on 5 years of monthly prices, July 2007 through July 2012, and movement relative to the S&P500 Market Index during this period.

Wal-Mart and IBM Dominate

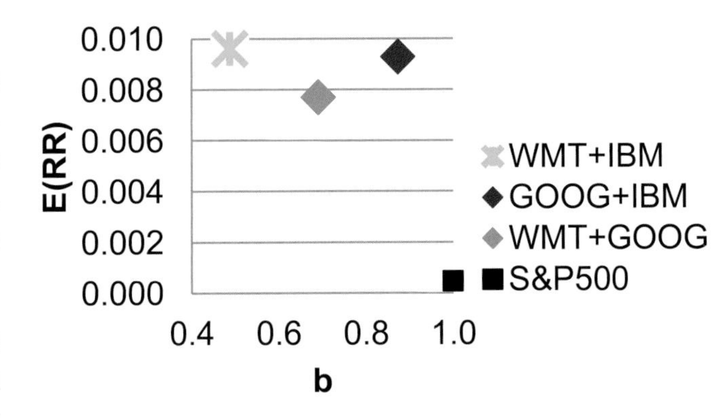

Expected monthly rates of return range from .8 % to 1.0 %. While all three portfolios outperform the Market, those with Google stock yield lower expected returns. The Wal-Mart and IBM combination yields the highest expected return, 1.0 %.

In response to a 1 % change in the S&P500, the Wal-Mart + IBM combination is expected to move less, .5 %, dampening Market movement. This is a conservative choice. Other combinations either mirror or dampen the Market and are expected to move more, .7–.9 %, in response to a 1 % change in the Market.

Choose Wal-Mart + IBM for highest expected return and lowest risk

The choice of Wal-Mart with IBM promises the highest expected return, as well as the lowest risk. Adding Google to either reduces expected rate of return and increases risk.

A larger number of stocks are suggested

You may wish to consider a portfolio with a larger number of stocks to increase your diversification and reduce your risk.

8.6 Portfolio Risk Depends on Correlations with the Market and Stock Variability

Both the expected rate of return of a portfolio and its risk, measured by its beta, depend on the expected rates of return and betas of the individual stocks in the portfolio. Individual stock betas are direct functions of

- The correlation between a stock's rate of return and the Market index rate of return, and
- The standard deviation of a stock's rate of return

Beta for a stock or a portfolio is estimated by regressing the stock or portfolio monthly rates of return against monthly Market rates of return. The resulting simple linear regression slopes are estimates of the stock or portfolio beta.

Excel 8.1 Estimate Portfolio Expected Rate of Return and Risk

Three Portfolios with Wal-Mart, IBM and Google. Monthly rates of return for each of the three stocks and the S&P500 index of the Market are in **Excel 8.1 Three Portfolios.xls**.

Correlations between stocks and the Market. Correlations between rates of return of pairs of stocks and the Market sometimes suggest combinations which might reduce risk through diversification.

Find the pairwise correlations

Alt A Y3 C

For **Input Range**, select the rates of return of the three stocks and the S&P500.

	A	B	C	D	E
1		*S&P500*	*WMT*	*IBM*	*GOOG*
2	S&P500	1			
3	WMT	0.352174	1		
4	IBM	0.654811	0.130101	1	
5	GOOG	0.585391	0.174391	0.475441	1

Monthly portfolio returns formula. Insert three new columns for each portfolio containing equally weighted pairs of the three stocks, which will be the average of rates of return of each of the two stocks in the portfolio.

In the first row of each new column enter a formula for the average of two stocks, using the Excel function **AVERAGE**(*number1,number2*).

C2 fx =AVERAGE(F2:G2)

	A	B	C	D	E
1	Date	S&P500	WMT+ IBM	WMT+ GOOG	IBM+ GOOG
2	8/1/2007	[illegible]13	0.002	-0.020	0.032
3	9/4/2007	0.036	0.005	0.051	0.055
4	10/1/2007	0.015	0.011	0.141	0.116
5	11/1/2007	-0.044	-0.017	0.020	-0.057
6	12/3/2007	-0.009	0.010	-0.005	0.013

Expected monthly rates of return. Find the expected monthly return for the three portfolios in the first row following the data with

Alt MUA:

B62 =AVERAGE(B2:B61)

	A	B	C	D	E
1	Date	S&P500	WMT+ IBM	WMT+ GOOG	IBM+ GOOG
60	6/1/2012	0.040	0.037	0.029	0.006
61	7/2/2012	0.002	-0.002	0.004	0.008
62	*E(RR)*	0.000	0.010	0.008	0.009

Estimate betas from simple regression. To find the Market specific risk, *beta,* find the simple regression slope of each portfolio rate of return with *S&P500.*

For the first portfolio, *WMT+IBM,* run regression with *WMT+IBM* in the **Input Y Range**, and *S&P500* in the **Input X Range**:

16		*Coefficients*	*andard Err*	*t Stat*	*P-value*	*Lower 95%*	*Upper 95%*
17	Intercept	0.00939	0.003754	2.50144	0.015213	0.001876	0.016904
18	S&P500	0.488689	0.06839	7.145641	1.67E-09	0.351792	0.625586

Sheet2 | WFM+IBM | Sheet1

Excel 8.2 Plot Return by Risk to Identify Dominant Portfolios and the Efficient Frontier

To compare the expected rates of return and estimated risk of the three portfolios, plot the portfolio rates of return against their betas to identify the Efficient Frontier.

Create a summary of the portfolio betas and expected returns below the data:

Copy the row containing portfolio labels and paste below the data rows using **Alt H I E**.

Add a row below this label row and above the *E(RR)* row, and then use the Excel function

=**slope**(*y array, x array*)

to add betas below each of the three portfolio labels, locking the *S&P500* column references by pressing **fn 4** three times:

C63 =SLOPE(C2:C61,$B2:$B61)

	A	B	C	D	E	F	G	H
1	Date	S&P500	WMT+ IBM	WMT+ GOOG	IBM+ GOOG	WMT	IBM	GOOG
62	Date	S&P500	WMT+ IBM	WMT+ GOOG	IBM+ GOOG	WMT	IBM	GOOG
63	beta		0.489	0.6921	0.8738			

Copy the three sets of labels, betas and expected rates of return and paste in a new row without formulas, transposing the rows and columns

Alt H V S U E

62		S&P500	WMT+ IBM	WMT+ GOOG	IBM+ GOOG
63	beta	1	0.489	0.6921	0.8738
64	*E(RR)*	0.000	0.010	0.008	0.009
65		beta	E(RR)		
66	S&P500	1	0.000		
67	WMT+IBM	0.4887	0.010		
68	WMT+GOO	0.6921	0.008		
69	IBM+ GOO(	0.8738	0.009		

Select and plot the beta and expected return for the *S&P500*,

Alt N D

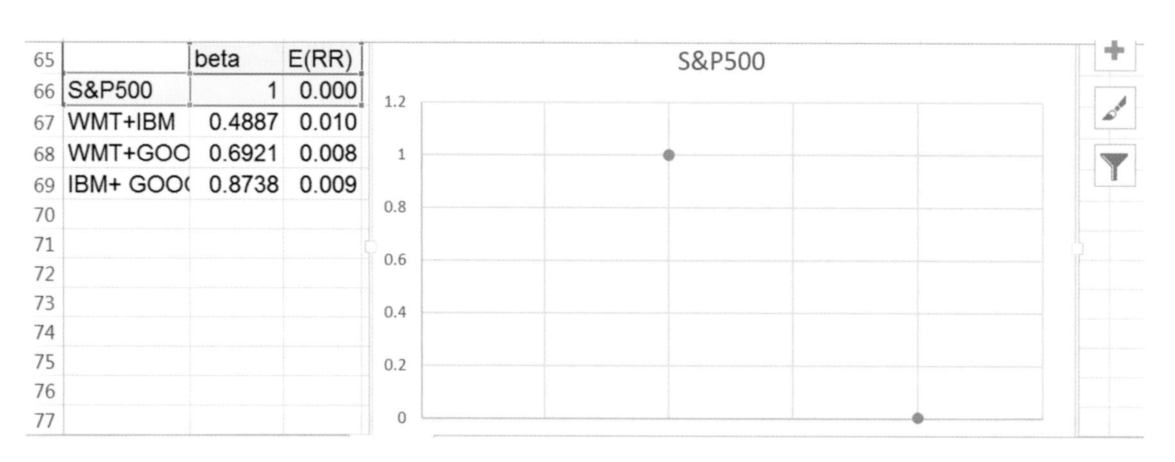

When we are plotting a single point, Excel will read both *b* and the *E(RR)* as series, plotting two points. To correct, edit the series,

Alt JC E Tab Tab E

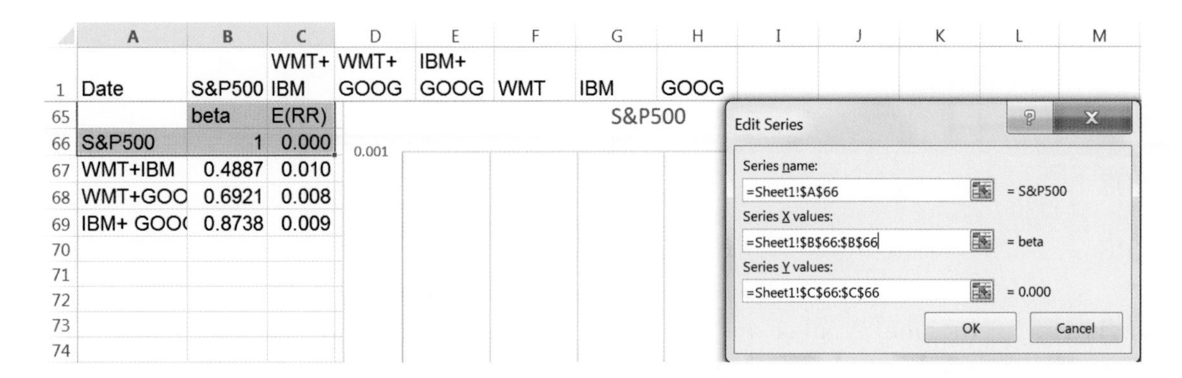

Add each of the three portfolio betas and expected returns as separate series,

Alt JC E Tab Tab A

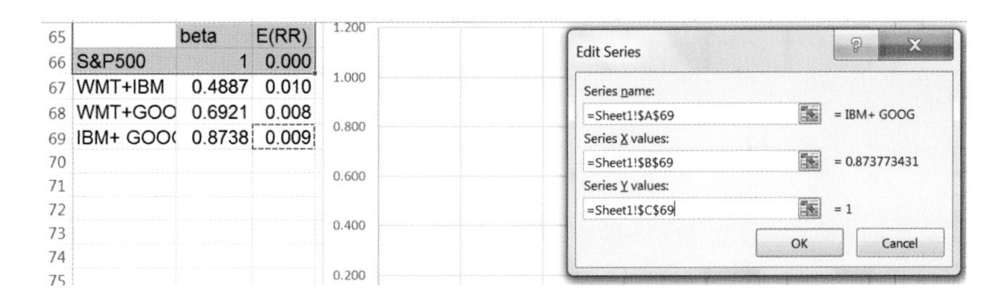

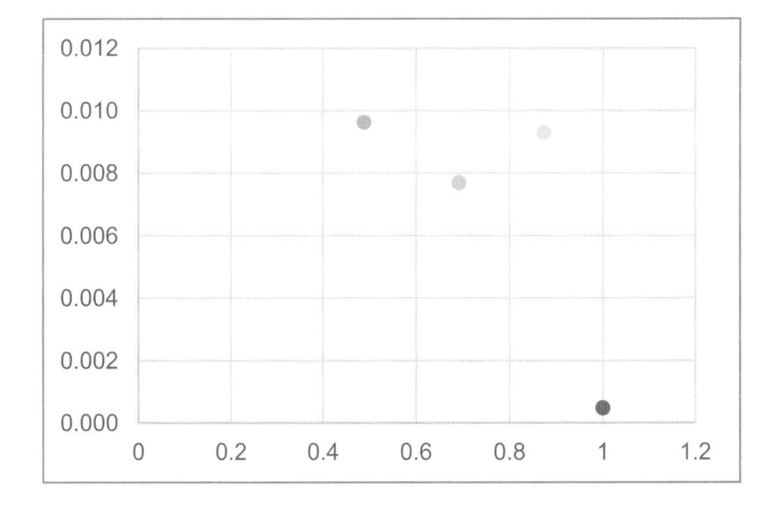

Add chart and axes titles and adjust axes:

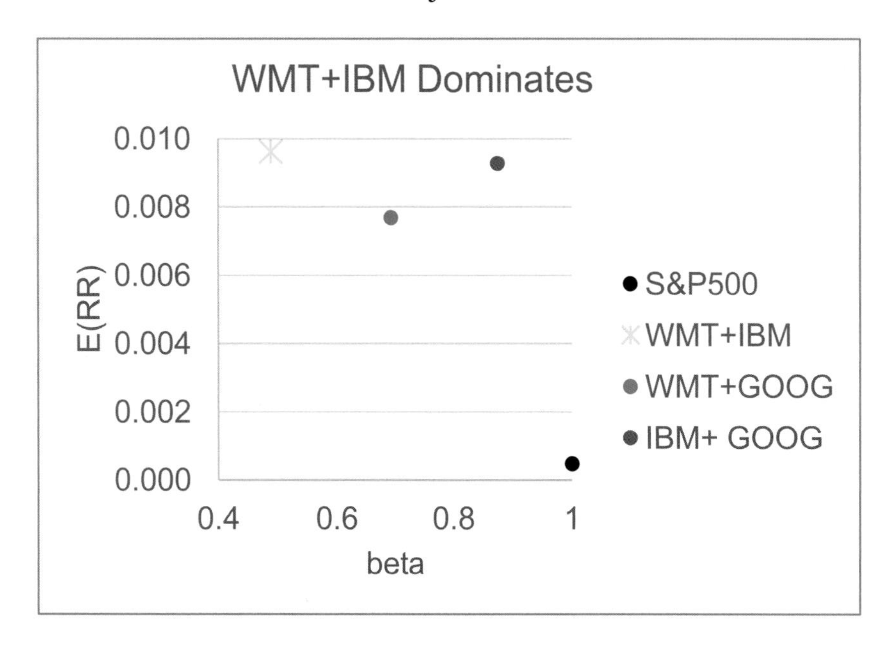

Lab 8 Portfolio Risk and Return

8 stocks 2012.xlsx contains a 5 year times series of monthly prices for 18 stocks and the S&P500 Market index for months July 2007 through July 2012. Stocks include ten "blue chips," as well as several others that have been in recent news.

Find the *rate of return* for each of the stocks using **Alt M U A**.

Find the *beta* for each of the stocks using **=SLOPE(***y array, x array***)**.

Which stocks performed worse than the Market, with lower expected rates of return and a higher expected *beta?*

____

Use logic to choose two stocks to combine in an equally weighted portfolio, and then add a column with portfolio rates of return.

The expected rate of return of my portfolio: ____________

Use regression to find the 95 % confidence interval for beta of your portfolio

The Market specific risk, beta, of my portfolio: _____ to _____

My portfolio ___dampens, ___mirrors, ___exaggerates
Market swings.

Create an alternative portfolio with two other stocks.

Find the expected rate of return with **Alt M U A** and expected *beta* with **=SLOPE(***y array, x array***)**.

Plot *E(RR)* by *b* for the S&P500 and your portfolios to see the *Efficient Frontier*, then compare the two portfolios:

Portfolio______ dominates Portfolio______ OR ___neither dominates the other

Chapter 9
Association Between Two Categorical Variables: Contingency Analysis with Chi Square

Categorical variables, including nominal and ordinal variables, are described by tabulating their frequencies or probability. If two categorical variables are associated, the frequencies of values of one will depend on the frequencies of values of the other. Chi square tests the hypothesized association between two categorical variables and contingency analysis quantifies their association.

9.1 When Conditional Probabilities Differ from Joint Probabilities, There Is Evidence of Association

Contingency analysis begins with the crosstabulation of frequencies of two categorical variables. Figure 9.1 shows a crosstabulation of sandwich spreads and topping combinations chosen by 40 students:

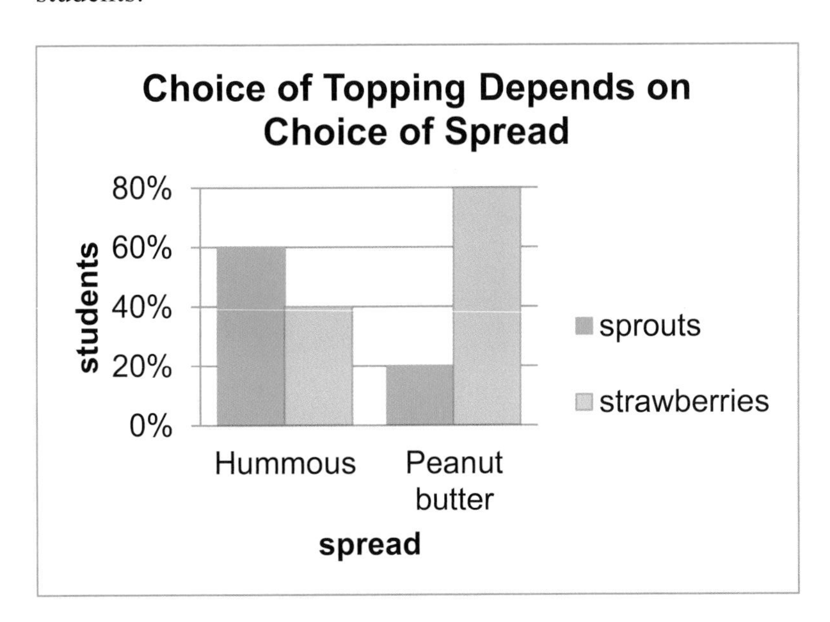

Counts	*Sprouts*	*Strawberries*	*Total*
Hummus	12	8	20
Peanut butter	4	16	20
Total	20	20	40

%Row	*Sprouts*	*Strawberries*	*Total*
Hummus	60	40	100
Peanut butter	20	80	100
Total	50	50	100

Fig. 9.1 Crosstabulation: sandwich topping depends on spread

C. Fraser, *Business Statistics for Competitive Advantage with Excel 2013: Basics, Model Building, Simulation and Cases*, DOI 10.1007/978-1-4614-7381-7_9, © Springer Science+Business Media New York 2013

If the unconditional probabilities of category levels, such as sprouts versus strawberries topping, differ from the probabilities, conditional on levels of another category, such as hummus or peanut butter spread, we have evidence of association. In this sandwich example, sprouts were chosen by half the students, making its unconditional probability .5. If a student chose hummus spread, the conditional probability of sprouts topping was higher (.60). If a student chose peanut butter spread, sprouts was the less likely topping choice (.40).

Example 9.1 Recruiting Stars

Human Resource managers are hoping to improve the odds of hiring outstanding performers and to reduce the odds of hiring poor performers by targeting recruiting efforts. Management believes that recruiting at the schools closer to firm headquarters may improve the odds of hiring stars. Students familiar with local customs may feel more confident at the firm. Removing schools far from headquarters may reduce the odds of hiring poor performers. Management's hypotheses are:

> H_0: Job performance is not associated with undergraduate program location.
> H_1: Job performance is associated with undergraduate program location.

To test these hypotheses, department supervisors throughout the firm sorted a sample of forty recent hires into three categories based on job performance: poor, average, and outstanding. The sample employees were also categorized by the proximity to headquarters: Home State, Same Region, and Outside Region. These cross-tabulations are shown in the PivotChart and PivotTable in Fig. 9.2.

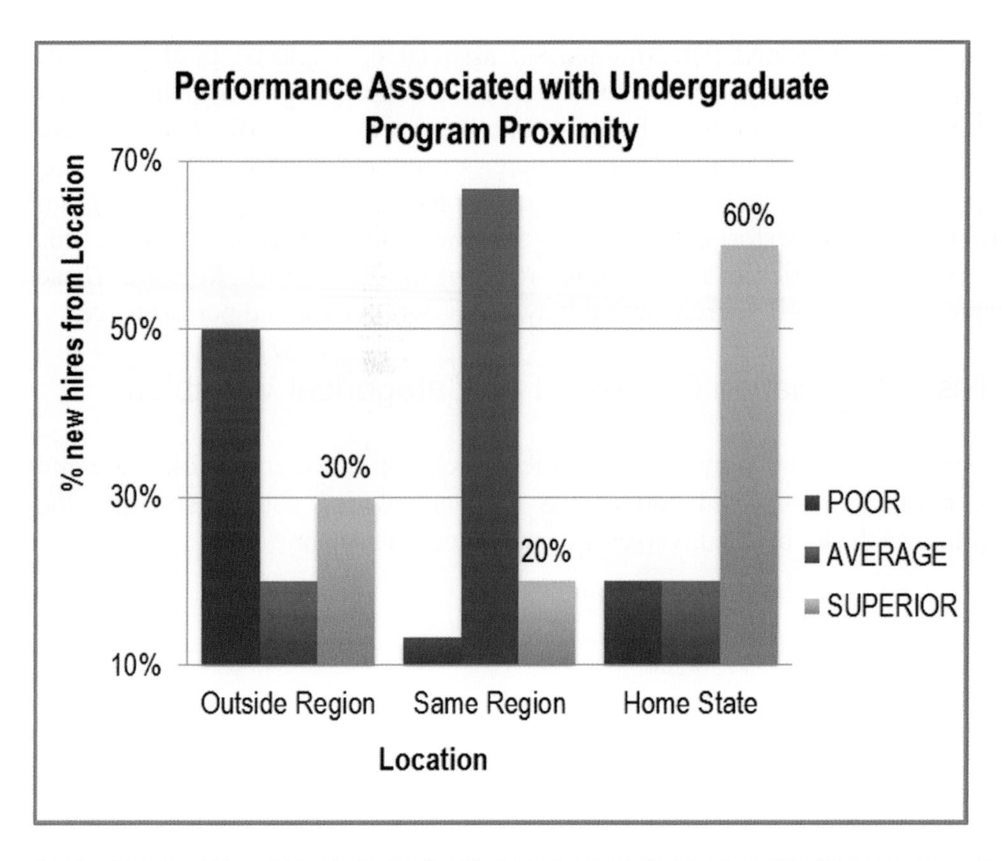

Count	*Performance*			
Location	*Poor*	*Average*	*Outstanding*	*Total*
Outside region	5	2	3	10
Same region	2	10	3	15
Home state	3	3	9	15
Total	10	15	15	40

% of Row	*Performance*			
Location	*Poor (%)*	*Average (%)*	*Outstanding (%)*	*Total (%)*
Outside region	50	20	30	100
Same region	13	67	20	100
Home state	20	20	60	100
Total	25	38	38	100

χ_4^2	12.5	*p value*	.02

Fig. 9.2 Job performance depends on program location

The crosstabs indicate that a quarter of the firm's new employees are *Poor* performers, about 40 % are *Average* performers, and about 40 % are *Outstanding* performers. From the PivotChart we see that more than a quarter of employees from programs *Outside Region* are *Poor* performers, and more than 40 % of employees from *Home State* programs are *Outstanding* performers. Were program location and performance *not* associated, a quarter of the recruits from each location would be *Poor* performers. We would, for example, expect a quarter of ten employees recruited from *Outside Region* to be Poor performers, or 2.5 (=.25(10)). Instead, there are actually five (*Outside Region, Poor*) employees. There is a greater chance, 50 %, of *Poor* performance, given *Outside Region*, relative to *Same Region* or *Home State*. Ignoring program location, the probability of poor performance is .25; acknowledging program location, this probability of poor performance varies from .13 (*Same Region*) to .50 (*Outside Region*). These differences in row percentages suggest an association between program rank and performance.

9.2 Chi Square Tests Association Between Two Categorical Variables

The chi square (χ^2) statistic tests the significance of the association between performance and program location, by comparing expected cell counts with actual cell counts, squaring the differences, and weighting each cell by the inverse of expected cell frequency.

$$\chi^2_{(R-1),(C-1)} = \sum_{ij}^{RC} (e_{ij} - n_{ij})^2 / e_{ij},$$

Where R is the number of row categories,
C is the number of column categories,
n is the number in the i'th row and j'th column,
e is the number expected in the i'th row and j'th column.

χ^2 gives more weight to the least likely cells. χ^2 distributions are skewed and with means equal to the number of degrees of freedom. Several χ^2 distributions with a range of degrees of freedom are shown in Fig. 9.3.

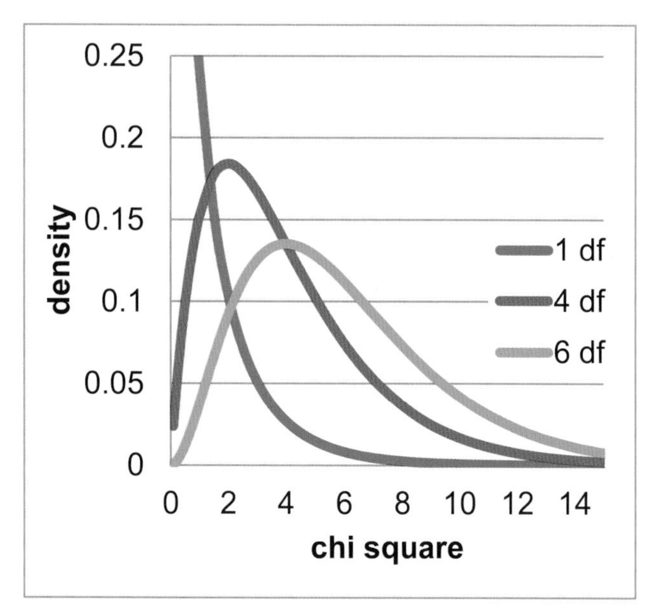

Fig. 9.3 Chi square distributions for a range of degrees of freedom

In the **Recruiting Stars** example, **Fig. 9.2**, chi square, χ_4^2, is 12.5, which can be verified using the formula:

$$\begin{aligned}\chi^2 = {} & (2.5 - 5)^2/2.5 + (3.8 - 2)^2/3.8 + (3.8 - 3)^2/3.8 \\ & + (3.8 - 2)^2/3.8 + (5.6 - 10)^2/5.6 + (5.6 - 3)^2/5.6 \\ & + (3.8 - 3)^2/3.8 + (5.6 - 3)^2/5.6 + (5.6 - 9)^2/5.6 \\ = {} & 2.5 + .9 + .2 \\ & + .9 + 3.5 + 1.2 \\ & + .2 + 1.2 + 2.0 = 12.5\end{aligned}$$

From a table of χ_4^2 distributions, we find that for a crosstabulation of this size, with three rows and three columns, $(df = (Rows-1) \times (Columns-1) = 2 \times 2 = 4)$, $\chi_4^2 = 12.5$ indicates that the p-value is .02. Two percent of the distribution lies right of 12.5. There is little chance that of observing the sample data were performance and program tier not associated. The null hypothesis of lack of association is rejected.

Those cells which contribute more to chi square indicate the nature of association. In this example, we see in Table 9.1 that these are the (*Outside Region*, *Poor*), (*Same Region*, *Average*), and (*Home State*, *Outstanding*) cells:

$$\begin{aligned}\chi^2 = {} & \mathbf{2.5} + .9 + .2 \\ & + .9 + \mathbf{3.5} + 1.2 \\ & + .2 + 1.2 + \mathbf{2.0} = 12.5\end{aligned}$$

Table 9.1 Contribution to chi square by cell

	Poor	*Average*	*Outstanding*
Outside region	**2.5**	.9	.2
Same region	.9	**3.5**	1.2
Home state	.2	1.2	**2.0**

Poor performance is more likely if a new employee came from a program *Outside Region*, *Average* performance is more likely if a new employee came from a program in the *Same Region*, and *Outstanding* performance is more likely if a new employee came from a *Home State* program. Job performance is associated with program location.

9.3 Chi Square Is Unreliable if Cell Counts Are Sparse

There are two possible reasons why the chi square statistic is large and apparently significant. The first reason is the likely actual association between program location and performance. The second reason is that there are few (fewer than five) expected employees in five of the nine cells, shown in Table 9.2.

Table 9.2 Expected counts by cell

	Poor	*Average*	*Outstanding*
Outside region	2.5	3.8	3.8
Same region	3.8	5.6	5.6
Home state	3.8	5.6	5.6

Since the chi square components include expected cell counts in the denominator, *sparse* (with expected counts less than five) cells inflate chi square. When sparse cells exist, we must either combine categories or collect more data.

In the **Recruiting Stars** example, management was most interested in increasing the chances of hiring *Outstanding* performers. Since some believed that *Outstanding* performers were recruited from programs in the *Home State*, these categories were preserved. *Same Region* and *Outside Region* program locations were combined. *Poor* and *Average* performance categories were combined. We are left with a 2 × 2 contingency analysis, Fig. 9.4.

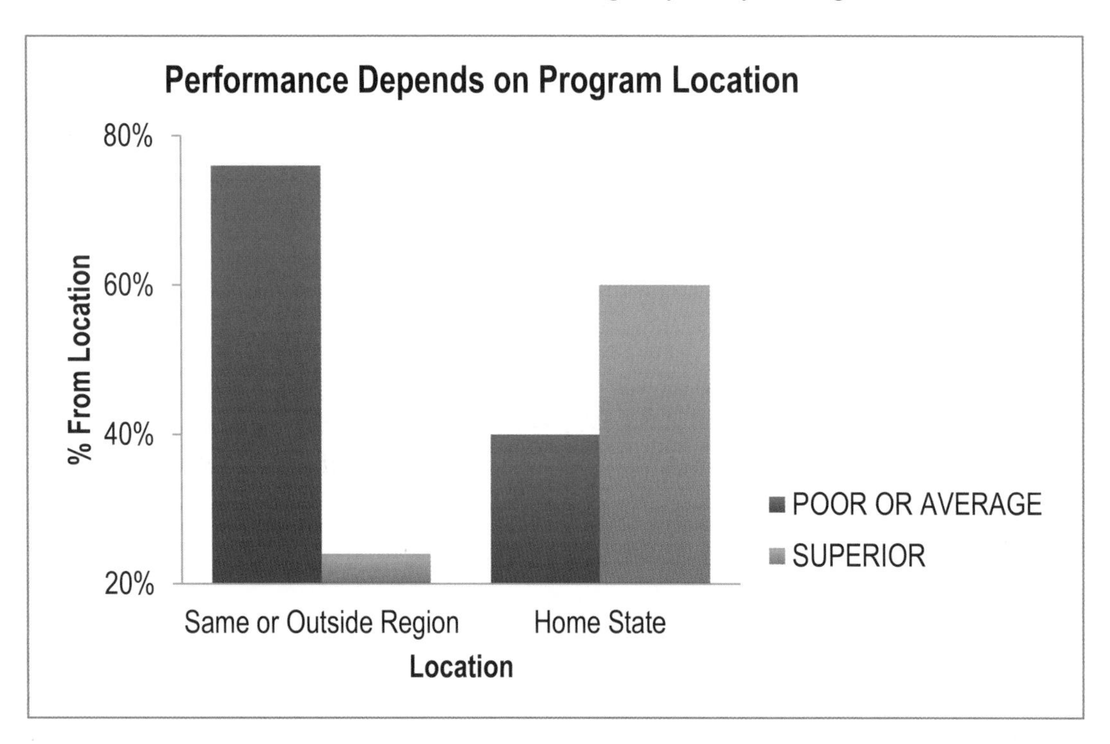

Count	*Performance*			*% Row*	*Performance*		
Location	*Poor/ average*	*Outstanding*	*Total*	*Location*	*Poor/ average (%)*	*Outstanding (%)*	*Total (%)*
Same or outside region	19	6	25	*Same or outside region*	76	24	100
Home state	6	9	15	*Home state*	40	60	100
Total	25	15	40	*Total*	63	38	100

Chi square	5.2
df	1
p value	.02

Fig. 9.4 PivotChart of performance by program location with fewer categories

With fewer categories, all expected cell counts are now greater than five, providing a reliable $\chi_1^2 = 5.2$, which remains significant (*p value* = .02). The PivotChart continues to suggest that the incidence of *Outstanding* performance is greater among employees recruited from *Home State* programs. The impact of program location on *Poor* performance is unknown, since *Poor* and *Average* categories were combined. Also unknown is the difference between employees from *Same* and *Outside Regions* programs, since these categories were likewise combined.

Recruiters would conclude:

"Job performance of newly hired employees is associated with undergraduate program location. Twenty-four percent of our new employees recruited from Same or Outside Region undergraduate programs have been identified as Outstanding performers. Within the group recruited from Home State undergraduate programs, more than twice this percentage, 60%, are Outstanding performers, a significant difference. Results suggest that in order to achieve a larger percent of Outstanding performers, recruiting should be focused on Home State programs."

9.4 Simpson's Paradox Can Mislead

Using contingency analysis to study the association between two variables can be potentially misleading, since all other related variables are ignored. If a third variable is related to the two being analyzed, contingency analysis may indicate that they are associated, when they may not actually be. Two variables may appear to be associated because they are both related to a third, ignored variable.

Example 9.2 American Cars

The CEO of American Car Company was concerned that the oldest segments of car buyers were avoiding cars that his firm assembles in Mexico. Production and labor costs are much cheaper in Mexico, and his long term plan is to shift production of all models to Mexico. If older, more educated and more experienced buyers avoid cars produced in Mexico, American Car could lose a major market segment unless production remained in The States.

The CEO's hypotheses were:

H_0: Choice between cars assembled in the U.S. and cars assembled in Mexico is not associated with age category.

H_1: Choice between cars assembled in the U.S. and cars assembled in Mexico is associated with age category.

He asked Travis Henderson, Director of Quantitative Analysis, to analyze the association between age category and choice of U.S. made versus Mexican made cars. The research staff drew a random sample of 263 recent car buyers, identified by age category. After preliminary analysis, age categories were combined to insure that all expected cell counts in an [Age Category × Origin Choice] crosstabulation were each at least five. Contingency analysis is shown in the PivotChart and Pivot Tables in Fig. 9.5.

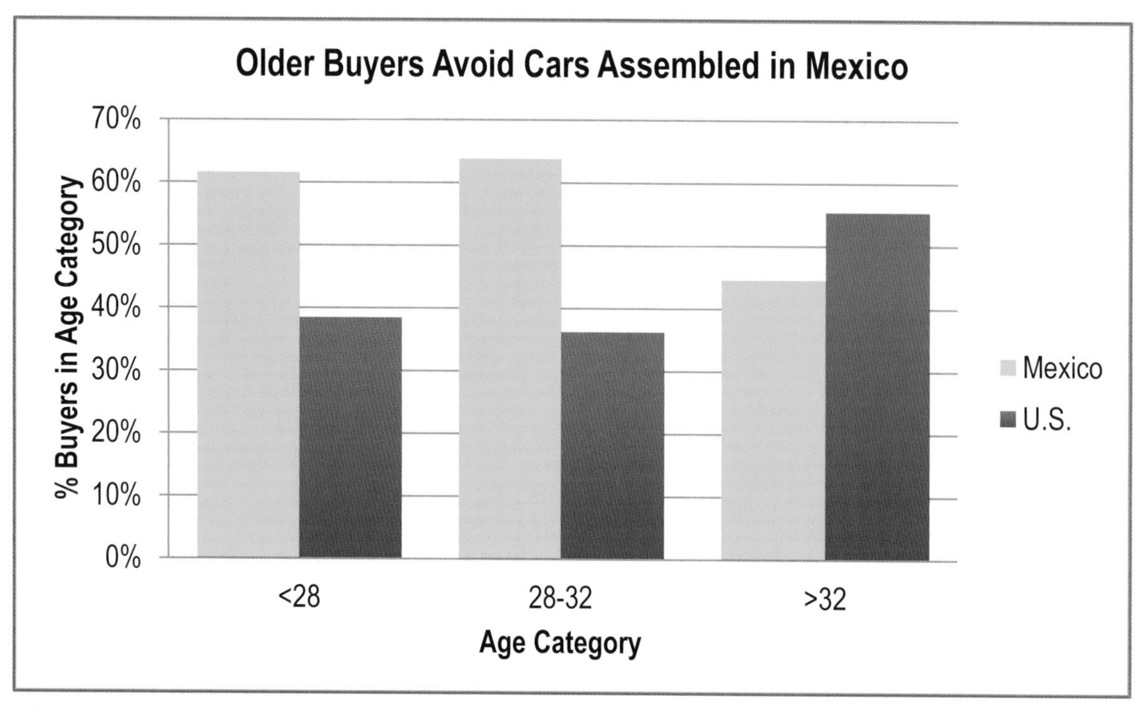

Count	*Assembled in*				*% Rows*		
Age	*Mexico*	*U.S.*	*Total*	*Age*	*Mexico (%)*	*U.S. (%)*	*Total (%)*
Under 28	56	35	91	*Under 28*	62	38	100
28 to 32	51	29	80	*28 to 32*	64	36	100
33 Plus	41	51	92	*33 Plus*	45	55	100
Total	148	115	263	*Total*	56	44	100

Chi square	8.0
df	2
p value	.02

Fig. 9.5 Contingency analysis of U.S. versus Mexican made car choices by age

A glimpse of the PivotChart confirmed suspicions that older buyers did seem to be rejecting cars assembled in Mexico. The *p value* for chi square was .02, indicating that the null hypothesis, lack of association, ought to be rejected. Choice between U.S. and Mexican made cars seemed to be associated with age category. Fifty six percent of the entire sample across all ages chose cars assembled in Mexico. Within the oldest segment, however, the Mexican assembled car share was lower: 45 %. While nearly two thirds of the younger segments chose cars assembled in Mexico, less than half of the oldest buyers chose Mexican made cars.

The CEO was alarmed with these results. His company could lose the business of older, more experienced buyers if production were shifted South of the Border. Brand managers were about to begin planning "Made in the U.S.A." promotional campaigns targeted at the oldest car buyers. Emily Ernst, the Director of Strategy and Planning, suggested that age was probably not the correct basis for segmentation. She explained that the older buyers shop for a particular *type* of car – a family sedan or station wagon – and few family sedans or wagons were being assembled in Mexico. Models assembled at home in the U.S. tended to be large sedans and station wagons – styles sought by older buyers. She proposed that it was *style* that influenced the U.S. versus Mexican assembled choice, and not age, and that it was *style* that was dependent on age. Her hypotheses were:

H_0: Choice of car style is not associated with age category.
H_1: Choice of car style is associated with age category.

To explore this alternate hypothesis, the research team ran contingency analysis of style choice (SUV, Sedan/Wagon and Coupe) by age category, Fig. 9.6.

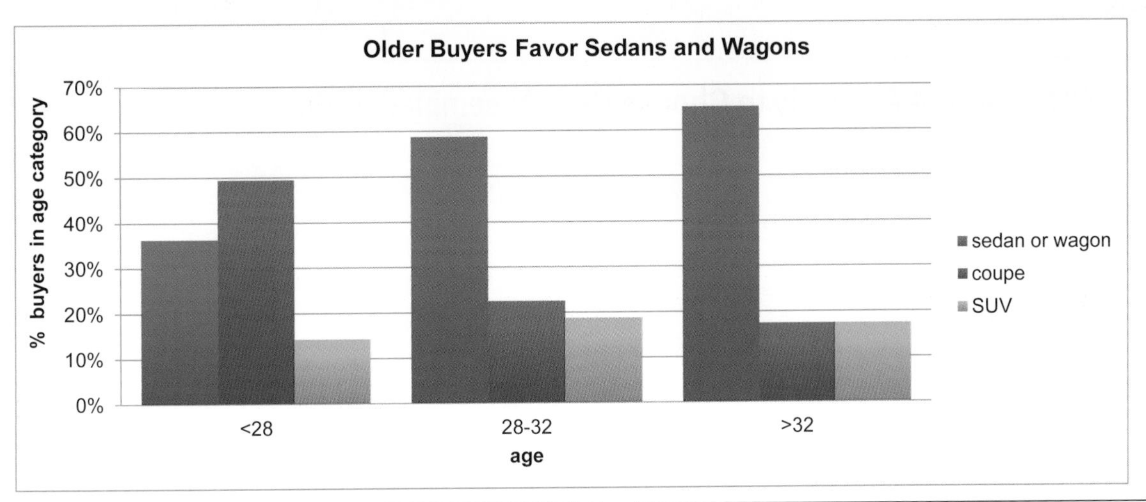

Count	*Style*			
Age	*Sedan/ wagon*	*Coupe*	*SUV*	*Total*
<28	33	45	13	91
28–32	47	18	15	80
33+	60	16	16	92
Total	140	79	44	263

Row%	*Style*			
Age	*Sedan/ wagon (%)*	*Coupe (%)*	*SUV (%)*	*Total (%)*
< 28	36	49	14	100
28–32	59	23	19	100
33+	65	17	17	100
Total	53	30	17	100

χ^2_4	**26.2**	***p value***	**.0000**

Fig. 9.6 Contingency analysis of car style choice by age category

Contingency analysis of this sample indicates that choice of style is associated with age category. More than half (53 %) of the car buyers chose a sedan or wagon, though only about a third (36 %) of the younger buyers chose a sedan or wagon, and nearly twice as many (65 %) older buyers chose a sedan or wagon. Thirty percent of the sample bought a coupe, and just nearly half (49 %) of the younger buyers chose a coupe. Only 17 % of the oldest buyers bought a coupe. These are significant differences supporting the conclusion that style of car chosen is associated with age category.

This is the news that the CEO was looking for. If older car buyers are choosing U.S.-made cars because they desire family styles, sedans and wagons, which tend to be assembled in the U.S., then perhaps these older buyers aren't shunning Mexican-made cars. His hypotheses were:

> H_0: Given choice of a sedan or wagon, choice of U.S. versus Mexican assembled is not associated with age category.
> H_1: Given choice of a sedan or wagon, choice of U.S. versus Mexican assembled is associated with age category.

To test these hypotheses, the analysis team conducted three contingency analyses of origin choice (U.S. vs. Mexican assembled) by age category, looking at each style separately in Fig. 9.7 and Table 9.3.

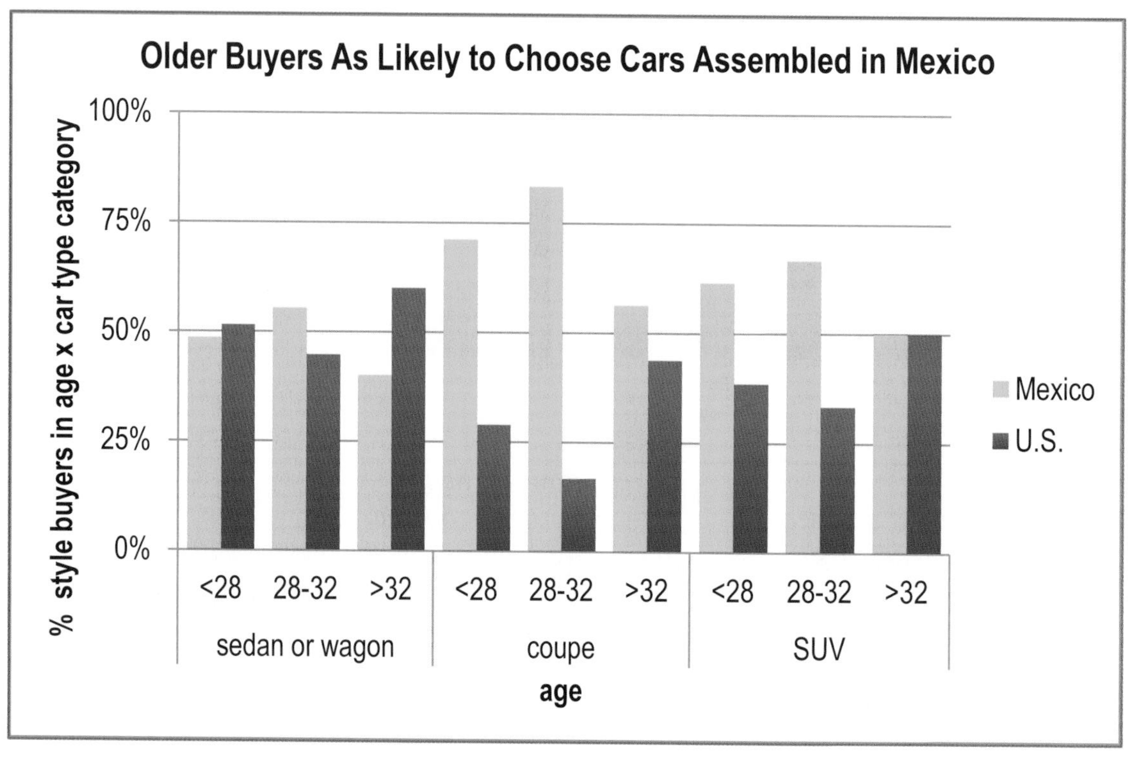

Fig. 9.7 Contingency analysis: origin choice given age by style

Table 9.3 Contingency analysis: origin choice by age given style

		% Age by style assembled in					
					χ^2	*df*	*p value*
Style	*Age*	*Mexico (%)*	*U.S. (%)*	*Total (%)*			
Sedan or wagon	*Under 28*	48	52	100			
	28–32	55	45	100			
	33 plus	40	60	100			
Total		47	53	100	2.5	2	.29
Coupe	*Under 28*	71	29	100			
	28–32	83	17	100			
	33 plus	56	44	100			
Total		71	29	100	3.0	2	.22
SUV	*Under 28*	62	38	100			
	28–32	67	33	100			
	33 plus	50	50	100			
Total		59	41	100	.9	2	.63
Grand total		56	44	100			

Controlling for style of car by looking at each style separately reveals lack of association between origin preference for U.S. versus Mexican made cars and age category. Across all three car styles, *p values* are greater than .05. There is not sufficient evidence in this sample to reject the null hypothesis. We conclude from this sample that the U.S. versus Mexican assembled choice is not associated with age category. The domestic automobile manufacturer should therefore not alter plans to move production South.

Simpson's Paradox describes the situation where two variables appear to be associated only because of their mutual association with a third variable. If the third variable is ignored, results are misleading. Because contingency analysis focuses upon just two variables at a time, analysts should be aware that apparent associations may come from confounding variables, as the **American Cars** example illustrates.

The Research Team summarized these results in this memo:

MEMO

Re.: **Country of Assembly Does Not Affect Older Buyers' Choices**
To: CEO, American Car Company
Emily Ernst, Director of Planning and Strategy
Brand Management
From: Travis Hendershott, Director of Quantitative Analysis

Analysis of a sample of new car buyers reveals that styles of car drive the choices of distinct age segments. Choices of all ages of buyers are independent of country of manufacture.

Contingency Analysis. Choices of 263 new car buyers were analyzed to assess the dependence of choice on country of manufacture, U.S. or Mexico, and age category.

Choice between Mexican and U.S. cars does not depend on buyer age.

Car choices between Mexican and U.S. cars do not depend on buyer age, though choices between *styles* are age dependent.

Younger buyers are more likely to choose a sporty coupe. Older buyers are more likely to buy a sedan or wagon.

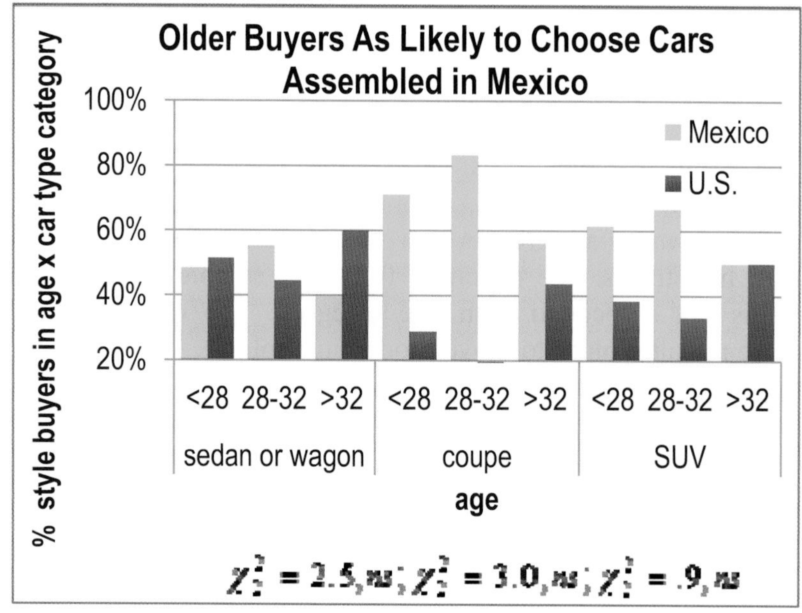

Production in Mexico will not influence car choices

Production in Mexico is not expected to affect car buyer choices, providing the opportunity to shift assembly South to take advantage of cheaper labor.

A larger, more diverse sample may provide additional information

A larger sample would enable examination of more representative age categories, and specifically, a broader middle segment and older oldest segment.

9.5 Contingency Analysis Is Demanding

Contingency analysis requires a large and balanced dataset to insure a stable chi square. Even large samples may contain small proportions of particular categories, forcing combinations that aren't ideal. In the **American Cars** example, a broad category was used for the oldest age segment, combining fairly different ages, 33 through 60, and a narrow category was defined for the middle age segment, ages 28 through 32. The sample, though large, was not balanced and contained a large proportion of car buyers ages 30 through 39. This group was split and combined with sparse younger and older age categories to allow expected cell counts greater than five. With smaller samples, just two categories for a variable may remain, which may limit hypothesis testing. In the **Recruiting Stars** example, final results could not be used to assess the association between recruiting and poor employee performance after Poor and Average performing employees were combined.

9.6 Contingency Analysis Is Quick, Easy, and Readily Understood

Despite the fairly demanding data requirements, contingency analysis is appealing because it is simple, and results are easily understood. For very large samples, sparse cells are not a problem and many categories may be used, increasing the specificity of results and allowing a range of hypothesis tests.

Excel 9.1 Construct Crosstabulations and Assess Association Between Categorical Variables with PivotTables and PivotCharts

American Cars. In order to explore the possible association between choice of U.S.-assembled and Mexican-assembled cars by age, begin by making a PivotTable to see the crosstabulation.

Open **Excel 9.1 American Cars.xls.**

Select filled cells in the *Age* and *Made In* columns and then insert a PivotTable.

Drag *Age* to **ROW**, *Made In* to **COLUMN**, and *Made In* to $\sum$ **Values**.

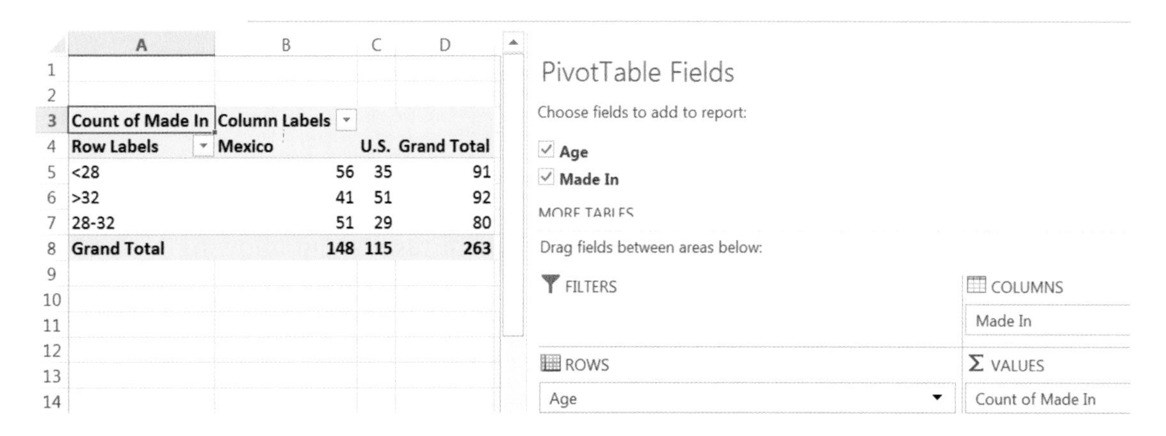

To see the conditional probabilities of choice of cars *Made In* the U.S. and Mexico given *age* category, convert cell counts to **% of row**. From a data cell in the table:

Alt JT G Tab > Tab down to **% of Row Total, Enter**

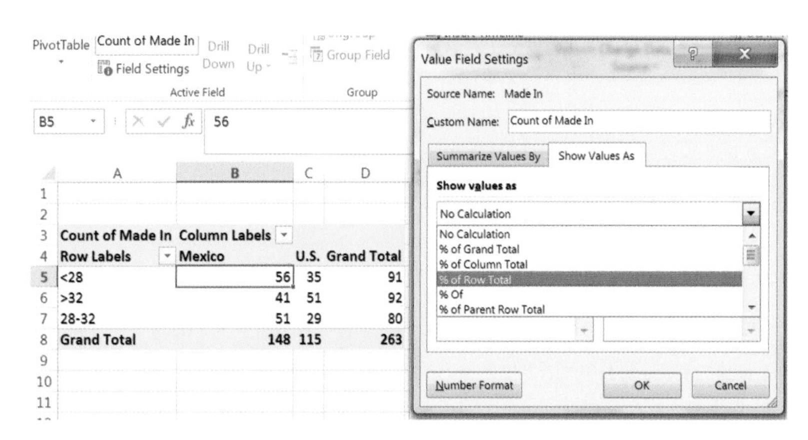

To put the age categories in order, select and right click the *>32* cell, **M**ove to **E**nd.

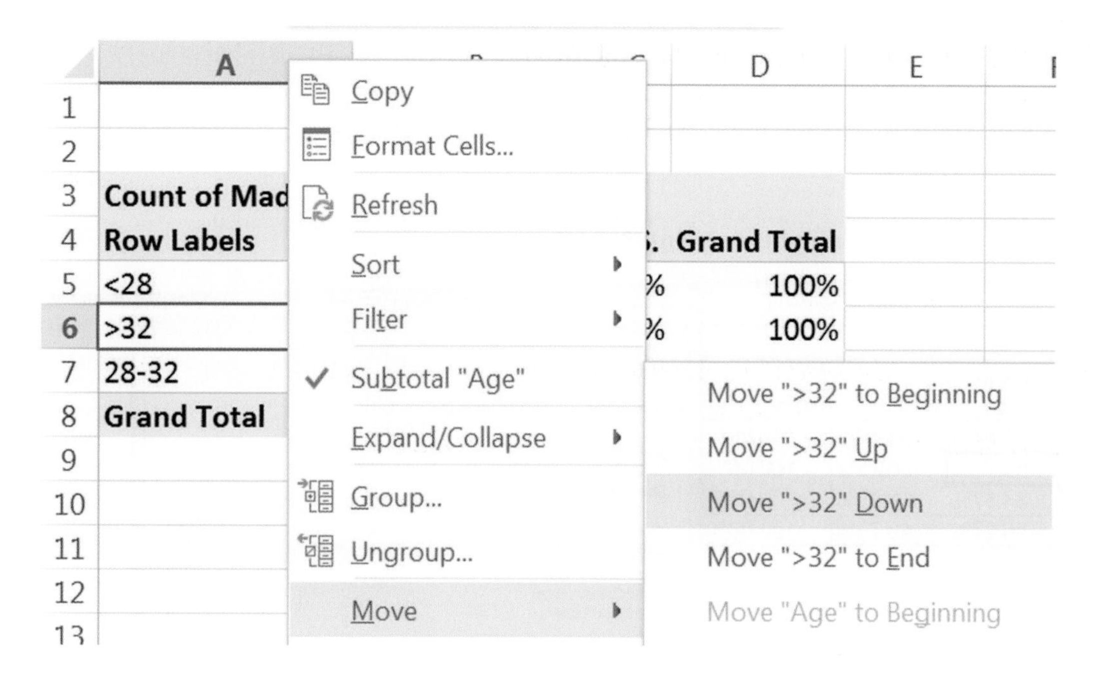

Make a PivotChart of *Made In* by *Age:*

Alt JT C

Choose a design and style, and add a chart title that reflects your conclusion.

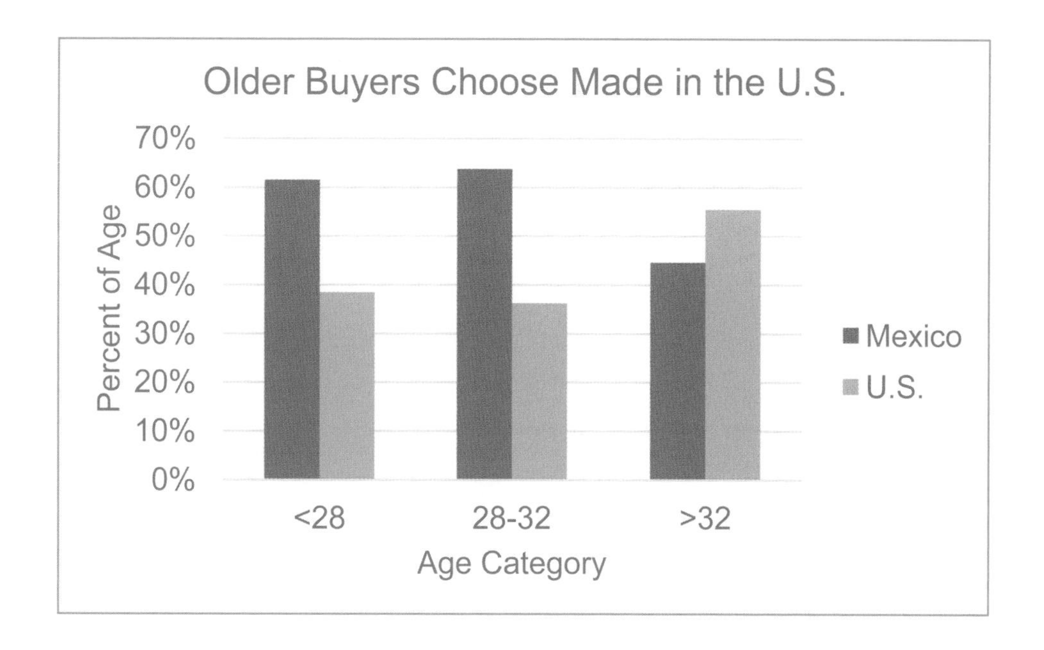

Excel 9.2 Use Chi Square to Test Association

To find the chi square statistic, change the PivotTable cells back to counts.
Select a cell in the PivotTable,

Alt JT G Tab > Tab up to **No Calculation, Enter**

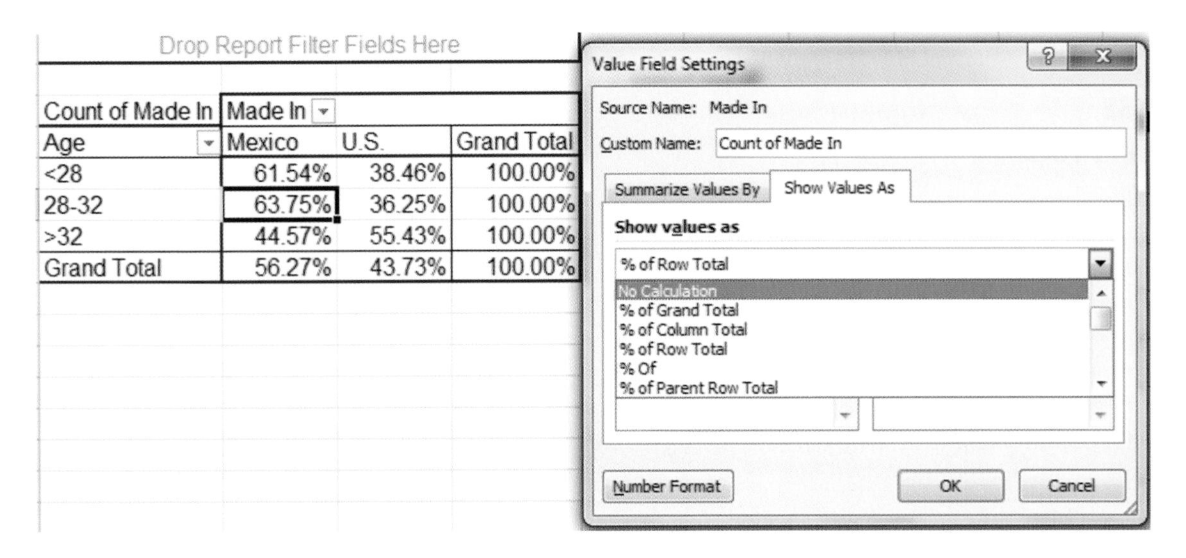

Count of Made In	Made In		
Age	Mexico	U.S.	Grand Total
<28	61.54%	38.46%	100.00%
28-32	63.75%	36.25%	100.00%
>32	44.57%	55.43%	100.00%
Grand Total	56.27%	43.73%	100.00%

For chi square, make a table of *expected* cell counts and a table of cell contributions to chi square.

Select the two empty rows above the PivotTable, plus the PivotTable, copy, and paste right of the PivotTable with values and formats, but not formulas,

Alt H V S U

Repeat to paste in a second copy.

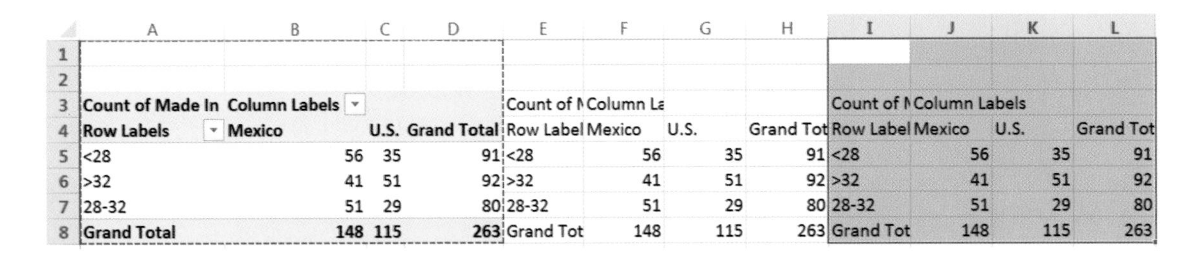

	A	B	C	D	E	F	G	H	I	J	K	L
1												
2												
3	Count of Made In	Column Labels			Count of M	Column La			Count of M	Column Labels		
4	Row Labels	Mexico	U.S.	Grand Total	Row Label	Mexico	U.S.	Grand Tot	Row Label	Mexico	U.S.	Grand Tot
5	<28	56	35	91	<28	56	35	91	<28	56	35	91
6	>32	41	51	92	>32	41	51	92	>32	41	51	92
7	28-32	51	29	80	28-32	51	29	80	28-32	51	29	80
8	Grand Total	148	115	263	Grand Tot	148	115	263	Grand Tot	148	115	263

In the first cell of the second, *expected* table, enter the formula for the expected count, multiplying cells containing the Grand Total of youngest buyers and the Grand Total of cars assembled in Mexico, and then dividing by the Grand Total:

=D5 fn4 fn4 fn4 * B8 fn4 fn4/D8 fn4.

Pressing **fn 4** three times locks the column, pressing **fn 4** twice locks the row, and pressing **fn 4** once locks both, so that we can fill in the remaining cells in the table with this formula.

Fill in the column, and then fill in the adjacent row.

F5 =$D5*B$8/$D$8

	A	B	C	D	E	F	G	H
1								
2								
3	**Count of Made In**	**Column Labels**			Count of M	Column La		
4	**Row Labels**	**Mexico**	**U.S.**	**Grand Total**	Row Label	Mexico	U.S.	Grand Tot
5	<28	56	35	91	<28	51.20913	39.79087	91
6	>32	41	51	92	>32	51.77186	40.22814	92
7	28-32	51	29	80	28-32	45.01901	34.98099	80
8	**Grand Total**	**148**	**115**	**263**	Grand Tot	148	115	263

In the third table, find each cell's contribution to chi square, the squared difference between expected counts, in the second table, and actual counts, in the first table, divided by expected counts in the second table.

In the first cell of the third table, contributions to chi square, **e**nter:

=(F5−B5)^2/F5.

Fill in the column and the rows:

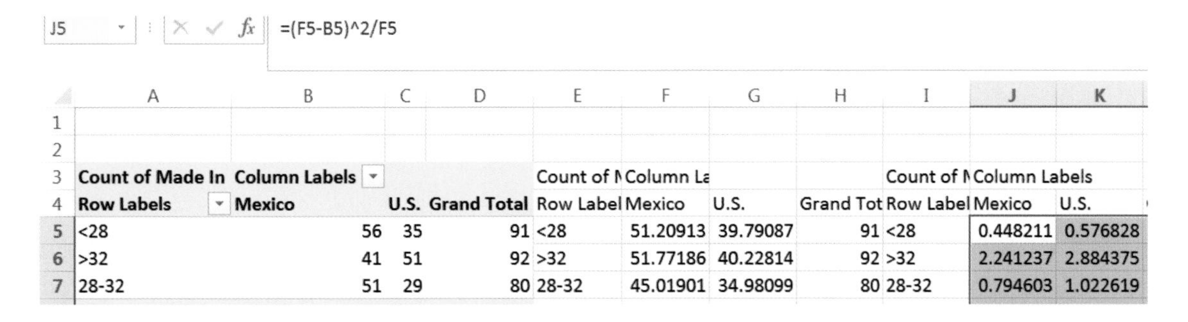

J5 =(F5-B5)^2/F5

	A	B	C	D	E	F	G	H	I	J	K
1											
2											
3	**Count of Made In**	**Column Labels**			Count of M	Column La			Count of M	Column Labels	
4	**Row Labels**	**Mexico**	**U.S.**	**Grand Total**	Row Label	Mexico	U.S.	Grand Tot	Row Label	Mexico	U.S.
5	<28	56	35	91	<28	51.20913	39.79087	91	<28	0.448211	0.576828
6	>32	41	51	92	>32	51.77186	40.22814	92	>32	2.241237	2.884375
7	28-32	51	29	80	28-32	45.01901	34.98099	80	28-32	0.794603	1.022619

Use the Excel function **SUM(**_array1,array2_) to add the cell contributions in the three rows:

In the Grand Total row, find the Mexico sum:

Alt M U S

Fill in the row to find chi square in the last cell:

	I	J	K	L	M
3	Count of N	Column Labels			
4	Row Label	Mexico	U.S.	Grand Total	
5	<28	0.448211	0.576828	1.025038	
6	>32	2.241237	2.884375	5.125612	
7	28-32	0.794603	1.022619	1.817222	
8	Grand Tot	3.484051	4.483822	7.967873	chi square

Find the *p value* for this chi square using the Excel function

CHISQ.DIST.RT(*chisquare,df***)** with degrees of freedom *df* of 2

L9 =CHISQ.DIST.RT(L8,2)

	I	J	K	L	M
3	Count of N	Column Labels			
4	Row Label	Mexico	U.S.	Grand Total	
5	<28	0.448211	0.576828	1.025038	
6	>32	2.241237	2.884375	5.125612	
7	28-32	0.794603	1.022619	1.817222	
8	Grand Tot	3.484051	4.483822	7.967873	chi square
9				0.018612	p value

Excel 9.3 Conduct Contingency Analysis with Summary Data

Sometimes data are in summary form. That is, we know the sample size, and we know the percent of the sample in each category.

Marketing Cereal to Children. Kooldogg expects that many Saturday morning cartoon viewers would be attracted to their sugared cereals. A heavy advertising budget for sugared cereals is allocated to Saturday morning television. We will use contingency analysis to analyze the association between Saturday morning cartoon viewing and frequent consumption of Kooldogg cereal with sugar added. From a survey of 300 households, researchers know whether or not children ages 2 through 5 *Watch Saturday Morning Cartoons* on a regular basis (at least twice a month) and whether or not those children *Eat Kooldogg Cereal with Added Sugar* (at least once a week).

Open **Excel 9 Kooldogg Kids Ads.xls**.

Select the summary data and make a PivotTable, with *Watches Saturday Morning Cartoons* in **Rows**, *Eats Kooldogg Sugary Cereal* in **Columns**, and *Number of Children* in **Σ Values**:

	A	B	C	D
1				
2				
3	Sum of Number of children	Column Labels		
4	Row Labels	0	1	Grand Total
5	0	36	4	40
6	1	4	256	260
7	Grand Total	40	260	300
8				
9				
10				
11				
12				
13				
14				

PivotTable Fields

Choose fields to add to report:

- ☑ Watches Saturday morning cartoons
- ☑ Eats Kooldog Sugary Cereal
- ☑ Number of children

Drag fields between areas below:

FILTERS

COLUMNS: Eats Kooldog Sugary Cereal

ROWS: Watches Saturday morning cartoons

Σ VALUES: Sum of Number of children

Copy rows **1** and **2** with the table and paste twice with formats and values, **Alt H V S U**.

	A	B	C	D	E	F	G	H	I	J	K	L
2												
3	Sum of Number of children	Column Labels			Sum of Number of children	Column Labels			Sum of Number of children	Column Labels		
4	Row Labels	0	1	Grand Total	Row Labels	0	1	Grand Total	Row Labels	0	1	Grand Total
5	0	36	4	40	0	36	4	40	0	36	4	40
6	1	4	256	260	1	4	256	260	1	4	256	260
7	Grand Total	40	260	300	Grand Tot	40	260	300	Grand Tot	40	260	300

In the second table, find the expected cell counts under the assumption that Kooldog cereal consumption is independent of Saturday morning TV viewing.

F5 =$D5*B$7/$D$7

	A	B	C	D	E	F	G
3	Sum of Number of children	Column Labels			Sum of Number of children	Column Labels	
4	Row Labels	0	1	Grand Total	Row Labels	0	1
5	0	36	4	40	0	5.3333	34.7
6	1	4	256	260	1	34.667	225
7	Grand Total	40	260	300	Grand Tot	40	260

In the third table, find cell contributions to chi square with squared differences between expected cell counts and actual cell counts, divided by expected cell counts.

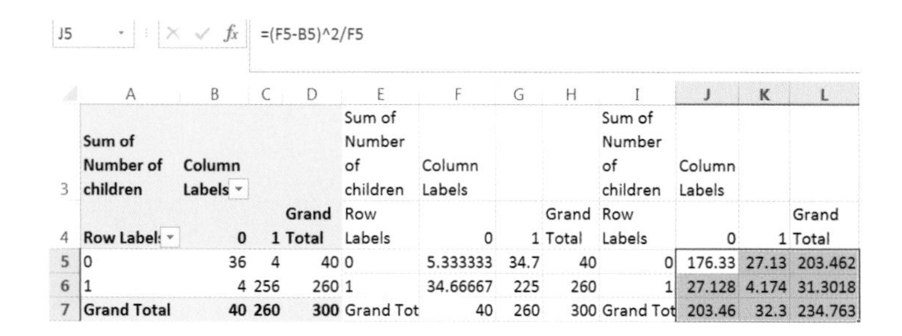

J5 | =(F5-B5)^2/F5

	A	B	C	D	E	F	G	H	I	J	K	L
3	Sum of Number of children	Column Labels			Sum of Number of children	Column Labels			Sum of Number of children	Column Labels		
4	Row Labels	0	1	Grand Total	Row Labels	0	1	Grand Total	Row Labels	0	1	Grand Total
5	0	36	4	40	0	5.333333	34.7	40	0	176.33	27.13	203.462
6	1	4	256	260	1	34.66667	225	260	1	27.128	4.174	31.3018
7	Grand Total	40	260	300	Grand Tot	40	260	300	Grand Tot	203.46	32.3	234.763

Sum the cell contributions to chi square in to find chi square.

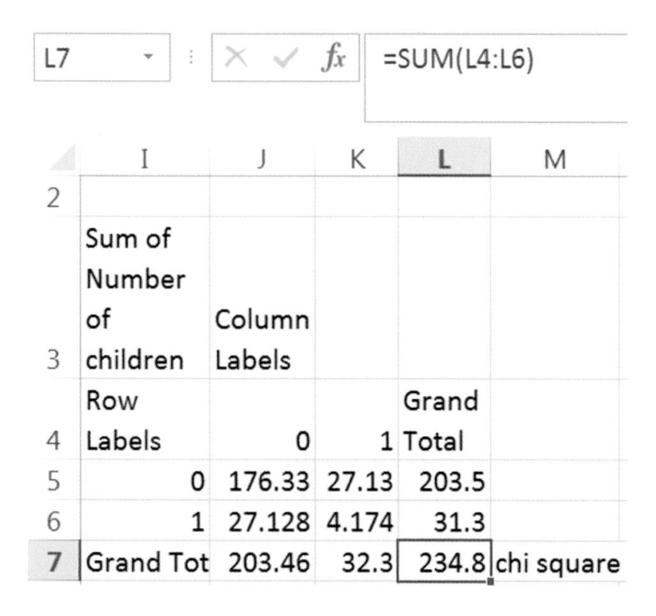

L7 | =SUM(L4:L6)

	I	J	K	L	M
2					
3	Sum of Number of children	Column Labels			
4	Row Labels	0	1	Grand Total	
5	0	176.33	27.13	203.5	
6	1	27.128	4.174	31.3	
7	Grand Tot	203.46	32.3	234.8	chi square

Use **CHISQ.DIST.RT()** to Find the *p value* of chi square:

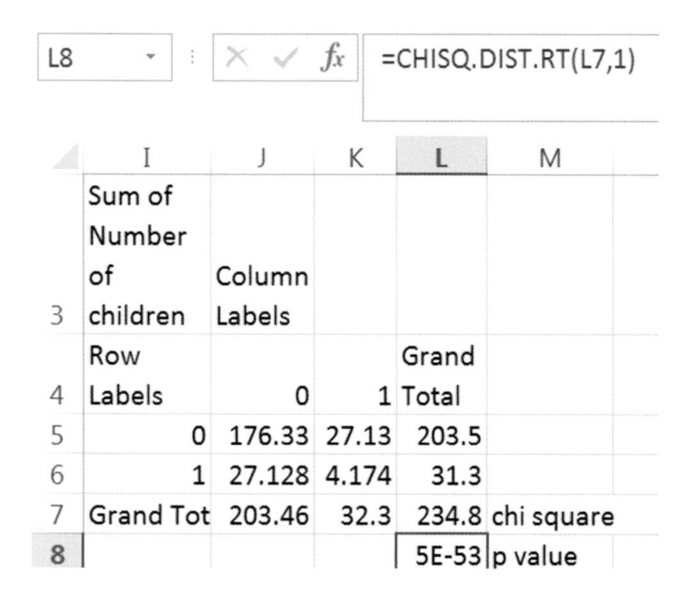

L8 | =CHISQ.DIST.RT(L7,1)

	I	J	K	L	M
3	Sum of Number of children	Column Labels			
4	Row Labels	0	1	Grand Total	
5	0	176.33	27.13	203.5	
6	1	27.128	4.174	31.3	
7	Grand Tot	203.46	32.3	234.8	chi square
8				5E-53	p value

Based on sample evidence, the null hypothesis of independence is rejected. Eating cereal with added sugar is associated with Saturday morning cartoon viewing.

To see the association, copy rows **1** and **2** with the PivotTable and paste below the original**,** this time with formulas.

Change the cell counts to percents of row: Right click a cell in the copied table,

Alt JT G Tab > **Tab** dn to **% Row Total**

Type in the labels *No* and *Yes* and *Don't Eat Kooldogs* and *Eat Kooldogs*.

	A	B	C	D
9	**Sum of Number of children**	**Column Lab**		
10	**Row Labels**	**Doesn't eat sugary cereal**	**Eats sugary cereal**	**Grand Total**
11	Doesn't watch Sat morn TV	90.00%	10.00%	100.00%
12	Watches Sat morn TV	1.54%	98.46%	100.00%
13	**Grand Total**	**13.33%**	**86.67%**	**100.00%**

Make a **PivotChart** with shortcuts **Alt JT C** to see the association.

Choose a design and style, add axes titles and a chart title which summarizes your conclusion:

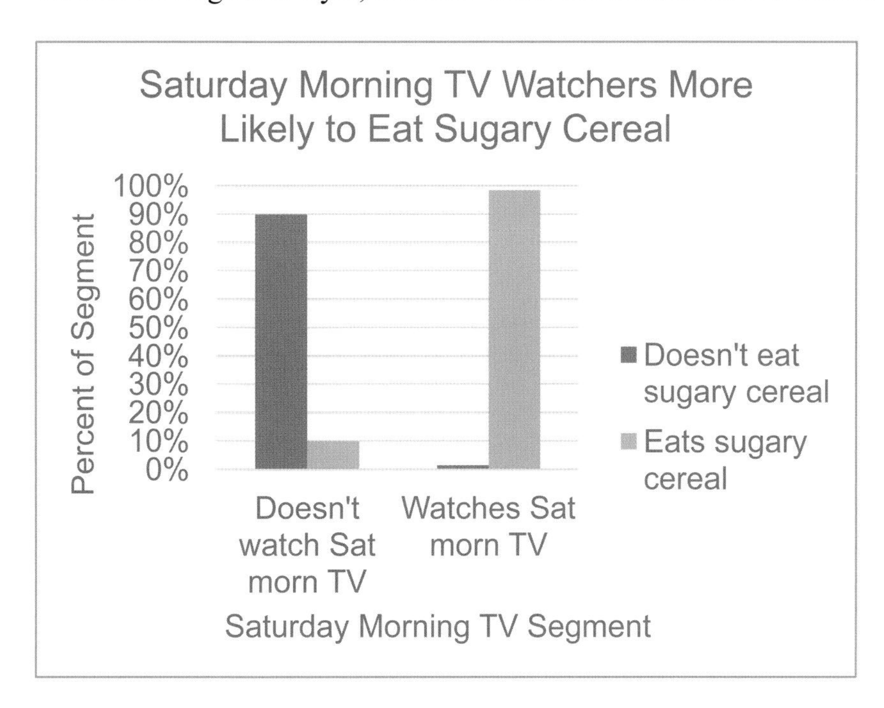

Lab 9 Skype Appeal

Following the launch of Google's Android phone, rumors surfaced that Google is considering a joint venture with Skype. Skype boasts 330 million users worldwide. Google management believes that Skype appeals most to younger consumers, who make long distance calls with Skype, instead of cell phones or land lines.

Google conducted a survey of 101 randomly chosen consumers, from ages 14 through 65. Consumers were asked which they relied on most for long distance: (i) cell phone, (ii) Skype, or (iii) landline.

A crosstabulation of the responses is in **Lab 9 Skype Appeal.xls**.

1. Make a PivotTable of long distance users by age and type of phone.
   Copy the PivotTable and paste in next to the first, and then change cell counts to percents of row.
   What percent of long distance users surveyed rely on Skype? ____________%

   What percent of 18–21 year olds rely on Skype? _________%

2. Make a table next to the second with the ***expected*** number of long distance users by age and type of phone ***given no*** *association* between age and type of phone,

expected count in row *i* column *j* = *number in row i x percent in column j*

$$e_{ij} = n_i p_j$$

OR *expected users in age segment i of phone j* = *number in age segment i x percent who use phone j*

How many of the fifteen 18–21 year olds would you expect to rely on Skype ***if*** long distance phone choice is ***not*** associated with age? _________

3. Mark (X) cells which are *sparse:*

Age	Cell phone	Land line	Skype
14–17			
18–21			
22–29			
30–39			
40–49			
50–59			
60–65			

   Group (circle) age segments so that no cells are sparse.
   Update your *expected* cell counts.

4. Make a table of cell contributions to chi square: $(e_{ij} - n_{ij})^2 / e_{ij}$

Sum the cell contributions to chi square to find chi square: ___________

5. Find the p value for your chi square with __ (=(rows−1) × (columns−1)) df: _________
6. Is choice of long distance phone type dependent on age? Y or N
7. *Type* of phone most dependent on age: ______________
8. Phone *type* choices depend most on *age segment*: __________
9. Make a PivotChart to illustrate your results.
10. Which *age segment*(s) are more likely than average to rely on *Skype*? ____________

Assignment 9-1 747s and Jets[1]

Boeing Aircraft Company management believes that demand for particular types of aircraft is associated with particular global region across their three largest markets, North America, Europe, and China. To better plan and set strategy, they have asked you to identify region(s) where demand is uniquely strong for 747 s and for regional jets.

Assignment 9-1 JETS747.xls contains Boeing's actual and projected deliveries 2005–2024 of each type of aircraft in each of the three regions:

a. Use contingency analysis to test the hypothesis that *demand* for particular aircraft is associated with *global region.*
b. If the association is significant, explain the nature of association.
c. Include a PivotChart and explain what it illustrates.

Assignment 9-2 Fit Matters

Procter & Gamble management would like to know whether intent to try their new preemie diaper concept is associated with the importance of fit. If Likely Triers value fit more than Unlikely Triers, fit could be emphasized in advertisements.

Assignment 9-2 Fit Matters.xls contains data from a concept test of 97 mothers of preemie diapers, including trial *Intention* and *Fit Importance*, measured on a 9-point scale.

You may decide to combine categories:

- Use contingency analysis to test the hypothesis that *intent* to try is associated with the *importance of fit.*
- If the association is significant, explain the nature of association.
- Include a PivotChart and explain what it illustrates.

Assignment 9-3 Allied Airlines

Rolls-Royce management has observed the growth in commercial airline alliances. Airline companies which are allied tend to purchase the same aircraft. Management would like to know whether or not alliance is associated with global region.

Data including the number of allied airline companies, *Allied*, and *Global Region* are contained in **Assignment 9-3 Allied Airlines.xls**.

You may decide to combine global regions:

- Use contingency analysis to test the hypothesis of association between *alliance* and *global region.*
- If the association is significant, describe the nature of association.
- Include a PivotChart and explain what it illustrates.

---

[1] This case is a hypothetical scenario using actual data.

Assignment 9-4 Netbooks in Color

Dell managers want to know whether college students' preferences for light weight netbooks and wide color selection are associated with major. Dell's netbook is lighter than many competing netbooks and comes in more colors than any other netbook. Managers believe that light weight and wide choice of colors may appeal to Arts & Sciences and Commerce students more than to Engineering students, which would give Dell an advantage to be promoted in those segments.

A sample of netbook and iPad owners was drawn from each of three schools on the UVA campus, Commerce, Arts & Sciences, and Engineering. Netbook or iPad brands owned by students were recorded. Those data, with number of *colors* available and *weight* are in **netbooks.xls**.

Determine whether preference for light weight and variety of colors are associated with college major:

1. State the hypotheses that you are testing.
2. What are your conclusions? (Include the statistical tests that you used to form your conclusions.)
3. What is the probability that:
   (a) A netbook or iPad owner will own a light weight brand?
   (b) An Arts & Science student will own a light weight brand?
   (c) A Commerce student will own a light weight brand?
   (d) An Engineering student will own a light weight brand?
4. What is the probability that:
   (a) A netbook or iPad buyer will own a brand available in at least six colors?
   (b) An Arts & Science student will own a brand available in at least six colors?
   (c) A Commerce student will own a brand available in at least six colors?
   (d) An Engineering student will own a brand available in at least six colors?

A Dell Intern believes that conclusions may differ if only netbooks owners are considered, excluding the unique segment of iPad owners.

5. Repeat your analyses using excluding iPad owners. Summarize your conclusions, including the statistics that you used.
6. Illustrate your netbook results (excluding iPad owners) with PivotCharts.

CASE 9-1 Hybrids for American Car

Rising gas prices and environmental concerns have led some customers to switch to hybrid cars.

American Car (AC) offers two hybrids, AC Sapphire and AC Durado, an SUV and a pickup. AC offers no hybrid automobiles. Major competitors, Ford, Toyota and Honda, offer hybrid automobiles. AC executives believe that with their hybrid SUV and pickup, they will be able to attract loyal AC customers who desire a hybrid. Shawn Green, AC Division Head, is worried that customers who were driving sedans, coupes or wagons may not want a truck or an SUV. They might switch from AC to Ford, Toyota or Honda in order to purchase a hybrid car.

To investigate further, Mr. Green commissioned a survey of car buyers. The new car purchases of a representative random sample of 4,000 buyers were sorted into eight groups, based on the type of car they had owned and *Traded* (Prestige Sport, Compact SUV, Large, and Full-size SUV) and whether or not they bought *Hybrid* or Conventional. These data are in **Case 9-1 Hybrid.xls**. The number of *Buyers* indicates popularity of each *Traded, Hybrid* combination.

Conduct contingency analysis with this data to determine whether *choice of hybrid vehicles* depends on *type of vehicle owned previously*.

Specifically,

1. Is there an association between the *type of car owned and Traded* and *choice of a Hybrid* instead of a Conventional car?
2. What is the probability that a new car buyer will choose a *hybrid*?
3. How likely is each of the segments to choose hybrids?
4. Illustrate your results with a PivotChart. Include a bottom-line title.
5. What are the implications of results for American Car Division? What is your advice to Mr. Green?

CASE 9-2 Tony's GREAT Advertising

Kellogg spends a hefty proportion of its advertising budget to expose children to ads for sweetened cereal on Saturday mornings. Kellogg brand ads feature cartoon hero characters similar to the cartoon hero characters that children watch on Saturday morning shows. This following press release is an example:

> ***Advertising Age***
> **Kellogg pounces on toddlers; Tiger Power to wrest tot monopoly away from General Mills' $500M Cheerios brand.** (News) *Stephanie Thompson.*
>
> Byline: STEPHANIE THOMPSON
>
> In the first serious challenge to General Mills' $500 million Cheerios juggernaut, Kellogg is launching a toddler cereal dubbed Tiger Power.
>
> The cereal, to arrive on shelves in January, will be endorsed by none other than Frosted Flakes icon Tony the Tiger and will be "one of our biggest launches next year," according to Kellogg spokeswoman Jenny Enochson. Kellogg will position the cereal-high in calcium, fiber and protein-as "food to grow" for the 2–5 set in a mom-targeted roughly $20 million TV and print campaign that begins in March from Publicis Groupe's Leo Burnett, Chicago.
>
> Cereal category leader Kellogg is banking on Tiger Power's nutritional profile as well as the friendly face of its tiger icon, a new shape and a supposed "great taste with or without milk" to make a big showing in take-along treats for tots.

Tony Grate, the brand manager for Frosted Flakes would like to know whether there is an association between Saturday morning cartoon viewing and consumption of his brand.

The Saturday morning TV viewing behaviors, *Saturday Morning Cartoons*, and consumption of Frosted Flakes, *Frosted Flake Eater*, are contained in **Case 9-2 Frosted Flakes.xls**. A random sample of 300 children ages 2 through 5 were sorted into four groups based on whether or not each watches at least 3 h of television on Saturday morning at least twice a month and whether or not each consumes Frosted Flakes at least twice times a week. The number of *Children* indicates popularity of each *Saturday Morning Cartoons*, *Frosted Flake Eater* combination.[2]

1. Is there an association between watching *Saturday morning cartoons* and consumption of Frosted Flakes?
2. What is the probability that a *cartoon watcher* consumes Frosted Flakes?
3. How likely is each segment to consume Frosted Flakes?
4. Illustrate your results with a properly labeled PivotChart. Include a bottom-line title.
5. What are the implications of results for Tony Grate?

CASE 9-3 Hybrid Motivations

American car executives have asked you to analyze data collected from a stratified sample of 301 car owners. The goal is to develop a profile of hybrid owners, which distinguishes them from conventional car owners. If differences are identified, those differences will be used to promote American's hybrid models.

One third of the owners surveyed own Priuses, one third own other hybrids, and one third own a conventional car. Car owners were asked to indicate the primary motivation which led to the last car choice. Possible responses included three *functional* benefits (fuel economy, lower emissions, tax incentives), *aesthetics/style* and *vanity* ("makes a statement about me"). Data are in **9 hybrid motivations.xlsx.**

It is thought that choice of a hybrid is associated with vanity…the desire to make a statement.

1. What is the probability that a car buyer was motived by vanity?
2. What is the probability that a buyer motivated by vanity chose a Prius?
3. What is the probability that a buyer motivated by vanity chose another hybrid model?
4. What is the probability that buyer motivated by vanity chose a conventional car?
5. Is choice of car type associated with motivation? Y or N
6. Statistic and p value you used to reach your conclusion in 5:
7. Embed a column chart which illustrates the association between motivation and car choice. Include axes labels with units and a title which describes what the conclusions which the audience should see:

Some American executives believe that Prius owners are unique.

8. Excluding conventional car owners, is choice of car type associated with motivation? Y or N
9. Statistic and p value you used to reach your conclusion in 8:

[2]These data are fictitious, though designed to reflect a realistic scenario.

Chapter 10
Building Multiple Regression Models

Models are used to accomplish *two* complementary goals: *identification of key drivers of performance* and *prediction of performance under alternative scenarios*. The variables selected affect both the explanatory accuracy and power of models, as well as forecasting precision. In this chapter, the focus is on variable selection, the first step in the process used to build powerful and accurate multiple regression models.

Multiple regression offers a major advantage over simple regression. Multiple regression enables us to account for the joint impact of multiple drivers. Accounting for the influence of multiple drivers provides a truer estimate of the impact of each one individually. In real world situations, multiple drivers together influence performance. Looking at just one driver, as we do with simple regression, we are very likely to conclude that its impact is much greater than it actually is. A single driver takes the credit for the joint influence of multiple drivers working together. For this reason, multiple regression provides a clearer picture of influence.

We use logic to choose variables initially. Some of the variables which logically belong in a model may be insignificant, either because they truly have no impact, or because their influence is part of the joint influence of a correlated set of predictors which together drive performance. *Multicollinear* predictors create the illusion that important variables are insignificant. *Partial F test(s)* are used to decide whether seemingly insignificant variables contribute to variance explained. Using *partial F tests* does not cure multicollinearity, but acknowledges its presence and helps us assess the incremental worth of variables that may be redundant or insignificant.

If an insignificant predictor adds no explanatory power, it is removed from the model. It is either not a performance driver, or it is a driver, but it is a redundant driver, because other variables reflect the same driving dimension. Correlations help to distinguish whether or not it is multicollinearity that is producing insignificance for a variable.

10.1 Multiple Regression Models Identify Drivers and Forecast

Multiple regression models are used to achieve two complementary goals: identification of key *drivers* of performance and prediction of performance under alternative scenarios. This prediction can be either what would have happened had an alternate course of action been taken, or what can be expected to happen under alternative scenarios in the future.

Decision makers want to know, given uncontrollable external influences, which controllable variables make a difference in performance. We also want to know the nature and extent of each of the influences when considered together with the full set of important influences. A multiple regression model will provide this information.

Once key drivers of performance have been identified, a model can be used to compare performance predictions under alternative scenarios. This *sensitivity analysis* allows managers to compare expected performance levels and to make better decisions.

C. Fraser, *Business Statistics for Competitive Advantage with Excel 2013: Basics, Model Building, Simulation and Cases*,
DOI 10.1007/978-1-4614-7381-7_10, © Springer Science+Business Media New York 2013

10.2 Use Your Logic to Choose Model Components

The first step in model building happens before looking at data or using software. Using logic, personal experience, and others' experiences, we first decide which of the potential influences ought to be included in a model. From the set of variables with available data, which could reasonably be expected to influence performance? In most cases, a reason is needed for including each independent variable in a model. Independent variables tend to be related to each other in our correlated world, and models are unnecessarily complicated if variables are included which don't logically affect the dependent performance variable. This complication from correlated predictors, *multicollinearity,* is explored later in the chapter.

Example 10.1 Sakura Motors Quest for Cleaner Cars

The new product development group at Sakura Motors is in the midst of designing a new line of cars which will offer reduced greenhouse gas emissions for sale to drivers in global markets where air pollution is a major concern. They expect to develop a car that will emit only 5 t of greenhouse gases per year.

What car characteristics drive emissions? The management team believes that smaller, lighter cars with smaller, more fuel efficient engines will be cleaner. The U.S. Government publishes data on the fuel economy of car models sold in the U.S. (fueleconomy.gov), which includes *manufacturer, model*, engine size (*cylinders)*, and gas mileage (*MPG*) for each category of car. This data source also includes *emissions* of tons of greenhouse gases per year. A second database, consumerreports.org, provides data on acceleration in *seconds* to go from 0 to 60 miles per hour, which reflects car model sluggishness, and two measures of size, *passengers* and curb *weight*. Management believes that responsiveness and size may have to be sacrificed to build a cleaner car.

The multiple linear regression model of *emissions* will include these car characteristics, *miles per gallon (MPG), seconds* to accelerate from 0 to 60, *horsepower, liters, cylinders, passenger* capacity, and weight in *pounds(K),* each thought to drive *emissions*:

$$\widehat{emissions}_i = b_0 + b_1 \times MPG_i + b_2 \times seconds_i + b_3 \times horsepower + b_4 \times liters$$

$$+ b_5 \times cylinders_i + b_6 \times passengers_i + b_7 \times pounds_i$$

Where $\widehat{emissions}_i$ is the expected tons of annual emissions of the *i*th car model,

b_0 is the intercept indicating expected emissions if *MPG, seconds, pounds(K), passengers, horsepower, cylinders* and *liters* were zero,

$b_1, b_2, b_3, b_4, b_5, b_6, b_7$ are the regression coefficient estimates indicating the expected marginal impact on emissions of a unit change in each car characteristic when other characteristics are at average levels, and

MPG_i, $seconds_i$, $horsepower_i$, $cylinders_i$, $liters_i$, $passengers_i$, $pounds(K)_i$ are characteristics of the *i*th car model.

When more than one independent variable in included in a linear regression, the coefficient estimates, or parameters estimates, are *marginal*. They estimate the marginal impact of each predictor on performance, given average levels of each of the other predictors.

The new product development team asked the model builder to choose a sample of car models which represents extremes of emissions, worst and best. Thirty-five car models were included in the sample. These included imported and domestic cars, subcompacts, compacts, intermediates, full size sedans, wagons, SUVs, and pickups. Within this set there are considerable differences in all of the car characteristics, shown in Table 10.1 and Fig. 10.1.

Table 10.1 Car characteristics in the Sakura motors sample

Car characteristic	*Minimum*	*Median*	*Maximum*
Emissions (tons)	5.2	8.7	12.5
MPG	15	22	34
Seconds (0–60)	7	9	12
Passengers	4	5	9
Pounds(K)	2.5	4.0	5.9
Horsepower	108	224	300
Cylinders	4	6	8
Liters	1.5	3.3	6.0

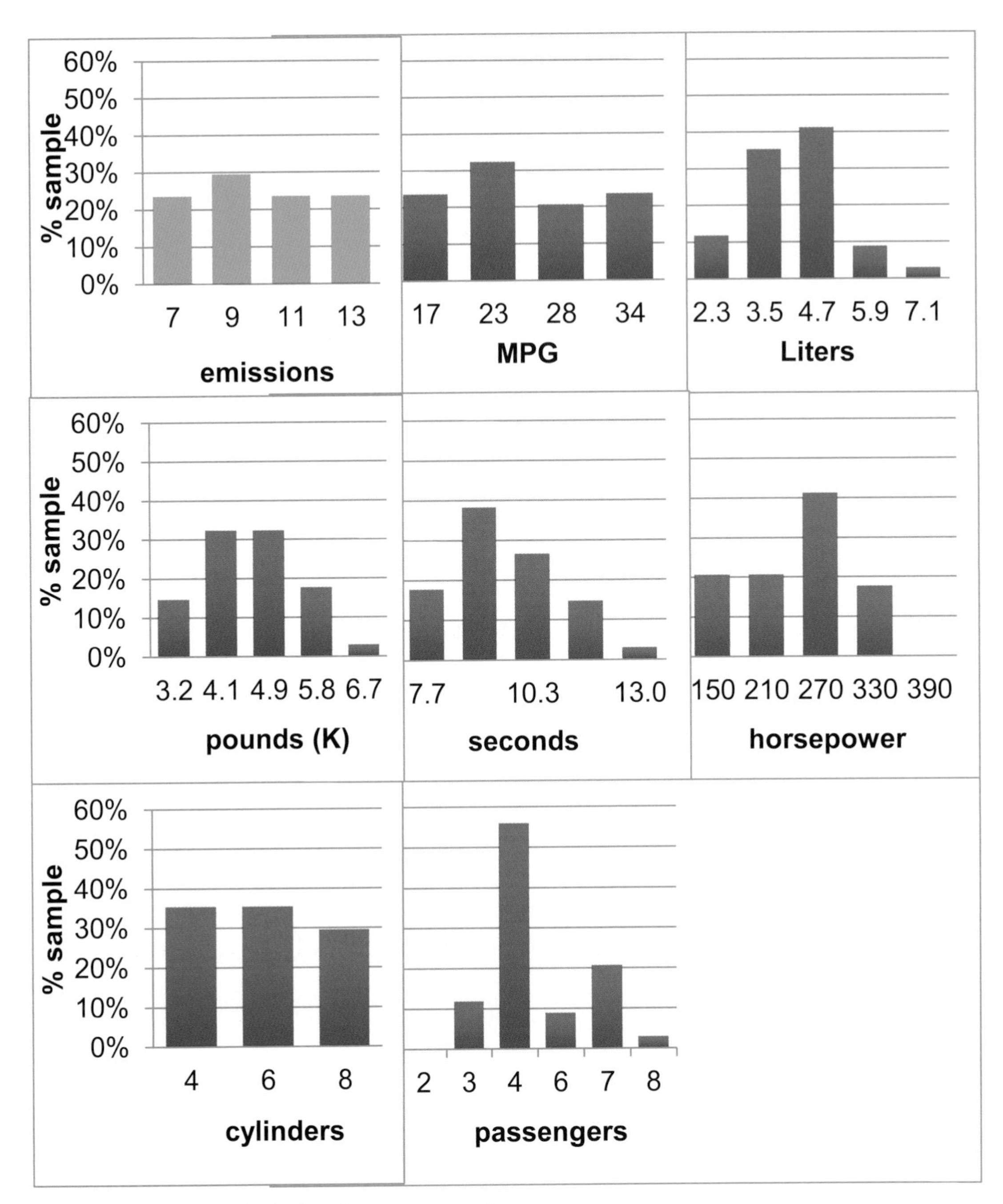

Fig. 10.1 Car characteristics in the Sakura motors sample

10.3 Multicollinear Variables Are Likely When Few Variable Combinations Are Popular in a Sample

Since these data come directly from the set of cars actually available in the market, many characteristic combinations do not exist. For example, there is no car with a 1.5 l engine that weighs 4,000 lb. The seven car characteristics to be related to each other and come in particular combinations in existing cars. We are knowingly introducing correlated independent variables, also called *multicollinear* independent variables, into our model, because the characteristic combinations which are not represented do not exist.

Results from Excel are shown in Table 10.2.

Table 10.2 Multiple linear regression of emissions with seven car characteristics

SUMMARY OUTPUT					
Regression Statistics					
R Square		.928			
Standard Error		.644			
Observations		34			
ANOVA	*df*	*SS*	*MS*	*F*	*Significance F*
Regression	7	138	19.8	47.7	.0001
Residual	26	11	.4		
Total	33	149			

RSquare is .928, or 93 %, indicating that, *together*, variation in the seven car characteristics accounts for 93 % of the variation in emissions. The *standard error* is .64, which indicates that forecasts of emissions would be within 1.3 t of average actual emissions for a particular car configuration.

10.4 *F* Tests the Joint Significance of the Set of Independent Variables

F tests the null hypothesis that *RSquare* is 0 %, or, equivalently, that all of the coefficients are zero:

$$\mathbf{H_0}\text{: } Rsquare = 0$$

Versus

$$\mathbf{H_1}\text{: } Rsquare > 0$$

or

$$\mathbf{H_0}\text{: All of the coefficients are equal to zero, } \beta_i = 0$$

Versus

$$\mathbf{H_1}\text{: At least one of the coefficients is not equal to zero.}$$

The F test compares explained to unexplained variation, which would be zero, under the null hypothesis. Sample evidence, shown in Fig. 10.2, will enable rejection of the null hypothesis if the explained slice is large enough.

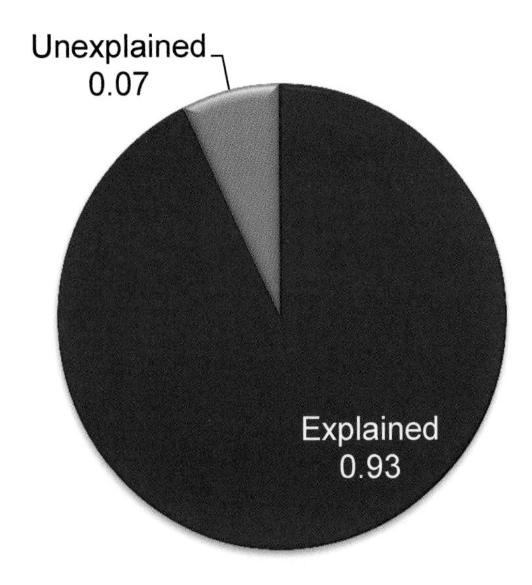

Fig. 10.2 Explained and unexplained variation

More drivers increase potential variation explained. The F test accounts for the number of drivers, as well as the sample size, with the comparison of explained variation per predictor (the *regression degrees of freedom*) with unexplained variation for a given sample and model size, the residual degrees of freedom. comparison for **Sakura** is shown in Fig. 10.3.

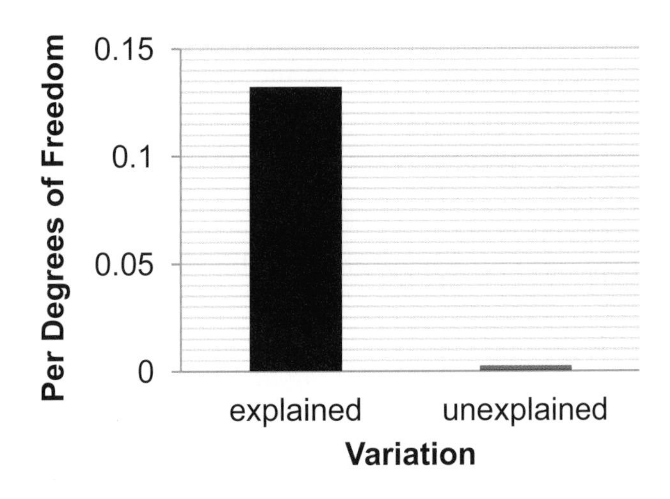

Fig. 10.3 Explained and unexplained variation

The ratio of these produces an F statistic with the regression and residual degrees of freedom:

$$F_{regression\ df, residual\ df} = \frac{RSquare/regression\ df}{(1 - RSquare)/residual\ df}$$

For **Sakura**, the F statistic is 48, with 7 and 26 degrees of freedom:

$$F_{7,26} = \frac{.928/7}{(1 - .928)/26} = \frac{.133}{.0028} = 48$$

The F statistic is compared to the F distribution with the same degrees of freedom. Figure 10.4 illustrates F distributions for 1, 2, 4, and 7 predictors with a sample of 30.

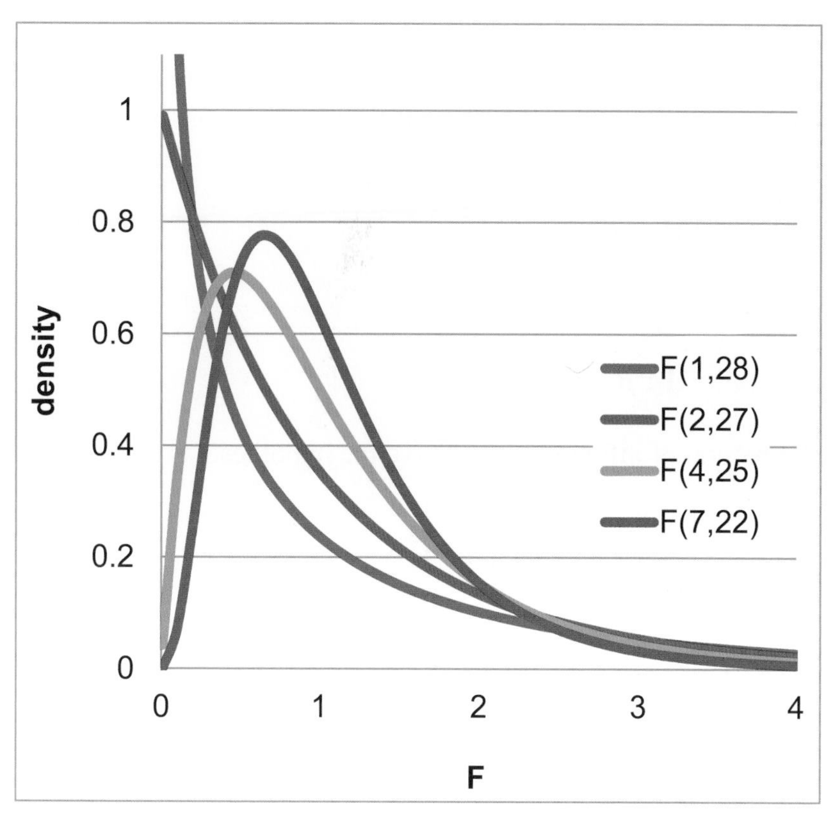

Fig. 10.4 A family of F distributions for a regression with sample of 30

The **Sakura** model F is 48, which lies to the extreme right of the $F_{7,\ 26}$ distribution. The *p value,* labeled *Significance F* in Excel, is .0001, indicating that it is unlikely that we would observe these data patterns, were none of the seven car characteristics driving emissions. It may be that just one of the seven characteristics drives emissions, or it may be that all seven are significant influences. With this set of seven predictors, some of the variation in *emissions* has been explained.

10.5 Insignificant Parameter Estimates Signal Multicollinearity

To determine which of the seven car characteristics are significant drivers of emissions, we initially look at the significance of *t tests* of the individual regression parameter estimates. A t statistic in multiple regression is used to test the hypothesis that a marginal coefficient is zero.

When we have no information about the direction of influence, a two tail test of each marginal slope is used:

$$\mathbf{H_0}\mathbf{:}\ \beta_i = 0$$

Versus

$$\mathbf{H_1}: \beta_i \neq 0$$

In the more likely case that, when, from theory or experience, we know the likely direction of influence, a one tail test is used. When the suspected direction of influence is positive, the null and alternate hypotheses are:

$$\mathbf{H_0}: \beta_i \leq 0$$

Versus

$$\mathbf{H_1}: \beta_i > 0.$$

Conversely, when the expected direction of influence is negative the hypotheses are:

$$\mathbf{H_0}: \beta_i \geq 0,$$

Versus

$$\mathbf{H_1}: \beta_i < 0.$$

Excel provides a two tail t statistic for each marginal slope by making calculating the number of standard errors each marginal slope is from zero:

$$t_{residual\ df,i} = b_i / s_{b_i}$$

Notice that a t statistic of a marginal slope in multiple regression is compared with the t distribution for the residual degrees of freedom. For each predictor in the model, we lose one degree of freedom. Excel provides the corresponding *p value* for the two tail t test of each marginal slope. In the case that we want to use a one tail test, the *p value* is divided by two. The t distribution used in the *emissions* model, with 26 degrees of freedom, is shown in Fig. 10.5.

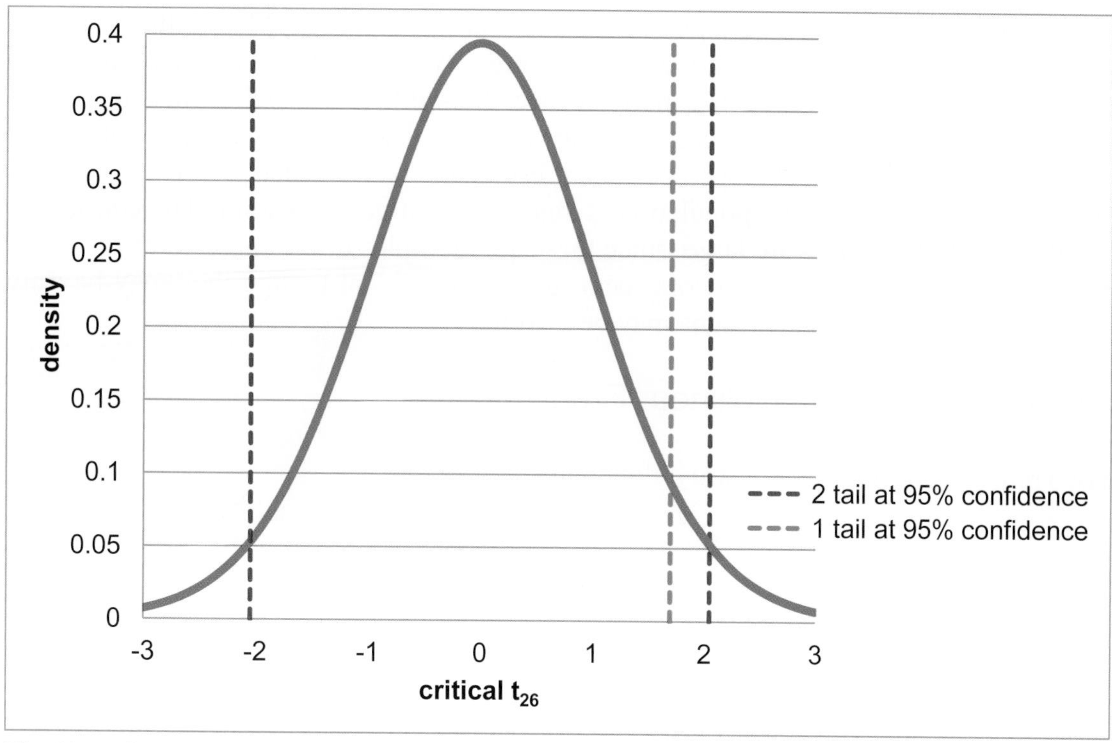

Fig. 10.5 *t* distribution with 26 degrees of freedom

In the emissions model, Sakura analysts were confident that the impact of *MPG* on emissions ought to be negative, and that each of the influences of *horsepower, cylinders, liters, weight,* and *passengers* on *emissions* ought to be positive. For these six potential drivers, one tail *t* tests could be used. Sakura managers were not sure of the direction of influence of acceleration on *emissions*, and so a two tail test would be used for the *seconds* slope.

Table 10.3 Marginal slopes and their *t* tests

	Coefficients	*Standard Error*	*t Stat*	*p value*	*1 tail p value*[a]
Intercept	9.2	1.90	4.8	<.0001	
Seconds	.23	.099	2.3	.03	
mpg	−.23	.037	−6.2	.0001	<.0001
Liters	.41	.29	1.4	.17	.08
Cylinders	−.035	.19	−.2	.85	.43
Horsepower	−.00052	.0037	−.1	.89	.44
Pounds (K)	.54	.30	1.9	.08	.04
Passengers	−.086	.12	−.7	.48	.24

[a]*p values* corresponding to one tail tests are not provided by Excel and have been added here

Excel *t* tests of the marginal slopes, shown in Table 10.3, suggest that only *seconds* to accelerate 0–60, *MPG,* and *pounds* drive differences in emissions. Neither engine size characteristics, *horsepower, cylinders* and *liters,* nor car size characteristic, *passengers,* appears to influence *emissions*. Coefficient estimates for *cylinders*, *horsepower,* and *passengers* have the "wrong signs." Cars with more cylinders, larger, more powerful engines and more passenger capacity are expected to emit more pollutants. These are surprising and nonintuitive results.

When predictors which ought to be significant drivers appear to be insignificant, or when parameter estimates are of the wrong sign, we suspect *multicollinearity.* Multicollinearity, the correlation between predictors, thwarts driver identification. When the independent variables are themselves related, they jointly influence performance. It is difficult to tell which individual variables are more important drivers, since they vary together. Because of their correlation, the standard errors s_{b_i} of the marginal slope coefficient estimates, b_i, are inflated. We are not very certain of each true influence in the population is since their influence is joint. The confidence intervals of the true partial slopes are large, since these are multiples of the standard errors of the partial slope estimates. Individual predictors seem to be insignificant though they may be truly significant. In some cases, coefficient signs may be "wrong."

10.6 Combine or Eliminate Collinear Predictors

We have two remedies for multicollinearity cloudiness:

- We can combine correlated variables, or
- We can eliminate variables that are contributing redundant information.

Correlations between the predictors are shown in Table 10.4.

Table 10.4 Pairwise correlations between predictors

	MPG	*Seconds*	*Liters*	*Horsepower*	*Cylinders*	*Pounds (K)*	*Passengers*
MPG	1						
Seconds	−.05	1					
Liters	−.81	−.17	1				
Horsepower	−.53	−.36	.76	1			
Cylinders	−.74	−.19	.92	.77	1		
Pounds (K)	−.77	−.01	.84	.72	.81	1	
Passengers	−.53	−.05	.59	.55	.60	.70	1

Some of correlated predictors can be eliminated, assuming that several reflect a common dimension. If *liters, horsepower,* and *cylinders* each reflect engine size, two are possibly redundant and may be represented by the third. The alternative is to combine correlated predictors, either by constructing an index from a weighted average of the correlated predictors, or by forming ratios of pairs of correlated predictors.

An index of engine size could be made from a weighted average of *liters, cylinders,* and *horsepower. Factor analysis* is a statistical procedure that would provide the weights to form such an index. The challenge associated with use of an index is in its interpretation. Sakura managers need to know how much difference particular car characteristics make, and they may not be satisfied knowing that an *engine size index* influences *emissions.* Factor analysis is beyond the scope of this text, but does enable construction of indices from correlated predictors.

Ratios of correlated predictors are used when they make intuitive sense. For example, economic models sometimes use the ratio of *GDP* and *population* to make *GDP per capita,* an intuitively appealing measure of personal wealth.

We will eliminate the seemingly redundant predictors to build a model for Sakura, though combining correlated predictors would be an acceptable alternative. This will not eliminate multicollinearity, but it will reduce multicollinearity by removing correlated predictors.

Cars with larger engines have more power. *Horsepower* and *cylinders* will be removed from the model, expecting that they are redundant measures of engine size. If explanatory power is not substantially reduced, we can designate *liters* as the measure of engine size which reflects *cylinders* and *horsepower*. *Liters* is the preferred predictor to retain, since its coefficient sign is as expected, while coefficient signs for both *cylinders* and *horsepower* are "wrong."

Passenger capacity is highly correlated with weight (*pounds(K)*): $r_{passengers,pounds} = .70$. Larger, more spacious cars weigh more. *Passengers* will be removed from the model, expecting that it is a redundant measure of car size. If explanatory power is not sacrificed, *pounds(K)* will reflect car size. *Pounds (K)* is chosen to represent car size, since its coefficient sign is as expected and significant, while the sign for the *passengers* coefficient is "wrong."

The revised *partial* model becomes:

$$emi\hat{s}sions_i = b_0 + b_1 \times MPG_i + b_2 \times seconds_i + b_3 \times liters_i + b_4 \times pounds_i$$

Regression results using this *partial* model are shown in Table 10.5.

Table 10.5 Regression of emissions with four car characteristics

SUMMARY OUTPUT						
Regression Statistics						
R Square		.926				
Standard Error		.617				
Observations		34				
ANOVA	*df*	*SS*	*MS*	*F*	*Significance F*	
Regression	4	138	34.5	90.8	.0000	
Residual	29	11	.4			
Total	33	149				
	Coefficients	*Standard Error*		*t Stat*	*p value*	*1 tail p value*
Intercept	9.0	1.8		5.0	<.0001	
seconds	.24	.087		2.8	.01	
mpg	−.23	.034		−6.7	<.0001	<.0001
liters	.36	.20		1.8	.08	.04
pounds (K)	.43	.24		1.8	.08	.04

The *partial* model *RSquare*, .926, is less than one percentage point lower than the *full* model *RSquare*, .929. With just four of the seven car characteristics, we can account for 93 % of the variation in emissions. Little explanatory power has been lost, and the standard error has dropped from .644 to .617, reducing the margin of error in forecasts by 5 % (= (1.32−1.26)/1.32). Model *F* is significant, suggesting that one or more of the four predictors influences emissions. All four of the predictors are significant drivers. All coefficient estimates have correct signs. As was the case in the full model, *emissions* are lower for smaller, responsive cars with higher fuel economy. By reducing multicollinearity, it can now also be concluded that *emissions* are lower for cars with smaller engines.

10.7 *Partial F* Tests the Significance of Changes in Model Power

Can *horsepower, cylinders* and *passengers* be eliminated without loss of explanatory and predictive power? Multicollinearity is reduced when we remove variables, increasing the certainty of parameter estimates for variables left in the model. With this small change in *RSquare* (less than 1 %), we do not need to test the significance of the change in *RSquare*. When *RSquare does* change by more than 1 %, we use a *Partial F* test to assess the significance of the decline. The *Partial F* tests the hypothesis that the explanatory power of the *partial* model equals that of the *full* model:

$$\mathbf{H_0}: SS_{regression_{partial}} = SS_{regression_{full}}$$

Versus

$$\mathbf{H_1}: SS_{regression_{partial}} < SS_{regression_{full}}$$

or

H$_0$: The variables removed add no explanatory power

Versus

H$_1$: The variables removed add explanatory power.

Under the null hypothesis, the (red) slice of additional variation explained by the full model, shown in Fig. 10.6, would be zero.

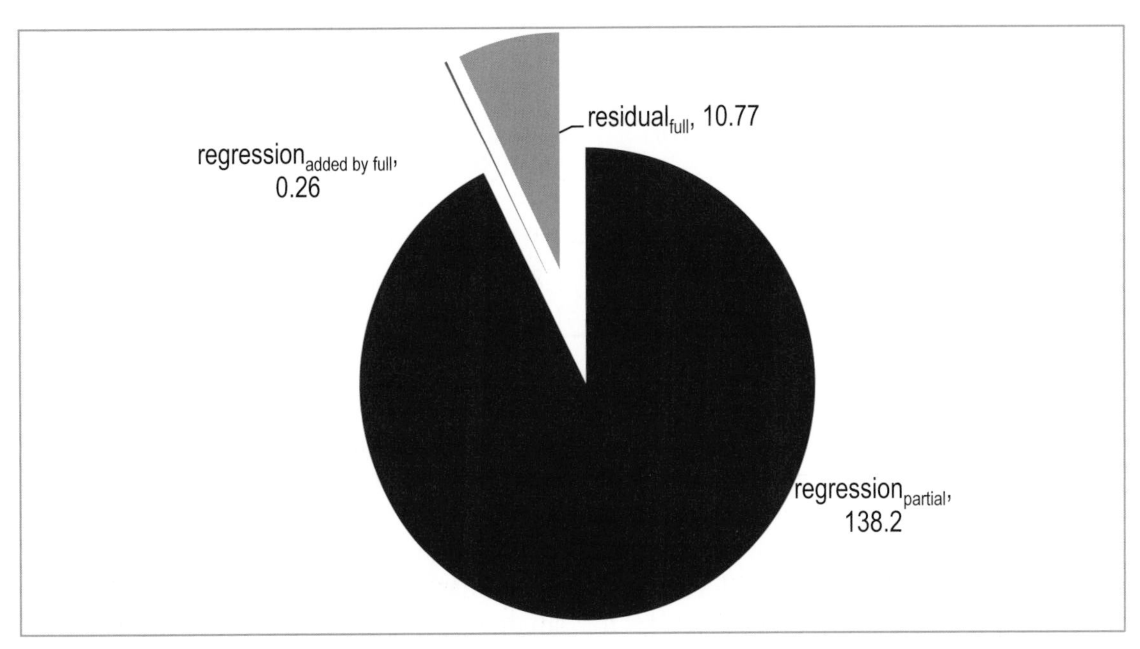

Fig. 10.6 Variation (SS) explained by partial and full models

A larger change in $SS_{regression}$ is expected if a larger number of variables are removed, so the change comparison is per driver removed, or the difference in the number of potential drivers between full and partial models. Unexplained variation, $SS_{residual_{full}}$, is also divided by the residual degrees of freedom to account for the sample and model size. These are shown in Fig. 10.7.

The ratio of these two *mean squares* provides the *Partial F* statistic:

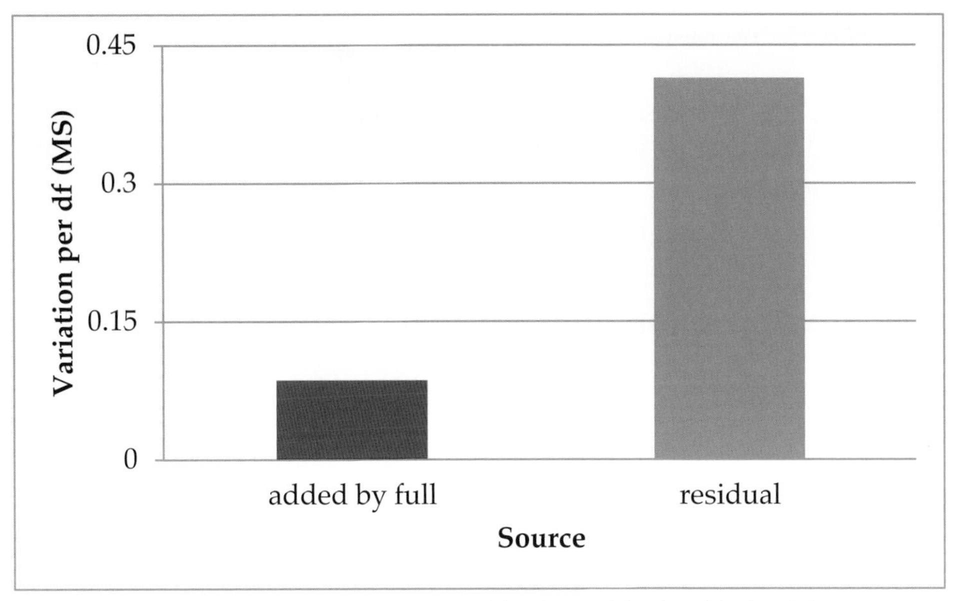

Fig. 10.7 Mean square variation added by full model and unexplained, residual variation

$$Partial\ F_{variables\ removed, residual\ df} = \frac{\dfrac{(SS_{regression_{full}} - SS_{regression_{partial}})}{(regression\ df_{full} - regression\ df_{partial})}}{\dfrac{SS_{residual_{full}}}{residual\ df_{full}}}$$

Which, for the **Sakura** models produces a *Partial F* of .2, with 3 and 26 *dfs*:

$$Partial\ F_{3,26} = \frac{(11.0 - 10.8)/3}{10.8/26} = \frac{.087}{.41} = .2$$

The *p value* for this *Partial F* of .2 with 3 and 26 *dfs* is .89. Sample evidence is consistent with the null hypothesis that the difference in variation explained by the full and partial models is zero. Variation explained did *not* change significantly when the three presumably redundant variables were eliminated. *Horsepower, cylinders* and *passengers* do *not* add sufficient explanatory power to the model and can be removed. The partial model now becomes the full model.

If the null hypothesis were rejected, we would return the variables removed, since they contribute significant explanatory power to the model. Because the null hypothesis cannot be rejected, the variables can be removed, since they add no additional explanatory power.

The final multiple linear regression model of emissions is:

$$\hat{emissions_i} = 9.0^a + .24^a \times seconds_i - .23^a \times MPG_i + .36^a \times liters_i$$
$$+ .43^a \times pounds_i$$

RSquare[a] = .93
[a]*Significant at a .05 level or better.*

To determine whether or not our model satisfies the assumptions of linear regression, the distribution of residuals is examined, just as with a simple regression model. In Fig. 10.8, the residuals are approximately *Normal*.

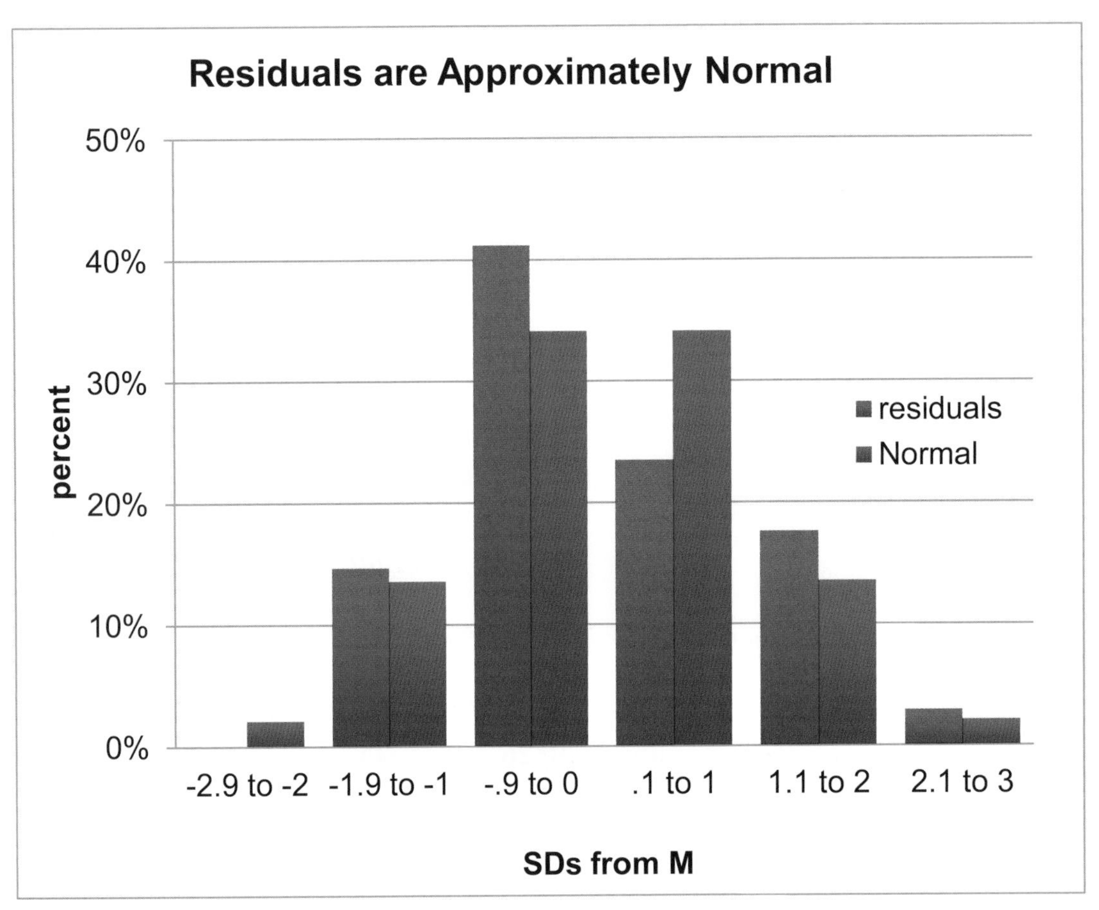

Fig. 10.8 Distribution of residuals

10.8 Decide Whether Insignificant Drivers Matter

It is typical to find that potential drivers, which logically ought to matter, are insignificant and fail to improve a model. The model is improved by removing those insignificant drivers. However, those apparently insignificant drivers should not be forgotten. They may not be performance drivers. Or, they may be performance drivers that do matter, but their correlation with other drivers makes them redundant. Management will want to know if potential drivers that were removed from a model are drivers, nonetheless.

Potential drivers which do matter, but are insignificant and appear not to matter, can be identified with correlations and simple regression. If multicollinearity has diminished the significance of a driver,

- That driver will be correlated with drivers remaining in the model, and
- That driver will be significant in a simple regression

Using simple regression to assess significance of a variable that appears not to matter gives that variable an "unfair" chance to claim credit for the joint impact of the set of correlated drivers. Simple regression will overstate a driver's importance, and we would not use the simple regression model in sensitivity analysis or forecasts. However, simple regression can provide important evidence that a potential driver matters, but is redundant.

In **Sakura,** three variables were removed from the multiple regression model: horsepower, cylinders, and passengers. All three were correlated with drivers that remained in the model. It seemed likely that all three were drivers of emissions that mattered, but simply could not be included in the multiple regression model, because of redundancy. The modeling team ran three simple regressions to decide whether or not this was the case.

Results of the three simple regressions, shown in Tables 10.6, 10.7 and 10.8 provide evidence that all three of the car characteristics drive emissions: All three simple regression models are significant, and all three slopes are in the expected positive direction. Sakura cannot ignore horsepower, cylinders, nor passengers in their car designs.

Table 10.6 Simple regression with cylinders alone

SUMMARY OUTPUT

Regression Statistics	
Multiple R	.76
R Square	.58
Standard Error	1.39
Observations	34

ANOVA

	df	*SS*	*MS*	*F*	*Significance F*
Regression	1	87.1	87.1	44.9	.0000
Residual	32	62.1	1.9		
Total	33	149.2			

	Coefficients	*Standard Error*	*t Stat*	*p value*	*Lower 95 %*	*Upper 95 %*
Intercept	3.16	.90	3.5	.0013	1.33	5.00
Cylinders	.99	.15	6.7	.0000	.69	1.30

Table 10.7 Simple regression with horsepower alone

SUMMARY OUTPUT

Regression Statistics	
Multiple R	.553
R Square	.306
Standard Error	1.798
Observations	34

ANOVA

	df	*SS*	*MS*	*F*	*Significance F*
Regression	1	45.7	45.7	14.1	.001
Residual	32	103.5	3.2		
Total	33	149.2			

	Coefficients	*Standard Error*	*t Stat*	*p value*	*Lower 95 %*	*Upper 95 %*
Intercept	4.53	1.22	3.7	.0008	2.04	7.02
Horsepower	.021	.006	3.8	.0007	.010	.032

Table 10.8 Simple regression with passengers alone

SUMMARY OUTPUT

Regression Statistics	
Multiple R	.540
R Square	.291
Standard Error	1.818
Observations	34

ANOVA

	df	*SS*	*MS*	*F*	*Significance F*
Regression	1	43.5	43.5	13.2	.001
Residual	32	105.7	3.3		
Total	33	149.2			

	Coefficients	*Standard Error*	*t Stat*	*p value*	*Lower 95 %*	*Upper 95 %*
Intercept	4.12	1.38	3.0	.005	1.31	6.92
passengers	.87	.24	3.6	.001	.38	1.35

10.9 Sensitivity Analysis Quantifies the Marginal Impact of Drivers

We want to compare influences of the significant drivers to identify those which make the greatest difference. We will forecast emissions at average levels of each of the car characteristics. Then, we will compare forecasts at minimum and maximum levels of each, holding the other three constant at mean levels. The sensitivity analysis is summarized in Table 10.9, below:

Table 10.9 Emissions response to car characteristics

MPG	*Seconds to accelerate 0–60*	*Pounds (K)*	*Liters*	*Expected emissions*	*Improvement (reduction) in expected emissions*
15.0	9	3.5	4.1	10.7	
33.5	9	3.5	4.1	6.5	**4.2**
22.6	**11.9**	3.5	4.1	9.7	
22.6	**6.7**	3.5	4.1	8.4	1.2
22.6	9	**6.0**	4.1	9.9	
22.6	9	**1.5**	4.1	8.3	1.6
22.6	9	3.5	**5.9**	9.8	
22.6	9	3.5	**2.5**	8.3	1.5

MPG. Within a representative range of values for each of the car characteristics, fuel economy makes the largest difference in emissions, shown in Fig. 10.9.

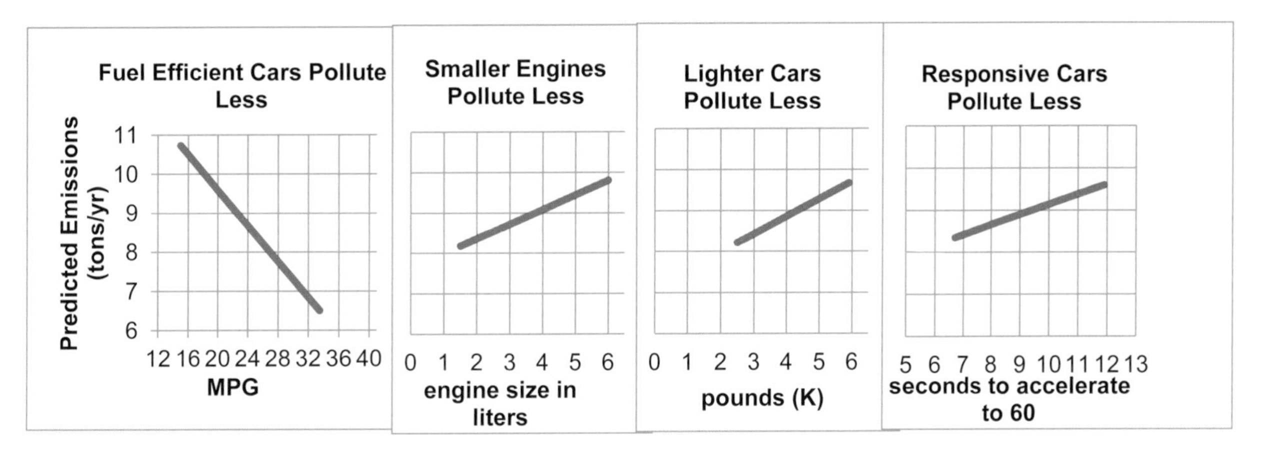

Fig. 10.9 Predicted emissions by car characteristic

Improving fuel economy by 19 MPG, the sample range, is associated with an expected reduction in emissions of 4.2 t per year. This is a large improvement, though not enough alone to meet the 5.0 t per year goal. Fuel economy improvements will need to be made in conjunction with improvements in one or more of the other car characteristics.

The linear model suggests that improving average fuel economy by 4 MPG, from 25 to 29, would produce an expected average improvement in emissions of about 1 t (.60–1.20 t) per year, assuming other car characteristics were at mean levels, which is shown in Fig. 10.10

$$\Delta MPG \times (b_{MPG} - 2 \times s_{b_{MPG}}) \leq \Delta MPG \times \beta_{MPG} \leq \Delta MPG \times (b_{MPG} + 2 \times s_{b_{MPG}})$$

$$(29 - 25) \times (-.23 - 2 \times .034) \leq (29 - 25) \times \beta_{MPG} \leq (29 - 25) \times (-.23 + 2 \times .034)$$

$$4 \times -.30 \leq 4 \times \beta_{MPG} \leq 4 \times -.16$$

$$-1.20 \leq 4 \times \beta_{MPG} \leq -.60$$

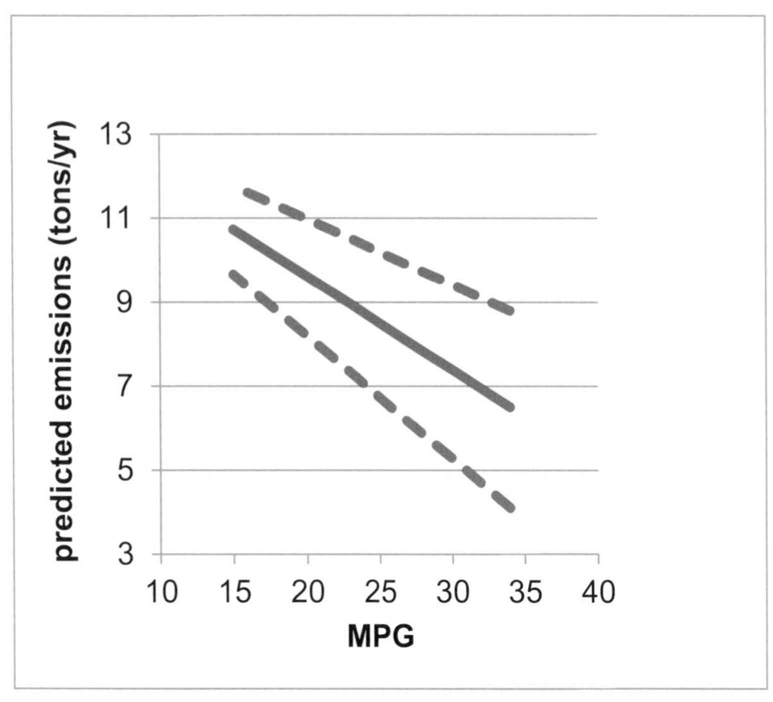

Fig. 10.10 Ninety-five percent prediction intervals by MPG

Pounds(K) and Liters. Reducing car weight by 2,500 pounds or reducing engine size by 3 l improves expected emissions by about 1 t per year. Even the combination of a lighter car with a smaller engine is probably not enough to reach the emissions goal of 5 t per year. In combination with fuel economy improvements, either car weight or engine size improvements could make the goal attainable.

Seconds. Improving car responsiveness by reducing the time to accelerate from 0 to 60 by 4 s could improve expected emissions about 1 t. Combined with any of the other car characteristics, responsiveness could help Sakura achieve their emissions goal, though acceleration alone makes the least difference in emissions.

The model provides clear indications for the new product development team. To improve emissions, they will need to design more responsive, lighter-weight cars with smaller engines and superior fuel economy. Changing just one car characteristic will not be enough to meet the goal of 5 t per year.

The Quantitative Analysis Director summarized model results in the following memo to Sakura Management:

MEMO

Re: Light, responsive, fuel efficient cars with smaller engines are cleanest
To: Sakura Product Development Director
From: Phil Levin, Quantitative Analysis Director
Date: June 2010

Improvements in gas mileage and responsiveness, with reductions in weight or engine size will allow Sakura to achieve the emissions target of 5 t per year.

A regression model of emissions was built from a representative sample of 34 diverse car models, considering fuel economy, acceleration, engine size and car size.

Fuel economy matters most.

Differences in fuel economy, weight, engine size, and acceleration account for 93 % of the variation in emissions. Forecasts from car characteristics are expected to be no further than 1.2 t from actual average emissions for a particular car.

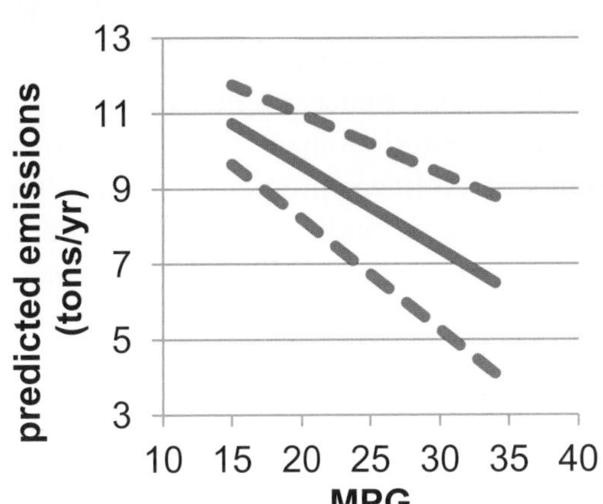

Fuel economy is the most powerful driver of emissions. Increasing gas mileage by four MPG is expected to reduce annual emissions by .6–1.2 t per year.

Car and engine size and responsiveness matter, but make less of a difference. A 2.5 K pound reduction in weight is expected to reduce emissions by about 1 t per year. Reducing engine size by 3 l reduces expected emissions by about 1 t per year.

$$\widehat{emissions} = 9.0^a + .24^a \times seconds$$
$$-.23^a \times MPG + .36^a \times liters$$
$$+.43^a \times pounds(K)$$
$$RSquare^a = .93$$

[a]*Significant at a .05 level or better.*

Passenger capacity, horsepower, and cylinders matter, but were not included in the model, since these characteristics are similar to weight and engine size in liters. Reduction in passenger capacity, horsepower or cylinders is expected to reduce emissions.

Responsiveness makes the smallest difference in emissions. Reducing acceleration from 0 to 60 by 4 s would improve emissions about 1 t per year.

Cleaner cars are more fuel efficient and more responsive

To achieve emissions of 5 t per year within existing characteristic ranges, more than one car characteristic must be changed. Improvements in fuel economy and responsiveness, with reductions in weight or engine size will enable Sakura to meet the emissions target of 5 t per year.

New technology may help to reduce emissions

Model results assume existing engine technology. With development of cleaner, more fuel efficient, responsive technologies, even lower emissions could possibly be achieved.

10.10 Model Building Begins With Logic and Considers Multicollinearity

Novice model builders sometimes mistakenly think that the computer can choose those variables which belong in a model. Computers have no experience making decisions and can never replace decision makers' logic. (Have you ever tried holding a conversation with a computer?) The first step in superior model building is to use your head. Use logic and experience to identify independent variables which ought to influence the performance variable which you are interested in explaining and forecasting. Both your height and GDP increased over the past 10 years. Given data on your annual height and annual GDP, the computer could churn out a significant parameter estimate relating variation in your height to variation in GDP (or variation in GDP to variation in your height). Decision makers must use their logic and experience to select model variables. Software will quantify and calibrate the influences that we know, from theory or experience, ought to exist.

It is a multicollinear world. Sets of variables together jointly influence performance. Using ratios of collinear predictors reduces multicollinearity. *Partial F tests* are used to confirm that eliminating redundancies does not reduce model power. Removing redundant predictors allows us to more accurately explain performance and forecast. Correlations and simple regressions are used to determine whether insignificant variables matter, but simply look as though they don't because of multicollinearity—or whether they simply do not matter.

From the logically sound set of variables, pruned to eliminate redundancies and reduce multicollinearity, we have a solid base for superior model building. To this we will consider adding variables to account for seasonality or cyclicality in time series in Chap. 11 and the use of indicators to build in influences of segment differences, structural shifts and shocks in Chap. 12. In Chaps. 13 and 14, alternative nonlinear models are considered, for situations where response is not constant.

Excel 10.1 Build and Fit a Multiple Linear Regression Model

Sakura Motors Quest for a Clean Car. Assist Sakura Motors in their quest for a less polluting car model, using data from bea.gov and consumerreports.org, which together provide information on individual car models. The dataset, Excel 10 Sakura Motors.xls contains data on 35 car models, representing U.S., European, and Asian manufacturers and a variety of sizes and styles.

Management is unsure which car characteristics influence *emissions*, but they suspect that fuel economy, *MPG,* acceleration capability, measured as *seconds* to accelerate from 0 to 60 mph, engine size, *cylinders, liters,* and *horsepower,* car *passenger* capacity, and weight in *pounds (K)* may be significant influences. Smaller, lighter models with smaller, less powerful engines are expected to be cleanest.

Open the dataset and run multiple regression with the dependent variable *emissions* in **Input Y Range** and the independent variables, *MPG, seconds, cylinders, liters, horsepower, passengers,* and *pounds* in the **Input X Range**.

Add a column for one tail *t* tests of *MPG, liters, horsepower, cylinders, pound (K)* and *passengers*, and the fill in *p values* by dividing Excel's two tail *p values* by 2:

	A	B	C	D	E	F	G	H
1	SUMMARY OUTPUT							
2								
3	*Regression Statistics*							
4	Multiple R	0.963233						
5	R Square	0.927817						
6	Adjusted F	0.908383						
7	Standard I	0.643631						
8	Observati(	34						
9								
10	ANOVA							
11		*df*	*SS*	*MS*	*F*	*gnificance F*		
12	Regressior	7	138.4448	19.77783	47.74239	3.05E-13		
13	Residual	26	10.7708	0.414261				
14	Total	33	149.2156					
15								
16		*Coefficients*	*andard Err*	*t Stat*	*P-value*	*Lower 95%*	*Upper 95%*	*1 tail pvalue*
17	Intercept	9.164994	1.903614	4.814523	5.48E-05	5.252059	13.07793	
18	MPG	-0.22612	0.036559	-6.18514	1.53E-06	-0.30127	-0.15097	7.6495E-07
19	seconds	0.229146	0.09864	2.323068	0.028265	0.02639	0.431903	
20	liters	0.414272	0.293367	1.412128	0.16977	-0.18875	1.017296	0.08488491
21	pounds (K	0.544133	0.294646	1.846736	0.076198	-0.06152	1.149786	0.03809903
22	cylinders	-0.03513	0.188284	-0.18656	0.853456	-0.42215	0.351897	0.42672815
23	horsepow	-0.00052	0.003688	-0.14111	0.888875	-0.0081	0.00706	0.44443729
24	passenger	-0.08552	0.119738	-0.7142	0.481466	-0.33164	0.160607	0.24073323

Multicollinearity symptoms. While the model is significant (*Significance F* < .0001), *only three of the car characteristics are significant* (*p value* < .05). We are not certain that *liters, cylinders, passengers,* and *horsepower* are influential, since their *p values* ≥ .05. *Horsepower, cylinders* and *passengers* have "incorrect" negative signs. Cars with greater horsepower, more cylinders, and more passenger space ought to be bigger polluters. Together, the lack of significance of seemingly important predictors and the three sign reversals signal multicollinearity.

Look at correlations to confirm suspicions that *liters, horsepower* and *cylinders* are correlated (and together reflect car power) and that *pounds(K)* and *passengers* are correlated (and together reflect car size). This may allow elimination of two of the power variables and one of the size variables to reduce multicollinearity.

Run correlations between the car characteristics:

	A	B	C	D	E	F	G	H
1		*MPG*	*seconds*	*liters*	*pounds (K)*	*cylinders*	*ıorsepowe*	*ɒassengers*
2	MPG	1						
3	seconds	-0.04901	1					
4	liters	-0.81008	-0.17065	1				
5	pounds (K	-0.76895	-0.01287	0.835189	1			
6	cylinders	-0.74357	-0.18889	0.924148	0.807687	1		
7	horsepow	-0.52917	-0.36347	0.762617	0.71753	0.770781	1	
8	passenger	-0.52581	-0.04871	0.592526	0.702565	0.602062	0.545275	1

Eliminating two of the three measures of power and one of the two measures of size will reduce multicollinearity.

Use *Partial F* to test significance of contribution to *RSquare.* Eliminate potentially redundant characteristics that appear to add little explanatory power. This does not mean that they are not important. More likely, they are closely related to other important characteristics and contribute redundant information. Characteristics with "wrong" signs in the full regression are removed first.

Run the partial model regression, changing the **Input X Range**, and add one tail *p values:*

	A	B	C	D	E	F	G	H
1	SUMMARY OUTPUT							
2								
3	*Regression Statistics*							
4	Multiple R	0.962329						
5	R Square	0.926077						
6	Adjusted F	0.915881						
7	Standard I	0.616733						
8	Observati	34						
9								
10	ANOVA							
11		*df*	*SS*	*MS*	*F*	*gnificance F*		
12	Regressior	4	138.1852	34.54629	90.82543	5.71E-16		
13	Residual	29	11.03042	0.380359				
14	Total	33	149.2156					
15								
16		*Coefficients*	*andard Err*	*t Stat*	*P-value*	*Lower 95%*	*Upper 95%*	*1 tail pvalue*
17	Intercept	8.98999	1.802313	4.988029	2.62E-05	5.303846	12.67613	
18	MPG	-0.2284	0.03409	-6.69989	2.38E-07	-0.29812	-0.15868	1.191E-07
19	seconds	0.239516	0.086938	2.755019	0.010033	0.061708	0.417325	
20	liters	0.360546	0.197403	1.826446	0.078094	-0.04319	0.764281	0.0390472
21	pounds (K	0.426995	0.235862	1.810361	0.080614	-0.0554	0.909388	0.0403068

This partial model, with three fewer predictors, is significant, and all predictors are now significant with "correct" signs. The standard error is smaller than that in the full model, though *RSquare* is lower.

Partial F compares reduction in *variation explained, SSR,* per variable removed, to unexplained variation, divided by the *residual degrees of freedom, MSE,* in the larger model.

Copy *SSR* in **C12** from the partial model (with only *MPG* and *seconds*) and paste it into the original full model output sheet.

11		*df*	*SS*	*MS*	*F*	*Significance F*	$SS_{partial}$
12	Regressior	7	138.444793	19.7778276	47.742392	3.04538E-13	138.19

Find the change in *SSR* due to removal of the three predictors:

11		*df*	*SS*	*MS*	*F*	*Significance F*	$SS_{partial}$	SS_{change}
12	Regressior	7	138.444793	19.7778276	47.742392	3.04538E-13	138.19	0.25962

Three variables were removed to build the partial regression model, reducing the *regression dfs* by three. Find the change in *SSR* per variable removed (3), which will be the numerator of *Partial F*.

	H	I	J
11	SS change	regression df change	SS change / regression df change
12	0.25962	3	0.0865409

Find the *Partial F* statistic by dividing the *change in SSR per change in regression dfs* by *MSE*:

K12 =J12/D13

	D	E	F	G	H	I	J	K
11	MS	F	Significance F	SS partial	SS change	regression df change	SS change / regression df change	Partial F
12	19.7778276	47.742392	3.04538E-13	138.19	0.25962	3	0.0865409	0.208904
13	0.414261348							

To find the level of significance of this *F* value, with 3 (variables omitted) and 26 (*residual df* in the larger model) degrees of freedom, use the Excel function **F.DIST.RT(***F,numerator df, denominator df***)**.

L12 =F.DIST.RT(K12,I12,B13)

	B	C	D	E	F	G	H	I	J	K	L
11	df	SS	MS	F	Significance F	SS partial	SS change	regression df change	SS change / regression df change	Partial F	p value
12	7	138.444793	19.7778276	47.742392	3.04538E-13	138.19	0.25962	3	0.0865409	0.208904	0.889306
13	26	10.770795	0.414261348								

The *Partial F p value* is greater than the *critical p value,* .05. The null hypothesis that the change in *RSquare* from removal of three predictors is zero cannot be rejected. The three car characteristics removed were redundant and were not adding explanatory power to the model.

Look at residuals to check model assumptions. Excel gives us the residuals (predicted minus actual) in the regression output sheet.

To be sure that the model residuals are *Normally* distributed, find skewness of the residuals:

C62 =SKEW(C28:C61)

	A	B	C	D
58	31	10.9469	0.453105	
59	32	11.35381	0.546186	
60	33	5.454643	1.145357	
61	34	7.529488	-1.12949	
62		skew	0.457325	

The residuals are approximately Normal.

Excel 10.2 Use Correlation and Simple Regression to Decide Whether Potential Drivers Removed Matter

Horsepower, cylinders and *passengers* were removed from the multiple regression model due to lack of significance and "incorrect" signs. *Horsepower* and *cylinders* were correlated with *liters,* all three reflecting engine size, leaving *liters* in the multiple regression model to represent engine size, as well as *horsepower* and *cylinders.* Run simple regression with *cylinders* alone to see if, given the chance to claim all of the impact of engine size, *cylinders* would be a significant driver.

	A	B	C	D	E	F	G
1	SUMMARY OUTPUT						
2							
3	*Regression Statistics*						
4	Multiple R	0.76416					
5	R Square	0.583941					
6	Adjusted F	0.570939					
7	Standard I	1.392867					
8	Observati(	34					
9							
10	ANOVA						
11		*df*	*SS*	*MS*	*F*	*gnificance F*	
12	Regressior	1	87.13306	87.13306	44.91212	1.45E-07	
13	Residual	32	62.08253	1.940079			
14	Total	33	149.2156				
15							
16		*Coefficients*	*andard Err*	*t Stat*	*P-value*	*Lower 95%*	*Upper 95%*
17	Intercept	3.164112	0.90002	3.515601	0.001335	1.330831	4.997393
18	cylinders	0.993569	0.148257	6.701651	1.45E-07	0.691579	1.295559

Cylinders is a significant driver of emissions. Its correlation with *liters* prevents it from reaching significance in the multiple regression model. Standing alone, its slope is positive, as expected. Multicollinearity is responsible for the "incorrect" sign when both *cylinders* and *liters* are included in a regression.

Excel 10.3 Use Sensitivity Analysis to Compare the Marginal Impacts of Drivers

For sensitivity analysis, to determine the impact of improvements in each of the four car characteristics, begin with the "worst" car, with lowest MPG, slowest acceleration, largest engine, and greatest weight. Then compare that baseline "worst" car with four cars that are "best" along one of the four car characteristics.

Insert five new rows of hypothetical cars. For the first baseline car, set all four characteristics to their least desirable levels, either the maximum (if the corresponding coefficient is negative) or the minimum (if the corresponding coefficient is positive).

Make the second hypothetical car identical, except for maximum improvement in MPG.

Make the third hypothetical car identical, except for maximum improvement in acceleration.

Make the fourth and fifth hypothetical cars identical, except improve one to have the smallest engine, and improve the other to be lightest in weight.

Add a row with labels above these new data cells

G41 =MIN(G2:G35)

	C	D	E	F	G
1	*emissions*	*MPG*	*seconds*	*liters*	*pounds (K)*
36		*MPG*	*seconds*	*liters*	*pounds (K)*
37	"worst"	15	11.9	6	5.9
38	best MPG	33.5	11.9	6	5.9
39	best acceleration	15	6.7	6	5.9
40	smallest engine	15	11.9	1.5	5.9
41	lightest weight	15	11.9	6	2.485

To find *Emissions* predicted by the model for each hypothetical car, first select the labels and new data cells, copy, and paste without formulas,

Alt H V S U

into the partial regression sheet, next to residuals:

	C	D	E	F	G	H
27	*Residuals*		MPG	seconds	liters	pounds (K)
28	0.107058	"worst"	15	11.9	6	5.9
29	0.225012	best MPG	33.5	11.9	6	5.9
30	-0.22103	best acceleration	15	6.7	6	5.9
31	0.209116	smallest engine	15	11.9	1.5	5.9
32	-0.20328	lightest weight	15	11.9	6	2.485

Use the regression equation formula to find *predicted emissions* using the car characteristic data and coefficient estimates.

I32 =$B$17+$B$18*E32+$B$19*F32+$B$20*G32+$B$21*H32

	D	E	F	G	H	I
27		MPG	seconds	liters	pounds (K)	*predicted emissions*
28	"worst"	15	11.9	6	5.9	13.096816
29	best MPG	33.5	11.9	6	5.9	8.8714552
30	best acceleration	15	6.7	6	5.9	11.851331
31	smallest engine	15	11.9	1.5	5.9	11.474358
32	lightest weight	15	11.9	6	2.485	11.638627

The difference between *predicted emissions* of each from the baseline provides an estimate of the expected difference that a improvement in each characteristic could make:

J29 =$I$28-I29

	D	E	F	G	H	I	J
27		MPG	seconds	liters	pounds (K)	*predicted emissions*	*impact*
28	"worst"	15	11.9	6	5.9	13.096816	
29	best MPG	33.5	11.9	6	5.9	8.8714552	4.225361
30	best acceleration	15	6.7	6	5.9	11.851331	1.245485
31	smallest engine	15	11.9	1.5	5.9	11.474358	1.622458
32	lightest weight	15	11.9	6	2.485	11.638627	1.458189

Scatterplots of marginal response. To see the impact of each driver, plot predicted *emissions* of hypotheticals.

First, focus on MPG.

Rearrange columns so that *MPG* and *Predicted* emissions are adjacent. Select the *predicted emissions* cells, cut, and past ("drop it in here") with **Alt H I E:**

Select *MPG* and *predicted emissions* in the first two hypothetical rows and insert a scatterplot.

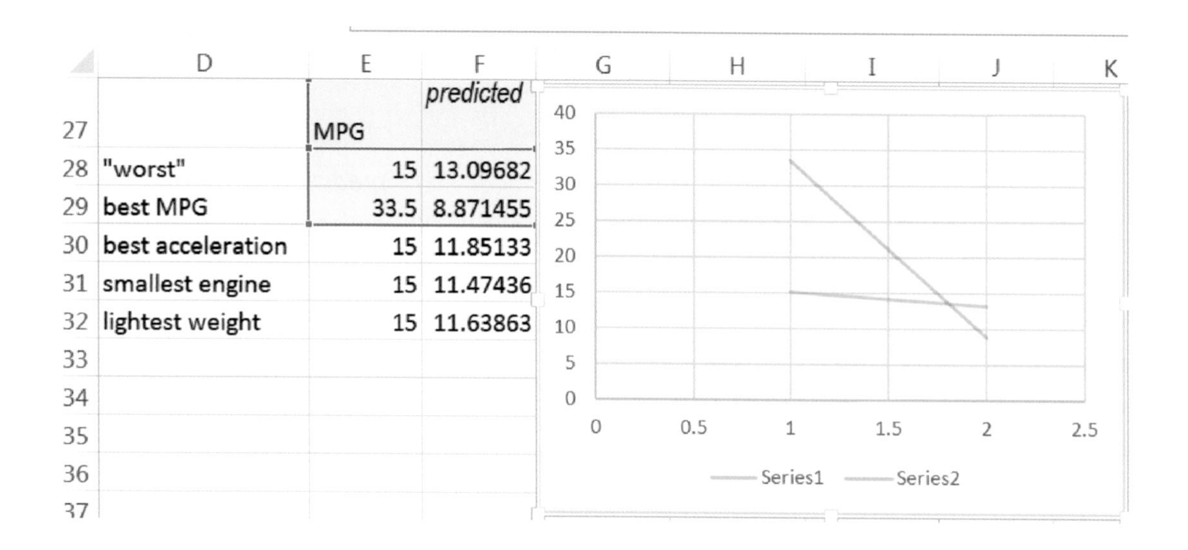

Excel sometimes reverses our intended axes, plotting x on the y axis, and vice versa. To correct this, switch row/column

Alt JC W

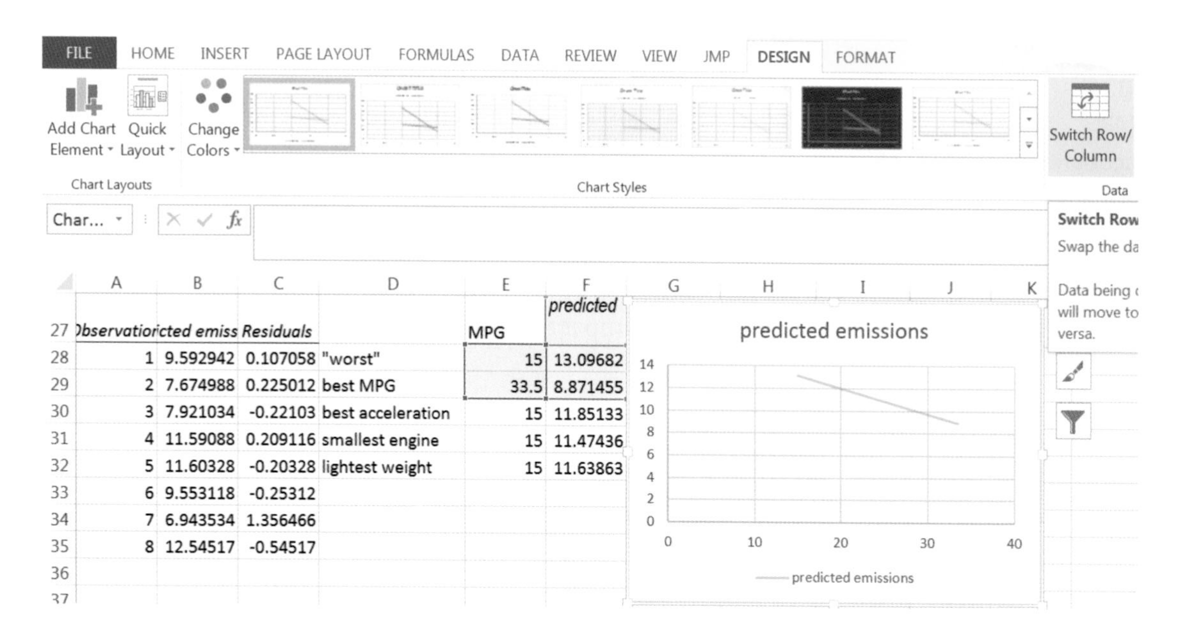

Choose a style and layout, and adjust axes.

Repeat this process see the impacts of acceleration, engine size, and car size. Choose the same minimum and maximum axes values for *predicted emissions* to make comparisons across car characteristics easiest:

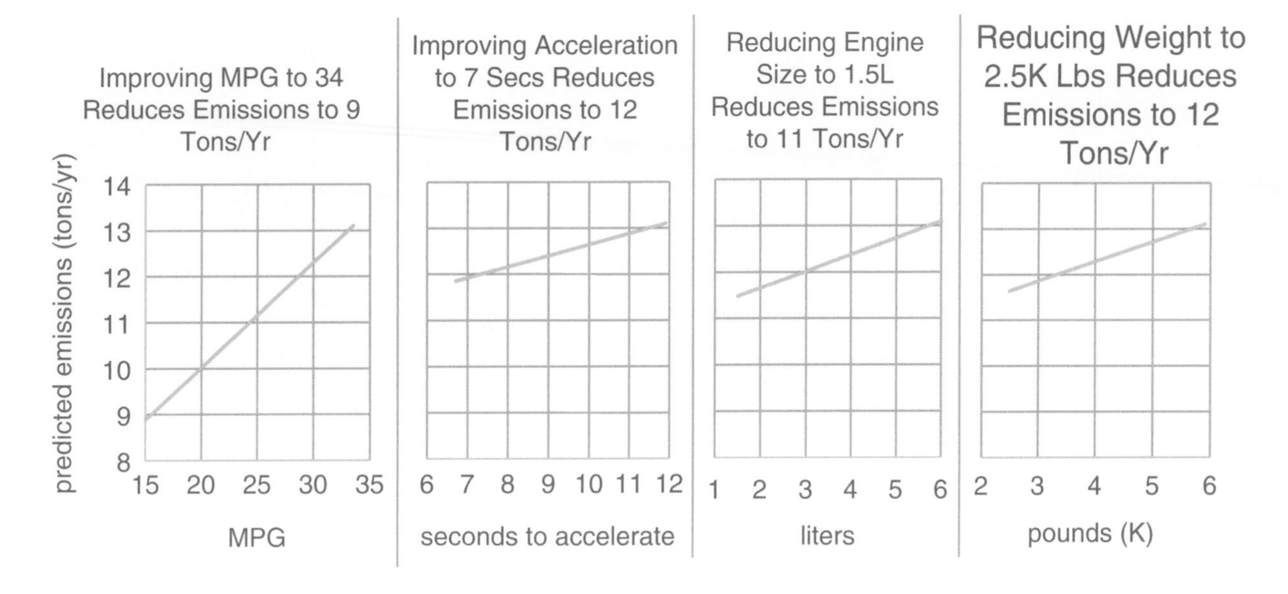

Lab 10 Model Building with Multiple Regression: Pricing Dell's Navigreat

Dell has experience selling GPS systems built by other firms and plans to introduce a Dell system, the Navigreat. They would like information that will help them set a price.
The Navigreat has

- An innovative, *highly portable* design, *weighing only 5 oz*, with a *state-of the art display*
- A *3.5" screen*, neither large, nor small, relative to competitors.
- Innovative technology which guarantees precise *routing time* estimates,

Dell executives believe that these features, *portability, weight, display* quality*, screen size,* and *routing time* precision, drive the price that customers are willing to pay for a GPS system.

Recent ratings by *Consumer Reports* provide data on the retail *price* of 18 competing brands, as well as

- *Portability* (1–5 scale), *weight* (ounces), and *display quality* (1–5 scale),
- *Screen size* (inches)
- *Routing time precision* (1–5 scale),

These data are in **Lab 10 Dell Navigreat.xls**. Also in the file, in row 21, are the attributes and expected ratings of the Navigreat.

Build a multiple regression model of GPS system *price*, including the characteristics thought by management to be drivers of *price*.

Regression results. Is the model *RSquare* significantly greater than 0? Y N

Evidence: *Significance F*=__________

Which of the potential drivers have slopes significantly different from 0?

	Portability	*Weight*	*Display*	*Screen size*	*Routing time*
Slope different from zero	Y or N	Y or N	Y or N	Y or N	Y or N
Evidence (*p value*)					

Which of the drivers have slopes of unexpected sign?

	Portability	*Weight*	*Display*	*Screen size*	*Routing time*
Slope sign unexpected	Y or N	Y or N	Y or N	Y or N	Y or N

Confirm suspected multicollinearity. The GPS system physical design determines its *screen size, display quality, weight* and *portability.* Run correlations to see if these characteristics are highly correlated.

	Highly correlated ($r_{\times1,\times2} > .5$)
Portability, weight	Y or N
Portability, display	Y or N
Portability, screen size	Y or N
Weight, display	Y or N
Weight, screen size	Y or N
Display, screen size	Y or N

Choose one of the set of correlated characteristics to represent the set, eliminating the other potentially redundant characteristics, and re-run the regression.

Is this partial model *RSquare* significantly greater than 0? Y N

Evidence: *Significance F*=____________

Which of the potential drivers in this partial model have slopes significantly different from 0? (Cross out characteristics that you excluded in this reduced model.)

	Portability	*Weight*	*Display*	*Screen size*	*Routing time*
Slope different from zero	Y or N	Y or N	Y or N	Y or N	Y or N
Evidence (*p value*)					

Which of the drivers have slopes of unexpected sign? (Cross out characteristics that you excluded in this partial model.)

	Portability	*Weight*	*Display*	*Screen size*	*Routing time*
Slope sign unexpected	Y or N	Y or N	Y or N	Y or N	Y or N

Find *Partial F* to decide whether the partial model's explanatory power is significantly lower than in the full model.

Full model $SS_{Regression}$ (1)	Partial model $SS_{Regression}$ (2)	Change in $SS_{Regression}$ (3) =(1)–(2)	Change per *g* predictors excluded (4) =(3)/g	$MS_{residuals}$ full model (5)	*Partial F* (6) =(4)/(5)	*p value* with *g* and (*N*-1-*k*) *dfs*

Conclusion:

__________partial model $SS_{Regression}$ is significantly lower than full model $SS_{Regression}$, and potentially redundant variables are jointly significant and cannot be excluded

or __________partial model $SS_{Regression}$ is equivalent to the full model $SS_{Regression}$, excluded variables are redundant or unimportant, and can remain excluded.

Determine the improvement in predictive accuracy:

	Full model (1)	Reduced model (2)	Improvement in *margin of error* (3)=(2)–(1)
Standard error	$	$	
Approximate margin of error in 95 % predictions	$	$	$

Assess residuals.
Are residuals approximately *Normal*? Y or N

Predict prices. Use the regression equation to find *expected prices* for each of the GPS systems, including the Navigreat.

Find the *critical t* value for 95 % prediction intervals with your model *residual degrees of freedom*.

Find the margin of error from the critical t and regression standard error.
Find the *lower* and *upper 95 % prediction intervals* for each model, including the Navigreat.

Will Dell be able to charge a retail price of $650 for the Navigreat? Y or N

Do variables removed from the model matter? Run simple regressions with each of the variables removed to decide whether each is a driver of price.

Driver removed	*Driver?*
Portability	Y or N
Display quality	Y or N
Weight	Y or N

Sensitivity analysis: Identify the most important driver of prices by comparing the differences in *expected prices* between four hypothetical GPS systems.
Add these four hypotheticals, and then extend *expected price* to include these.

Screen size	*Route time rating*	*Expected price*	*Difference due to*
Largest (5")	Average (4="Good")	$	Screen size: $__________
Smallest (3.4")	Average(4="Good")	$	
Average (3.8")	Best (5="Excellent")	$	Route time rating: $__________
Average (3.8")	Worst (2="poor")	$	

If Dell wants to charge a retail price of $650 for the Navigreat, what product design modification ought to be made?______________________________

Assignment 10-1 Sakura Motor's Quest for Fuel Efficiency

The new product development team at Sakura Motors has decided that the new car which they are designing will have superior gas mileage on the highway.

Use the data in **Assignment 10 Sakura Motors.xls** to build a model to help the team. Variables in the dataset include:

MPGHwy
manufacturer's suggested retail base price
engine size (*liters*)
engine *cylinders*
engine *horsepower*
curb *weight*
acceleration in *seconds* to go from 0 to 60
percent of owners *satisfied* who would buy the model again

Use your logic to choose car characteristics which ought to influence highway gas mileage.
Determine which car characteristics influence *highway gas mileage*. Use *partial F test(s)* to decide whether to remove apparently insignificant variables.
With sensitivity analysis, find the relative importance of significant influences on *highway fuel economy*
Find the car characteristic levels which could be expected to achieve **40 miles per gallon** in highway driving. (Sakura is not limited to existing designs.)

1. Write a **one page, single spaced** memo presenting your model, sensitivity analysis and design recommendations.

Present your final model in standard format

What is the margin of error of model forecasts of MPG?

Discuss the relative importance of significant influences, including the expected difference in *fuel economy* that differences in each could be expected to make if other characteristics were held at mean values

Conduct a sensitivity analysis comparing expected fuel economy with best and worst levels of each predictor in your final model when other characteristics are at average levels. Discuss the relative importance of significant influences, referring to

i. *A table of Fuel Response to Car Characteristics which you have added to the second page of Attachments*
ii. *A scatterplot of the impact of each significant driver on fuel economy. This plot shows predicted fuel economy on the vertical axis by values of a driver.*

There is a two page limit:

One single spaced page for your memo text with a single embedded scatterplot (of *predicted MPGHwy* by the most important car characteristic).
A second page of Exhibits showing your sensitivity analysis table and plots.

Please use Times New Roman 12 pt font and round your statistics to two or three significant digits.
You do not need to include description of your partial F test.

Assignment 10-2 Identifying Promising Global Markets

Harley-Davidson would like to identify the most promising global markets for motorcycle sales.
Some managers believe that motorcycle sales potential is greater in developed countries with higher GDP. Others believe that per capita GDP may a better indicator.
Management believes that motorcycle sales potential will necessarily be greater in more populated countries. Some believe that population density may matter more, since motorcycles may be preferred to cars for parking and commuting in larger cities.
Build a model to identify the drivers of motorcycle market potential. **10 Global Moto**.xls contains measures of

Motorcycle sales, GDP, per capita GDP, population, and *population density*

in 2009 for 20 countries with the highest motorcycle sales.

1. Identify outliers and list. Which countries have unusually high motorcycle sales?
2. Which economic and population variables drive motorcycle sales?

___ GDP ___ per capita GDP ___ population ___ population density

Explain how you reached your conclusions, including statistics that you used to decide:

3. Present your regression equation, including slope significance levels:
4. Illustrate the impact of the two most influential drivers, using the same scale for the y axis on both plots.
5. Compare predicted sales with actual sales in all countries in the sample to identify two markets with the greatest unrealized potential that Harley-Davidson should target:
6. H-D is considering expansion in the BRIC countries (Brazil, Russia, India, and China). The column *BRIC* distinguishes these four countries from other countries in the sample.

   What economic or demographic characteristic(s) distinguish the BRIC countries from other countries in the sample? (Include the statistics that you used to test hypothesized differences between BRIC and other countries.)

7. In light of your model results and the distinguishing characteristic(s) of the BRICs, should Harley-Davidson make expansion in the BRICs a high priority? Y or N

   Explain the logic of your answer:

Case 10-1 Chasing Whole Foods' Success

Trader Vics management is considering an attempt to replicate Whole Foods' success by implementing a process for selecting new store locations that mirrors Whole Foods' selection process, detailed in their website:

> If you have a retail location you think would make a good site for Whole Foods Market, Inc., please review the following guidelines carefully for consideration:
>
> **200,000 people or more in a 20-min drive time**
> 25,000–50,000 Sq Ft
> **Large number of college-educated residents**
> Abundant parking available for our exclusive use
> Stand alone preferred, would consider complementary
> Easy access from roadways, lighted intersection
> Excellent visibility, directly off of the street
> Must be located in a high traffic area (foot and/or vehicle)

James Sud, executive vice president of growth and development at WFM, describes WFM's store location selection success:

> "We're still very proud of the fact that in our 30-plus year history, we've never had a store that we opened ourselves ever fail. So we're really determined to keep that track record alive,"

Some executives at Trader Vic's believe that States in the Heartland may be the most promising locations for new stores, since residents of Heartland States are thought earn more college *degrees per person.*

Data in **10 WFM Stores.xlsx** contains these data for each State:

Heartland State or not
WFM stores in 2011
colleges
college *degrees* conferred in 2010
2010 population
colleges per person in 2010
square miles
colleges per sq mi
college *degrees* conferred *per person* in 2010
2010 people per sq mi

Help Trader Vics' strategists by identifying drivers of WFM's store selection (*stores* in a State).

1. Identify outlier State(s) and list:
2. What drives number of *stores* in a State? ___ *colleges* ___*college degrees*

   ___*population* ___*people per sq mi* ___*colleges per sq mi* ___*colleges per person*

   ___*degrees per person*

3. Write the equation for your final model of *stores* in a State.
4. Embed a graph illustrating the impact on *stores* of the most important driver.
5. Test managers' suspicion that *Heartland* State residents earn more college *degrees per person* than residents of other States. Present and interpret statistics that you used.
6. Using the data, identify differences between *Heartland* States and other States. Present and interpret the statistics that you used:
7. Based on the WFM store selection process revealed by your model results, which two States ought Trader Vics consider first for new store locations?
8. Based on your results, explain why Trader Vics should or should not focus on Heartland States for new store locations:

Case 10-2 Fast Food Nations

Fast food restaurants have been sprouting in diverse global regions. McDonalds has announced plans to open 700+ new restaurants in China. There is some concern that other global regions may be more fruitful locations.

Executives believe that countries with larger urban populations, and more populations under 15 years are likely the most fruitful locations for potential fast food establishments. (They are not sure whether education and economic productivity favors or discourages the fast food industry.)

Fast Food Nations.xlsx contains data on number of fast food establishments, demographics, and economic productivity of 47 countries. Build a model linking number of fast food establishments to demographic and economic characteristics of countries to help McDonalds choose the most promising locations for new restaurants.

1. Identify the outlier(s), with respect to fast food establishments and list, below:
2. Which variables drive fast food sales in a country?

___ Population under 15 ___ Urban population ___ Expected years in school

___ GDP per capita

3. Present your model equation linking demographics and economic productivity to fast food establishments in a country. (Be sure to specify units and significance levels.) Format matters.
4. Illustrate the impact of the most influential driver in your model and embed below:
5. Describe the power of your model, referring to the appropriate statistic:
6. Describe the precision of your model, referring to the appropriate statistic:
7. Comparing predicted with actual fast food establishments, which country would be the most fruitful choice for future McDonalds restaurants?
8. How do the "Fast Food Nations" differ with respect to fast food, demographics, or in terms of economics, from the other countries? (For each difference that you identify, include reference to the statistic that you used to conclude that a difference exists. You do NOT need to embed graphics to illustrate.)

Chapter 11
Model Building and Forecasting with Multicollinear Time Series

A regression model from time series data allows us to identify performance drivers and forecast performance given specific driver values, just as regression models from cross sectional data do. When decision makers want to forecast *future* performance, a time series of past performance is used to identify drivers and fit a model. A time series model can be used to identify drivers whose variation over time is associated with later variation in performance over time.

Three differences in the model building process distinguish cross sectional and time series models:

- The use of lagged predictors,
- Addition of trend, seasonality and cyclical variables, and
- The model validation process.

In time series models, the links between drivers and performance are stronger if changes in the drivers precede change in performance. Therefore, lagged predictor variables are often used. Patterns of change in drivers that also occur in the dependent variable in later time periods are identified to choose driver lags. Time series models are built using predictor values from past periods to explain and forecast later performance. Figure 11.1 illustrates the differences in model building processes between cross sectional and time series models.

Most business performance variables and economic indicators are cyclical. Economies cycle through expansion and recession, and performance in most businesses fluctuates following economic fluctuation. Business and economic variables are also often seasonal. Cyclicality and seasonality are accounted for by adding cyclical and seasonal predictors.

Before a time series model is used to forecast future performance, it is validated:

- The two most recent observations are hidden while the model is built,
- The model equation is used to forecast performance in those two most recent periods,
- Model prediction intervals are compared with actual performance values in those two most recent periods, and if the prediction intervals contain actual performance values, this is evidence that the model has *predictive validity* and can be reliably used to forecast unknown performance in future periods.

C. Fraser, *Business Statistics for Competitive Advantage with Excel 2013: Basics, Model Building, Simulation and Cases*, DOI 10.1007/978-1-4614-7381-7_11,

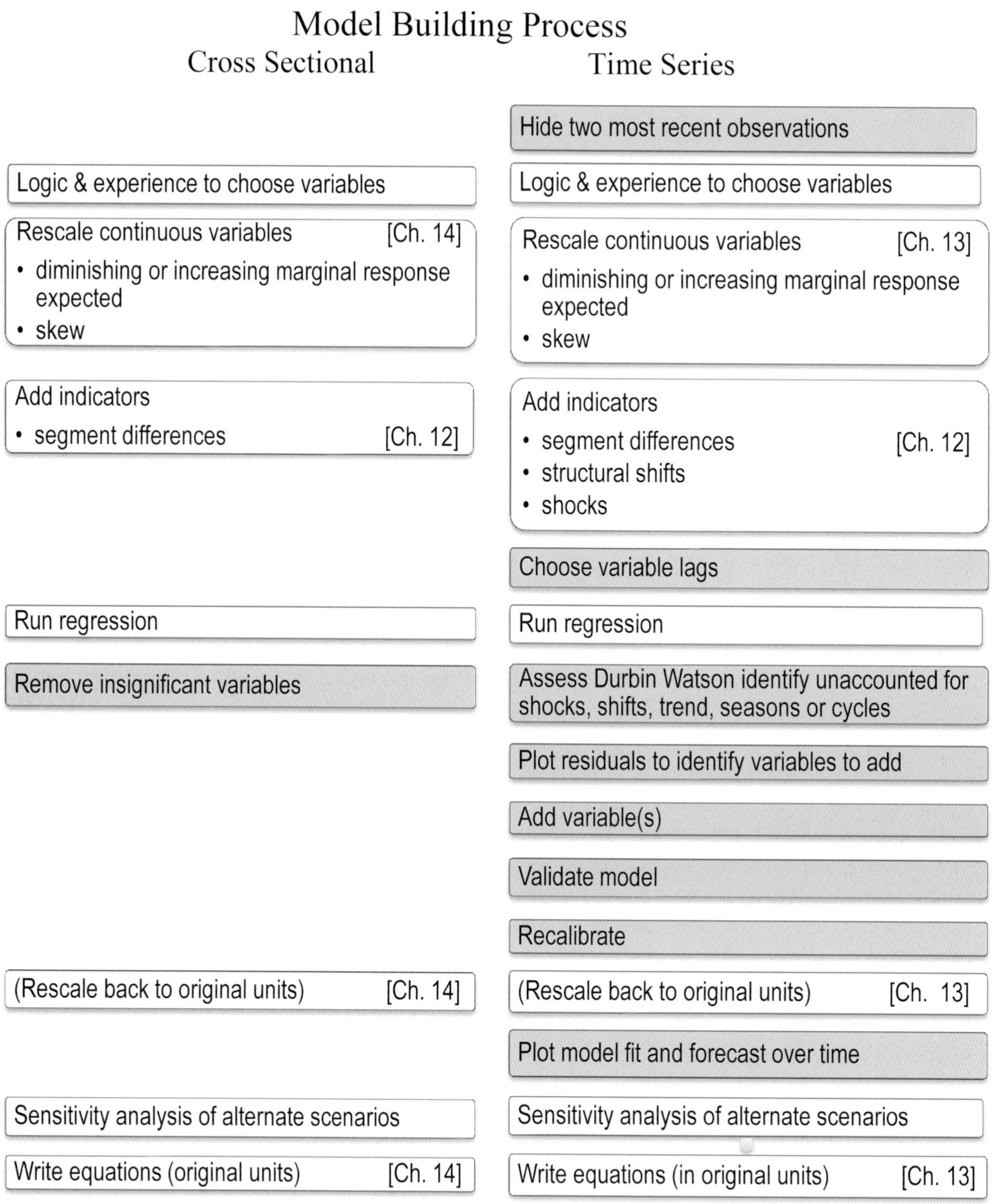

Model Building Process

Cross Sectional	Time Series
	Hide two most recent observations
Logic & experience to choose variables	Logic & experience to choose variables
Rescale continuous variables [Ch. 14] • diminishing or increasing marginal response expected • skew	Rescale continuous variables [Ch. 13] • diminishing or increasing marginal response expected • skew
Add indicators • segment differences [Ch. 12]	Add indicators • segment differences [Ch. 12] • structural shifts • shocks
	Choose variable lags
Run regression	Run regression
Remove insignificant variables	Assess Durbin Watson identify unaccounted for shocks, shifts, trend, seasons or cycles
	Plot residuals to identify variables to add
	Add variable(s)
	Validate model
	Recalibrate
(Rescale back to original units) [Ch. 14]	(Rescale back to original units) [Ch. 13]
	Plot model fit and forecast over time
Sensitivity analysis of alternate scenarios	Sensitivity analysis of alternate scenarios
Write equations (original units) [Ch. 14]	Write equations (in original units) [Ch. 13]

Fig. 11.1 Model building processes with cross sectional and time series data

11.1 Time Series Models Include Decision Variables, External Forces, and Leading Indicators

Most successful forecasting models logically assume that performance in a period, y_t, depends upon

- Decision variables under the management control,
- External forces, including
  - Shocks such as 9/11, Hurricane Katrina, change in Presidential Party
  - Market variables,
  - Competitive variables,
- Inertia from past performance
- Leading indicators of the economy, industry or the market
- Seasonality
- Cyclicality

Ultimately, the multiple regression forecasting models contain several of these components, which together account for variation in performance. This chapter introduces inertia and leading indicator components of regression models built from time series.

Performance across time depends on decision variables and the economy. Decision variables, such as spending on advertising, sales effort and research and development tend to move together. In periods of prosperity, spending in all three areas may increase; in periods where performance is sluggish, spending in all three areas may be cut. Firm strategy guides resource allocation to the various firm functions. As a result, it is common for spending and investment variables to be correlated in time series data.

Many economic indicators also move together across time. In times of economic prosperity, GDP is growing faster, consumer expectations increase, and investments increase. Increasing wealth filters down from the economy to consumers and stock holders, where some proportion of gains are channeled back into consumption of investments.

It is common for decision variables, past performance, and leading indicators to be correlated in time series data. This inherent correlation of performance drivers in time series data makes logical choice of drivers a critical component of good model building.

It is also often more promising to build models by adding variables, one at a time, looking at residuals for indications of the most promising variables to add next. Multicollinearity, including its consequences, diagnosis and alternate remedies, is further considered in this chapter.

Example 11.1 Home Depot Revenues[1]

Home Depot executives were concerned in early 2012 that revenues were not yet recovering from the economic recession of 2008–2009. Quarterly revenues had been down the first quarter of 2011, relative to the first quarter in 2010. The U.S. economy had recovered, though growth was slow. The financial crisis had reduced lending, and *new home sales* had slowed. Traditionally, Home Depot Revenues have grown following growth in *New Home Sales*, since builders and homeowners buy construction materials, flooring, and appliances at Home Depot.

[1] This example is a hypothetical scenario based on actual data.

Lowes' business was similar to Home Depot's business. Both firms' revenues were seasonal and linked to the housing market, though Lowes offered installation services and Home Depot had not. Amanda was not sure whether Lowes revenues had a positive or a negative impact on Home Depot revenues. Whenever either firm advertised or promoted home improvement items, later sales at both tended to be higher. Nonetheless, the two firms were competing for the business of many of the same customers. It was not clear whether *Lowes*, Home Depot's major competitor, was helping to expand the home improvement market, or taking business from Home Depot.

11.2 Indicators of Economic Prosperity Lead Business Performance

A *leading indicator* model links changes in a leading indicator and later performance:

$$re\widehat{ve}nue(\$B)_q = b_0 + b_1 \times GDP(\$T)_{q-l} + b_2 \times New\ Home\ Sales(K)_{q-l}$$

where l denotes the length of lag, or delay from change in GDP or new home sales to change in revenues.

Amanda, a recent business school graduate with modeling expertise, was asked to build a model of Home Depot Revenues, which would both explain revenue fluctuations and forecast revenues in the next four quarters.

Home Depot executives wanted to know how strongly

- Growth in past GDP, and
- Growth in past *New Home Sales*
- Growth in *Lowes revenues*

influenced revenues.

After being briefed by the executives, Amanda created a model reflecting their logic. She considered as possible drivers in her model:

- *New home sales*$(K)_{q-l}$
- *GDP*$(\$B)_{q-l}$
- *Lowes revenues*$_{q-l}$

11.3 Hide the Two Most Recent Datapoints to Validate a Time Series Model

Amanda used datapoints for quarterly revenue in 2004 through 2011, including quarters before, during and after the financial crisis and recession.

Before Amanda proceeded further, she excluded the two most recent observations from third and fourth quarter 2011. These *hold out* observations would allow her to compare forecasts for the two most recent periods with actual revenues to *validate* her model. If the 95 % prediction intervals from the model contained the actual revenues for both quarters, she would be able to conclude that her model is valid. She could then use the model to forecast with confidence.

11.4 Compare Scatterplots to Choose Driver Lags: Visual Inspection

The potential drivers each reflect economic conditions and move together over time. Consequently, they are highly correlated. Including all of the drivers in a multiple regression model at once would introduce a high degree of multicollinearity and make it difficult to identify each of their marginal impacts. To most effectively build a time series model, start with one driver, and then add additional drivers, one at a time.

Amanda began by plotting *Home Depot revenues* quarter. She focused on pattern in recent quarters, since her goal was to build a model which produced valid forecasts. She added a trend line for reference. (The trend is the average linear growth over the series.) She noted that revenues grew through mid 2006, then began declining late in 2006 through mid 2010. Revenue losses seemed to be linked to the housing bubble of 2007–2008 and to the recession of 2008–2009. In recent quarters, from mid 2010, revenues had again begun growing. Revenues were also seasonal. Her scatterplot is shown in Fig. 11.2.

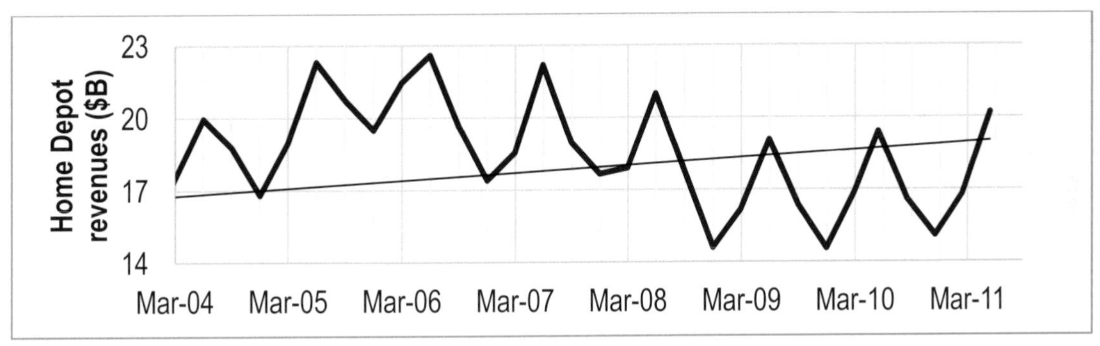

Fig. 11.2 Home Depot revenues by quarter

To account for the 2008–2009 recession, Amanda focused on quarterly GDP. She considered both 8 and 12 quarter lags, thinking that the impact of changes in the U.S. economy on home improvement was most likely 2 or 3 years later. This made the relevant series of quarters those which began 3 years before the Home Depot series began, March 2001, through June 2009, 2 years before the Home Depot revenue series ended.

Amanda plotted this quarterly GDP series, and then added a trendline. GDP growth was slowing through mid 2003, then growth accelerated through 2007, before slowing in more recent quarters. The plot of quarterly GDP ($T) is shown in Fig. 11.3.

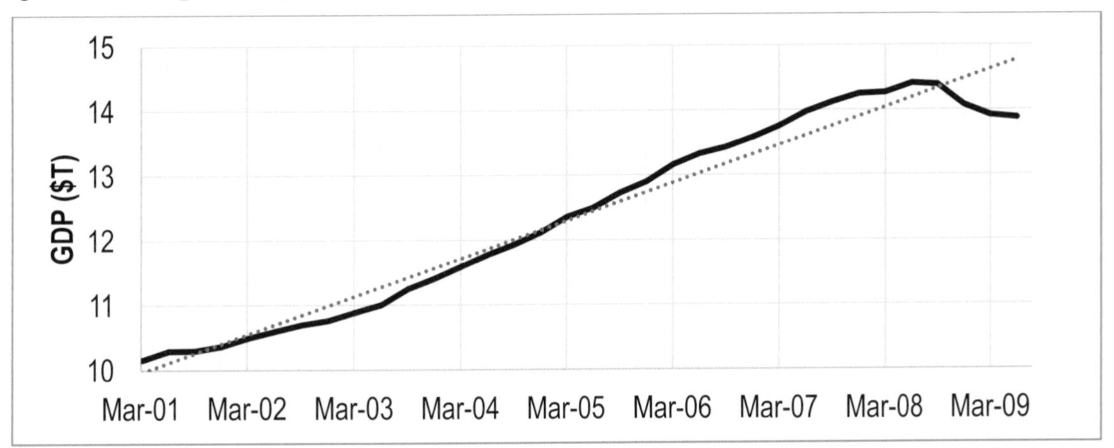

Fig. 11.3 GDP by quarter

In order to choose between the 2 or the 3 year lag, Amanda lined up Home Depot revenues with each. The 2 year lag is shown in Fig. 11.4.

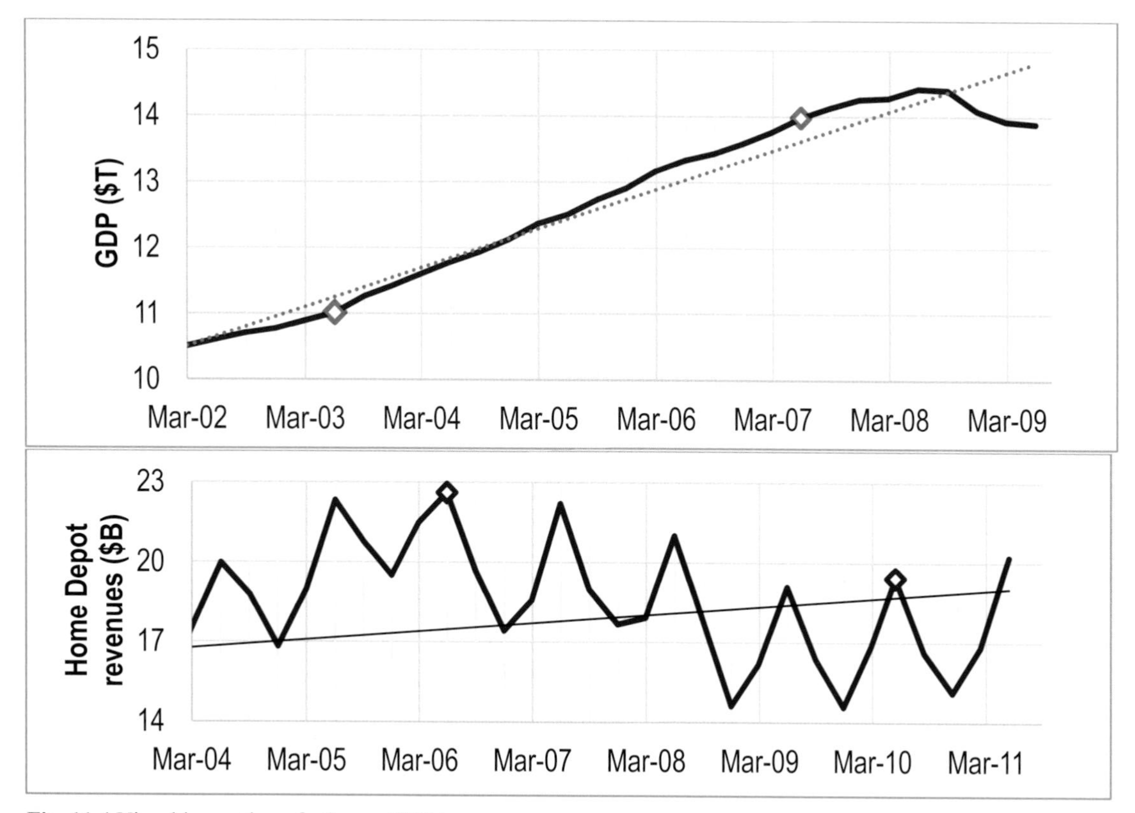

Fig. 11.4 Visual inspection of a 2 year GDP lag

Pivotal quarters in GDP (noted with markers) precede pivotal quarters in Home Depot revenues by about four quarters. The 2 year lag is not a good match. A 3 year lag seems promising.

The 3 year GDP lag with Home Depot revenues in shown in Fig. 11.5.

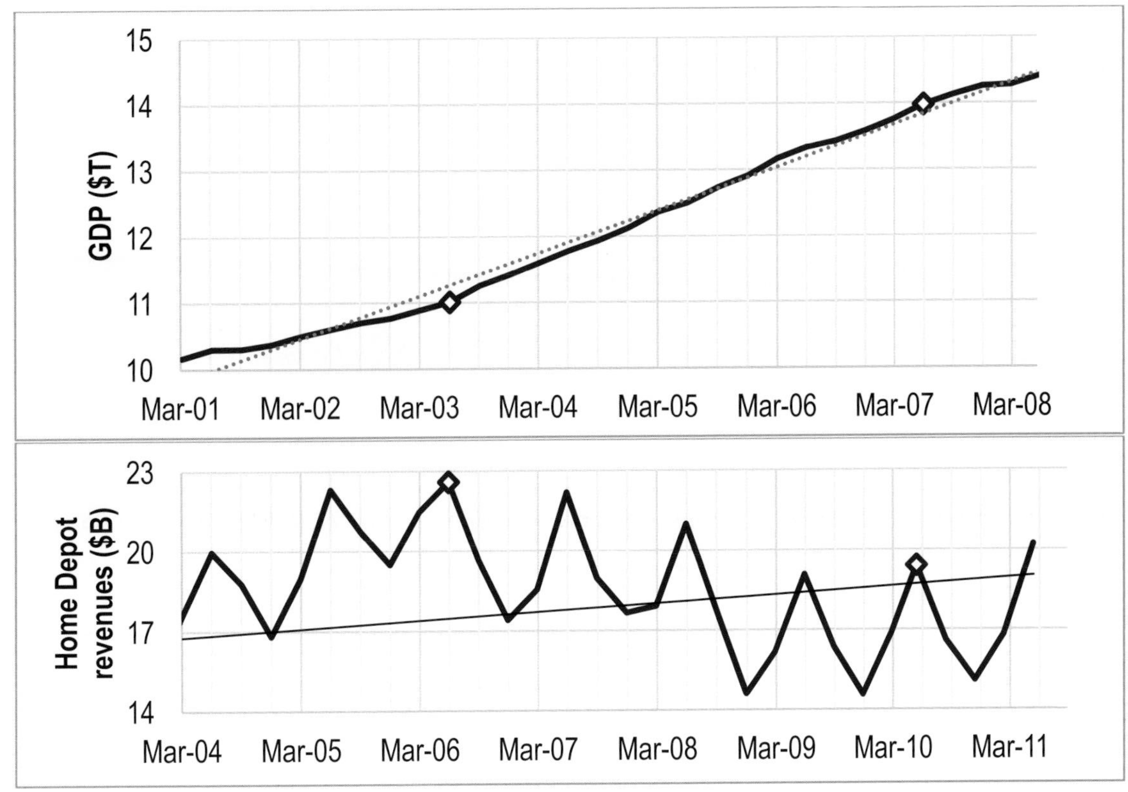

Fig. 11.5 Visual inspection of a 3 year GDP lag

The 3 year lag pairs pivotal quarters in the two series. Home depot revenues improve following a slow down in the U.S. economy, perhaps because home owners elect to improve, rather than to sell and buy a larger or newer home.

To confirm her conclusions from visual inspection, Amanda ran correlations between *Home Depot revenues* and 8 and 12 quarter *GDP* lags, shown in Table 11.1.

Table 11.1 Correlations between Home Depot revenues and past GDP

	Home Depot revenues (B$) $_q$
Gross domestic product ($T) $_{q-8}$	−.475
Gross domestic product ($T) $_{q-12}$	−.482

The GDP lags are equally correlated with revenues, since correlation weights each quarter equally. The longer lag pattern provides a better match and allows for longer, 12 quarter forecasts. Amanda ran simple regression using the 12 quarter lag. Results are shown in Table 11.2

Table 11.2 Regression with past GDP

SUMMARY OUTPUT

Regression Statistics						
R Square	.23					
Standard Error	1.97					
Observations	30					
ANOVA						
	df	*SS*	*MS*	*F*	*Significance F*	
Regression	1	33	32.9	8.5	.007	
Residual	28	109	3.9			
Total	29	142				
	Coefficients	*Standard Error*	*t Stat*	*P-value*	*Lower 95%*	*Upper 95%*
Intercept	27.60	3.14	8.8	.000	21.17	34.04
$GDP(T\$)q_{-12}$	−.75	.26	−2.9	.007	−1.27	−.22

Regression results suggest that past *GDP* drives *Home Depot revenues* 12 quarters later, accounting for 23 % of quarterly variation over the past 8 years.

11.5 The Durbin Watson Statistic Identifies Positive Autocorrelation

The Durbin Watson (*DW*) statistic incorporates correlation between residuals across adjacent time periods which allows assessment of the presence of unaccounted for trend or cycles in the residuals. If there is an unaccounted for cycle or trend, higher residuals are likely to be followed by similar higher residuals, and lower residuals are likely to be followed by similar lower residuals.

In Amanda's model, past *GDP* accounts for trend and the economic cycle in *Home Depot revenues.* If the Home Depot trend differs from the trend in past GDP, or if cyclicality in revenues differs from cyclicality in the economy, there may be unaccounted for pattern in the residuals.

DW indicates such *positive autocorrelation,* the correlation of residuals over time, which signals that a trend or cycle has been ignored. The Durbin Watson statistic compares the sum of squared differences between pairs of adjacent residuals with the sum of squared residuals:

$$DW = \frac{\sum_2^N (e_q - e_{q-1})^2}{\sum_1^N e_q^2}.$$

If all of the trend, seasons, and cycles in the data have been accounted for *DW* will be "high." Exactly how high depends on the length of time series, which is the number of observations used in the regression, and the number of independent variables, including the intercept. *DW* critical values are available online at stanford.edu/~clint/bench/dwcrit.htm, found by googling "Durbin Watson critical values." (In this online table, sample size is indexed by *T*, and the number of independent variables, plus intercept, is indexed by *K*.)

There are two relevant critical values, a lower value and an upper value, dL and dU.

DW below the lower critical value, dL, indicates presence of positive autocorrelation from unaccounted for trend, cycle, or seasons which we would then attempt to identify and incorporate into the model.

DW above the upper critical value, dU, indicates lack of autocorrelation and freedom from unaccounted for trend, cycle or seasons, which is the goal.

DW between dL and dU is the gray area, indicative of possible autocorrelation and presence of unaccounted for trend, cycle or seasons. When DW is in the gray area, we look for pattern in the residuals from unaccounted for trend, cycle or seasons, knowing that there is a reasonable chance that pattern may not be identified.

Figure 11.6 illustrates critical values for several sample and model sizes:

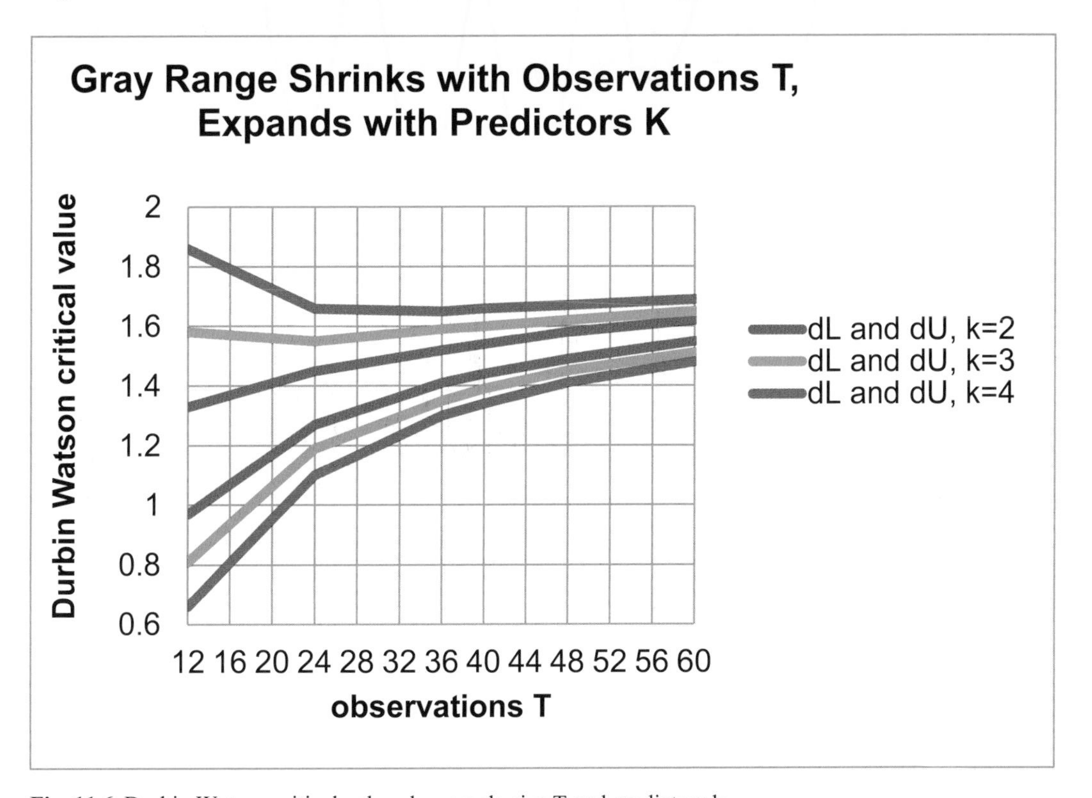

Fig. 11.6 Durbin Watson critical values by sample size T and predictors k

Notice that the gray area, dL to dU, shrinks as sample size increases, but expands as the number of predictors increases.

Amanda's initial model, with one driver, plus intercept, and a sample size of 30, has *DW* critical values of $dL = 1.35$ and $dU = 1.49$. The model *DW* statistic is 1.52, leading to the conclusion that the residuals are free of positive autocorrelated. The trend and cycle in past GDP match well the trend and cycle in Home Depot revenues.

11.6 Assess Residuals to Identify Unaccounted for Trend or Cycles

RSquare is just 23 %. There remains variation in revenues to be explained. Visual inspection of the residuals can provides clues which may suggest which potential driver to add to the model. Figure 11.7 contains a scatterplot of residuals by quarter.

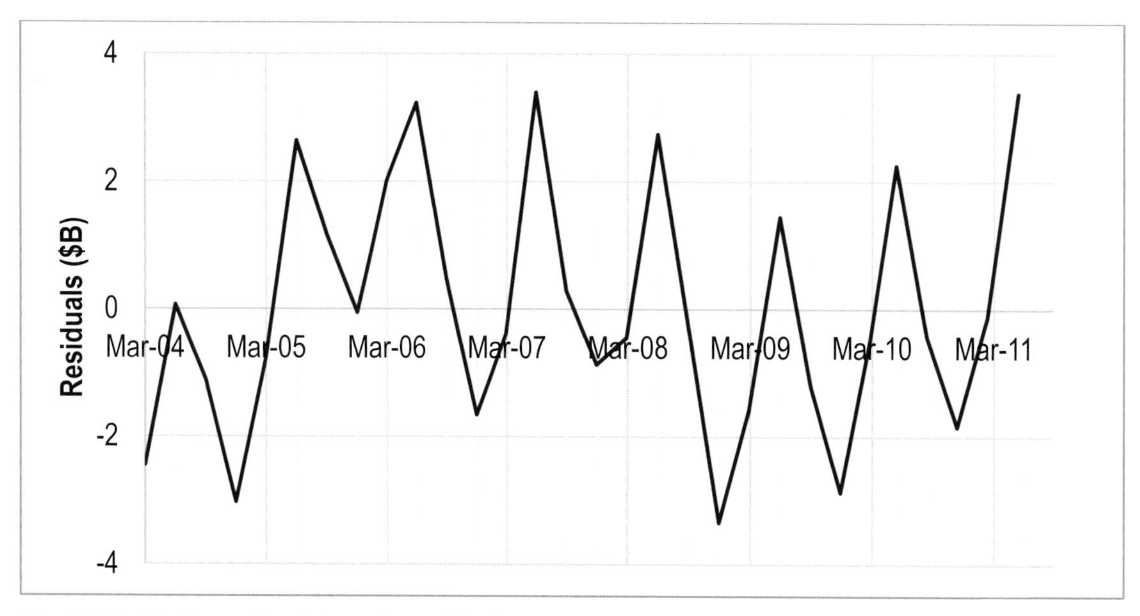

Fig. 11.7 Residuals reveal variation to be explained

Seasonality (*negative autocorrelation*) is apparent, particularly in more recent quarters. This pattern is similar to *Lowes revenue* patterns. Lining up the two plots, an eight quarter lag in *Lowes revenues* promises to improve the model (Fig. 11.8).

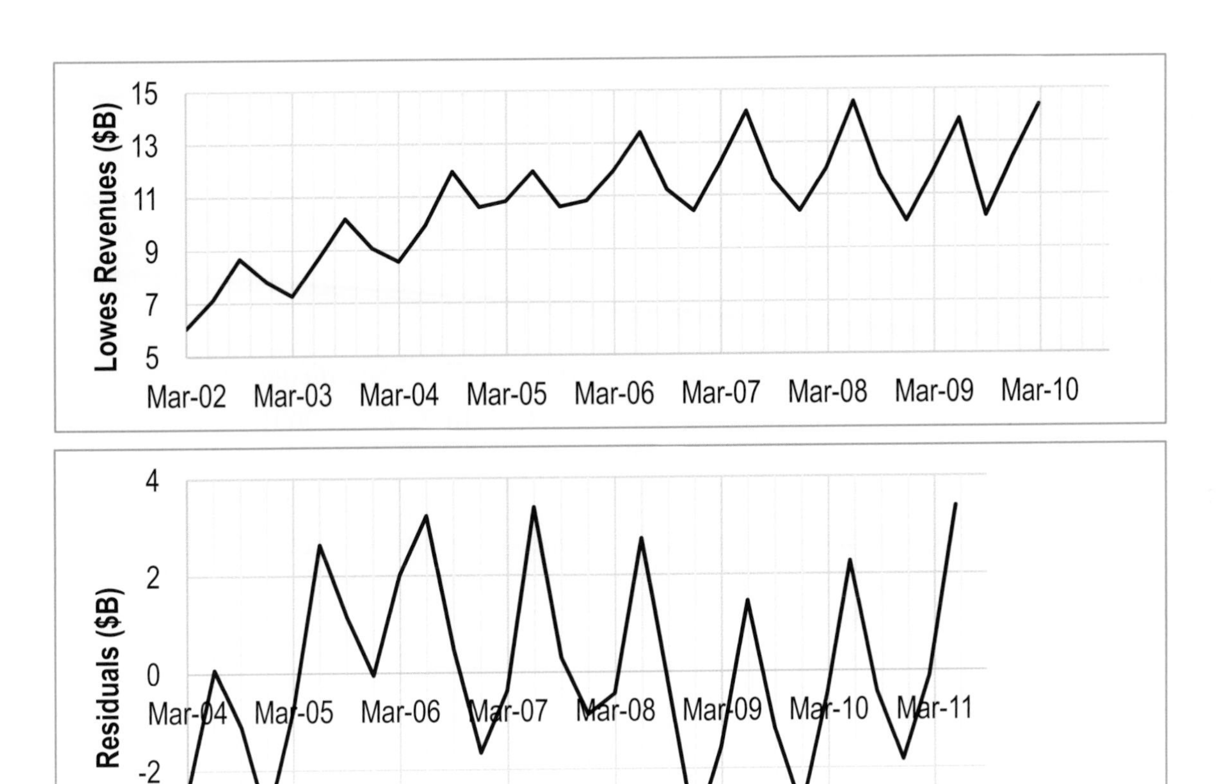

Fig. 11.8 Lowes revenues ($B) and residuals ($B)

To backup her conclusions from visual inspection, Amanda compared correlations between the residuals and six- and eight-quarter *Lowes revenue* lags, shown in Table 11.3.

Table 11.3 Correlations between residuals and past *Lowes revenue*

	Lowes revenues (B$) $_{q-8}$	*Lowes revenues (B$)* $_{q-6}$
Residuals	.40	−.23

The residual correlation with the eight quarter lag is stronger. Amanda added *Lowes Revenues* with an eight quarter lag to the model and ran multiple regression. Results are shown in Table 11.4.

Table 11.4 Home Depot revenue regression with past GDP and past Lowes revenue

SUMMARY OUTPUT

Regression Statistics	
R Square	.51
Standard Error	1.60
Observations	30

ANOVA

	df	*SS*	*MS*	*F*	*Significance F*	
Regression	2	73	36.3	14.2	.0001	
Residual	27	69	2.6			
Total	29	142				
	Coefficients	*Standard Error*	*t Stat*	*p value*	*Lower 95 %*	*Upper 95 %*
Intercept	30.03	2.62	11.4	.0000	24.65	35.42
GDP (\$T) $_{q-12}$	−1.69	.32	−5.3	.0000	−2.34	−1.04
Lowes Revenues(\$B) $_{q-8}$	.85	.22	3.9	.0005	.41	1.29

dw	T	k	dL	dU
1.31	30	3	1.28	1.57

Together, past *GDP* and *Lowes Revenues* drive *Home Depot revenues. RSquare* is now 51 %. The model *standard error* has been reduced from $2.0B to $1.6B, reducing the forecast margin of error to $3.3B.

The Durbin Watson statistic, 1.31, is now in the grey area, which may be due to differences in trend or cycles between the two home improvement businesses. *RSquare* is just 51 %. Variation remains to be explained.

Amanda examined a plot of the residuals, this time with a plot of *new home sales*, shown in Fig. 11.9. In more recent quarters, *new home sales* were declining, and had not recovered from the housing bubble of 2007–2008, the financial crisis and recession of 2008–2009. The residuals, in contrast, show recovery, increasing from late 2008, perhaps because the recession motivated home improvement rather than new home purchases.

To complement visual inspection, Amanda viewed correlations between eight and ten quarter lags of *new home sales* and residuals, shown in Table 11.5.

Table 11.5 Correlations between past new home sales and residuals

	New home sales $_{q-10}$	*New home sales (K)* $_{q-8}$
Residuals	−.45	.00

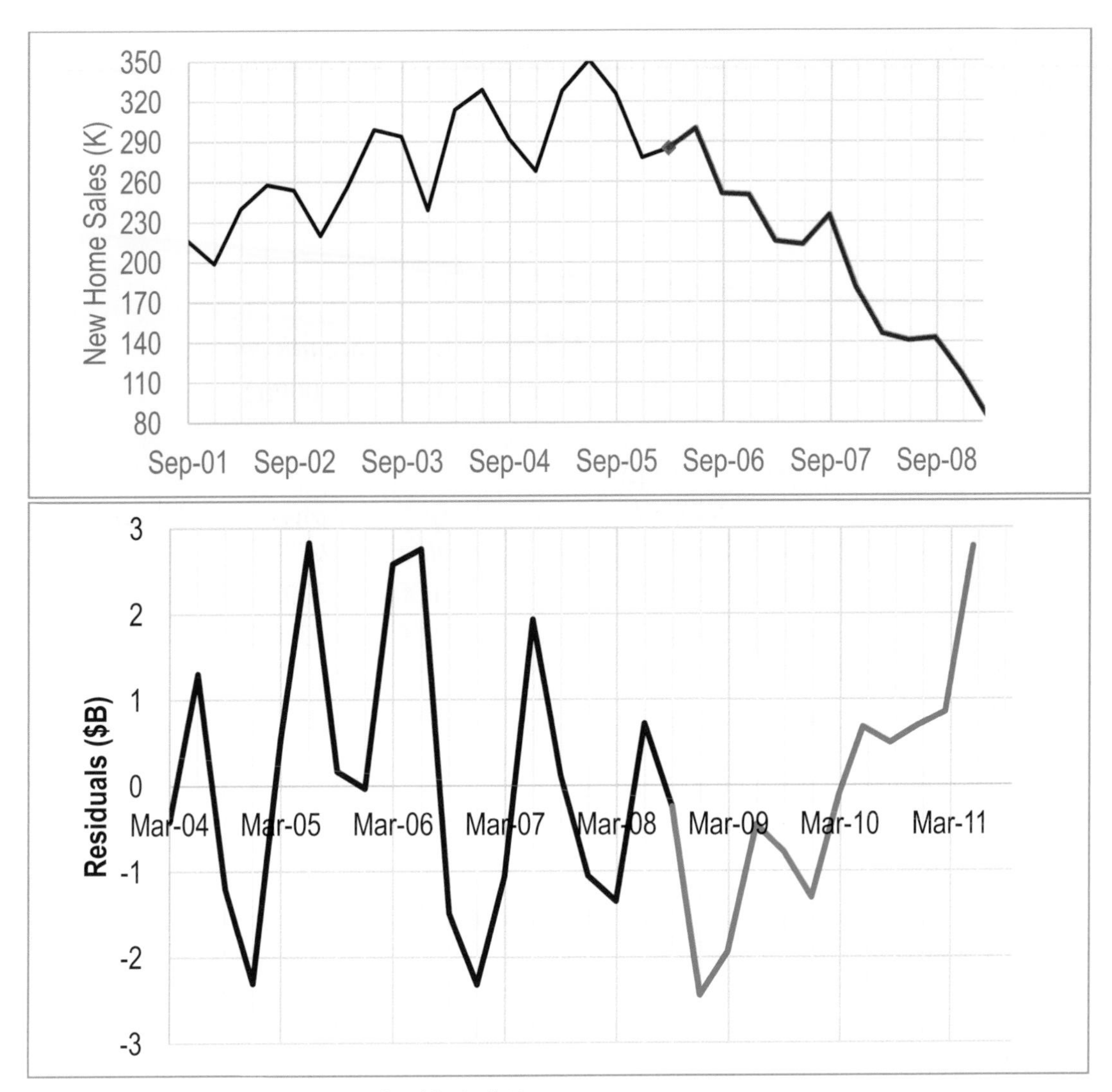

Fig. 11.9 Past new home sales (K) and residuals ($B).

Amanda added the ten quarter lag of *new home sales* to the model. Regression results are shown in Table 11.6.

Table 11.6 Regression with leading indicators and competition

SUMMARY OUTPUT

Regression Statistics							
R Square	.67						
Standard Error	1.34						
Observations	30						
ANOVA							
	df	*SS*	*MS*	*F*	*Significance F*		
Regression	3	95	31.6	17.6	.000002		
Residual	26	47	1.8				
Total	29	142					
	Coefficients	*Standard Error*	*t Stat*	*p value*	*Lower 95 %*	*Upper 95 %*	
Intercept	40.052	3.590	11.2	.0000	32.672	47.431	
GDP (B\$) $_{q-12}$	−2.421	.337	−7.2	.0000	−3.114	−1.728	
Lowes Revenues (B\$) $_{q-6}$	1.169	.202	5.8	.0000	.753	1.585	
new home sales (K) $_{q-4}$	−.018	.005	−3.5	.0016	−.029	−.008	
DW	*dL*	*dU*					
1.28	1.21	1.65					

The addition of past new home sales improved the model. *RSquare* is now 67 %. New home sales increased explanatory power. The model *standard error,* \$1.3B is now smaller, reducing the margin of error in predictions to \$2.8B.

All three marginal slopes differ from zero. The marginal slope for *Lowes revenues* is positive, indicating that the competitor's promotions and marketing expand the home improvement market and benefit *Home Depot revenues.* The coefficients for GDP and *new home sales* are negative. The home improvement business apparently improves when new homes are unaffordable.

The Durbin Watson statistic, $DW = 1.28$, remains in the grey area, suggesting that unaccounted for trend or cycles may exist in the residuals.

11.7 Forecast the Recent, Hidden Points to Assess Predictive Validity

With a significant model, logically correct coefficient signs, and residuals possibly free of autocorrelation, Amanda could proceed to assess the predictive validity of her model by comparing actual *Home Depot revenues (\$B)* in the two most recent quarters with the model's 95 % prediction intervals. (Recall that those two most recent quarters were hidden and not used in the regression to fit the model.) Validation evidence is shown in Table 11.7.

Table 11.7 Model predictions include actual values

Quarter	*Lower 95 % prediction*	*Home Depot revenues ($B)*	*Upper 95 % prediction*
Sep-11	12.8	17.3	18.3
Dec-11	15.8	16.0	21.3

11.8 Add the Most Recent Datapoints to Recalibrate

With evidence of predictive validity, Amanda used the model to forecast revenues in the next four quarters. Before making the forecast, she added the two most recent observations that were hidden to validate. The recalibrated model became:

$$revênues(\$B)_q = 39^a - 2.28^a \times GDP(\$B)_{q-12}$$

$$+1.06^a \times Lowes\ revenue(\$B)_{q-8}$$

$$-\ .017^a \times New\ Home\ Sales(K)_{q-10}$$

RSquare: $.63^a$
[a] *Significant* at .01.

Variation in past U.S. GDP, *Lowes revenue,* and *new home sales*, together account for 63 % of the quarterly variation in *Home Depot revenues*. Using this multiple linear regression model, forecast quarterly revenues are expected to fall within $2.9B of predictions.

The economy is the strongest driver of revenues. Revenues follow longer economic cycles. For a typical recent increase in U.S. GDP of $80B in a quarter, revenues are expected to decline by about $190M (= −2.28($B/$T) × $.08T) 3 years later.

Changes in the housing market influence revenues to a smaller degree. Revenues follow the changes in the housing market ten quarters later. In recent quarters, a decline of 5,000 (K) new homes sold would be typical. Following such a typical decline in a quarter, Home Depot revenues are expected to improve by about $90M (= −.017($B/K) × (−5 (K))) ten quarters later.

Revenues are higher when revenues of Lowes, a major competitor, grow. Lowes revenues have declined by an average of about $100 million in recent quarters. Following similar declines by competitor Lowes, Home Depot revenues are expected to decline by about $100M (= 1.06 ($B/$B) × −$.1B) eight quarters later.

Model forecasts are shown in Fig. 11.10 and Table 11.8.

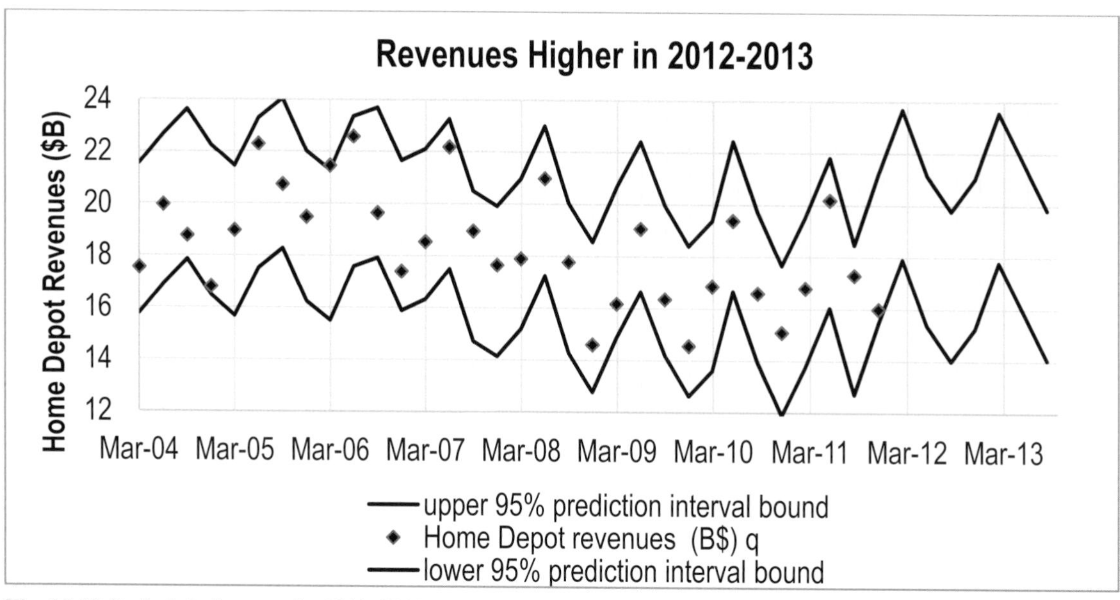

Fig. 11.10 Optimistic forecast for 2012–2013

Table 11.8 Quarterly revenue forecast

Quarter	*95 % lower prediction (B$)*	*95 % upper prediction ($B)*	*Prior year Revenues ($B)*	*Forecast annual growth for quarter*
Mar-12	17.9	23.7	16.8	24 %
Jun-12	15.4	21.2	20.2	−9 %
Sep-12	14.0	19.8	17.3	−2 %
Dec-12	15.3	21.0	16.0	13 %
Mar-13	17.8	23.6	20.8[a]	−1 %
Jun-13	14.1	19.8	20.7	24 %

[a]Forecast growth for March 2013–Sep 2013 is relative to revenues forecasts

Annual quarterly growth (from same quarter in the past year) has been 0–6 % in the past six quarters. Home Depot revenues are expected to vary more in the next six quarters, declining as much as 9 %, and increasing as much as 24 %, with average quarterly growth of 3.2 % over same quarter in the past year, on average.

Amanda summarized her model results for Management:

MEMO

Re: Revenue Recovery Forecast for 2012 through 2013
To: Home Depot Management
From: Amanda Chanel
Date: June 2010

Following past growth in the U.S. economy, new home sales, and competitor sales, quarterly revenues are expected to increase an average of 3 % annually over the next six quarters.

A regression model of quarterly revenues was built from past U.S. GDP, new home sales, and Lowes revenues. The model accounts for 63 % of the variation in revenues and produces valid forecasts within $2.9 billion of actual revenues.

Revenues are driven by past economic productivity

Revenues have benefitted from slowed economic growth and declining new home sales, as customers substitute home improvement for home purchases.

Declining new home sales are expected to improve quarterly revenues by an average of $90M in future quarters; however, slowed past annual growth in quarterly GDP, the strongest drives, is expected to reduce quarterly revenues by an average of $190M in future quarters.

Declining Lowes revenue is expected to reduce quarterly revenues by an average of $130 M in future quarters.

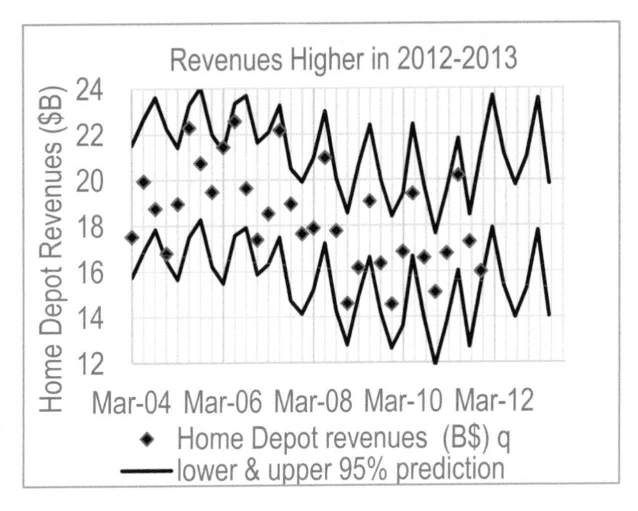

With the negative impacts of economic growth and reduced competitor sales and the positive impact of slowed new home sales, average annual revenue growth of 3 % is expected in future quarters.

$$\begin{aligned} \widehat{revenues}(\$B)_q &= 39^a - 2.3^a \times GDP(\$B)_{q-12} \\ &+1.1^a \times Lowes\ revenue(\$B)_{q-8} \\ &- .017^a \times New\ Home\ Sales(K)_{q-10} \end{aligned}$$

RSquare: $.63^a$
^{a}*Significant* at .01.

Recovering growth is forecast

Revenues are driven by past economic performance. Recovering growth is forecast in late 2012 and 2013.

Revenue response may be nonlinear

This model assumes constant response, though nonlinear response is possible.

Quarter	*Forecast ($B)*	*Forecast growth*
12-I	17.9 to 23.7	24 %
12-II	15.4 to 21.2	−9 %
12-III	14.0 to 19.8	−2 %
12-IV	15.3 to 21.0	13 %
13-I	17.8 to 23.6	−1 %
13-II	14.1 to 19.8	24 %

11.9 Leading Indicator Components Are Powerful Drivers and Often Multicollinear

Like cross sectional models, time series models allows identification of performance drivers and forecasts of performance. However, time series models differ from cross sectional models, and the model building process with time series contains additional steps.

- Often lagged predictors are used to make driver identification more certain and to enable forecasts.
- Lagged predictors tend to move together across time and are often highly correlated. Consequently, to minimize multicollinearity issues, model building begins with one predictor, and then others are added, considering their joint influence and incremental model improvement.
- Forecasting accuracy of time series models is tested, or validated, before they are used for prediction of future performance.

Predictors in time series models tend to be highly correlated, since most move with economic variables and most exhibit predictable growth (*trend*). Model building with time series begins with the strongest among logical predictors, and additional predictors are added which improve the model.

Time series typically contain trend, business cycles, and seasonality that are captured with these components. Unaccounted for trend, cycles, or seasonality are detected through inspection of the residual plot and the Durbin Watson statistic. Leading indicators are often stable and predictable performance drivers. Competitive variables may account for trend, seasonality or cycles common to a market.

Useful forecasting models must be valid. Holding out the two most recent performance observations allows a test of the model's forecasting capability. With successful prediction of the most recent performance, the model is validated and the recalibrated model can be used with confidence to forecast performance in future periods.

Excel 11.1 Build and Fit a Multiple Regression Model with Multicollinear Time Series

WFM Revenues. Build a model of WFM quarterly revenues which potentially includes past economic growth, growth in the housing market, and variation in a competitor's revenues. The data are in 11 WFM quarterly revenues.xls.

Plot *WFM revenues* by quarter to see the pattern of movement over time. (Hide or ignore the two most recent datapoints, September 2011 and December 2011.)

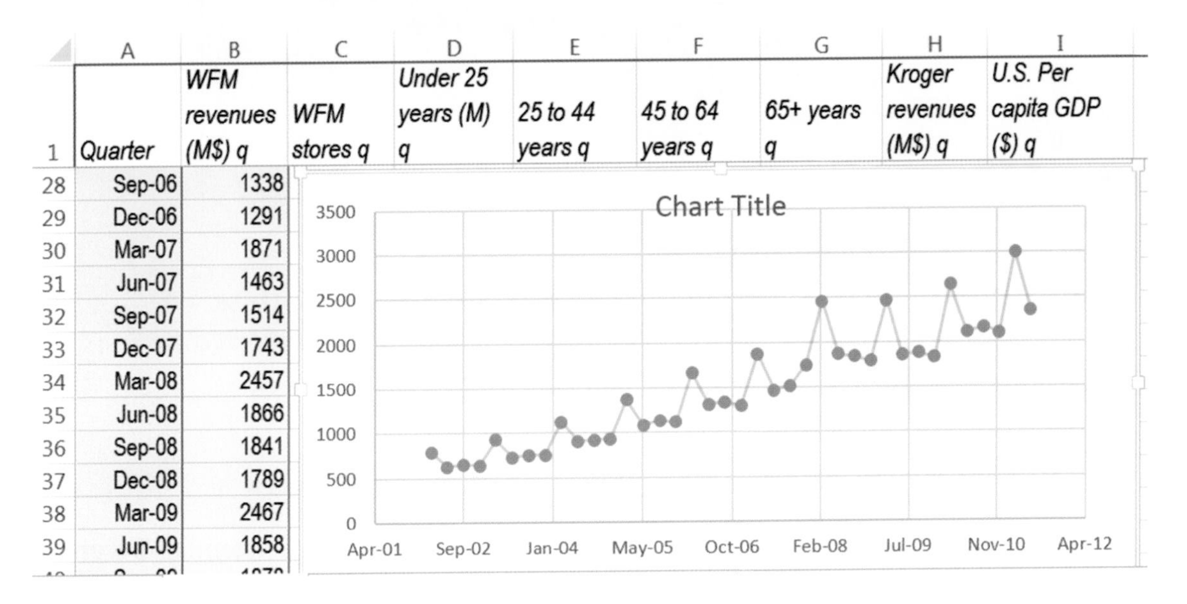

	A	B	C	D	E	F	G	H	I
1	Quarter	WFM revenues (M$) q	WFM stores q	Under 25 years (M) q	25 to 44 years q	45 to 64 years q	65+ years q	Kroger revenues (M$) q	U.S. Per capita GDP ($) q
28	Sep-06	1338							
29	Dec-06	1291							
30	Mar-07	1871							
31	Jun-07	1463							
32	Sep-07	1514							
33	Dec-07	1743							
34	Mar-08	2457							
35	Jun-08	1866							
36	Sep-08	1841							
37	Dec-08	1789							
38	Mar-09	2467							
39	Jun-09	1858							

In Excel scatterplots, time is measured in days. To set the quarter axis beginning and end points, format the axis, with minimum 373,200, major units 366 (1 year), and minor units 91.5 (one quarter).

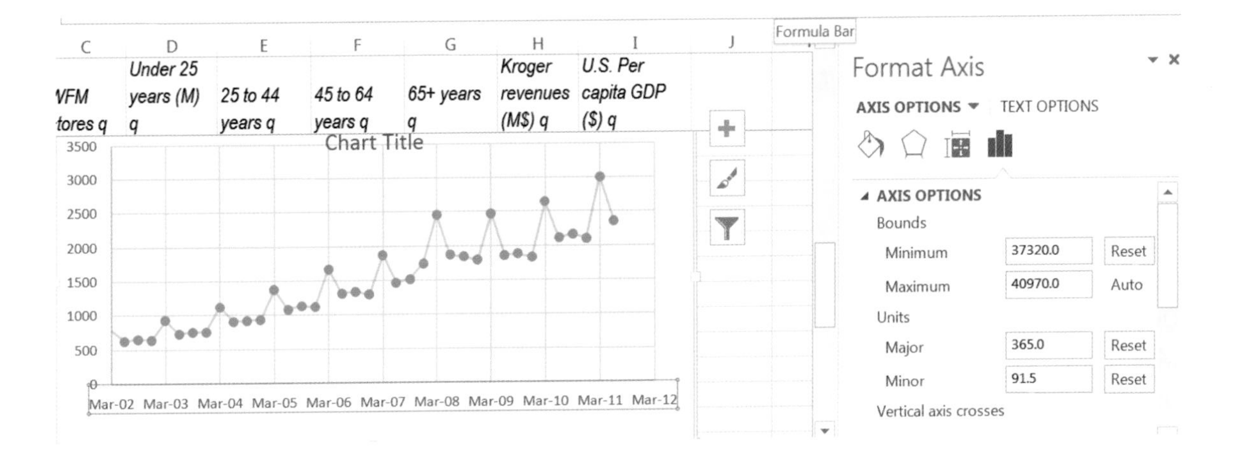

Add a trendline:

Alt JC A T L

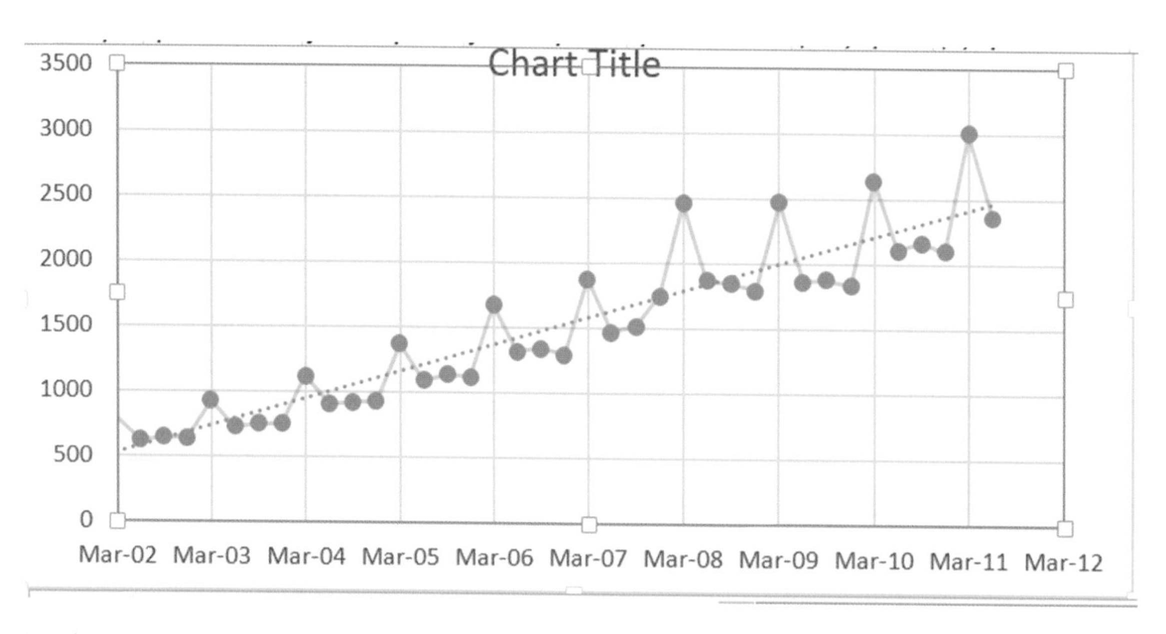

Notice that revenues grew show a positive trend with obvious seasonality.

Excel 11.2 Create Driver Lags

Demographic data are available through December 2010. In order use these potential drivers to make a forecast through December 2012, each must be lagged by eight quarters.

	A	B	C	D	E	F	G	H
1	Quarter	WFM revenues (M$) q	Under 25 years (M) q	25 to 44 years (M) q	45 to 64 years (M) q	65+ years (M) q	Kroger revenues (M$) q	U.S. Per capita GDP ($) q
44	Sep-10	2163	105.0	82.24	82.5	46.5	18694	36208
45	Dec-10	2097	105.1	82.24	82.9	46.7	19928	36436
46	Mar-11	3004					27461	36834
47	Jun-11	2351					20913	37020
48	Sep-11	2400						36964
49	Dec-11	2354						
50	Mar-12							
51	Jun-12							
52	Sep-12							
53	Dec-12							

Create working columns that will contain the lagged drivers by copying columns **A** through **H** and then pasting into column **I:**

	H	I	J	K	L	M	N	O	P
1	U.S. Per capita GDP ($) q	Quarter	WFM revenues (M$) q	Under 25 years (M) q	25 to 44 years (M) q	45 to 64 years (M) q	65+ years (M) q	Kroger revenues (M$) q	U.S. Per capita GDP ($) q
2	25506	Mar-00		99.7	84.97	62.4	39.3	14329	25506
3	25817	Jun-00		99.9	84.86	62.9	39.4	11017	25817
4	26170	Sep-00		100.2	84.75	63.5	39.5	10962	26170
5	26282	Dec-00		100.4	84.64	64.0	39.5	12692	26282
6	26602	Mar-01		100.7	84.52	64.5	39.6	15102	26602
7	26591	Jun-01		100.9	84.39	65.0	39.7	11485	26591
8	27200	Sep-01		101.0	84.26	65.6	39.7	11382	27200
9	26827	Dec-01		101.2	84.12	66.1	39.8	12129	26827
10	27552	Mar-02	791	101.4	83.99	66.7	39.9	15667	27552
11	27856	Jun-02	623	101.6	83.84	67.2	40.0	11927	27856
12	27829	Sep-02	649	101.7	83.69	67.8	40.1	11696	27829

Create the eight quarter lags on each of the demographic driver columns. Selecting the first eight cells of each,

	H	I	J	K	L	M	N	O
1	U.S. Per capita GDP ($) q	Quarter	WFM revenues (M$) q	Under 25 years (M) q	25 to 44 years (M) q	45 to 64 years (M) q	65+ years (M) q	Kroger revenues (M$) q
2	25506	Mar-00		99.7	84.97	62.4	39.3	14329
3	25817	Jun-00		99.9	84.86	62.9	39.4	11017
4	26170	Sep-00		100.2	84.75	63.5	39.5	10962
5	26282	Dec-00		100.4	84.64	64.0	39.5	12692
6	26602	Mar-01		100.7	84.52	64.5	39.6	15102
7	26591	Jun-01		100.9	84.39	65.0	39.7	11485
8	27200	Sep-01		101.0	84.26	65.6	39.7	11382
9	26827	Dec-01		101.2	84.12	66.1	39.8	12129
10	27552	Mar-02	791	101.4	83.99	66.7	39.9	15667

then insert cells and shift the selected cells down with

Alt H I I D

	H	I	J	K	L	M	N	O
1	U.S. Per capita GDP ($) q	Quarter	WFM revenues (M$) q	Under 25 years (M) q	25 to 44 years (M) q	45 to 64 years (M) q	65+ years (M) q	Kroger revenues (M$) q
2	25506	Mar-00						14329
3	25817	Jun-00						11017
4	26170	Sep-00						10962
5	26282	Dec-00						12692
6	26602	Mar-01						15102
7	26591	Jun-01						11485
8	27200	Sep-01						11382
9	26827	Dec-01						12129
10	27552	Mar-02	791	99.7	84.97	62.4	39.3	15667
11	27856	Jun-02	623	99.9	84.86	62.9	39.4	11927

Population values from March 2000 now appear in the March 2002 row, eight quarter lags.

Focusing on the Kroger revenues driver, notice that a minimum lag of five quarters is required in order to use this driver for the December 2012 forecast:

	A	B	C	D	E	F	G
1	Quarter	WFM revenues (M$) q	Under 25 years (M) q	25 to 44 years (M) q	45 to 64 years (M) q	65+ years (M) q	Kroger revenues (M$) q
44	Sep-10	2163	105.0	82.24	82.5	46.5	18694
45	Dec-10	2097	105.1	82.24	82.9	46.7	19928
46	Mar-11	3004		-	-		27461
47	Jun-11	2351		-	-		20913
48	Sep-11	2400		-	-		
49	Dec-11	2354		-	-		
50	Mar-12			-	-		
51	Jun-12			-	-		
52	Sep-12			-	-		
53	Dec-12			-	-		

Since both WFM revenues and Kroger revenues are seasonal, use an eight quarter lag so that the four seasons match, with December Kroger revenues in the same row as December WFM revenues.

Insert eight cells in rows 2–9 to create the eight quarter lag:

	I	J	K	L	M	N	O
1	Quarter	WFM revenues (M$) q	Under 25 years (M) q	25 to 44 years (M) q	45 to 64 years (M) q	65+ years (M) q	Kroger revenues (M$) q
2	Mar-00						
3	Jun-00						
4	Sep-00						
5	Dec-00						
6	Mar-01						
7	Jun-01						
8	Sep-01						
9	Dec-01						
10	Mar-02	791	99.7	84.97	62.4	39.3	14329
11	Jun-02	623	99.9	84.86	62.9	39.4	11017

A minimum lag of five quarters is required for GDP in order to forecast Dec 2010 revenues.

	A	B	C	D	E	F	G	H
1	Quarter	WFM revenues (M$) q	Under 25 years (M) q	25 to 44 years (M) q	45 to 64 years (M) q	65+ years (M) q	Kroger revenues (M$) q	U.S. Per capita GDP ($) q
44	Sep-10	2163	105.0	82.24	82.5	46.5	18694	36208
45	Dec-10	2097	105.1	82.24	82.9	46.7	19928	36436
46	Mar-11	3004		-	-		27461	36834
47	Jun-11	2351		-	-		20913	37020
48	Sep-11	2400		-	-			36964
49	Dec-11	2354		-	-			
50	Mar-12			-	-			
51	Jun-12			-	-			
52	Sep-12			-	-			
53	Dec-12			-	-			

The delay between declining income, GDP per capita, and changes in food spending is probably not more than 1 year. Create six and eight quarter lags for per capita GDP, and adjust driver labels to reflect lag lengths:

	I	J	K	L	M	N	O	P	Q
1	Quarter	WFM revenues (M$) q	Under 25 years (M) q-8	25 to 44 years (M) q-8	45 to 64 years (M) q-8	65+ years (M) q-8	Kroger revenues (M$) q-8	U.S. Per capita GDP ($) q-6	U.S. Per capita GDP ($) q-8
2	Mar-00								
3	Jun-00								
4	Sep-00								
5	Dec-00								
6	Mar-01								
7	Jun-01								
8	Sep-01							25506	
9	Dec-01							25817	
10	Mar-02	791	99.7	84.97	62.4	39.3	14329	26170	25506
11	Jun-02	623	99.9	84.86	62.9	39.4	11017	26282	25817

Finish creating your lags by selecting cells in rows that contain blanks:

	I	J	K	L	M	N	O	P	Q
1	Quarter	WFM revenues (M$) q	Under 25 years (M) q-8	25 to 44 years (M) q-8	45 to 64 years (M) q-8	65+ years (M) q-8	Kroger revenues (M$) q-8	U.S. Per capita GDP ($) q-6	U.S. Per capita GDP ($) q-8
2	Mar-00								
3	Jun-00								
4	Sep-00								
5	Dec-00								
6	Mar-01								
7	Jun-01								
8	Sep-01							25506	
9	Dec-01							25817	
10	Mar-02	791	99.7	84.97	62.4	39.3	14329	26170	25506
11	Jun-02	623	99.9	84.86	62.9	39.4	11017	26282	25817

Delete these cells from rows with incomplete data and shift remaining cells up

Alt H D D U

	I	J	K	L	M	N	O	P	Q
1	Quarter	WFM revenues (M$) q	Under 25 years (M) q-8	25 to 44 years (M) q-8	45 to 64 years (M) q-8	65+ years (M) q-8	Kroger revenues (M$) q-8	U.S. Per capita GDP ($) q-6	U.S. Per capita GDP ($) q-8
2	Mar-02	791	99.7	84.97	62.4	39.3	14329	26170	25506
3	Jun-02	623	99.9	84.86	62.9	39.4	11017	26282	25817
4	Sep-02	649	100.2	84.75	63.5	39.5	10962	26602	26170
5	Dec-02	638	100.4	84.64	64.0	39.5	12692	26591	26282
6	Mar-03	924	100.7	84.52	64.5	39.6	15102	27200	26602
7	Jun-03	725	100.9	84.39	65.0	39.7	11485	26827	26591
8	Sep-03	749	101.0	84.26	65.6	39.7	11382	27552	27200
9	Dec-03	751	101.2	84.12	66.1	39.8	12129	27856	26827
10	Mar-04	1118	101.4	83.99	66.7	39.9	15667	27829	27552
11	Jun-04	902	101.6	83.84	67.2	40.0	11927	27957	27856
12	Sep-04	917	101.7	83.69	67.8	40.1	11696	28201	27829

Excel 11.3 Examine Patterns in More Recent Time Periods

To increase the chances that your model will be valid, focus on the eight most recent quarters (that aren't hidden). To more easily see recent patterns in the data, mark cells in which a value is less than the previous value:

	I	J	K	L	M	N	O	P	Q
1	Quarter	WFM revenues (M$) q	Under 25 years (M) q-8	25 to 44 years (M) q-8	45 to 64 years (M) q-8	65+ years (M) q-8	Kroger revenues (M$) q-8	U.S. Per capita GDP ($) q-8	U.S. Per capita GDP ($) q-6
31	Jun-09	1858	103.9	82.48	77.5	43.1	16139	34344	35042
32	Sep-09	1878	104.1	82.45	77.8	43.4	16135	34579	35810
33	Dec-09	1829	104.2	82.43	78.2	43.7	17235	35042	36854
34	Mar-10	2639	104.3	82.40	78.6	44.0	23137	35810	36311
35	Jun-10	2106	104.4	82.35	79.0	44.2	18088	36854	35690
36	Sep-10	2163	104.5	82.31	79.4	44.5	17615	36311	35127
37	Dec-10	2097	104.6	82.26	79.9	44.7	17308	35690	35245
38	Mar-11	3004	104.66	82.21	80.3	45.0	22789	35127	34935
39	Jun-11	2351	104.72	82.22	80.6	45.2	17728	35245	35044
40	Sep-11	2400	104.8	82.22	81.0	45.5	17662	34935	35557
41	Dec-11	2354	104.8	82.22	81.4	45.7	18554	35044	36001
42	Mar-12		104.9	82.23	81.8	46.0	24779	35557	36208

Find matching patterns. Compare the pattern in each potential driver with the dependent variable pattern. For each match in a quarter, change the font to blue. In September 2009, for example, WFM revenues have increased from the previous quarter. The number under 25 also increased from the previous quarter:

	I	J	K	L	M	N	O	P	Q
1	Quarter	WFM revenues (M$) q	Under 25 years (M) q-8	25 to 44 years (M) q-8	45 to 64 years (M) q-8	65+ years (M) q-8	Kroger revenues (M$) q-8	U.S. Per capita GDP ($) q-8	U.S. Per capita GDP ($) q-6
31	Jun-09	1858	103.9	82.48	77.5	43.1	16139	34344	35042
32	Sep-09	1878	104.1	82.45	77.8	43.4	16135	34579	35810
33	Dec-09	1829	104.2	82.43	78.2	43.7	17235	35042	36854
34	Mar-10	2639	104.3	82.40	78.6	44.0	23137	35810	36311
35	Jun-10	2106	104.4	82.35	79.0	44.2	18088	36854	35690
36	Sep-10	2163	104.5	82.31	79.4	44.5	17615	36311	35127
37	Dec-10	2097	104.6	82.26	79.9	44.7	17308	35690	35245
38	Mar-11	3004	104.66	82.21	80.3	45.0	22789	35127	34935
39	Jun-11	2351	104.72	82.22	80.6	45.2	17728	35245	35044
40	Sep-11	2400	104.8	82.22	81.0	45.5	17662	34935	35557
41	Dec-11	2354	104.8	82.22	81.4	45.7	18554	35044	36001
42	Mar-12		104.9	82.23	81.8	46.0	24779	35557	36208

The single potential driver that moves in the same direction as WFM revenues more than half of the time (at least three out of the last four quarters, and at least five out of the last eight quarters) is Kroger revenues.

Now consider inverse relationships, which remain in black font. Potential drivers that move in the opposite direction as WFM revenues more than half of the time are: the six and eight quarter per capita GDP lags.

Move columns that match well to columns following WFM revenues. Within a column, select that column

Cntl+spacebar

Cut,

Cntl+X

Move to the desired column, and drop in there:

Alt H I E

	I	J	K	L	M
1	Quarter	WFM revenues (M$) q	Kroger revenues (M$) q-8	U.S. Per capita GDP ($) q-8	U.S. Per capita GDP ($) q-6
31	Jun-09	1858	16139	34344	35042
32	Sep-09	1878	16135	34579	35810
33	Dec-09	1829	17235	35042	36854
34	Mar-10	2639	23137	35810	36311
35	Jun-10	2106	18088	36854	35690
36	Sep-10	2163	17615	36311	35127
37	Dec-10	2097	17308	35690	35245
38	Mar-11	3004	22789	35127	34935
39	Jun-11	2351	17728	35245	35044
40	Sep-11	2400	17662	34935	35557
41	Dec-11	2354	18554	35044	36001
42	Mar-12		24779	35557	36208

To select the first driver for your model, compare correlations with WFM revenues:

	A	B
1		WFM revenues (M$) q
2	WFM revenues (M$) q	1
3	Kroger revenues (M$) q-8	0.915572
4	U.S. Per capita GDP ($) q-8	0.909519
5	U.S. Per capita GDP ($) q-6	0.902758

Run regression with Kroger revenues:

	A	B	C	D	E	F	G
1	SUMMARY OUTPUT						
2							
3	*Regression Statistics*						
4	Multiple R	0.915572					
5	R Square	0.838273					
6	Adjusted F	0.83378					
7	Standard I	257.7033					
8	Observati(	38					
9							
10	ANOVA						
11		*df*	*SS*	*MS*	*F*	*gnificance F*	
12	Regressior	1	12392078	12392078	186.5969	8.22E-16	
13	Residual	36	2390795	66410.97			
14	Total	37	14782873				
15							
16		*Coefficients*	*andard Erro*	*t Stat*	*P-value*	*Lower 95%*	*Upper 95%*
17	Intercept	-1293.55	208.6567	-6.1994	3.77E-07	-1716.72	-870.371
18	Kroger rev	0.184661	0.013518	13.66005	8.22E-16	0.157245	0.212077
19							
20							
21							
22	RESIDUAL OUTPUT						
23							
24	*)bservatior*	*VFM reven*	*Residuals*				
25	1	1352.461	-561.461				
26	2	740.8636	-117.864				

Excel 11.4 Plot Residuals to Identify Unaccounted for Trend, Cycles, or Seasonality and Assess Autocorrelation

Unaccounted for trend, cycles, or seasonality produce positive autocorrelation in the residuals. The Durbin Watson statistic will allow assessment of positive autocorrelation in the residuals.

Next to the residuals in the regression page, find the Durbin Watson statistic using the two Excel functions, **SUMXMY2(***array1,array2***)** and **SUMSQ(***array***).**

SUMXMY2 sums the squared differences between adjacent residuals. For *array1***,** enter all but the *last* residual, and for *array2*, enter all but the *first* residual. **SUMSQ** sums the squared residuals. Enter all of the residuals in this array.

=SUMXMY2(*array1,array2***)/SUMSQ(***array***).**

Consult the online DW critical value table. Google “Durbin Watson critical values” to find the Stanford University site: stanford.edu/~clint/bench/dw05a.htm.

Copy and paste the critical values for sample size, T = 38, and two coefficient estimates (including the intercept), K = 2, next to the Durbin Watson statistic, and plot the residuals:

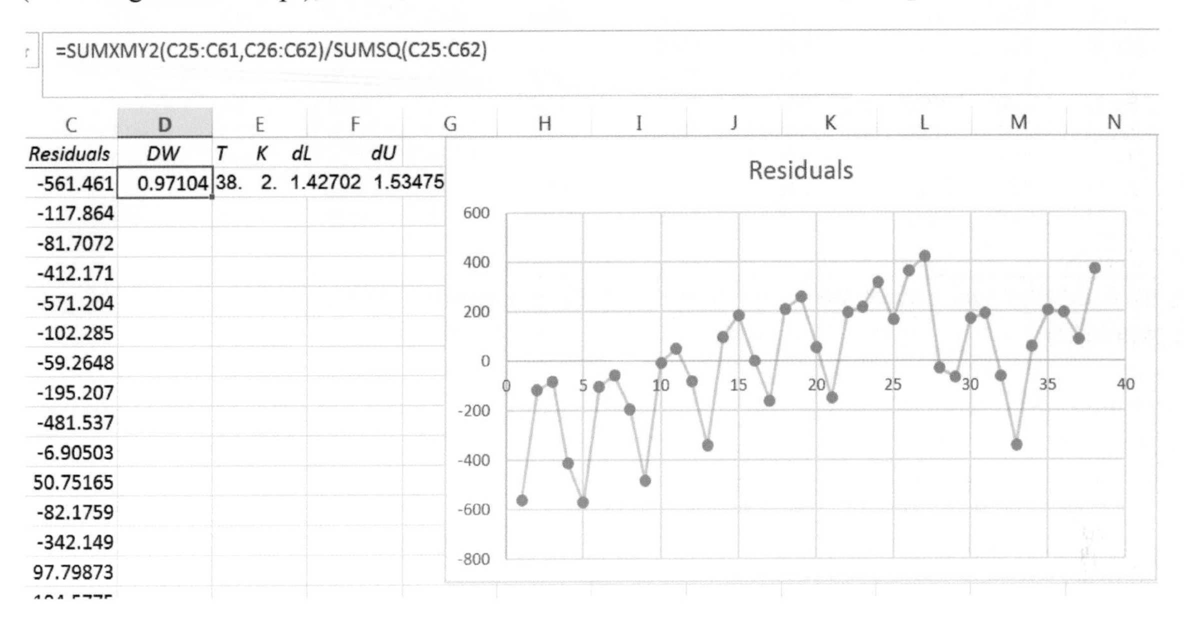
=SUMXMY2(C25:C61,C26:C62)/SUMSQ(C25:C62)

C	D	E		F	
Residuals	*DW*	*T*	*K*	*dL*	*dU*
-561.461	0.97104	38.	2.	1.42702	1.53475
-117.864					
-81.7072					
-412.171					
-571.204					
-102.285					
-59.2648					
-195.207					
-481.537					
-6.90503					
50.75165					
-82.1759					
-342.149					
97.79873					

The residuals contain positive autocorrelation. Trend has not yet been accounted for, and evidence of seasonality remains.

Copy the residuals and paste into the data sheet. Rearrange columns, so that Kroger revenues follows WFM revenues, and residuals follow Kroger revenues.

Mark the residual pattern in the 8 most recent quarters:

	I	J	K	L
1	*Quarter*	*WFM revenues (M$) q*	*Kroger revenues (M$) q-8*	*Residuals*
31	Jun-09	1858	16139	171
32	Sep-09	1878	16135	192
33	Dec-09	1829	17235	-60
34	Mar-10	2639	23137	-340
35	Jun-10	2106	18088	59
36	Sep-10	2163	17615	204
37	Dec-10	2097	17308	194
38	Mar-11	3004	22789	89
39	Jun-11	2351	17728	371
40	Sep-11	2400	17662	
41	Dec-11	2354	18554	
42	Mar-12		24779	

Compare potential driver patterns with the residual pattern, using blue font to designate movement in the same direction:

	I	J	K	L	M	N	O	P	Q	R
1	Quarter	WFM revenues (M$) q	Kroger revenues (M$) q-8	Residuals	U.S. Per capita GDP ($) q-8	U.S. Per capita GDP ($) q-6	Under 25 years (M) q-8	25 to 44 years (M) q-8	45 to 64 years (M) q-8	65+ years (M) q-8
31	Jun-09	1858	16139	171	34344	35042	103.9	82.48	77.5	43.1
32	Sep-09	1878	16135	192	34579	35810	104.1	82.45	77.8	43.4
33	Dec-09	1829	17235	-60	35042	36854	104.2	82.43	78.2	43.7
34	Mar-10	2639	23137	-340	35810	36311	104.3	82.40	78.6	44.0
35	Jun-10	2106	18088	59	36854	35690	104.4	82.35	79.0	44.2
36	Sep-10	2163	17615	204	36311	35127	104.5	82.31	79.4	44.5
37	Dec-10	2097	17308	194	35690	35245	104.6	82.26	79.9	44.7
38	Mar-11	3004	22789	89	35127	34935	104.66	82.21	80.3	45.0
39	Jun-11	2351	17728	371	35245	35044	104.72	82.22	80.6	45.2
40	Sep-11	2400	17662		34935	35557	104.8	82.22	81.0	45.5

Potential drivers that match well are: the two U.S. per capita GDP lags, and the 25–44 year old population.

Rearrange columns so that these three follow the residual column, and run correlations with residuals:

	A	B
1		Residuals
2	Residuals	1
3	U.S. Per capita GDP ($) q-8	0.546004
4	U.S. Per capita GDP ($) q-6	0.547403
5	25 to 44 years (M) q-8	-0.6512

The 25–44 year population is most strongly correlated with the residuals.

Add this driver to your regression model:

	A	B	C	D	E	F	G
1	SUMMARY OUTPUT						
2							
3	*Regression Statistics*						
4	Multiple R	0.980542					
5	R Square	0.961462					
6	Adjusted F	0.95926					
7	Standard I	127.5821					
8	Observati(	38					
9							
10	ANOVA						
11		*df*	*SS*	*MS*	*F*	*gnificance F*	
12	Regressior	2	14213171	7106585	436.5974	1.79E-25	
13	Residual	35	569702.1	16277.2			
14	Total	37	14782873				
15							
16		*Coefficients*	*andard Err*	*t Stat*	*P-value*	*Lower 95%*	*Upper 95%*
17	Intercept	28731.79	2840.528	10.11495	6.28E-12	22965.21	34498.36
18	Kroger rev	0.121495	0.00897	13.54522	1.75E-15	0.103286	0.139704
19	25 to 44 y(	-349.429	33.03566	-10.5773	1.91E-12	-416.495	-282.363
20							
21							
22							
23	RESIDUAL OUTPUT						
24							
25	*)bservatior*	*VFM reven*	*Residuals*				
26	1	780.5268	10.47317				
27	2	417.4523	205.5477				

Reuse your DW formula, look up dL and dU for T = 39 and K = 3, and plot the residuals:

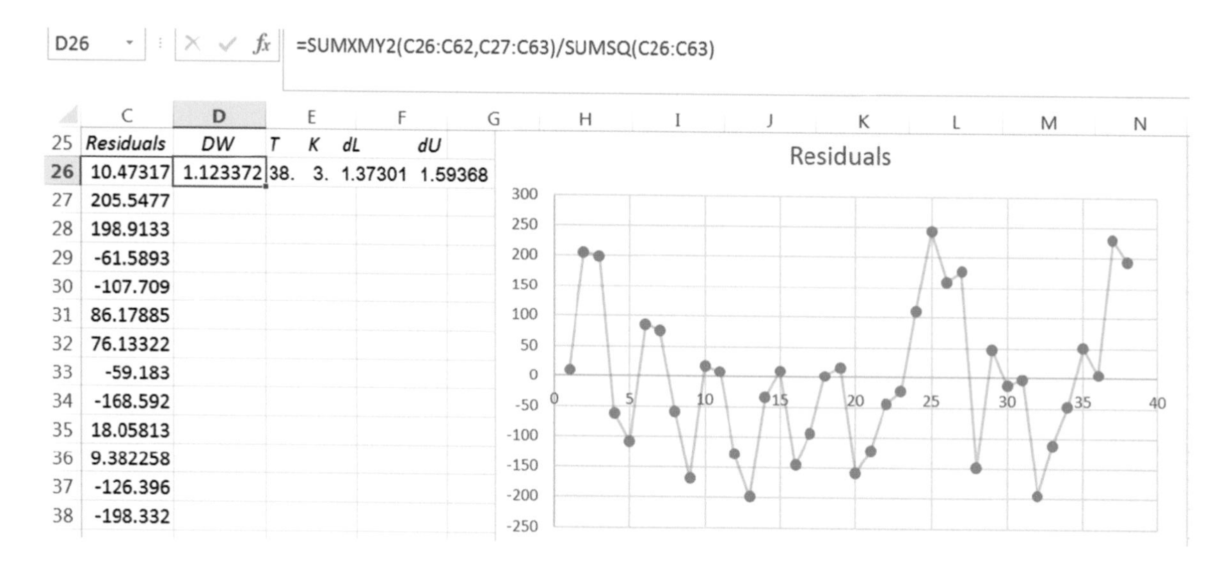
D26 =SUMXMY2(C26:C62,C27:C63)/SUMSQ(C26:C63)

	C	D	E	F	G
25	*Residuals*	*DW*	*T K*	*dL*	*dU*
26	10.47317	1.123372	38. 3.	1.37301	1.59368
27	205.5477				
28	198.9133				
29	-61.5893				
30	-107.709				
31	86.17885				
32	76.13322				
33	-59.183				
34	-168.592				
35	18.05813				
36	9.382258				
37	-126.396				
38	-198.332				

DW for the model is less than the lower critical value. The residuals contain positive autocorrelation.

Focusing on the more recent time periods to match patterns increases the chance you're your model will be valid. However, trend or cycle that are more long term are also important to capture. To choose the third driver, give equal weight to the entire series by looking at correlations with residuals. Copy the residuals and paste into the data sheet. Rearrange so that the residuals follow residuals from the one driver model. Run correlations to select the third driver for your model.

	A	B
1		*Residuals*
2	Residuals	1
3	U.S. Per capita GDP ($) q-8	0.095888
4	U.S. Per capita GDP ($) q-6	0.059542
5	Under 25 years (M) q-8	0.038741
6	45 to 64 years (M) q-8	0.094675
7	65+ years (M) q-8	0.166378

The population of 65+ years is most strongly correlated with residuals. Add this driver to your model:

	A	B	C	D	E	F	G
1	SUMMARY OUTPUT						
2							
3	*Regression Statistics*						
4	Multiple R	0.983638					
5	R Square	0.967544					
6	Adjusted F	0.96468					
7	Standard I	118.7921					
8	Observati(	38					
9							
10	ANOVA						
11		*df*	*SS*	*MS*	*F*	*'gnificance F*	
12	Regressior	3	14303079	4767693	337.857	2.3E-25	
13	Residual	34	479793.5	14111.57			
14	Total	37	14782873				
15							
16		*Coefficients*	*andard Err(*	*t Stat*	*P-value*	*Lower 95%*	*Upper 95%*
17	Intercept	18696.01	4775.256	3.915184	0.000413	8991.519	28400.49
18	Kroger rev	0.10898	0.009712	11.22073	5.64E-13	0.089242	0.128718
19	25 to 44 y(	-259.309	47.12614	-5.50245	3.82E-06	-355.081	-163.537
20	65+ years	65.63629	26.00345	2.524138	0.016435	12.79092	118.4817
21							
22							
23							
24	RESIDUAL OUTPUT						
25							
26	*)bservatior*	*VFM reven*	*Residuals*				
27	1	804.8162	-13.8162				
28	2	477.4929	145.5071				

Reuse your formula to find DW, look dL and dU, and plot residuals:

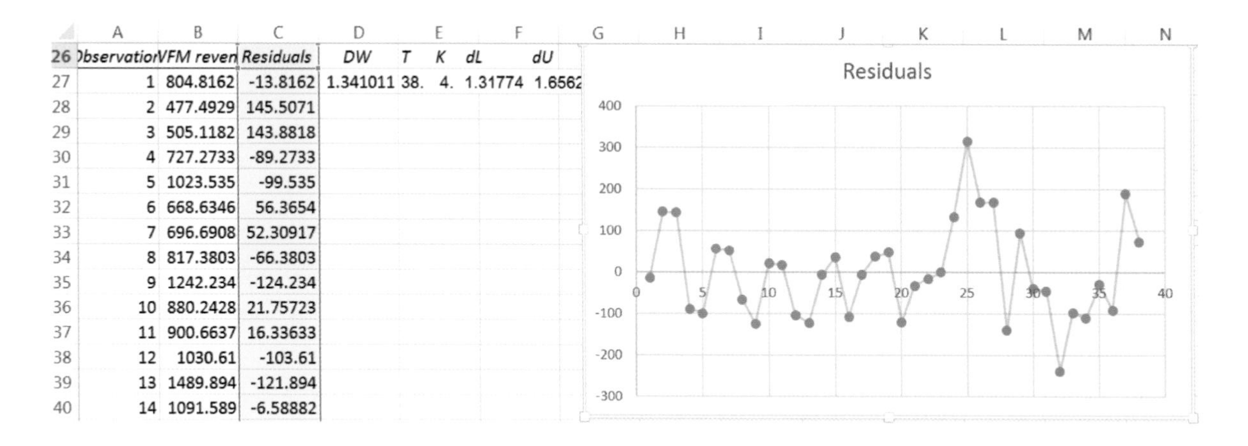

	A	B	C	D	E	F
26	ObservatiorWFM reven		Residuals	DW	T K dL	dU
27	1	804.8162	-13.8162	1.341011	38. 4. 1.31774	1.6562
28	2	477.4929	145.5071			
29	3	505.1182	143.8818			
30	4	727.2733	-89.2733			
31	5	1023.535	-99.535			
32	6	668.6346	56.3654			
33	7	696.6908	52.30917			
34	8	817.3803	-66.3803			
35	9	1242.234	-124.234			
36	10	880.2428	21.75723			
37	11	900.6637	16.33633			
38	12	1030.61	-103.61			
39	13	1489.894	-121.894			
40	14	1091.589	-6.58882			

DW is now in the gray area, indicating that there remains possible positive autocorrelation.

Use residual correlations with the remaining potential drivers to identify the most promising:

None of the correlations are large.

	A	B
1		*Residuals*
2	Residuals	1
3	U.S. Per capita GDP ($) q-8	-0.0159
4	U.S. Per capita GDP ($) q-6	-0.024
5	Under 25 years (M) q-8	-0.0194
6	45 to 64 years (M) q-8	0.014306

Excel 11.5 Test the Model's Forecasting Validity

Given a significant model with high RSquare, significant drivers, and DW that exceeds the lower critical level, test the model's validity.

Copy Quarter, WFM revenues, and the three drivers from the data sheet and paste next to residuals, without formulas

Alt H V S U

	A	B	C	D	E	F	G	H	I	J
26	ObservatiorWFM reven		Residuals	DW	T K d	Quarter	WFM reve	Kroger rev	25 to 44 y	65+ years
27	1	804.8162	-13.8162	1.341011	38. 4. 1.	Mar-02	791	14329	84.97	39.3
28	2	477.4929	145.5071			Jun-02	623	11017	84.86	39.4
29	3	505.1182	143.8818			Sep-02	649	10962	84.75	39.5
30	4	727.2733	-89.2733			Dec-02	638	12692	84.64	39.5

Find the *margin of error* from *critical t and the standard error*:

D7 =B7*C7

	A	B	C	D
3	*Regression Statistics*			
4	Multiple R	0.983638		
5	R Square	0.967544		
6	Adjusted F	0.96468	critical t	me
7	Standard I	118.7921	2.032245	241.4147

Use the regression equation to find predicted WFM revenues from past Kroger revenues and populations 25–44 and 65+:

K27 =$B$17+$B$18*H27+$B$19*I27+$B$20*J27

	A	B	C	D	E	F	G	H	I	J	K
16		*Coefficients*	*andard Err*	*t Stat*	*P-value*	*Lower 95%*	*Upper 95%*	*ower 95.0%*	*pper 95.0%*		
17	Intercept	18696.01	4775.256	3.915184	0.000413	8991.519	28400.49	8991.519	28400.49		
18	Kroger rev	0.10898	0.009712	11.22073	5.64E-13	0.089242	0.128718	0.089242	0.128718		
19	25 to 44 y	-259.309	47.12614	-5.50245	3.82E-06	-355.081	-163.537	-355.081	-163.537		
20	65+ years	65.63629	26.00345	2.524138	0.016435	12.79092	118.4817	12.79092	118.4817		
21											
22											
23											
24	RESIDUAL OUTPUT										
25											
26	*Observation*	*VFM reven*	*Residuals*	*DW*	*T K di*	Quarter	WFM reve	Kroger rev	25 to 44 y	65+ years	predicted
27	1	804.8162	-13.8162	1.341011	38. 4. 1.	Mar-02	791	14329	84.97	39.3	805
28	2	477.4929	145.5071			Jun-02	623	11017	84.86	39.4	477
29	3	505.1182	143.8818			Sep-02	649	10962	84.75	39.5	505
30	4	727.2733	-89.2733			Dec-02	638	12692	84.64	39.5	727

Add and subtract the *margin of error* from *predicted* values to find the *lower* and *upper 95 % prediction interval bounds*:

L27 =K27-$D$7

	F	G	H	I	J	K	L	M
26	Quarter	WFM reve	Kroger rev	25 to 44 y	65+ years	predicted	lower	upper
27	Mar-02	791	14329	84.97	39.3	805	563	1,046
28	Jun-02	623	11017	84.86	39.4	477	236	719
29	Sep-02	649	10962	84.75	39.5	505	264	747
30	Dec-02	638	12692	84.64	39.5	727	486	969
31	Mar-03	924	15102	84.52	39.6	1,024	782	1,265

Compare actual revenues with the 95 % prediction interval in the two most recent quarters to assess your model's validity for forecasting:

	F	G	H	I	J	K	L	M
64	Jun-11	2351	17728	82.22	45.2	2,278	2,036	2,519
65	Sep-11	**2400**	17662	82.22	45.5	2,286	**2,044**	**2,527**
66	Dec-11	**2354**	18554	82.22	45.7	2,398	**2,156**	**2,639**
67	Mar-12		24779	82.23	46.0	3,091	2,850	3,333
68	Jun-12		18788	82.23	46.2	2,453	2,212	2,695
69	Sep-12		18694	82.24	46.5	2,458	2,216	2,699
70	Dec-12		19928	82.24	46.7	2,607	2,366	2,849

The model correctly forecast the two hidden datapoints from the most recent quarters, providing confidence in its validity for forecasting.

Excel 11.6 Recalibrate to Forecast

Recalibrate the model to update the coefficients by rerunning regression, this time with all 41 rows of data. Reuse your formulas for *critical t, margin of error, predicted WFM revenues,* and *95 % prediction interval bounds*:

	A	B	C	D	E	F	G	H	I	J	K	L	M
1	SUMMARY OUTPUT												
2													
3	*Regression Statistics*												
4	Multiple R	0.984683											
5	R Square	0.969601											
6	Adjusted F	0.967068	critical t	me									
7	Standard I	117.1352	2.028094	237.5612									
8	Observati	40											
9													
10	ANOVA												
11		*df*	*SS*	*MS*	*F*	*gnificance F*							
12	Regressior	3	15754930	5251643	382.7546	2.37E-27							
13	Residual	36	493943.6	13720.65									
14	Total	39	16248874										
15													
16		*Coefficients*	*andard Err*	*t Stat*	*P-value*	*Lower 95%*	*Upper 95%*	*ower 95.0%*	*pper 95.0%*				
17	Intercept	18206.88	4365.15	4.170963	0.000183	9353.945	27059.81	9353.945	27059.81				
18	Kroger rev	0.10792	0.009316	11.58418	1.06E-13	0.089026	0.126814	0.089026	0.126814				
19	25 to 44 y	-255.437	44.02212	-5.80247	1.28E-06	-344.718	-166.156	-344.718	-166.156				
20	65+ years	70.06931	21.74306	3.222606	0.002698	25.97234	114.1663	25.97234	114.1663				
21													
22													
23													
24	RESIDUAL OUTPUT												
25													
26	*bservatio*	*VFM reven*	*Residuals*			Quarter	WFM reve	Kroger rev	25 to 44 y	65+ years	predicted	lower	upper
27	1	803.9006	-12.9006			Mar-02	791	14329	84.97	39.3	804	566	1,041
28	2	479.954	143.046			Jun-02	623	11017	84.86	39.4	480	242	718
29	3	507.5021	141.4979			Sep-02	649	10962	84.75	39.5	508	270	745

Excel 11.7 Illustrate the Fit and Forecast

To see the model fit and forecast, move the *lower* and *upper 95 % prediction interval bounds* next to *WFM revenues ($B)* and plot the prediction intervals with actual revenues through June 2013:

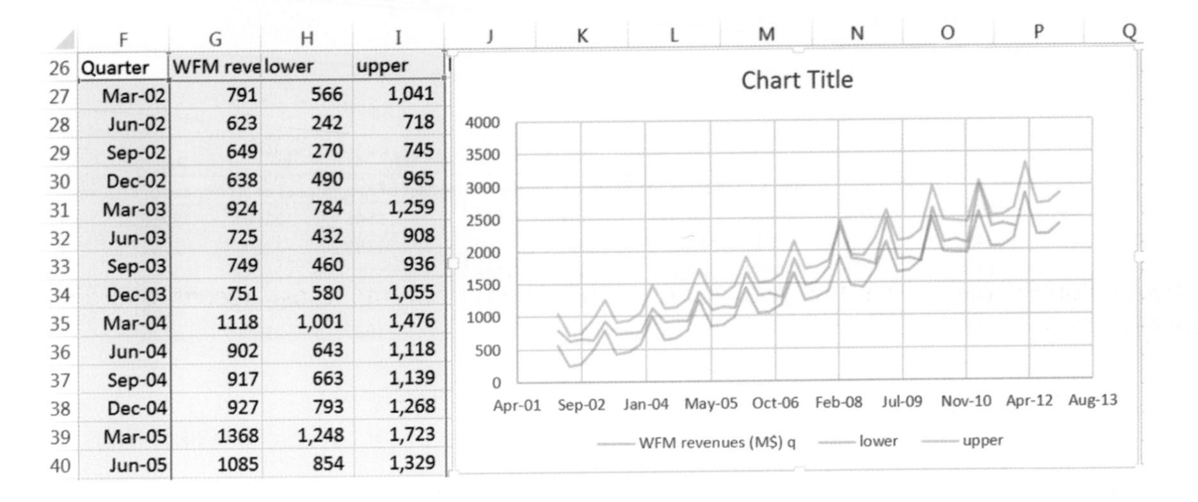

	F	G	H	I
26	Quarter	WFM reve	lower	upper
27	Mar-02	791	566	1,041
28	Jun-02	623	242	718
29	Sep-02	649	270	745
30	Dec-02	638	490	965
31	Mar-03	924	784	1,259
32	Jun-03	725	432	908
33	Sep-03	749	460	936
34	Dec-03	751	580	1,055
35	Mar-04	1118	1,001	1,476
36	Jun-04	902	643	1,118
37	Sep-04	917	663	1,139
38	Dec-04	927	793	1,268
39	Mar-05	1368	1,248	1,723
40	Jun-05	1085	854	1,329

Recolor one of the 95 % prediction intervals so that both are the same color.
Change the *WFM revenues ($B)* series from a line to markers.
Format axes to reduce white space and add gridlines.
Select a style and design and add a chart title that summarizes your conclusion:

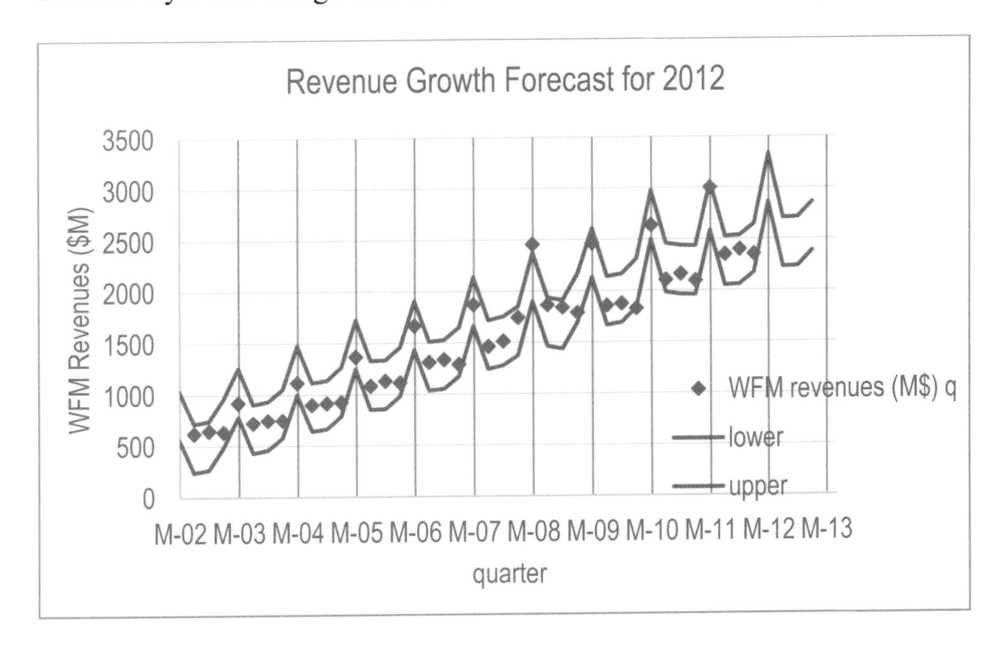

Excel 11.8 Assess the Impact of Drivers

Use the regression equation to compare the impacts of each of the drivers on model forecasts.

Impact of drivers. To see the impact of each of the drivers, add three columns, *impact of Kroger revenues, impact of 25–44, impact of 65+*. Fill these columns with the product of the corresponding driver slope times the driver value:

N27 =$B$18*J27

	F	G	H	I	J	K	L	M	N	O	P
26	Quarter	WFM reve	lower	upper	Kroger rev	25 to 44 y	65+ years	predicted	impact of Kro	impact of 25	impact of (
27	Mar-02	791	566	1,041	14329	84.97	39.3	804	1546	-21705	2756
28	Jun-02	623	242	718	11017	84.86	39.4	480	1189	-21677	2761
29	Sep-02	649	270	745	10962	84.75	39.5	508	1183	-21648	2765
30	Dec-02	638	490	965	12692	84.64	39.5	728	1370	-21619	2770
31	Mar-03	924	784	1,259	15102	84.52	39.6	1,021	1630	-21590	2775

From the max(*array)* and min(*array)* of each of the three impacts, find the range.

Past Kroger revenues drive an increase of 1491 ($M) over the time series, making twice the impact of populations.

N71 =MAX(N27:N70)

	M	N	O	P
69	2,711	2017.452	(21,006.41)	3255.439
70	2,860	2150.625	(21,007.52)	3272.603
71	max	2674	-21000	3273
72	min	1183	-21705	2756
73	range	1491	706	517

Plot each of the three impacts by quarter, choosing a maximum vertical axis value equal to the minimum vertical axis value plus 1,600 ($M), for easy comparison of the three impacts:

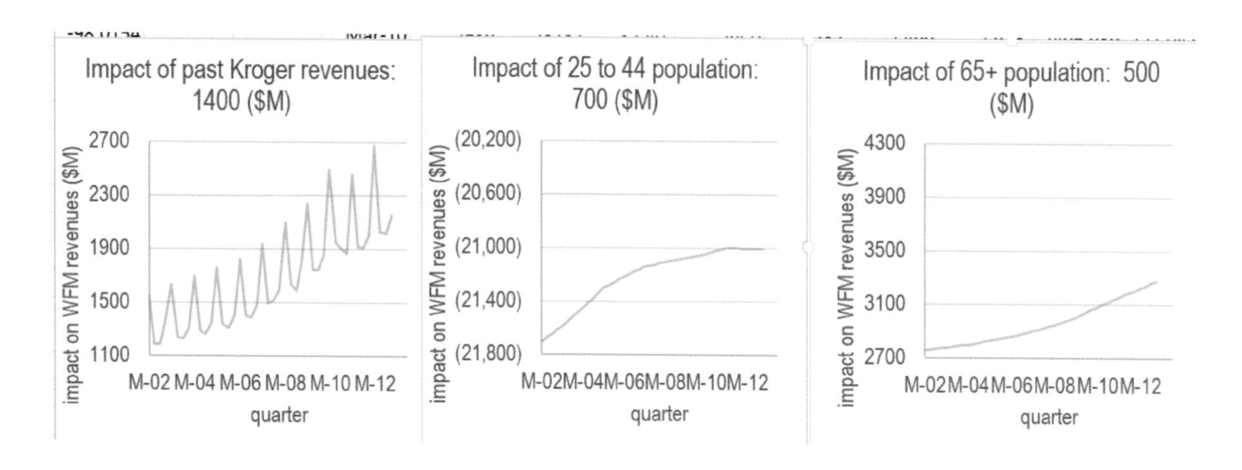

Lab Practice 11 Starbucks in China

Read the ***first page*** of the case description on page 363 and then build a *valid* Leading Indicator model of Starbucks revenues to forecast revenues in 2007–2009 from data in **Case 11 Starbucks Revenue.xls**. Instead of writing a memo, answer the questions below.

I. Assess your model and its implications.

1. What percent of variation in Starbucks *revenues* can be explained with variation in past *Chinese per capita GDP* and past *revenues?*_________

2. What is the margin of error for your forecasts?__________________

3. Is your model is free of unaccounted for trend and cycles? Y Maybe N

   Evidence: ____

4. Is your model valid? Y or N

   Evidence:____

II. Recalibrate if your model is valid.

5. Following each increase of **$.3K (three hundred dollars)** in *Chinese per capita GDP*, the expected change in *revenues* is __________ _____ years later.

6. Is there evidence of Starbucks customer loyalty? N or Y
   If yes, what is the extent of this loyalty…what range of increase in *revenues* is expected following each *revenue* increase of **$1B?**

   __________ ______ years later.

7. Is Starbucks' likely to make its claim of *revenues* of **$13 billion by '09?** Y or N

   Evidence:____

III. Embed a scatterplot of your fit and forecast by year with your regression equation, *RSquare* and *significance* levels.

IV. By what percents has annual *revenue* grown in the 5 years, 2002–2006, and what are the expected annual growth percents in the years 2007–2009?

	% revenue growth		Expected % revenue growth
2001–2002	____________	2006–2007	____________
2002–2003	____________	2007–2008	____________
2003–2004	____________	2008–2009	____________
2004–2005	____________		
2005–2006			

Lab 11-1 What Is Driving WFM Revenues…and What Revenues Can WFM Expect Next Year?

Management believes that some of WFM's success may be due to demographics. Stores are often located near colleges. Is this location strategy successful because a growing number of WFM shoppers are college students? Is the college age population increasing? Identify demographic drivers of revenues.

Competition with the larger chain stores has intensified in recent years, as Wal-Mart, Safeway, Kroger, and others add organic lines. Revenues of competitors may be hurting WFMs. Or demand for organic products may simply be increasing, allowing revenues of WFM and competitors to increase together.

WFM is regarded as expensive by some shoppers and organic items might be considered luxuries. Revenues may be sensitive to per capita GDP.

11 WFM revenues.xls contains a 10 year time series of quarterly data on

WFM Revenues
WFM stores
Population by age groups
Kroger revenues
U.S. per capita GDP

Build a model to identify drivers of WFM revenues and forecast WFM revenues in the four quarters of 2012.

1. Identify drivers of WFM revenues:

___*Under 18* ___*18–24 years* ___*25–44 years* ___*45–64 years* ___*over 65* ___*over 85*

___*Kroger revenues* ___*U.S. per capita GDP*

2. In your final model, can you be 95 % certain that at least one of the potential drivers influences *revenues? Y N*

Statistic(s) that you used to decide? _________

3. Present your model equation. (Note: Format matters.)

4. Present and interpret the statistical evidence that you used to evaluate the explanatory power of your model:

5. Present and interpret the statistical evidence that you used to determine whether or not you accounted for all of the trend, seasonality and cycles in quarterly revenues:

6. Present and interpret the statistical evidence that you used to confirm that your model is valid for forecasting:

7. Present and interpret the statistical evidence that you used to assess the precision of your model:

8. Embed an illustration of the relative influence of each of the drivers in your model.

9. Which driver has had the most impact on WFM revenues in the past 2 years?

____

10. Present your forecast for WFM revenues for the four quarters of 2012:

Year	Quarter	Forecast
2012	I	
2012	II	
2012	III	
2012	IV	

11. Qualitatively describe the short term outlook for WFMs and offer two useful suggestions to improve revenues based on your model results:

12. Embed an illustration of your model fit and forecast.

CASE 11-1 McDonalds Revenue Drivers and Future Prospects

McDonalds executives believe that success depends on demographics, economic productivity, and the growing fast food industry. They require identification of drivers of revenues, as well as a forecast of revenues over the next five quarters.

McDs QR.xlsx contains data on quarterly revenues, global GDP per capita among high and upper middle income countries, the global fast food market value, competitor Yum Foods (Taco Bell, Kentucky Fried Chicken and others) revenues, and global population of the youngest and oldest segments for the 8 year period 2003–2011. Each potential driver differs in data availability, with data from some drivers being available as early as March 2003, and data for other drivers being available only since March 2004.

It is a widely held belief that people's incomes in developing global regions are growing, and more families are electing to dine out, motivating new restaurants to open. McDonalds is a favorite choice for these new restaurant diners. Experts believe that there is a lag of about 12 quarters before increases in incomes are felt in the fast food restaurant industry.

Yum Foods competes with McDonalds, though McDonalds is substantially larger, both in terms of revenues and number of restaurants. Following increases in Yum Foods revenues, executives have observed slower growth in McDonalds revenues, about 6 quarters later.

The global fast food restaurant business is growing. As new restaurant diners enter the market, new restaurants appear, and advertising and promotion levels increase. Following expansion in the industry, McDonalds benefits later, from this growth, and this delayed benefit is thought to occur 8 or 9 quarters later.

Executives are convinced that demographics are key drivers of revenues. There is particular interest in learning whether the youngest and oldest population segments are particularly strong drivers. Both global population segments are growing. Young children are thought to first become fast food restaurant diners between 18 and 24 months of age (e.g., a 6–8 quarter delay). The over 65, retired segment is thought to alter fast food dining habits about 18–24 months after retirement (e.g. a 6–8 quarter delay).

Build a valid model of quarterly revenues, identifying drivers, and forecasting revenues in the next five quarters.

1. Identify the potential driver lags which move in the *same* direction as McDonalds revenues in at least 6 of the last 8 quarters that you are using to build your model and list those below:
2. Identify the potential driver lags which move in the *opposite* direction as McDonalds revenues in at least 6 of last 8 quarters that you are using to build your model and list those below:

Choose one of the potential drivers identified in 1. or 2. to run your first regression.

3. What pattern(s) do you see in the residuals?
4. Are your residuals free of positive autocorrelation (trends and cycles)?

   Explain how you know:

5. Are your residuals free of negative autocorrelation (seasonality)? ___ no ___ yes

Explain how you know:

6. Identify the potential driver lags which move in the *same* direction as the residuals in at least six of the last eight quarters that you are using to build your model and list those below:
7. Identify the potential driver lags which move in the *opposite* direction as residuals in at least six of last eight quarters that you are using to build your model and list those below:

Choose a second driver from potential drivers identified in 6. or 7. to add to the model and then run your second regression.

8. Are your residuals free of positive autocorrelation (trends and cycles)?

   Explain how you know:

9. Are your residuals free of negative autocorrelation (seasonality)? ___ no ___ yes

   Explain how you know:

Choose one or two more drivers that improve your model and finish building your model.

10. Which variables drive quarterly revenues:

___ Global GDP per capita$_{\text{hi\&up mid inc countries}}$ ___ Global fast food market value

___ Yum brands revenues ___ Global population under 15 ___ Global population 65+

11. Explain how you know that your model is valid for forecasting:
12. Describe the power of your model, referring to the appropriate statistic:
13. Describe the precision of your model, referring to the appropriate statistic:
14. Present your equation for McDonalds quarterly revenues. (Be sure to specify units, time periods, and significance levels.) Format matters.
15. Illustrate your fit and forecast and embed, below. (Be sure to choose a Stand Alone title, format axes to make good use of space, and label axes correctly.) Format matters.
16. What is your forecast for quarterly revenues in the first quarter of 2013?
17. Illustrate the influence of each of the drivers in your model, rescaling the vertical axes so that each has the same vertical axis range. Embed below, side by side:
18. Which driver is the most influential?
19. One of the McDonalds executives has observed a sizeable number of older customers. She is convinced that the growing number of retired customers constitute a key segment. Is she correct? Explain how you reached your conclusion.
20. Offer one suggestion to management which is motivated by your results, and which could improve McDonalds revenues:

An intern overheard you discussing Monte Carlo simulation with another intern. He produced a five quarter forecast by

i) Finding the range of quarterly revenue growth from same quarter previous year, −10 to 16 %,
ii) Generating a random sample of 1,000 uniformly distributed quarterly revenue growth values within the range −10 to 16 %,
iii) Multiplying the 97.5 % and the 2.5 % values from the random sample of growth rates, −9 % and 15 %, times revenues in the four most recent quarters to find *predicted quarterly revenue* in the next five quarters.

His predictions, shown below, fall outside your 95 % prediction intervals in some quarters:

	Quarterly revenue forecast ($B)				
	Mar-12	Jun-12	Sep-12	Dec-12	Mar-13
Lower	5.54	6.26	6.49	6.18	5.02
Upper	7.04	7.95	8.25	7.86	8.11

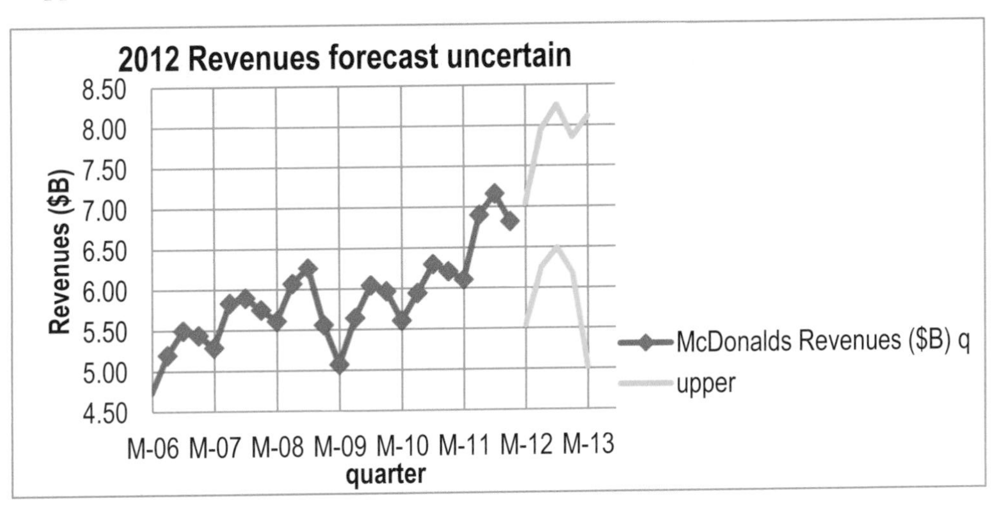

He is arguing that his forecast is more accurate than yours, since it takes into account the uncertainty in quarterly revenue growth using Monte Carlo simulation.

21. Provide three reasons why your forecast is superior:
22. The intern has asked you for feedback regarding his purported Monte Carlo simulation. Explain why his analysis was NOT Monte Carlo simulation:

CASE 11-2 Starbucks in China

Despite recent press that their revenue growth is stagnating, Starbucks management is claiming that revenues will grow by 20 % annually, reaching $13 billion by 2009.

Starbucks management is counting on the growing coffee consumption in China to fuel revenues. In China, Starbucks coffee is considered a luxury. More and more Chinese will be able to afford the treat, as per capita GDP continues to grow. Two recent articles explain:

A Tall Espresso Con Panna costs $1.63, while a small coffee of the day is $1.50. And a Mocha Frappuccino Grande sets you back a substantial 3.63 at the crowded Starbucks stores of Beijing, Shanghai, and Tianjin. Wait a second – isn't the mainland better known for leaves steeped in water, as demonstrated by the phrase "all the tea in China?" There's no shortage of tea in the country that invented it, but the fact is that java beans are a new sensation for the relatively well-off urban Chinese, who now earn on average $1,312 per year, up 9.6 % this year. [Rural Chinese won't likely be drinking Seattle's finest anytime soon, however; rural incomes, still less than a third of their urban counterparts, this year grew 6.2 % to $407.]

In the seven years since H&Q Asia—the former controlling shareholder of Beijing Mei Da Coffee—opened the first Starbucks shop in Beijing in 1999, the Seattle phenomenon has grown to 190 stores in 19 cities in mainland China . "It's not just a drink in China. It's a destination. It's a place to be seen and a place to show how modern one is," adds Technomic Asia's Kedl. And with China's economy growing in double digits, there are likely to be lots more young urban and modern Chinese ready to sip java in a sleek new Starbucks. (Business Week Online, October 26, 2006)

Starbucks Corp. executives have forecast that about 20 percent of its international growth will occur in China this year, which has the potential for more than 200 million customers. There already are more than 500 Starbucks Coffee outlets in China, about 300 of which have opened in the past two years, and Martin Coles, president of Starbucks ' international division, told a telephone conference of financial analysts that the chain would add 200 more there by 2008. Chairman Howard Schultz, emphasizing Starbucks ' current presence in Beijing and 17 provinces, said he anticipates the brand will continue to do well in Hong Kong and gain strength in Taiwan. "We are dreaming very big in China ," he said. (Nation's Restaurant News, May 21, 2007)

Starbucks managers also believe that their loyal customers will continue to return to purchase their favorite coffees, in spite of growing competition.

Build a *valid* Leading Indicator model of Starbucks revenues to forecast revenues in 2007 through 2009 from data in **Case 11 Starbucks Revenue.xls**. The dataset contains *Starbucks Revenues* (B$) in billion dollars, and *China GDP per capita ($T)* in trillion dollars for years 1988 through 2006, with estimates of *China GDP per capita* through 2008.

Use years 1991 through 2004 to build your model.

First, choose Chinese per capita GDP from 2 to 3 years prior.

Next, choose Starbucks revenues from 2 to 3 years prior. (Prior revenues reflect inertia in consumer behavior, or the tendency for Starbucks customers to remain loyal, rather than switch to other coffee sources.)

Write a one page memo presenting your results to management. Be sure to include in your memo:

i. Percent of variation in Starbucks *revenues* which can be explained with variation in past Chinese per *capita GDP* and past *Starbucks revenues*
ii. The margin of error for *your forecasts*
iii. *Following* each *increase of $1K* (one thousand dollars) in *Chinese per capita GDP*, the expected change in *revenues*.
(Be sure to specify units and the expected time of the change)
iv. Evidence of Starbucks customer loyalty and the extent of this loyalty…the range of increase in *Starbucks revenues* expected, following each *revenue* increase of **$1B (one billion dollars)**
v. Whether or not your model is free of unaccounted for trend and cycles
(Use a footnote to include the statistic that you used to draw your conclusion.)
vi. The *range* in revenues forecast in 2007, 2008, and 2009 with 95 % confidence
vii. Likelihood that Starbucks' will match its claim to achieve *revenues* of $13 billion by 2009
viii. Average annual *revenue* growth percent in the past 5 years, 2002 through 2006 and expected annual growth percent the next 3 years
ix. Model validity

Embed a scatterplot of your fit and forecast with your regression equation, *RSquare* and *significance* levels.

Chapter 12
Indicator Variables

In this chapter, 0-*1 indicator* or "*dummy*" variables are used to incorporate segment differences, shocks, or structural shifts into models. In cross sectional data, indicators can be used to incorporate the unique responses of particular groups or segments. In time series data, indicators can be used to account for external shocks or structural shifts. Indicators also offer one option to account for seasonality or cyclicality in time series.

Analysis of variance sometimes is used as an alternative to regression when potential drivers are categorical, or when data are collected to assess the results of an experiment. In this case, the categorical drivers could be represented with indicators in regression, or analyzed directly with analysis of variance.

This chapter introduces the use of indicators to analyze data from conjoint analysis experiments. Conjoint analysis is used to quantify customer preferences for better design of new products and services.

Model variable selection begins with the choice of potential drivers from logic and experience. Indicators are added to account for segment differences, shocks, shifts or seasonality, and, in time series models, if autocorrelation remains, trend, inertia, a leading indicator or an indicator variable may be added to remedy the autocorrelation. The addition of indicators in the variable selection process is considered in this chapter.

12.1 Indicators Modify the Intercept to Account for Segment Differences

To compare two segments, a 0-1 indicator can be added to a model. One segment becomes the baseline, and the indicator represents the amount of difference from the base segment to the second segment. Indicators are like switches that turn on or off adjustments in a model intercept.

Example 12.1 Hybrid Fuel Economy

In a model of the impact of car characteristics on fuel economy:

$$\begin{aligned} M\hat{P}G &= b_0 + b_1 Hybrid + b_2 Emissions + b_3 Horsepower \\ &= 48 + 8.8 Hybrid - 2.3 Emissions - .025 Horsepower \end{aligned}$$

The coefficient estimate of 8.8 MPG for the *hybrid* indicator modifies the intercept. For conventional cars, the *hybrid* indicator is 0, making the intercept for conventional cars 48 MPG:

$$\begin{aligned} M\hat{P}G &= 48 + 8.8(0) - 2.3 Emissions - .025 Horsepower. \\ &= 48 - 2.3 Emissions - .025 Horsepower \end{aligned}$$

C. Fraser, *Business Statistics for Competitive Advantage with Excel 2013: Basics, Model Building, Simulation and Cases*,
DOI 10.1007/978-1-4614-7381-7_12, © Springer Science+Business Media New York 2013

For hybrids in the sample, the *hybrid* indicator is 1, which adjusts the intercept for hybrids to 56.8 MPG by adding 8.8 MPG to the baseline 48 MPG:

$$\begin{aligned}\hat{MPG} &= 48 + 8.8(1) - 2.3Emissions - .025Horsepower \\ &= 56.8 - 2.3Emissions - .025Horsepower\end{aligned}$$

The adjustment is switched on when *hybrid* = 1, but remains switched off if *hybrid* = 0. The parameter estimate for the indicator tells us that on average, hybrid gas mileage is 8.8 MPG higher than conventional gas mileage.

Example 12.2 Yankees v Marlins Salaries[1]

The Yankees General Manager has discovered that the hot rookie whom the Yankees are hoping to sign is also considering an offer from the Marlins. The General Manager would like to know whether there is a difference in salaries between the two teams. He believes that, in addition to a possible difference between the two teams, *Runs* by players ought to affect salaries.

We will build a model of baseball salaries, including *Runs* and an indicator for Team. This variable, *Yankees,* will be equal to 1 if a player is on the Yankees Team, and equal to 0 if the player is a Marlin. The Marlins is the baseline team. Data are shown in Table 12.1, and regression results are shown in Table 12.2.

From the regression output, the model is:

$$Sa\hat{l}ary(\$M) = -3.90^{a} + 6.31^{b} \times Yankee + .104^{b}Runs$$

RSquare: $.57^{b}$

^{a}*Significant* at .05
^{b}*Significant* at .01

The coefficient estimate for the Yankee indicator is \$6.3 M. The intercept for Yankees is \$6.3 M greater than the intercept for Marlins. The rookie can expect to earn \$6.3 million more if he signs with the Yankees.

His expected salary, with 40 runs last season, is:

- As a Marlin, setting the *Yankee* indicator to zero:

$$Sa\hat{l}ary(\$M) = -3.90 + .104 \times 40 = -3.90 + 4.16 = .26(\$M) = \$260{,}000$$

[1] This example is a hypothetical scenario based on actual data

Table 12.1 Baseball team salaries

Player	*Team*	*Yankee*	*Runs*	*Salary (M$)*
Castillo	Marlin	0	72	5.2
Delgado	Marlin	0	81	4.0
Pierre	Marlin	0	96	3.7
Gonzalez	Marlin	0	45	3.4
Easley	Marlin	0	37	.8
Cabrera	Marlin	0	106	.4
Aguila	Marlin	0	11	.3
Treanor	Marlin	0	10	.3
Rodriguez	Yankee	1	111	21.7
Jeter	Yankee	1	110	19.6
Sheffield	Yankee	1	94	13.0
Williams	Yankee	1	48	12.4
Posada	Yankee	1	60	11.0
Matsui	Yankee	1	97	8.0
Martinez	Yankee	1	41	2.8
Womack	Yankee	1	46	2.0
Sierra	Yankee	1	13	1.5
Giambi	Yankee	1	66	1.3
Flaherty	Yankee	1	8	.8
Crosby	Yankee	1	10	.3
Phillips	Yankee	1	7	.3

Table 12.2 Multiple regression of baseball salaries

SUMMARY OUTPUT						
Regression Statistics						
R Square	0.57					
Standard Error	4.2					
Observations	35					
ANOVA	*df*	*SS*	*MS*	*F*	*Significance F*	
Regression	2	754	377	21.3	.0000	
Residual	32	566	18			
Total	34	1,320				
	Coefficients	*Standard Error*	*t Stat*	*p value*	*Lower 95 %*	*Upper 95 %*
Intercept	−3.90	1.56	−2.5	.02	−7.06	−0.73
Yankee	6.31	1.43	4.4	.0001	3.40	9.22
runs	.104	.020	5.1	.0000	.062	.15

- As a Yankee, setting the *Yankee* indicator to 1:

$$\hat{Salary}(\$M) = -3.90 + 6.31 + .104 \times 40 = 6.57(\$M) = \$6{,}570{,}000$$

The *Yankee* indicator modifies the intercept of the regression line, increasing it by $6.31 M.

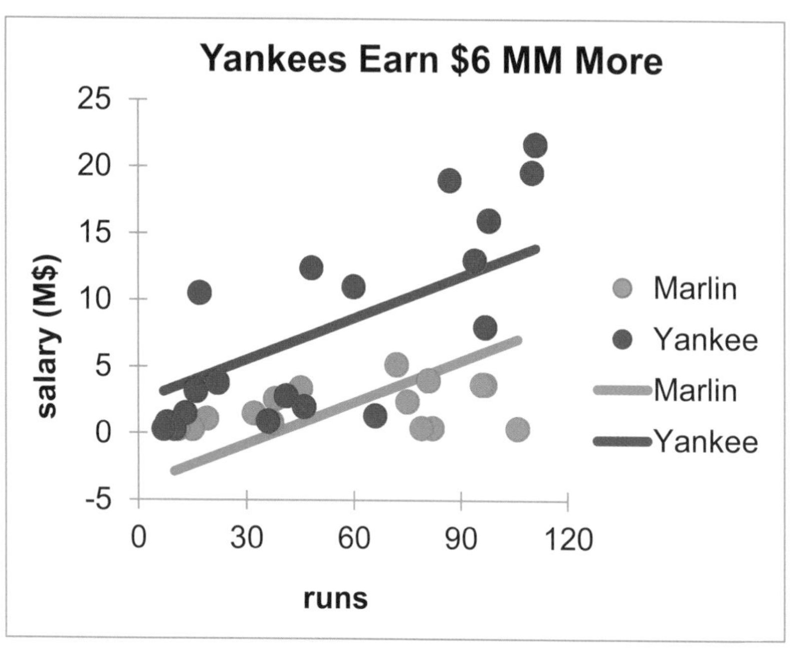

Fig. 12.1 Yankees expect to earn $6 million more

In Fig. 12.1, the intercept represents the baseline Marlins segment; the indicator adjusts the intercept to reflect the difference between Yankees and Marlins.

It does not matter which team is the designated baseline. The model will provide identical results either way.

12.2 Indicators Estimate the Value of Product Attributes

New product development managers sometimes use *conjoint analysis* to identify potential customers' most preferred new product design and to estimate the relative importance of product attributes. The conjoint analysis concept assumes that customers' preferences for a product are the sum of the values of each of the product's attributes, and that customers *trade off* features. A customer will give up a desired feature if another, more desired feature is offered. The offer of a more desired feature compensates for the lack of a second, less desired feature.

Example 12.3 New PDA Design

As an example, consider preferences for PDAs. Management believes that customers choose PDAs based on desired size, design, keypad, and price. For a new PDA design, they are considering

- Three sizes: bigger than shirt-pocket, shirt-pocket, and ultra thin shirt-pocket
- Three designs: single unit, clamshell, and slider
- Three keypads: standard, touch screen, and QWERTY
- Three prices: $150, $250 and $350

Management believes that price is a quality signal, and that customers suspect the quality of less expensive PDAs.

The least desirable, baseline configuration is expected to be:

Bigger than shirt pocket, single unit, with standard keypad at the lowest price.

To find the *part worth utilities,* or the value of each cell phone feature, indicators are used to represent features that differ from the baseline. The conjoint analysis regression model is:

$$
\begin{aligned}
PDA\ pr\hat{e}ference_i = b_0 &+ b_1 \times shirt\ pocket\ size_i &&+ b_2 \times ultra\ thin\ shirt\ size_i \\
&+ b_3 \times clam\ shell_i &&+ b_4 \times slider_i \\
&+ b_5 \times touch\ screen_i &&+ b_6 \times QWERTY_i \\
&+ b_7 \times \$250_i &&+ b_8 \times \$350_i
\end{aligned}
$$

for the *i'th* PDA configuration, where

b_0 is the intercept, which reflects preference for the baseline configuration,
b_1, b_2, b_3, b_4, b_5, b_6, b_7, and b_8 are estimates of the *part worth utilities* of features.

The conjoint analysis process assumes that it is easier for customers to rank or rate products or brands, rather than estimating the value of each feature. For price preferences, this may be particularly true. It will be easier to customers to rate hypothetical PDA designs than it would be for customers to estimate the value of a $250 PDA, relative to a $150 PDA.

The four PDA attributes could be combined in 81 ($=3^4$) unique ways. Eighty one hypothetical PDAs would be too many for customers to accurately evaluate. From the 81, a set of nine are carefully chosen so that the chance of each feature is equally likely (33 %), and each feature is uncorrelated with other features. Slider designs, for example, are equally likely to be paired with each of the three sizes, each of the three keypads, and each of the three prices. This will eliminate multicollinearity among the indicators used in the regression of the conjoint model. Such a subset of hypothetical combinations is an *orthogonal array* and is shown in Table 12.3.

Table 12.3 Nine hypothetical PDA designs in an orthogonal array

Size	*Shape*	*Keypad*	*Price*
Bigger than shirt-pocket	Single unit	Standard	$150
Bigger than shirt pocket	Clamshell	Touch screen	$250
Bigger than shirt pocket	Slider	QWERTY	$350
Shirt pocket	Single unit	Touch screen	$350
Shirt pocket	Clamshell	QWERTY	$150
Shirt pocket	Slider	Standard	$250
Ultra thin shirt pocket	Single unit	QWERTY	$250
Ultra thin shirt pocket	Clamshell	Standard	$350
Ultra thin shirt pocket	Slider	Touch screen	$150

Three customers rated the nine hypothetical PDAs after viewing concept descriptions with sketches. The configurations judged extremely attractive were rated 9 and those judged not at all attractive were rated 1. The regression with eight indicators is shown in Table 12.4.

Table 12.4 Regression of PDA preferences

R Square	.747					
Standard error	1.644					
Observations	27					
ANOVA	*df*	*SS*	*MS*	*F*	*Significance F*	
Regression	8	143	17.9	6.6	.0004	
Residual	18	49	2.7			
Total	26	192				
	Coefficients	*Standard error*	*t Stat*	*p value*	*Lower 95 %*	*Upper 95 %*
Intercept	1.00	.95	1.1	.31	−.99	2.99
Shirt pocket	.78	.78	1.0	.33	−.85	2.41
Ultra thin shirt pocket	1.89	.78	2.4	.03	.26	3.52
clamshell	−1.56	.78	−2.0	.06	−3.18	.07
Slider	−1.44	.78	−1.9	.08	−3.07	.18
Touch screen	4.22	.78	5.4	.0000	2.59	5.85
QWERTY	3.78	.78	4.9	.0001	2.15	5.41
$250	1.67	.78	2.2	.05	.04	3.30
$350	1.67	.78	2.2	.05	.04	3.30

PDA size, keypad, and price features influence preferences, while design options do not. The preferred PDA is *ultra thin* and fits in a *shirt pocket,* features a *touch screen* or *QWERTY keypad,* and is priced at $250 or $350.

The *coefficients* estimate the part worth utilities of the PDA features since the part that each feature is worth is its contribution, with indicator turned on, relative to its lack of contribution, with indicator turned off: $b_i \times (1-0) = b_i$. Expected preference for the ideal design is the sum of the part worth utilities for features included. Of all possible configurations of the four attributes, an ultra thin PDA that fits in a shirt pocket, with the simplest single unit design, with a touch screen, at the highest price is the ideal PDA. Design does not affect preferences, so the least expensive option would be used, and the two higher prices are equivalent to customers, so the higher, more profitable price would be charged:

$$\begin{aligned} PDApr\hat{e}ference_i = 1.00 &+ 0.78 \times shirtpocket_i + 1.89 \times ultra\ thin\ shirtpocket_i \\ &- 1.56 \times clamshell_i - 1.44 \times slider_i \\ &+ 4.22 \times touch\ screen_i + 3.78 \times QWERTY_i \\ &+ 1.67 \times \$250_i + 1.67 \times \$350_i \end{aligned}$$

$$\begin{aligned} = 1.00 &+ 0.78\,(0) + 1.89\,(1) \\ &- 1.56\,(0) - 1.44\,(0) \\ &+ 4.22\,(1) + 3.78\,(0) \\ &+ 1.67\,(0) + 1.67\,(1) \end{aligned}$$

$$=8.78$$

The part worth utilities from coefficient estimates are shown in Fig. 12.2 and Table 12.5.

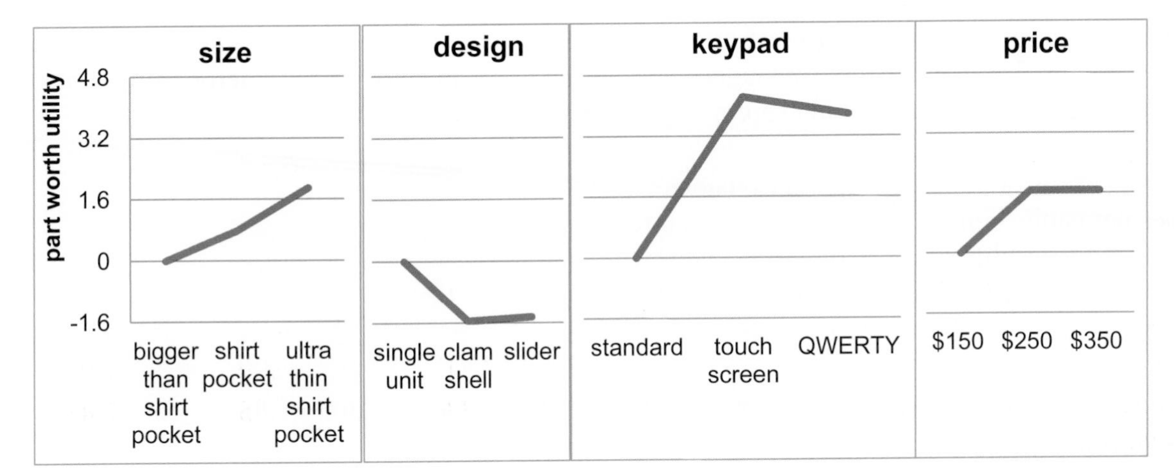

Fig. 12.2 PDA part worth utilities

Preferred *ultraslim shirt pocket size* adds 1.89 (= 1.89 × (1−0)) to the preference 125 rating, a *touch screen* adds 4.22 (= 4.22 × (1− 0)), and a price of *$250* adds 1.67 (= 1.67 × (1− 0)). The preferred design makes no significant difference, 1.56 (= 1.56 × (1−0)).

The range in part worth utilities for each attribute is an indication of that attribute's importance. Preference depends most on the keypad configuration, which is more than twice as important as size or price.

Table 12.5 Relative importance of PDA attributes

Attribute	*Part worth utility of least preferred*	*Part worth utility of most preferred*	*Part worth utility range*	*Attribute importance*
Size	0	1.9	1.9	1.9/9.4 = .20
Shape	−1.6	0	1.6	1.6/9.4 = .17
Keypad	0	4.2	4.2	4.2/9.4 = .18
Price	0	1.7	1.7	1.7/9.4 = .45
	Sum of part worth utility ranges:		9.4	

Conjoint analysis been used to improve the designs of a wide range of products and services, including:

- Seating, food service, scheduling and prices of airline flights
- Offer of outpatient services and prices for a hospital
- Container design, fragrance and design of a aerosol rug cleaner,
- Digital camera pixels, features and prices

Conjoint analysis is versatile and the attributes studied can include characteristics that are difficult to describe, such as fragrance, sound, feel, or taste. It is difficult for customers to tell us how important color, package design, or brand name is in shaping preferences, and conjoint analysis often provides believable, valid estimates.

12.3 Indicators Estimate Segment Mean Differences

Indicators are used in regression to test hypotheses regarding equivalence of segment, group, or category means. With indicators, managers can compare mean performance across categories. The following are questions that managers might use indicators to address:

Does job satisfaction differ across divisions?
Does per capita demand differ across global regions?
Do preferences differ across flavors?
Do rates of return differ across portfolios?
Does customer loyalty differ across brands?

Where differences exist, regression with indicators enables estimation of the extent of those differences.

In each of these scenarios, the question concerns performance differences across categories or groups: divisions, global regions, flavors, portfolios, or brands.

Regression with indicators compares performance variation across groups with performance variation within groups, and more across group variation is evidence that the group performance levels differ.

Example 12.4 Background Music to Create Brand Interest

A brand manager suspects that the background music featured in a brand's advertising may affect the level of interest in the advertised brand. Several background options are being considered, and those options differ along two categories, or *factors*.

Three vocals options are:

(i) Backgrounds which feature vocals,
(ii) Backgrounds with brand related vocals substituted for original vocals, and
(iii) Backgrounds with vocals removed.

Three orchestration options are:

(i) Saxophone,
(ii) Saxophone and percussion, and
(iii) Saxophone and piano.

The hypotheses that the brand manager would like to test are:

$H_{vocals0}$: Mean interest ratings following exposure to ads with alternate vocals options are equivalent.

$$\mu_{original} = \mu_{brand\_specific} = \mu_{no\_vocals}$$

Versus

$H_{vocals1}$: At least one mean interest rating following exposure to ads with alternate vocals differs.

And

$H_{orchestration0}$: Mean interest ratings following exposure to ads with alternate orchestrations are equivalent.

$$\mu_{saxophone} = \mu_{saxophone+percussion} = \mu_{saxophone+piano}$$

Versus

$H_{orchestration1}$: At least one mean interest rating following exposure to ads with alternate orchestrations differs.

To determine whether vocals and orchestration of backgrounds affect brand interest ratings, the ad agency creative team designed nine backgrounds for a brand ad. Since the ad message, visuals, and length of ad could also influence interest, the agency creatives were careful to make those ad features identical across the nine versions. By using ads that were identical, except for their musical backgrounds, any difference in resulting brand interest could be attributed to the difference in backgrounds.

Nine consumers were randomly selected and then randomly assigned to one of the nine background *treatments,* or combination of *vocals* and *orchestration.* Each viewed the brand advertisement with one of the nine backgrounds, and then rated their interest in the brand using a scale from 1 ("not at all interested") to 9 ("very interested"). The data are shown in Table 12.6 and Fig. 12.3.

Table 12.6 Brand interest ratings by vocals and orchestration levels

Brand interest ratings by ad background music

	Orchestration			
Vocals option	Sax	Sax and Percussion	Sax and Piano	Mean
None	9	6	7	7.3
Original	6	4	5	5.0
Brand specific	5	4	3	4.0
Mean	6.7	4.7	5.0	5.4

To set up the data for regression analysis, zeros and ones are used to distinguish *levels* of both *factors*. Each factor in this experiment has three levels. Two indicator variables, shown in Tables 12.7 and 12.8, are needed to distinguish two of the three levels from the third *baseline* level. Regression results do not depend on which level is designated as the *baseline.*

Only the two indicator variables, *original* and *brand,* are included in the regression, since together with the baseline, the three form an identity matrix. The value of the baseline will be reflected in the intercept.

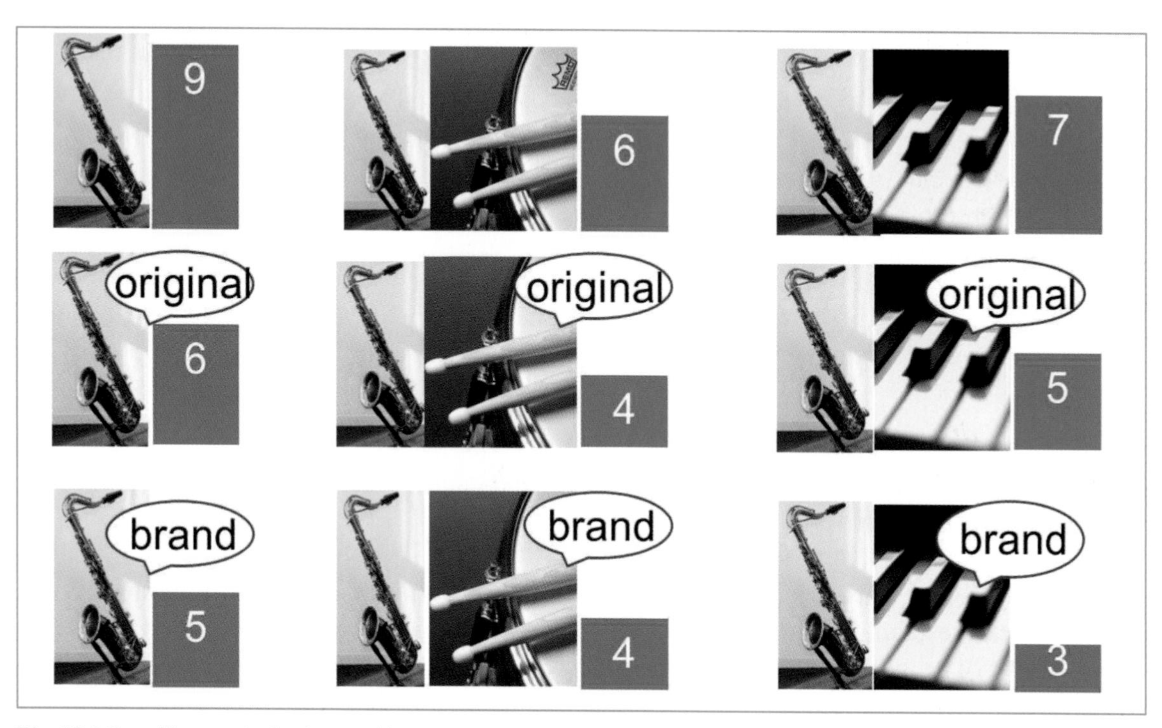

Fig. 12.3 Brand interest by background instrumentation and vocals

Table 12.7 Vocals indicators

	Baseline	*Indicator variables*	
Vocals factor levels	*No vocals*	*Original*	*Brand*
No vocals	1	0	0
Original	0	1	0
Brand	0	0	1

Table 12.8 Orchestration indicators

	Baseline	*Indicator variables*	
Orchestration factor levels	*Sax*	*Saxperc*	*Saxpiano*
Sax	1	0	0
Sax + percussion	0	1	0
Sax + piano	0	0	1

Regression enables us to determine whether at least one of the factors, either one or both, matters. Regression also identifies particular levels which produce higher or lower expected performance relative to the baseline. To illustrate, a regression model of *vocals* and *orchestration* background influences on *brand interest* is shown below.

$$\widehat{Interest} = b_0 + b_{original} \times original + b_{brand} \times brand + b_{sax+perc} \times saxperc + b_{sax+piano} \times saxpiano$$

where *no vocals* with orchestration for *saxophone* are the baseline levels.

Regression results are below in Table 12.9:

Table 12.9 Multiple regression with indicators

SUMMARY OUTPUT

Regression Statistics						
R Square	.932					
Standard Error	.667					
Observations	9					
ANOVA	*df*	*SS*	*MS*	*F*	*Significance F*	
Regression	4	24.4	6.1	13.8	.01	
Residual	4	1.8	0.4			
Total	8	26.2				
	Coefficients	*Standard Error*	*t Stat*	*P-value*	*Lower 95 %*	*Upper 95 %*
Intercept	8.6	.50	17.2	.0001	7.2	9.9
original	−2.3	.54	−4.3	.01	−3.8	−.8
brand	−3.3	.54	−6.1	.004	−4.8	−1.8
saxperc	−2.0	.54	−3.7	.02	−3.5	−.5
saxpiano	−1.7	.54	−3.1	.04	−3.2	−.2

The model *F* statistic, 13.8, has a *p value* (=.01) less than the *critical p value* of .05. Sample evidence allows the conclusion that at least one of the *vocals* or *orchestration* options is driving the level of brand *interest.* From the model *RSquare,* we learn that differences in *vocals* and *orchestration* together account for 93 % of the variation in brand *interest* ratings.

Regression enables identification of indicators which differ from the baseline. The coefficient estimates for *original vocals* and *brand vocals* are significant. *Original* vocals reduces *interest* by 1–4 points, and *brand* vocals reduces *interest* by 2–5 points.

Both the coefficient estimates for *saxperc* and *saxpiano* are significant. Adding percussion to the background reduces expected brand *interest* ratings by 1–4 rating scale points. Adding piano to the background reduces *interest* by as much as 3 rating scale points.

When a regression model is built using indicators, part worth graphs can be used to illustrate results. In the background music example, *part worth interest ratings* can be compared, as Fig. 12.4 illustrates:

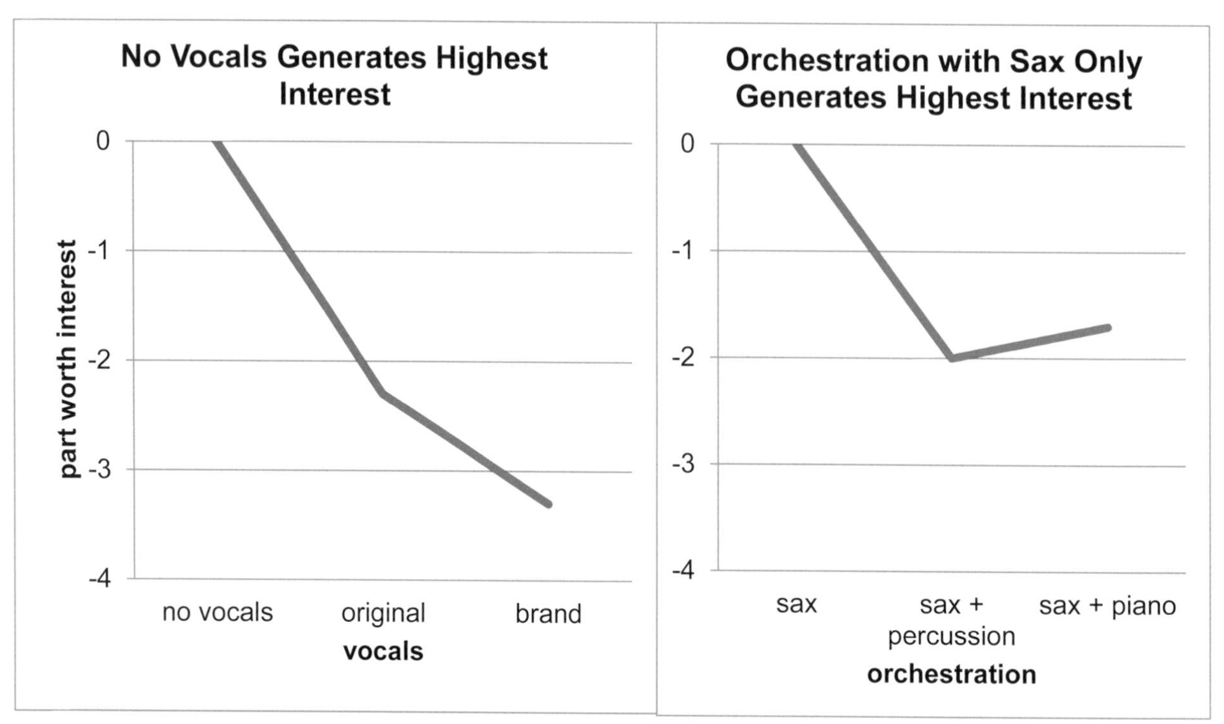

Fig. 12.4 Interest part worths

From regression, we learn that together, *vocals* and *orchestration* options account for 93 % of the variation in *interest* ratings, and that backgrounds with *no vocals* instead of *original* or *brand* vocals, and *saxophone* alone, instead of a combination with either piano or percussion, are expected to generate the highest ratings, as much as eight scale points higher than backgrounds with vocals and either piano or percussion.

12.4 Analysis of Variance Offers an Alternative to Regression with Indicators

ANalysis Of VAriance is an alternative to regression with indicators for situations in which all of the drivers are categorical. ANOVA also tests hypotheses regarding factor level means and provides individual F statistics for each factor but splitting the variation explained by factors, SSR in regression, into pieces explained by each factor.

In the background music example, variation across *vocals* levels, ignoring *orchestration* levels, is:

$$SSB_{vocals} = n_{no\ vocals} \times (\bar{X}_{no\ vocals} - \bar{\bar{X}})^2$$

$$+ n_{original} \times (\bar{X}_{original} - \bar{\bar{X}})^2$$

$$+ n_{brand} \times (\bar{X}_{brand} - \bar{\bar{X}})^2$$

$$= 3 \times (7.3 - 5.4)^2 + 3 \times (5.0 - 5.4)^2 + 3 \times (4.0 - 5.4)^2$$

$$= 17.6$$

There are three *vocals* levels. The degrees of freedom for variation across *vocals* levels is two, comparing two of the levels to the third baseline. Mean variation is:

$$MSB_{vocals} = SSB_{vocals}/df_{vocals}$$
$$= 17.6/2$$
$$= 8.8$$

Variation across *orchestration* levels is

$$SSB_{orchestration} = n_{sax} \times (\bar{X}_{sax} - \bar{\bar{X}})^2$$
$$+ n_{sax+perc} \times (\bar{X}_{sax+perc} - \bar{\bar{X}})^2$$
$$+ n_{sax+piano} \times (\bar{X}_{sax+piano} - \bar{\bar{X}})^2$$
$$= 3 \times (6.7 - 5.4)^2 + 3 \times (4.7 - 5.4)^2 + 3 \times (5.0 - 5.4)^2$$
$$= 6.9$$

And mean variation between *orchestration* levels is:

$$MSB_{orchestration} = SSB_{orchestration}/df_{orchestration}$$
$$= 6.9/2$$
$$= 3.4$$

To compare mean variation across *vocals* levels and *orchestration* levels with mean variation within *vocals* and *orchestration* levels, the variation within levels is calculated by subtracting SSB_{vocals} and $SSB_{orchestration}$ from total variation, SST:

$$SST = (9 - 5.4)^2 + (6 - 5.4)^2 + (7 - 5.4)^2$$
$$+ (6 - 5.4)^2 + (4 - 5.4)^2 + (5 - 5.4)^2$$
$$+ (5 - 5.4)^2 + (4 - 5.4)^2 + (3 - 5.4)^2$$
$$= 26.2$$

Of the total variation of 26.2, 17.6 has been explained by differences across *vocals* levels, and 6.9 has been explained by differences across *orchestration* levels, leaving 1.8 unexplained from variation within levels:

$$SSW = SST - SSB_{vocals} - SSB_{orchestration}$$
$$= 26.2 - 17.6 - 6.9$$
$$= 1.8$$

Mean unexplained variation is:

$$MSW = SSW / (N - df_{vocals} - df_{orchestration} - 1)$$
$$= 1.8 / 4$$
$$= .4$$

To test each of the two sets of hypotheses, the corresponding F statistic is calculated from the ratio of *mean squares between,* MSB_{vocals} or $MSB_{orchestration}$, and *mean square within, MSW:*

$$F_{vocals_{2,4}} = MSB_{vocals}/MSW$$

$$= 8.8/.4$$

$$= 19.8$$

$$F_{orchestration_{2,4}} = MSB_{orchestratio}/MSW$$

$$= 3.4/.4$$

$$= 7.8$$

With 2 and 4 degrees of freedom, the *critical F* for 95 % confidence is 6.9. Both F statistics exceed the *critical F*, and have *p values* of .008 and .04. Based on the sample data, there is evidence that the *vocals* alternatives are not equally effective in backgrounds, and that the *orchestration* alternatives are also not equally effective. Both null hypotheses are rejected.

Excel provides the F statistics and their *p values,* as well as factor level means (Table 12.10):

Table 12.10 ANOVA results from Excel

Anova: Two Factor Without Replication						
SUMMARY	*Count*	*Sum*	*Average*	*Variance*		
No vocals	3	22	7.3	2.3		
original	3	15	5.0	1.0		
brand	3	12	4.0	1.0		
sax	3	20	6.7	4.3		
sax & percussion	3	14	4.7	1.3		
sax & piano	3	15	5.0	4.0		
ANOVA						
Source of Variation	*SS*	*df*	*MS*	*F*	*p value*	*F crit*
Rows	17.6	2	8.8	19.8	.008	6.9
Columns	6.9	2	3.4	7.8	.04	6.9
Error	1.8	4	.4			
Total	26.2	8				

In the sample, ads with *no vocals* produced highest average brand *interest ratings,* $\bar{X}_{no\ vocals} = 7.3$, and ads with *brand* vocals, produced lowest average *interest ratings,* $\bar{X}_{brand} = 4.0$. The F_{vocals} test allows the conclusion that at least one of the *vocals* factor levels differs. Therefore, it is possible that (i) *no vocals* is more effective than either option with vocals, (ii) *brand vocals* are less effective than either *original* or *no vocals,* or (iii) all three levels may differ. To determine which of the three levels differ, *multiple comparisons,* which resemble t tests, would be used, though Excel does not offer this ability. (Other more specialized software packages, such as SPSS and SAS, do offer multiple comparisons.)

Ads with *sax* produced highest average brand *interest ratings,* $\bar{X}_{sax} = 6.7$, and ads with *sax+percussion orchestration* produced the lowest average *interest ratings,* $\bar{X}_{saxperc} = 4.7$ The $F_{orchestration}$ test allows the conclusion that at least one of the *orchestration* factor levels differs; however, from analysis of variance results, it is not possible to determine whether any of the three levels are statistically unique.

12.5 ANOVA and Regression with Indicators are Complementary Substitutes

The *F* statistics used to test hypotheses with analysis of variance and with regression are similar. Both compare variation explained by model drivers or factors with unexplained variation. Analysis of variance enables us to determine whether each factor matters. For example, both *vocals* and *instrumentation* in ad backgrounds matter and at least one *vocal* option and at least one *instrumental* option are more effective in generating *brand interest* following ad exposure. Regression enables us to determine whether at least one of the factors, either one or both, matters. Regression also identifies particular indicators which produce higher or lower expected performance relative to the baseline.

Regression *RSquare* provides a measure of the power of the model: differences in *vocals* and *orchestration* together account for 93 % of the variation in brand *interest* ratings. While analysis of variance does not explicitly provide *RSquare,* it can be easily found from analysis of variance output as the ratio of explained variation, the sum of squares due to the factors, and total variation. Variation explained by the two factors in analysis of variance is equivalent to variation explained by the model in regression:

$$SSB_{vocals} + SSB_{orchestration} = SSR$$
$$17.6 + 6.9 = 24.4$$

And

$$RSquare = (SSB_{vocals} + SSB_{orchestration})/SST$$
$$= (17.6 + 6.9)/26.2$$
$$= .932$$

From analysis of variance, we learn that both *vocals* and *orchestration* influence *interest* ratings. From regression, we learn that together, *vocals* and *orchestration* options account for 93 % of the variation in *interest* ratings, and that backgrounds with *no vocals* instead of *original* or *brand vocals,* and *saxophone* alone, instead of a combination with either piano or percussion, are expected to generate the highest ratings, as much as eight scale points higher than backgrounds with vocals and either piano or percussion.

Multiple regression with indicators and analysis of variance are substitutes, though they are each offer particular advantages. Multiple regression is designed to accommodate both categorical and continuous drivers, and interest is twofold: (i) identify performance drivers, including differences across groups, and (ii) forecast performance under alternate scenarios. Regression accounts for the impact of continuous drivers by building them into a model. Analysis of variance is designed to identify performance differences across groups. Where possible, continuous drivers are controlled by choosing groups that have equivalent profiles, often in the context of an experiment.

12.6 ANOVA and Regression in Excel

Regression's dual goals of (i) identification of drivers and quantification of their influence, plus (ii) forecasting performance under alternate scenarios, provides more information than analysis of variance in Excel, where output is primarily geared toward hypothesis tests of the factors. However, other, more specialized software packages, such as *SAS, JMP,* and *SPSS,* offer more powerful and versatile analysis of variance features, including multiple comparisons. Marketing researchers and psychometricians sometimes use *analysis of covariance* to account for variation in experiments that has not been controlled, and to compare factor levels to identify those that differ.

Analysis of variance is particularly well suited for use with experimental data, and, since experiments tend not to be routinely conducted by managers, experimental data collection and analysis are often outsourced to marketing research firms. Because Excel is targeted for use by managers, analysis of variance in Excel is basic. In Excel, there is the additional limitation that *replications,* the number of datapoints for each combination of factor levels, must be equivalent. In the background music experiment, for example, had 15 consumers been randomly selected to view one of the 9 ads, data from only 9 consumers could be used in analysis of variance with Excel. Six of the ads would have been viewed by two consumers each, and three of the ads would have been viewed by only one consumer. Data from six consumers would have to be ignored in order to use analysis of variance in Excel. Since all of the data could be used in regression with indicators, regression is a more useful choice in Excel, and allows both hypothesis tests and forecasts under alternate scenarios.

12.7 Indicators Quantify Shocks in Time Series

Example 12.5 Tyson's Farm Worker Forecast[2]

Tyson's Management would like to forecast quarterly self employed workers in agriculture. They believe that these self employed workers, family farmers, are leaving the farm to find more profitable work elsewhere, and that this hypothetical exodus may have accelerated by the Stimulus program of 2009. Stimulus legislation enacted in 2009 offered benefits to wage and salary workers, but not to self employed workers, encouraging some self employed to take wage and salary jobs. This might result in a permanent shrinking of the self employed worker segment, or the segment might switch back to self employment once Stimulus benefits expire.

Tyson's meet labor demand left unsatisfied by hiring agricultural workers. They have asked Mark, their master model builder, to build a model to forecast quarterly self employed agriculture workers. In months where the number of workers is expected to be down from the prior year, they will hire additional workers. If these gaps are large enough, they will implement a lobbying campaign to lesson restrictions on illegal immigrant workers who would work for lower wages.

Choice of the first predictor. Since Mark was working with a time series, he first chose a logically appealing leading indicator of *self employed workers*: *unpaid family workers* in agriculture. Self employed farmers often relied on unpaid family members. If *unpaid family workers* were leaving agriculture to work in paid jobs elsewhere, this might drive *self employed workers* to leave agriculture the following year. Mark began with this single predictor to minimize multicollinearity issues.

---

[2]This example is a hypothetical scenario based on actual data

Choice of lag. In order to forecast *self employed ag workers* from *unpaid family workers,* Mark needed to lag the leading indicator. He hid the two most recent observations, April and May 2010, to later validate the model, since he wanted to be sure that his model could be relied upon to produce solid forecasts. Then, to confirm that 12 months was the appropriate lag for *unpaid family workers*, he plotted *self employed workers* and *unpaid family workers*, using data from the Bureau of Labor, January, 2004 through May 2010, shown in Fig. 12.5.

The scatterplots confirmed that agricultural labor follows an annual cycle that corresponds to planting and harvesting. In 2006, there were fewer *unpaid family workers,* and all but one datapoints lie below the trendline. One year later, in 2007, there were fewer *self employed workers,* and all but two datapoints lie below the trendline. Twelve months is the traditional growing cycle in agriculture, and the year with an unusually low number of *self employed workers* lags by 1 year the year with an unusually lower number of *unpaid family workers.* Mark chose a 12 month lag for *unpaid family workers* for the regression model, using datapoints for *unpaid family workers* from April 2005 with datapoints for *self employed workers* from April 2006. His regression is in Table 12.11.

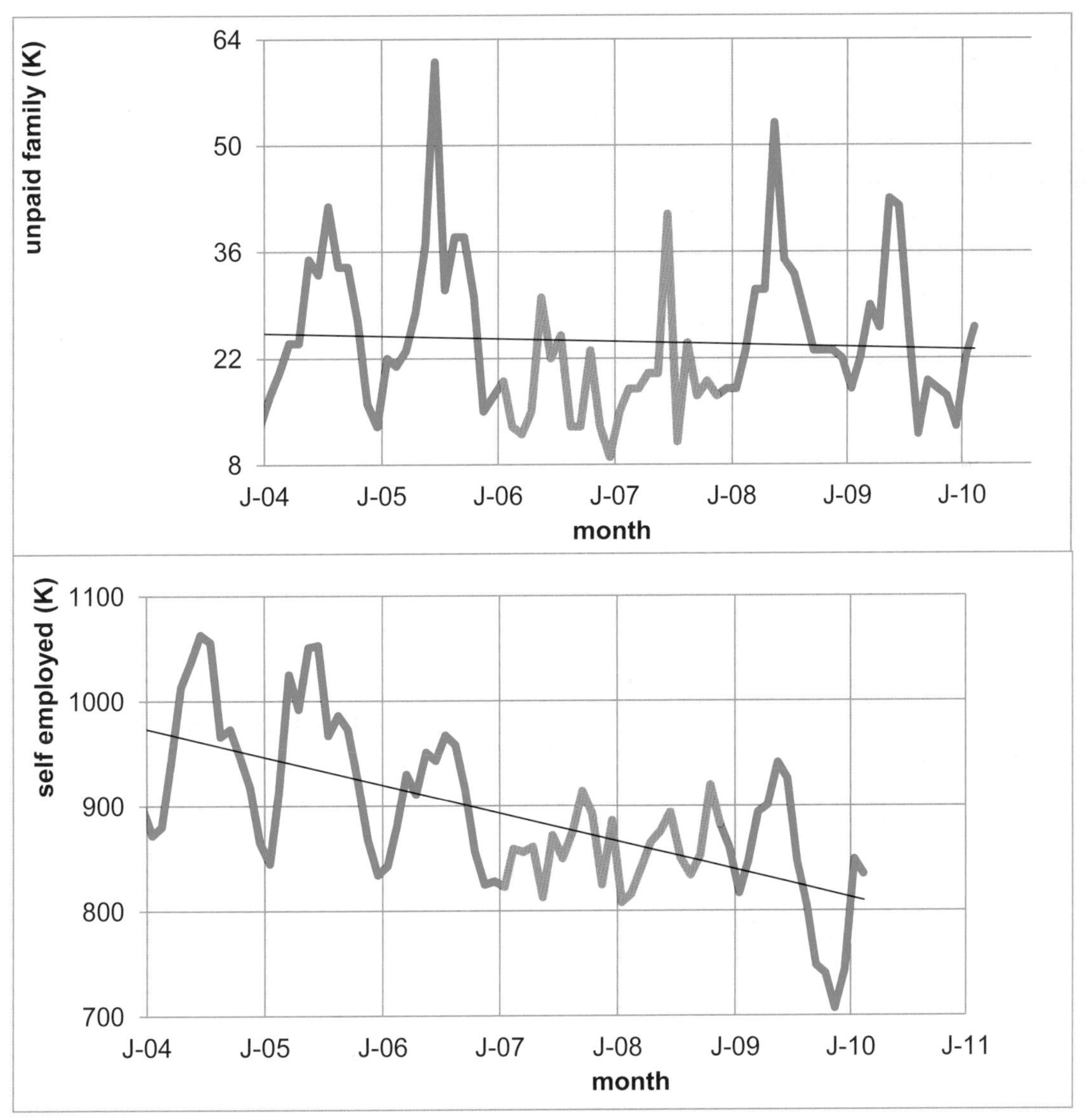

Fig. 12.5 Self employed and unpaid family workers in agriculture, January 04 through May 10

Table 12.11 Regression of self-employed workers in agriculture

R Square	.188					
Standard error	52.2					
Observations	48					
ANOVA	*df*	*SS*	*MS*	*F*	*Significance F*	
Regression	1	28,913	28,913	10.6	.002	
Residual	46	1,25,285	2,724			
Total	47	1,54,197				
	Coefficients	*Standard error*	*t Stat*	*p value*	*Lower 95 %*	*Upper 95 %*
Intercept	804	19	42.7	.0000	766	842
Unpaid family workers q-12	2.36	.72	3.3	.0021	.90	3.81
DW:	0.78					

The model is significant (*Significance F* = *.002)*, though the *RSquare*, .19, is low for time series data. The coefficient estimate is positive as expected: *self employed workers* leave agriculture following the exit of *unpaid family workers.*

Assessment of autocorrelation. Since time series often contain trend, cycles, and seasonality, those must be accounted for. If these systematic variations in the data are present, but unaccounted for, they will be present in the model residuals. The Durbin Watson statistic will identify presence of unaccounted for trend, cycles, or seasonality in the residuals. Mark found that the residuals are autocorrelated ($DW = .78 < dL_{48,2} = 1.49$). Trend, cycles or seasonality are present in the data and have not been accounted for. Mark plotted the residuals in Fig. 12.6 to identify potential trend, cycle or seasonality variables.

There was evidence that the Stimulus legislation of 2009 had affected *self employed workers*, as management had hypothesized. Late in 2009, the number of workers had fallen noticeably. However, workers appeared to have returned by early 2010.

Mark added an indicator of the financial industry *Stimulus 09* to his model, setting the indicator equal to one is months September 2009 through January 2010, and setting the indicator to zero in all other months. The expanded regression model, with the *Stimulus 09* indicator, is shown in Table 12.12.

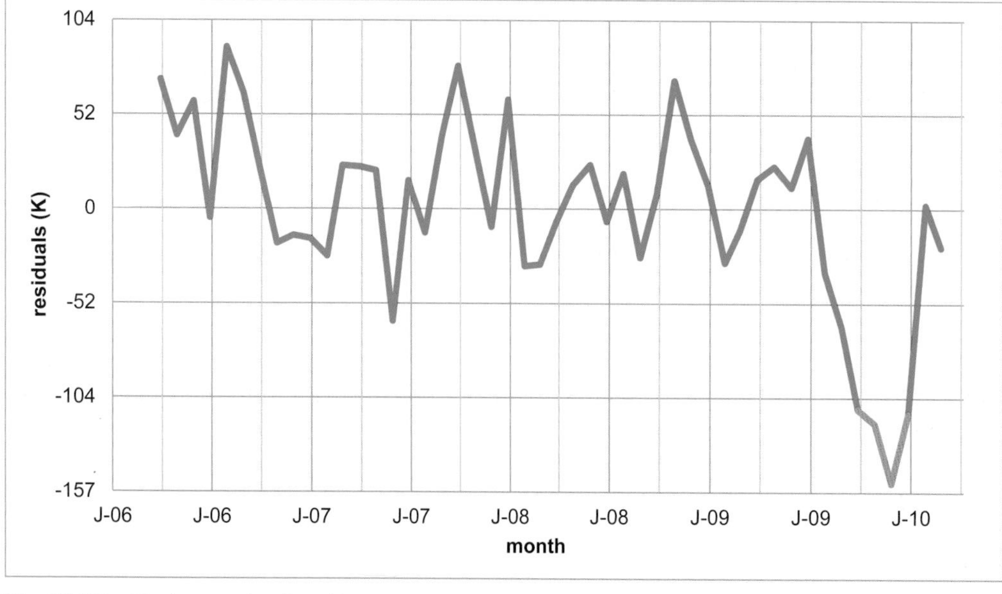

Fig. 12.6 Residuals are not pattern free

Table 12.12 Regression with credit crunch 09 indicator

R Square	.640					
Standard error	35.1					
Observations	48					
ANOVA	*df*	*SS*	*MS*	*F*	*Significance F*	
Regression	2	98,737	49,368	40.1	.0000	
Residual	45	55,461	1,232			
Total	47	1,54,197				
	Coefficients	*Standard error*	*t Stat*	*p value*	*Lower 95 %*	*Upper 95 %*
Intercept	818	13	63.9	.0000	792	843
Unpaid family $_{m-12}$	2.35	.49	4.8	.0000	1.37	3.33
Stimulus 09	−125	17	−7.5	.0000	−158	−91
DW: 1.66						

RSquare is now much higher, .64, and the standard error is now much smaller. Forecasts can be expected to fall within 71K (=2.0*35.1K) workers. The coefficient signs are as Mark expected. The number of *self employed workers* follows the number of *unpaid family workers* a year later. The *Stimulus 09* had a sizeable, though temporary, negative impact on the number of *self employed workers*. The residuals are now free of autocorrelation. *DW* is 1.66, which exceeds $dU_{48,3}$ = 1.62 for this sample of 48 months and a model with three variables, including intercept.

Model validity. To assess the model's validity, Mark compared the two most recent, hidden observations with the 95 % mean prediction intervals, shown in Table 12.13.

Table 12.13 Model validation

Month	*95 % lower prediction (K)*	*Self-employed workers (K)*	*95 % upper prediction (K)*
Apr-10	815	837	956
May-10	808	848	949

The model correctly predicts the number of *self employed workers* in the two most recent months.

With this evidence of model validity, Mark recalibrated the model by adding these two most recent months, which had been hidden to build the model and validate. The model became:

$$self\ emp\hat{l}oyed\ workers(K)_t = 817(K)^a - 123(K)^a \times Stimulus09_t$$

$$+2.29^a \times unpaid\ family\ workers(K)_{t-12}$$

RSquare: .62
[a]*significant at .01.*

In months before September 2009 and after January 2010, setting the *Stimulus09* indicator to 0, the expected number of *self employed workers* in agriculture is:

$$self\ emp\hat{l}oyed\ workers(K)_t = [817(K) - 123(K) \times 0]$$

$$+2.29 \times unpaid\ family\ workers\ (K)_{t-12}$$

$$= 817(K) + 2.29\ \times unpaid\ family\ workers(K)_{t-12}$$

In months September 2009 through January 2010, the *Stimulus 09* indicator is 1, and the expected number of *self-employed workers* is:

$$self\ emp\hat{l}oyed\ workers(K)_t = [817(K) - 123(K) \times 1]$$

$$+2.29 \times unpaid\ family\ workers\ (K)_{t-12}$$

$$= 694(K) + 2.29\ \times unpaid\ family\ workers(K)_{t-12}$$

The *Stimulus 09* indicator shifts the regression intercept and line down by 123(K) workers, as Fig. 12.7 illustrates.

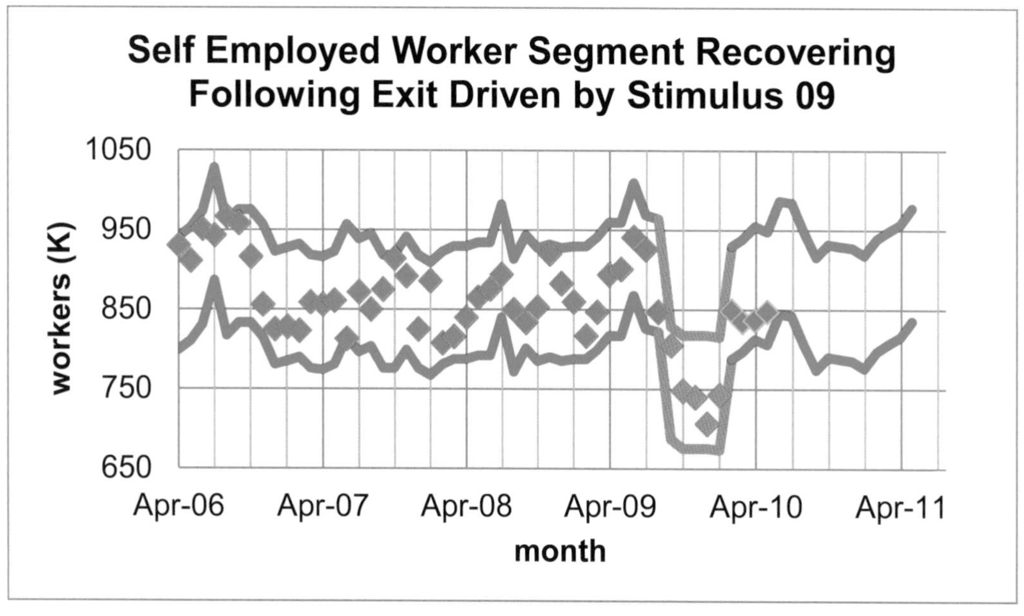

Fig. 12.7 Self employed workers segment recovering

After accounting for the temporary depressing impact of the 2009 Stimulus program, the model forecasts mild growth in the self employed segment in 2010 and 2011. Figure 12.8 shows the percent increase each month over the same month in the past year.

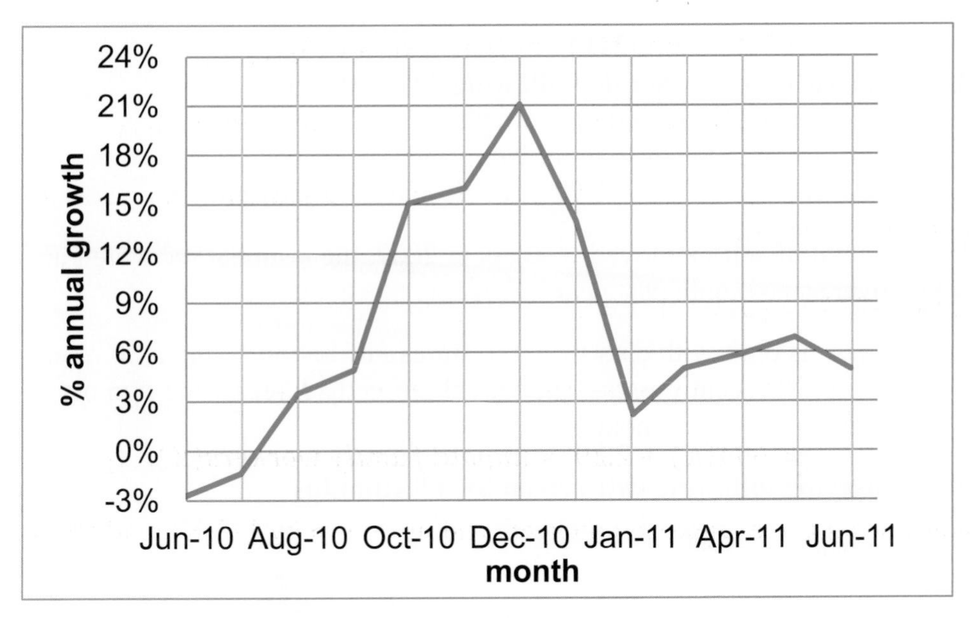

Fig. 12.8 Projected growth in *self employed worker* segment in '10 and '11

Mark would report to Management:

MEMO

Re: Supply of Self Employed Workers Stable Following 09 Contraction
To: Tyson Directors of Planning and Legal Affairs
From: Mark Weisselburg, Director, Econometric Forecasting and Analysis
Date: June 2010

Following an unusually large exit of self employed workers in 2009, the segment recovered and is expected to grow by 8 % over same month past year in 2011.

Econometric Model. A model was built with data from the Bureau of Labor. Using a 50 month series which excluded the two most recent months, the model correctly forecast the number of self employed workers in the two most recent months.

Self employed segment recovering following exit driven by '09 stimulus

Variation in past year unpaid family workers and Stimulus programs account for 62 % of the variation in monthly self employed workers. The model forecast margin of error is 71,000 workers.

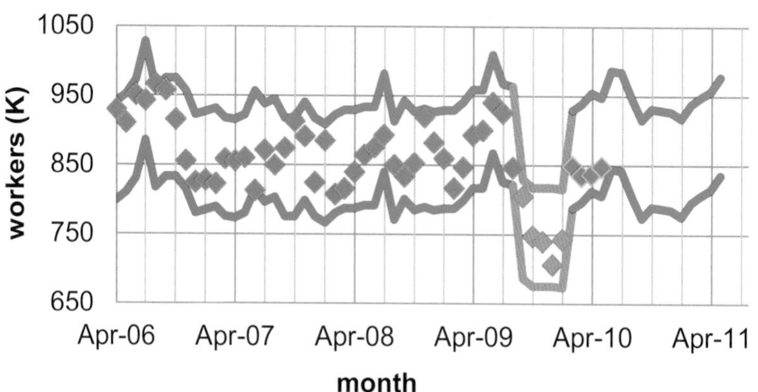

$$\widehat{Self\ employed\ (K)}_t = 694(K) + 2.29 \times unpaid\ family\ workers(K)_{t-12}$$

with '09 Stimulus

$$= 817(K) + 2.29 \times unpaid\ family$$

Following a decline of 1,000 unpaid family workers, the number of self employed workers is expected to decline by 2,000 the following year. '09 Stimulus programs led to the exit of 120 (K) self employed workers each month from September 09 through January 10. Self employed workers have since returned, and their numbers are expected to average 8 % above 09 levels. Forecasts for the next 12 months are:

Month	J-10	J-10	A-10	S-10	O-10	N-10	D-10	J-11	F-11	M-11	A-11	M-11
Lower	850	850	810	780	790	790	790	780	800	810	820	840
Upper	990	990	950	920	930	930	930	920	940	950	960	980

Modest growth forecast in the self employed segment

The number of self employed agriculture workers is expected to show modest growth in the next 12 months.

Farm prices, wages, and pool of workers may influence the self employed segment

This model accounts for less than two thirds of the monthly variation in self employed workers and does not account for changing prices, wages, or the pool of wage and salary workers, which may affect the self employed segment.

12.8 Indicators Allow Comparison of Segments and Scenarios, Quantify Shocks, and Offer an Alternative to Analysis of Variance

Indicators adjust the intercept in linear models to allow for differences in average levels of diverse segments or scenarios. Incorporating indicators in time series models allows us to gauge the impact of structural shifts and to estimate response levels that would have manifested had shocks not occurred. Similarly, if a shock is expected to recur, its indicator can be to one in future periods to forecast the expected change should the shock occur again.

Indicators are used to analyze conjoint analysis data, and estimate the part worth utilities, or the value of each product feature. The part worth utility estimates enable new product development managers to identify most preferred product designs and the most important attributes driving preferences

Analysis of variance enables a manager wants to determine whether or not group or category means differ. Building a regression model with indicators offers an alternative to analysis of variance that also allows identification of the particular groups or categories that differ, as well as the extent of the difference. In Excel, regression with indicators may provide more information for decision making than the basic analysis of variance alternative.

Excel 12.1 Use Indicators to Find Part Worth Utilities and Attribute Importances from Conjoint Analysis Data

Three customers from the target market rated nine hypothetical PDA designs, shown in **Table 12.3,** using a scale from 1 (=least preferred) to 9 (=most preferred). This data is in **Excel 12 PDA conjoint.xls.**

Use indicators to estimate the part worth utilities of *size, shape, keypad* and *price* attribute options for PDAs.

Baseline hypothetical. The baseline PDA is *bigger than shirt pocket,* with *single unit* design, *standard* keypad, at a retail price of *$150.*

The first hypothetical PDA design in **Table 12.3,** and in rows **2, 11,** and **20** of the file, corresponds to the baseline.

Add indicators for differences from baseline. Add four indicators, two for each PDA attribute: *shirt pocket, ultra slim shirt pocket, clamshell, slider, QWERTY, touch screen, $250,* and *$350.*

Enter a zero or a one in each of these columns for each of the nine hypotheticals.

The baseline hypothetical, for example, will have zeros in all eight columns, since it is not *shirt pocket* or *ultra slim shirt pocket size,* it does not feature a *clamshell* or *slider design,* it does not have a *QWERTY* or *touch screen keypad,* and it is not priced at *$250* or $350:

	C	D	E	F	G	H	I	J	K	L	M	N	O
1	*size*	*design*	*key pad*	*price*	*rating*	*shirt pocket*	*ultra thin shirt pocket*	*clamshell*	*slider*	*touch screen*	*QWERTY*	*$250*	*$350*
2	bigger than shirt pocket	single unit	standard	$150	1	0	0	0	0	0	0	0	0
3	bigger than shirt pocket	clamshell	touch screen	$250	5	0	0	1	0	1	0	1	0
4	bigger than shirt pocket	slider	QWERTY	$350	5	0	0	0	1	0	1	0	1
5	shift pocket	single unit	touch screen	$350	7	1	0	0	0	1	0	0	1
6	shift pocket	clamshell	QWERTY	$150	3	1	0	1	0	0	1	0	0
7	shift pocket	slider	standard	$250	3	1	0	0	1	0	0	1	0
8	ultra thin shirt pocket	single unit	QWERTY	$250	8	0	1	0	0	0	1	1	0
9	ultra thin shirt pocket	clamshell	standard	$350	5	0	1	1	0	0	0	0	1
10	ultra thin shirt pocket	slider	touch screen	$150	5	0	1	0	1	1	0	0	0

Select and copy the indicator values for the nine hypotheticals in the first customer's rows, and then paste into the other two customers' rows:

	H	I	J	K	L	M	N	O
1	shirt pocket	ultra thin shirt pocket	clamshell	slider	touch screen	QWERTY	$250	$350
2	0	0	0	0	0	0	0	0
3	0	0	1	0	1	0	1	0
4	0	0	0	1	0	1	0	1
5	1	0	0	0	1	0	0	1
6	1	0	1	0	0	1	0	0
7	1	0	0	1	0	0	1	0
8	0	1	0	0	0	1	1	0
9	0	1	1	0	0	0	0	1
10	0	1	0	1	1	0	0	0
11	0	0	0	0	0	0	0	0
12	0	0	1	0	1	0	1	0

Run a regression of *rating*, with the eight indicators:

	A	B	C	D	E	F	G
1	SUMMARY OUTPUT						
2							
3	*Regression Statistics*						
4	Multiple R	0.864018					
5	R Square	0.746528					
6	Adjusted F	0.633873					
7	Standard I	1.644294					
8	Observati(	27					
9							
10	ANOVA						
11		*df*	*SS*	*MS*	*F*	*gnificance F*	
12	Regressior	8	143.3333	17.91667	6.626712	0.000438	
13	Residual	18	48.66667	2.703704			
14	Total	26	192				
15							
16		*Coefficients*	*andard Err*	*t Stat*	*P-value*	*Lower 95%*	*Upper 95%*
17	Intercept	1	0.949334	1.05337	0.306107	-0.99448	2.994476
18	shirt pock(	0.777778	0.775128	1.003419	0.328958	-0.85071	2.406261
19	ultra thin :	1.888889	0.775128	2.436874	0.025423	0.260406	3.517372
20	clamshell	-1.55556	0.775128	-2.00684	0.060027	-3.18404	0.072927
21	slider	-1.44444	0.775128	-1.86349	0.078791	-3.07293	0.184039
22	touch scre	4.222222	0.775128	5.447131	3.56E-05	2.593739	5.850705
23	QWERTY	3.777778	0.775128	4.873749	0.000122	2.149295	5.406261
24	250	1.666667	0.775128	2.150183	0.04538	0.038184	3.29515
25	350	1.666667	0.775128	2.150183	0.04538	0.038184	3.29515

Coefficients for *shirt pocket, clamshell,* and *slider* are not significant. With experimental conjoint analysis data, the indicators reflect an orthogonal design in which the product features present or absent are *uncorrelated.* Since multicollinearity will not affect results, there is no need to remove the insignificant indicators.

Part worth utilities. The *coefficients* are estimates of the part worth utilities, the value of each feature: $b_i \times (1 - 0) = b_i$. Size, price, and keypad options drive preferences, while design options do not. The most preferred PDAs would be that combining the features with highest part worth utilities: ultrathin shirt pocket size, with a touch screen or QWERTY keypad, at a price of \$250 or \$350.

Attribute importances. To find the *attribute importances*, make a table of the part worth utilities, including the baselines. (Format cells in your *feature* column as text so that Excel will treat these cells as categories.)

J27 : fx =B24

	A	B	C	D	E	F	G	H	I	J
15		*Coeffi-cients*	*Standard Error*	*t Stat*	*P-value*	*Lower 95%*	*Upper 95%*	*Attribute*	*Attribute option*	*part worth utility*
16	Intercept	1.00	0.95	1.1	0.3061	-0.99	2.99	Size	bigger than shirt pocket	0
17	shirt pocket	0.78	0.78	1.0	0.3290	-0.85	2.41		shirt pocket	0.78
18	ultra thin shirt p	1.89	0.78	2.4	0.0254	0.26	3.52		ultra thin shirt pocket	1.89
19	clamshell	-1.56	0.78	-2.0	0.0600	-3.18	0.07	Design	single unit	0
20	slider	-1.44	0.78	-1.9	0.0788	-3.07	0.18		clam shell	-1.56
21	touch screen	4.22	0.78	5.4	0.0000	2.59	5.85		slider	-1.44
22	QWERTY	3.78	0.78	4.9	0.0001	2.15	5.41	Keypad	standard	0
23	250	1.67	0.78	2.2	0.0454	0.04	3.30		touch screen	4.22
24	350	1.67	0.78	2.2	0.0454	0.04	3.30		QWERTY	3.78
25								Price	$150	0
26									$250	1.67
27									$350	1.67

To see the difference that each feature makes, plot the part worth utilities for each attribute.

To see the preference difference due to alternate sizes, make a line plot, **Alt** N N.

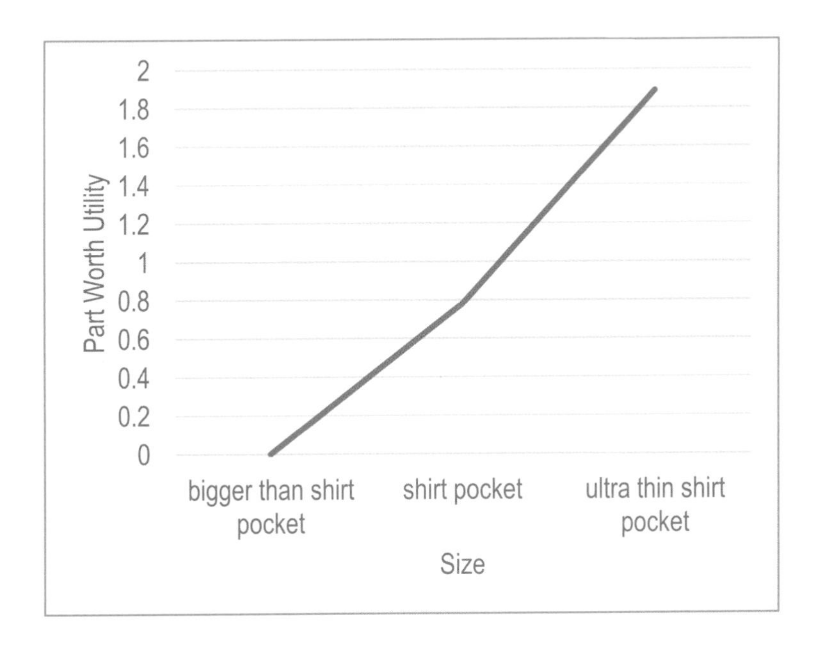

So that attributes can be compared, reformat the vertical axis range, from the most negative to the most positive part worth utility, −2 to 5, choosing a value for major unit, such as 1, and specify that the horizontal axis crosses at the axis value −2:

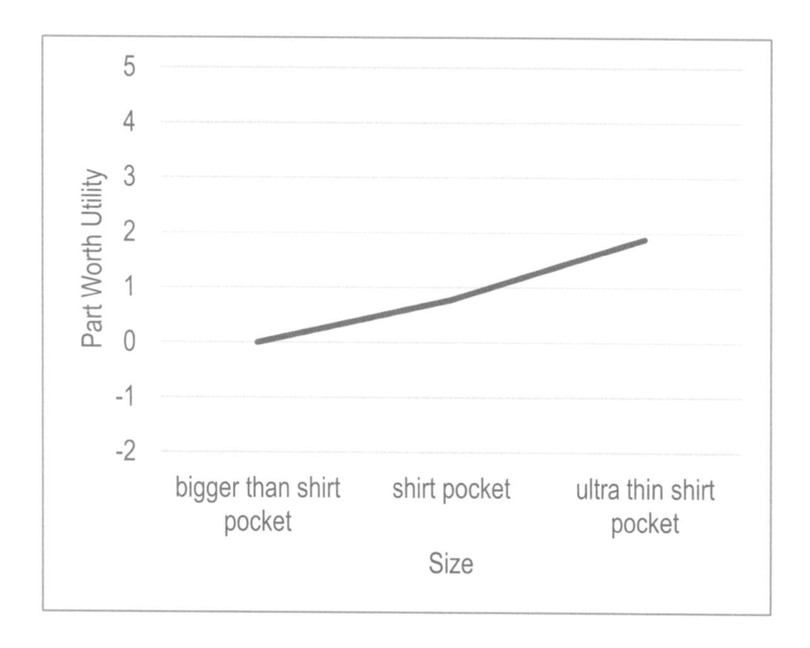

Make plots of part worth utilities for the other three attributes and reformat the vertical axes similarly.

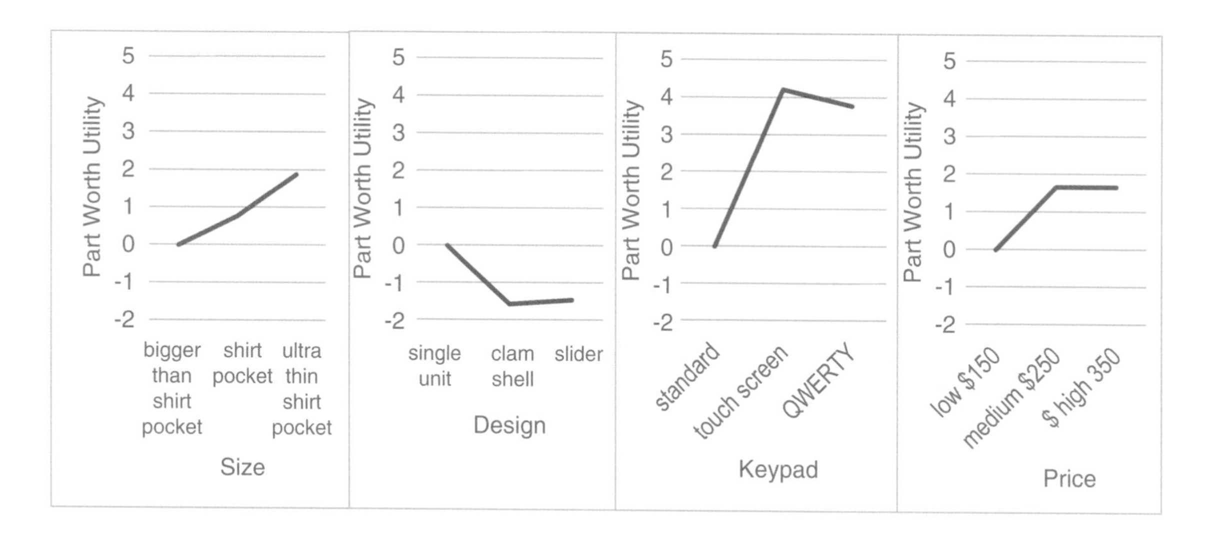

The importance of each attribute is the difference between the part worth utilities of the most and least preferred *attribute options:*

K25 =J26-J25

	H	I	J	K
15	*Attribute*	*Attribute option*	*part worth utility*	*Attribute importances*
16	Size	bigger than shirt pocket	0	1.89
17		shirt pocket	0.78	
18		ultra thin shirt pocket	1.89	
19	Design	single unit	0	1.56
20		clam shell	-1.56	
21		slider	-1.44	
22	Keypad	standard	0	4.22
23		touch screen	4.22	
24		QWERTY	3.78	
25	Price	low $150	0	1.67
26		medium $250	1.67	
27		$ high 350	1.67	

To find the *standardized attribute importances,* first find the total of all attribute importances, **SUM(***array***)**, and then divide the attribute importances by the SUM. (Use **fn 4** to lock the SUM cell reference.)

L16 =K16/$K$28

	H	I	J	K	L
15	*Attribute*	*Attribute option*	*part worth utility*	*Attribute importances*	*standardized attribute importances*
16	Size	bigger than shirt pocket	0	1.89	**0.20**
17		shirt pocket	0.78		**0.00**
18		ultra thin shirt pocket	1.89		**0.00**
19	Design	single unit	0	1.56	**0.17**
20		clam shell	-1.56		**0.00**
21		slider	-1.44		**0.00**

Excel 12.2 Add Indicator Variables to Account for Segment Differences or Structural Shifts

Tyson forecast of self employed workers in agriculture. Build a model to forecast self employed workers in agriculture to assist Tyson in the decision whether or not to employ lobbyists to argue for more lenient immigrant farm labor. If self employed workers are declining, Tyson sees an opportunity to replace self employed agricultural output by employing immigrant workers.

Data included in the time series, *month, self employed workers(K), unpaid family workers (K),* and past year *unpaid family works*$_{m-12}$, are in **Excel 12 Self Employed Ag Workers.xls.**

Build a simple regression of *self employed workers* with the leading indicator past year *unpaid family workers.* Find *DW* and compare with the lower and upper critical values.

	A	B	C	D	E	F	G
1	SUMMARY OUTPUT						
2							
3	*Regression Statistics*						
4	Multiple R	0.433018					
5	R Square	0.187504					
6	Adjusted F	0.169842					
7	Standard I	52.18791					
8	Observatio	48					
9							
10	ANOVA						
11		*df*	*SS*	*MS*	*F*	*gnificance F*	
12	Regressior	1	28912.67	28912.67	10.6157	0.002111	
13	Residual	46	125284.6	2723.578			
14	Total	47	154197.3				
15							
16		*Coefficients*	*andard Erro*	*t Stat*	*P-value*	*Lower 95%*	*Upper 95%*
17	Intercept	804.3563	18.84129	42.69114	1.18E-38	766.4308	842.281
18	unpaid far	2.356803	0.723351	3.258174	0.002111	0.900773	3.81283
19							
20							
21							
22	RESIDUAL OUTPUT						
23							
24	*Observation*	*f employed*	*Residuals*	*DW*	*T k*	*dL dU*	
25	1	858.5628	71.4372	0.778293	48. 2.	1.49275 1.57762	

Since RSquare is quite low and the Durbin Watson statistic indicates presence of unaccounted for trend, seasonality, cycles, shifts, or shocks, plot the residuals to identify pattern(s) which might suggest variables to add to the model.

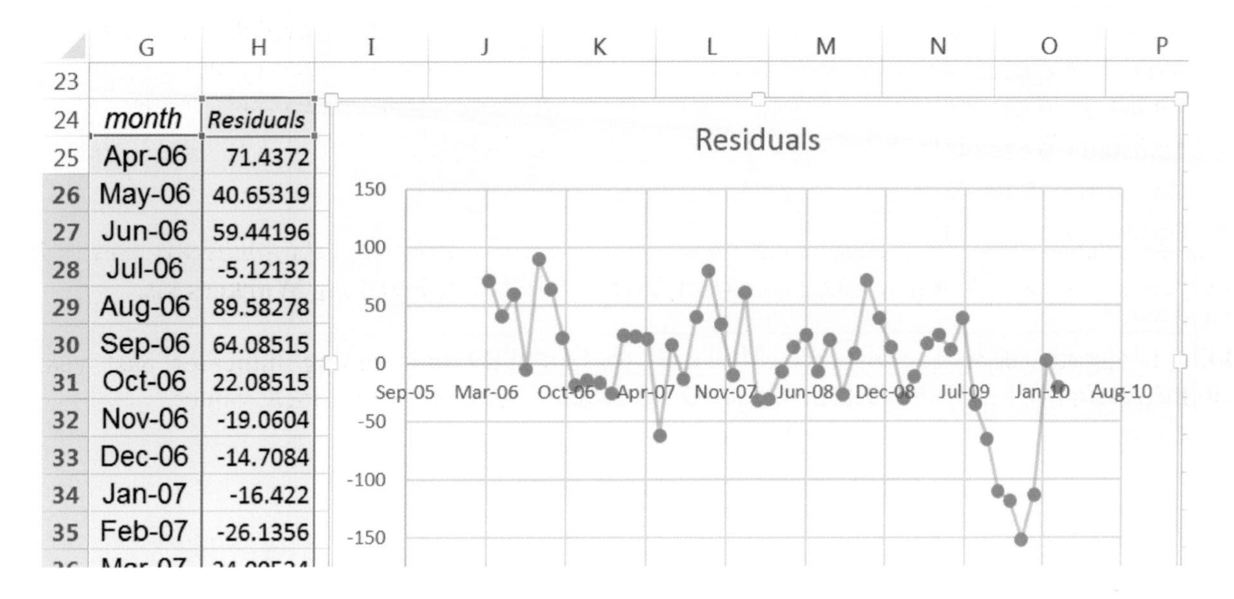

There was evidence that the Stimulus legislation of 2009 had affected *self employed workers*, as management had hypothesized. Late in 2009, the number of workers had fallen noticeably. However, workers appeared to have returned by early 2010.

Add an indicator of the financial industry *Stimulus 09* to the model, setting the indicator equal to 1 in months September 2009 through January 2010, and setting the indicator to 0 in all other months. Run an expanded regression model, with the *Stimulus 09* indicator, and reassess *DW*.

	A	B	C	D	E	F	G
1	SUMMARY OUTPUT						
2							
3	*Regression Statistics*						
4	Multiple R	0.800204					
5	R Square	0.640326					
6	Adjusted F	0.624341					
7	Standard I	35.10642					
8	Observati(	48					
9							
10	ANOVA						
11		*df*	*SS*	*MS*	*F*	*gnificance F*	
12	Regressior	2	98736.51	49368.26	40.05665	1.02E-10	
13	Residual	45	55460.74	1232.461			
14	Total	47	154197.3				
15							
16		*Coefficients*	*andard Erri*	*t Stat*	*P-value*	*Lower 95%*	*Upper 95%*
17	Intercept	817.5768	12.79552	63.89553	8.3E-46	791.8053	843.3483
18	unpaid far	2.347808	0.486594	4.82498	1.64E-05	1.367757	3.32786
19	Stimulus 0	-124.855	16.58782	-7.52689	1.7E-09	-158.264	-91.445
20							
21							
22							
23	RESIDUAL OUTPUT						
24							
25	*Observatior*	*f employec*	*Residuals*	*DW*	*T k*	*dL*	*dU*
26	1	871.5764	58.42364	1.661045	48. 3.	1.45004	1.62308

RSquare has improved substantially, and *DW* exceeds the upper critical value, allowing the conclusion that the residuals are free of unaccounted for trend, cycles, seasons, shifts and shocks.

Model validation. To test the model's validity, select and copy *month, self employed workers,* past *unpaid family workers,* and *Stimulus 09* and paste without formulas in the regression sheet, next to Durbin Watson limits.

	F		G	H	I	J
25	dL	dU	month	self emplc	unpaid far	Stimulus C
26	45004	1.6	Apr-06	930	23	0
27			May-06	911	28	0
28			Jun-06	951	37	0
29			Jul-06	943	61	0
30			Aug-06	967	31	0

Use the regression equation with the coefficients and data to find *predicted self employed workers (K):*

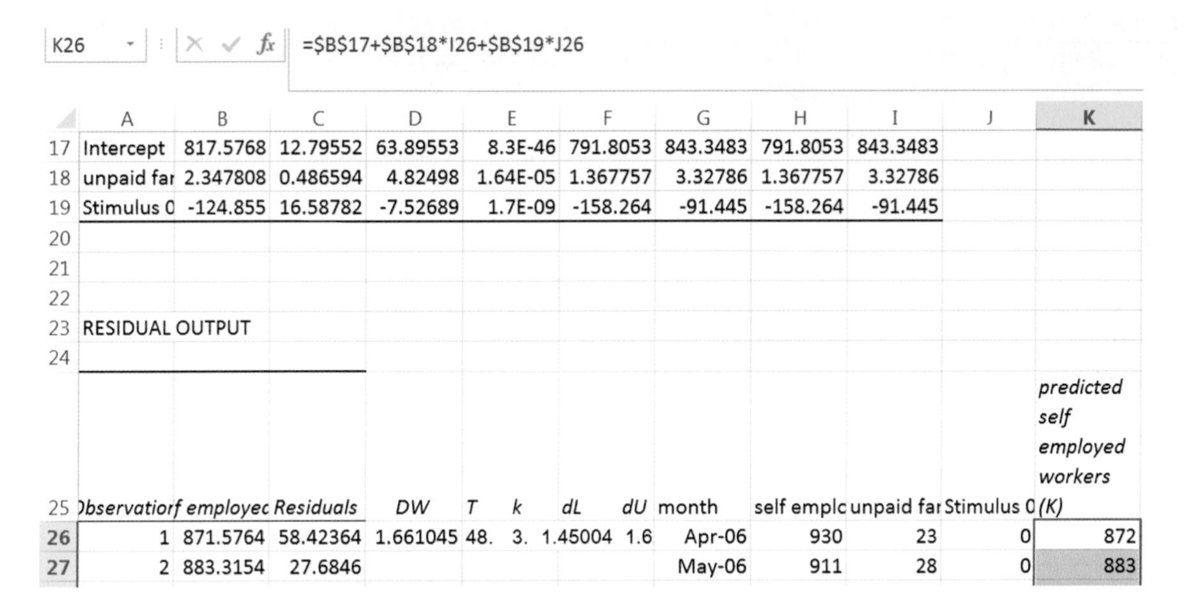

K26 =$B$17+$B$18*I26+$B$19*J26

	A	B	C	D	E	F	G	H	I	J	K
17	Intercept	817.5768	12.79552	63.89553	8.3E-46	791.8053	843.3483	791.8053	843.3483		
18	unpaid far	2.347808	0.486594	4.82498	1.64E-05	1.367757	3.32786	1.367757	3.32786		
19	Stimulus C	-124.855	16.58782	-7.52689	1.7E-09	-158.264	-91.445	-158.264	-91.445		
20											
21											
22											
23	RESIDUAL OUTPUT										
24											

	Observatio	f employec	Residuals	DW	T	k	dL	dU	month	self emplc	unpaid far	Stimulus C	predicted self employed workers (K)
26	1	871.5764	58.42364	1.661045	48.	3.	1.45004	1.6	Apr-06	930	23	0	872
27	2	883.3154	27.6846						May-06	911	28	0	883

To make the lower and upper prediction interval bounds, find the *critical t* value for the residual degrees of freedom in **B**13, and then find make the margin of error from the product of the *critical t* and the regression *standard error* (**B7** in the regression sheet**)**.

D7 =B7*C7

	A	B	C	D
3	*Regression Statistics*			
4	Multiple R	0.800204		
5	R Square	0.640326		
6	Adjusted F	0.624341	critical t	me
7	Standard I	35.10642	2.014103	70.70796

Add and subtract the *margin of error* from the *predicted* values to find the *lower* and *upper 95 % prediction interval bounds.* (Lock the margin of error cell reference, **D7** with **fn 4**.)

L26	fx	=K26-$D$7

	K	L	M
25	*predicted self employed workers (K)*	lower 95% prediction interval bound	upper 95% prediction interval bound
26	872	801	942
27	883	813	954
28	904	834	975
29	961	890	1032

Confirm that the model is valid by comparing actual *self employed workers* in with *lower* and *upper 95 % prediction* interval bounds for April and May 2010:

	G	H	I	J	K	L	M
74	Apr-10	837	29	0	886	815	956
75	May-10	848	26	0	879	808	949

Recalibrate by running the regression, this time including with the two most recent months of data:

3	*Regression Statistics*						
4	Multiple R	0.788418					
5	R Square	0.621602					
6	Adjusted F	0.6055					
7	Standard I	35.31301					
8	Observatio	50					
9							
10	ANOVA						
11		*df*	*SS*	*MS*	*F*	*gnificance F*	
12	Regressior	2	96279.08	48139.54	38.604	1.21E-10	
13	Residual	47	58609.42	1247.009			
14	Total	49	154888.5				
15							
16		*Coefficients*	*andard Err*	*t Stat*	*P-value*	*Lower 95%*	*Upper 95%*
17	Intercept	817.1976	12.86213	63.53519	3.17E-47	791.3224	843.0729
18	unpaid far	2.290301	0.488077	4.692499	2.36E-05	1.308417	3.272186
19	Stimulus C	-123.107	16.64714	-7.39507	2.06E-09	-156.597	-89.6171

Recalibrate forecasts. Update the *critical t*, and *margin of error* by pasting in the formulas from the validation regression:

	A	B	C	D
6	Adjusted F	0.6055	critical t	me
7	Standard I	35.31301	2.011741	71.04062

Copy and paste the data and predicted values from the validation regression to update the forecasts:

	G	H	I	J	K	L	M
25	month	self emplc	unpaid far	Stimulus C	*predicted self employed workers (K)*	lower 95% prediction interval bound	upper 95% prediction interval bound
26	Apr-06	930	23	0	870	799	941
27	May-06	911	28	0	881	810	952
28	Jun-06	951	37	0	902	831	973
29	Jul-06	943	61	0	957	886	1028

Plot the Fit and Forecast. Cut, **Cntl+X,** the *lower* and *upper 95 % prediction intervals* and paste next to *self employed workers,* **Alt H I E**.

Select *month, self employed workers, lower* and *upper 95 % prediction interval bounds* for all months, through May 2011, and insert a scatterplot:

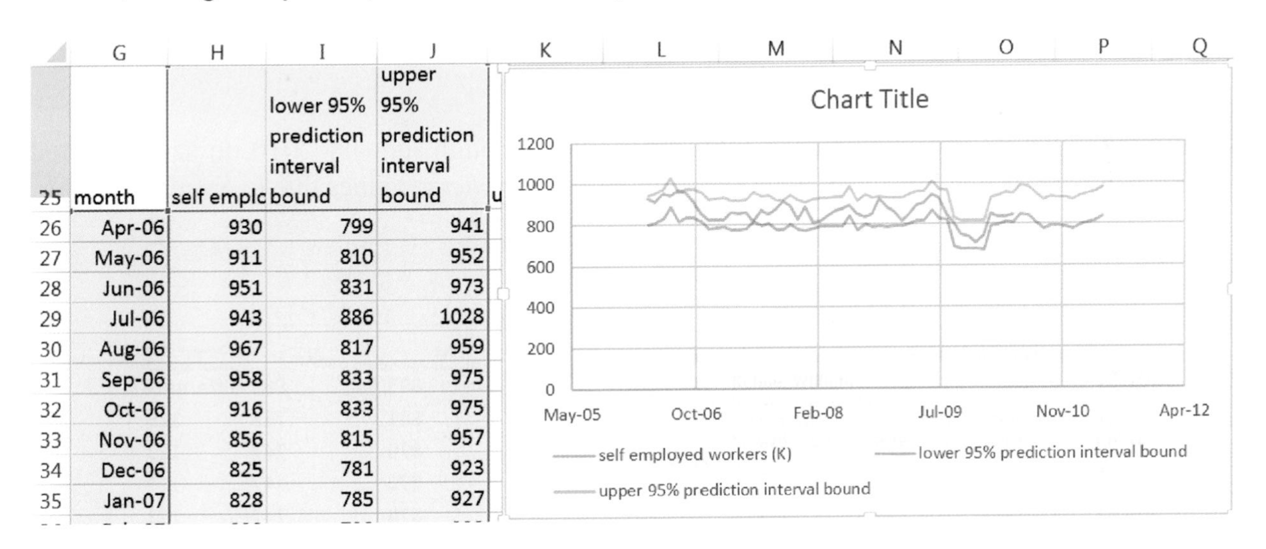

	G	H	I	J
25	month	self emplc	lower 95% prediction interval bound	upper 95% prediction interval bound
26	Apr-06	930	799	941
27	May-06	911	810	952
28	Jun-06	951	831	973
29	Jul-06	943	886	1028
30	Aug-06	967	817	959
31	Sep-06	958	833	975
32	Oct-06	916	833	975
33	Nov-06	856	815	957
34	Dec-06	825	781	923
35	Jan-07	828	785	927

Adjust axes, recolor one of the prediction interval bounds to match the other, change actual datapoints to markers, add axes labels and a title which describes your conclusion:

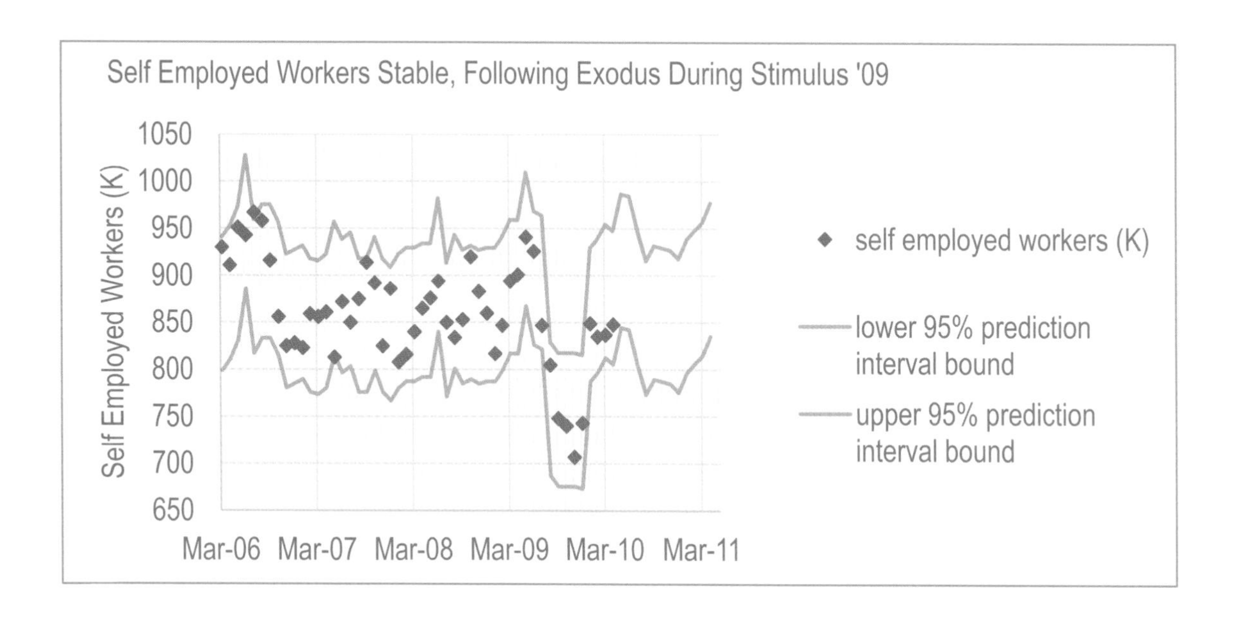

Sensitivity Analysis. Copy *predicted* values (which assume that the *Stimulus 09* was offered September 2009 through January 2010) and paste into a new column *predicted self employed with Stimulus 09* with **values and formats** (but not formulas):

Alt H V S U

	K	L	M	N
25	*predicted self employed workers (K)*	lower 95% prediction interval bound	upper 95% prediction interval bound	*predicted self employed (K) with Stimulus 09*
26	870	799	941	870
27	881	810	952	881
28	902	831	973	902
29	957	886	1028	957

Change *Stimulus 09* values to zero in September 2009 through January 2010, to see predicted values had there been no stimulus. Find the difference in *predicted* values due to *Stimulus 09:*

O71 =M71-N71

	G	H	I	J	K	L	M	N	O
66	Aug-09	847	822	964	33	0	893	893	difference
67	Sep-09	805	810	952	28	0	881	758	123
68	Oct-09	748	799	941	23	0	870	747	123
69	Nov-09	740	799	941	23	0	870	747	123
70	Dec-09	707	799	941	23	0	870	747	123
71	Jan-10	743	797	939	22	0	868	744	123

Rearrange columns so that predictions under the two alternative scenarios follow *year,* and then make a scatterplot to compare:

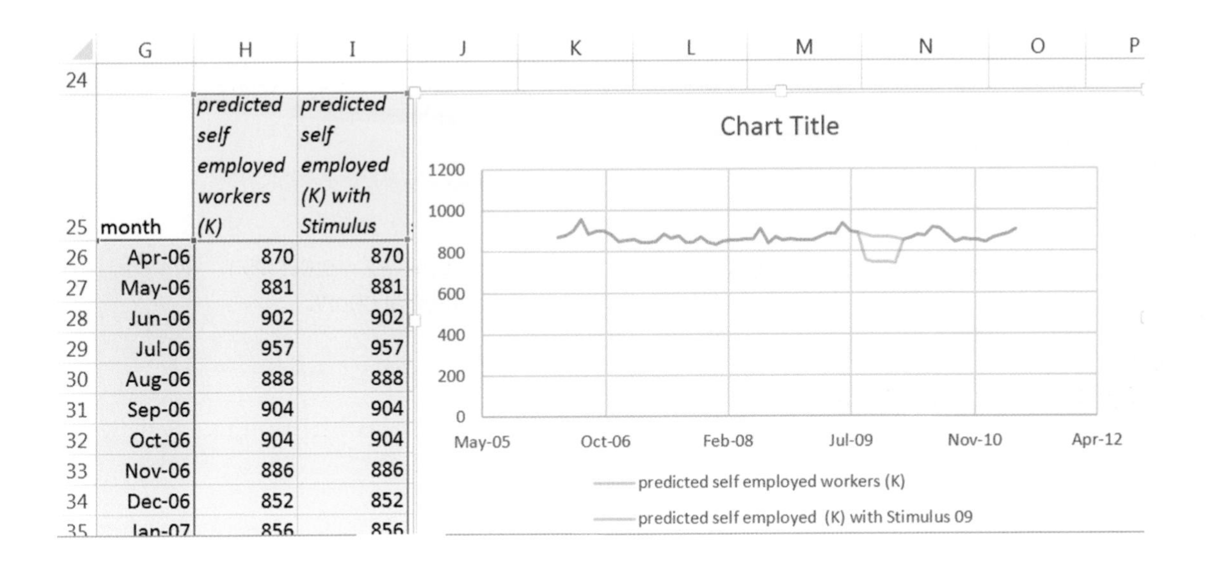

	G	H	I
24			
25	month	predicted self employed workers (K)	predicted self employed (K) with Stimulus
26	Apr-06	870	870
27	May-06	881	881
28	Jun-06	902	902
29	Jul-06	957	957
30	Aug-06	888	888
31	Sep-06	904	904
32	Oct-06	904	904
33	Nov-06	886	886
34	Dec-06	852	852
35	Jan-07	856	856

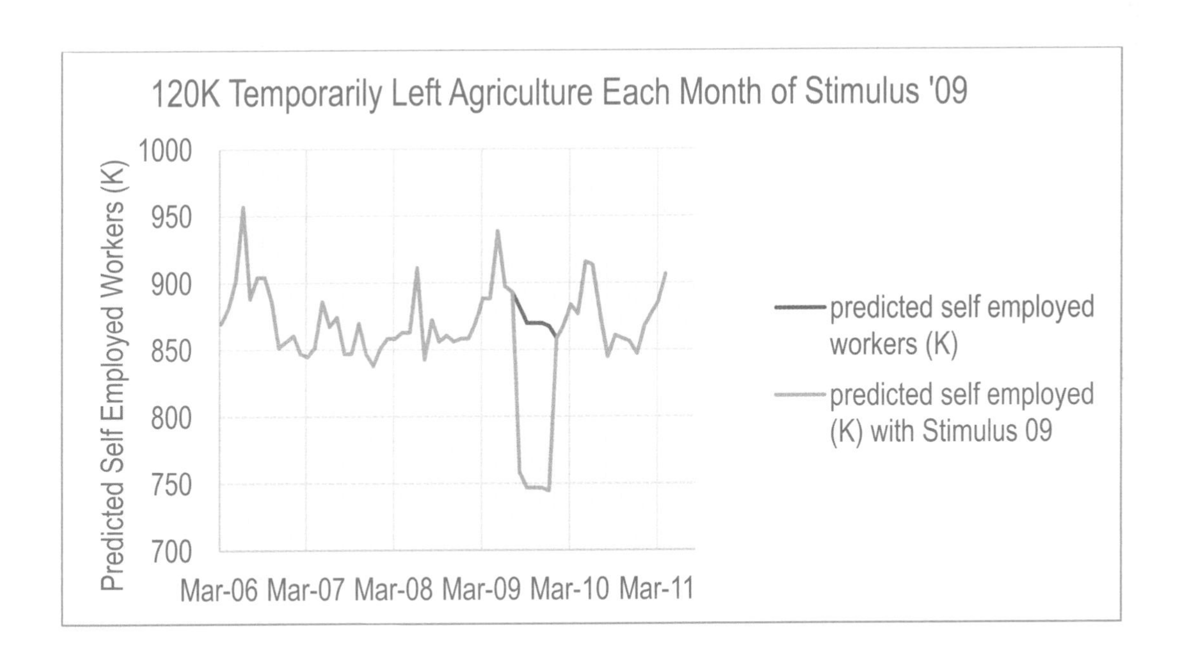

Lab 12 Power PowerPoints

Corporate Relations management is considering alternate designs for the firm's PowerPoint presentation. Content will focus on performance in firm divisions and new ventures by the firm in several global markets. Ideally, audience members would remember twelve key points made in the presentation.

Graphics. Some managers believe that graphics illustrating performance and market potential are easier for audience members to digest and remember. Others argue that tables with precise numbers are more effective. A third group prefers to use photographs with visual images that support conclusions from the numbers stated directly in text.

Text. There is also difference of opinion regarding the amount of text to include in each slide. Some favor a single sentence. Others prefer to use bullet points to remind presenters what material should be covered.

To identify the most effective design, Corporate Relations personnel created six PowerPoint sets:

Graphs, single sentence	Photographs, single sentence	Tables, single sentence
Graphs, bullet points	Photographs, bullet points	Tables, bullet points

A random sample of 12 shareholders not employed for the firm viewed one of the six PowerPoint sets, then answered questions about the content. For each of the 12 key points mentioned, one point was added, making scores from zero to 12 possible.

To account for the influence of experience, the shareholders were asked to report their years of professional experience. **Lab 12 Power PowerPoints.xlsx** contains this data.

One of the participants had to leave early, leaving responses incomplete, reducing the sample to 11.

I. Use regression to test whether visuals or text influence audience content recall and to identify the most effective combination(s).

1. State the hypotheses that you are testing:

2. Report your conclusions, noting the statistical tests that you used:

3. Find the part worth recall scores and attribute importances:

Attibute level		*Part worth recall score*	*Attribute importance*
Visuals	Graphics		
	Photographs		
	Tables		
Text	Single sentence		
	Bullet points		
		Total:	

4. Plot the part worth recall scores by level for visuals and text using the same scale for y axes.

5. Write regression equations for PowerPoints with

   Graphics and single sentences:

   Graphics and bullet points:

   Photographs and single sentences:

   Photographs and bullet points:

   Tables and single sentences:

   Tables and bullet points:

6. Identify the most effective visual, text combination(s): ________________________

7. Identify the least effective visual, text combination(s): ________________________

8. Find the part worth of experience:

Coefficient b (1)	*Max* (2)	*Min* (3)	*Range* (4) = (2) − (3)	*Part worth* = (1) * (4)

9. Which matters more: ___visuals ___text ___ audience experience

Lab 12 ANOVA and Regression with Indicators: Powerful PowerPoints

I. Use ANOVA to test whether type of visual or text format influences audience content recall. Data are in **Lab 12 PowerPoint.xls**.

1. State the hypotheses that you are testing.

2. Report your conclusions, noting the statistical tests that you used.

3. Illustrate average recall scores by type of visual and by text format, choosing the same y axis scale for each.

4. What proportion of variation in recall scores can be accounted for by PowerPoint design differences? ____________

5. Which visual type is most effective? ___________________

6. Which visual type should be avoided? _________________

7. Which text format is most effective? __________________

8. Which makes a bigger difference, the visual type or the text format?

____

II. Compare ANOVA with Regression

Use the same sample that you used for ANOVA, ignoring the influence of experience to investigate the impact of PowerPoint design on content recall.

1. Do visuals and/or text influence recall scores? Y N

   Evidence: *p* value from F test: ____

2. Report your conclusions, noting the statistical tests that you used:

Lab 12 The H-D Buell Blast

The Buell Motorcycle Company is an American motorcycle manufacturer based in East Troy, Wisconsin, founded by ex-Harley-Davidson engineer Erik Buell. H-D bought controlling shares of Buell in 1998 and began selling Buell cycles. Buell became a wholly owned subsidiary of H-D in 2003. Harley-Davidson assumed that new riders who learned on a Buell would later trade up to a H-D.

The Blast. Buell engines were designed to be fuel efficient, and the single cylinder Blast, introduced in 2000, achieved 84 mpg. Body parts of the Blast were made from the same plastic that is used to make the outside of golf balls, to protect Blast parts when dropped.

Buell Blasted. In the economic recession of 2008–2009, lower priced Buell motorcycles began cannibalizing Harley-Davidson bike sales. Keith Wandell, appointed CEO in 2009, began to question the fit of Buell with Harley-Davidson, mentioning "Erik's racing hobby", and asking "why anyone would even want to ride a sport bike". His team and concluded that the adrenaline sport bike segment would encounter high competition, offering low profits, while the cruiser segment could provide high returns.

On October 15, 2009, Harley-Davidson announced the discontinuation of the Buell product line as part of its strategy to focus on the Harley-Davidson brand. The last Buell motorcycle was produced on October 30, 2009.

Harley-Davidson is considering the return of the Blast, and management would like to know what impact the Blast would have on future revenues, should it be reintroduced.

There is some possibility that H-D may sell Buell. Management would like an estimate of the annual future contribution to revenues that could be attributed to the Buell acquisition.

In 2006, H-D changed its stock ticker from HDI to HOG. Management would like to assess the impact of this move on revenues. Some believe that the ticket change motivated some investors to purchase Harleys.

Lab 12 Harley-Davidson Buell Blast.xls contains annual observations on H-D revenues from 1996 to 2009, as well as demographic data thought to drive revenues. Add indicators for and then build a model in identify revenue drivers and to forecast revenues.

Blast availability, 2000 through 2009,
Buell acquisition in 2003,
Ticker change to *HOG* in 2006,

Hide the two most recent data points. To more easily identify demographic drivers, run regression with the three indicators, and then plot the residuals.

Focusing in the eight most recent years, highlight cells in which the residuals have decreased, relative to the previous year, and then highlight cells in which each of the demographic drivers has decreased, relative to the previous year.

1. Which demographic variable has the same pattern at least 3 out of 4 and at least 5 out of 8 of the most recent years?

2. Which demographic variable shows a pattern exactly opposite of the residuals in at least 3 out of 4 and at least 5 out of 8 of the most recent years of data?

If more than one variable was identified in steps 1 and 2, use correlations between residuals and the variables with same or opposite patterns to choose a driver to add to the model. (If a single variable was identified, select that variable to add to your model.)

3. Do either the indicators or demographics (or both) influence revenues? Y or N

4. Have you accounted for trend, cyclicality, and shifts with your model?

5. Is your model valid? Y or N

6. Recalibrate and then plot your fit and forecast for years 1996 through 2012.

7. If H-D were to re-introduce the Blast, how much could revenues in 2012 be expected to increase?

8. If H-D were to sell Buell, what potential revenues could the buyer expect in 2012?

9. What additional annual revenues can H-D attribute to the HOG ticker?

Assignment 12-1 Conjoint Analysis of PDA Preferences

Dell is considering introduction of a new PDA which would be sold at a competitive price through WalMarts. New product development managers believe that customers would choose brightly colored Dell PDAs at competitive prices.

Choose four attributes of PDAs that you believe to be influences on college students' preferences. Identify three alterative options for each attribute and fill in the orthogonal array table, below, to make nine hypothetical PDAs. You may use whichever attributes and attribute levels you believe matter to college students. Those shown below are used to illustrate the orthogonal design.

Hypothetical PDA	*Brand*	*Color*	*Keypad*	*Price*
1	Dell	Silver	Standard	$150
2	Dell	White	QWERTY	$250
3	Dell	Lime green	Touch screen	$350
4	Apple	Silver	QWERTY	$350
5	Apple	White	Touch screen	$150
6	Apple	Lime green	Standard	$250
7	Palm	Silver	Touch screen	$250
8	Palm	White	Standard	$350
9	Palm	Lime green	QWERTY	$150

Rate the nine hypothetical PDAs, using a scale from 1 ("undesirable") to 9 ("very desirable"). Ask three friends or classmates to rate the nine hypotheticals also.

Enter your ratings in the **Assignment 12 Dell PDA conjoint.xls.** The file contains 36 rows, 9 rows for each person in your sample, and 7 columns, *customer, hypothetical PDA, brand, color keypad, price*, and *rating*. Change the labels to match the attributes and attribute levels that you chose.

Identify the baseline PDA, and then make eight indicator variables to designate options other than baseline.

Run a regression to find the preferred PDA configuration, the *part worth utilities*, and the relative *importances* of attributes.

Deliverables: Write a paragraph to management, summarizing your results, with recommendations for the new product development team.

Attach a copy of your regression sheet with a table and plots of *part worth utilities*, and a table of *attribute importances*.

Assignment 12-2: Indicators with Time Series: Impact of the Commodity Price Bubble on World Oil Prices

Rolls-Royce is facing several decisions which hinge on future oil prices. They require a forecast of oil prices over the next 4 years, through 2015. They are particularly interested in learning

- The degree to which the Commodity Price Bubble shock of 2007–2008 affected oil prices, and
- The influence of past World GDP on oil prices.

The Commodity Price Bubble of 2007–2008 followed the shock of the collapse in housing prices in several Western nations, including the U.S. Commodities were seen as safer alternatives to other investments. The price of oil nearly tripled from $50 to $147 a barrel from early 2007 to 2008, before plunging with the financial crisis and global recession.

Demand for oil is thought to depend on past productivity, reflected in past World GDP. Higher wealth is thought to stimulate higher demand for energy, driving up oil prices.

The dataset **Assignment 12 World Oil Price Forecast.xls** contains time series of *World oil* $prices_t$, and past *World* GDP_{t-5} for years 2003 through 2011.

Build a model of World oil prices, including

i. *An indicator of the Commodity Bubble in 2007 and 2008,*
ii. *Past World* GDP_{t-5}

1. Since this is a time series model, assess the model Durbin Watson statistic to determine whether or not unaccounted for trend or cycles remain.

   Residuals ___ contain _____possibly contain OR ___ are free of unaccounted for trend or cycles.

2. Is your model valid? Y or N

3. Illustrate your fit and forecast and embed your plot.

4. Write your model equations for *World oil price*, turning the indicator on or off, adjusting the intercept and simplifying:

   - During **baseline years 2003–2006** and **2009–2015,**
   - During the Commodity Bubble in years **2007–2008.**

5. What is your prediction interval for oil prices in 2015?

6. If there were to be a similar Commodity Price Bubble in 2012 and 2013, perhaps due to increasing demand in emerging markets and perceived scarcity, similar to the Bubble of 2007–2008, what would be the estimated impact on oil prices in 2013?

	Predicted world oil price ($/barrel)		Expected influence of commodity price bubble ($/barrel)
Year	No commodity price bubble	Commodity price bubble	
2013			

7. To illustrate the potential impacts of the Commodity Price Bubble and past *World GDP* on *World oil prices*, plot *predicted oil prices* by year for years 2003 through 2015

   - Assuming no Commodity Price Bubble in 2012–2013, and
   - Assuming a Commodity Price Bubble in 2012–2013.

Embed your plot.

CASE 12-1 Modeling Growth: Procter & Gamble Quarterly Revenues

Procter & Gamble revenues are growing, as the company's managers innovate and forge into new markets, and as the company acquires complementary businesses. Procter & Gamble management want to quantify the impact on revenues of the acquisition of Gillette late in 2005. They have asked for a model which quantifies quarterly revenue drivers, including the Gillette acquisition, which can also be used to forecast. Quarterly revenues are in **Case 12 Procter & Gamble Revenues.xlsx.**

The Gillette Acquisition. Procter & Gamble acquired Gillette in 2005. The first quarter of the combination is December 2005. Revenues in that quarter were nearly $4 billion greater than in the preceding quarter.

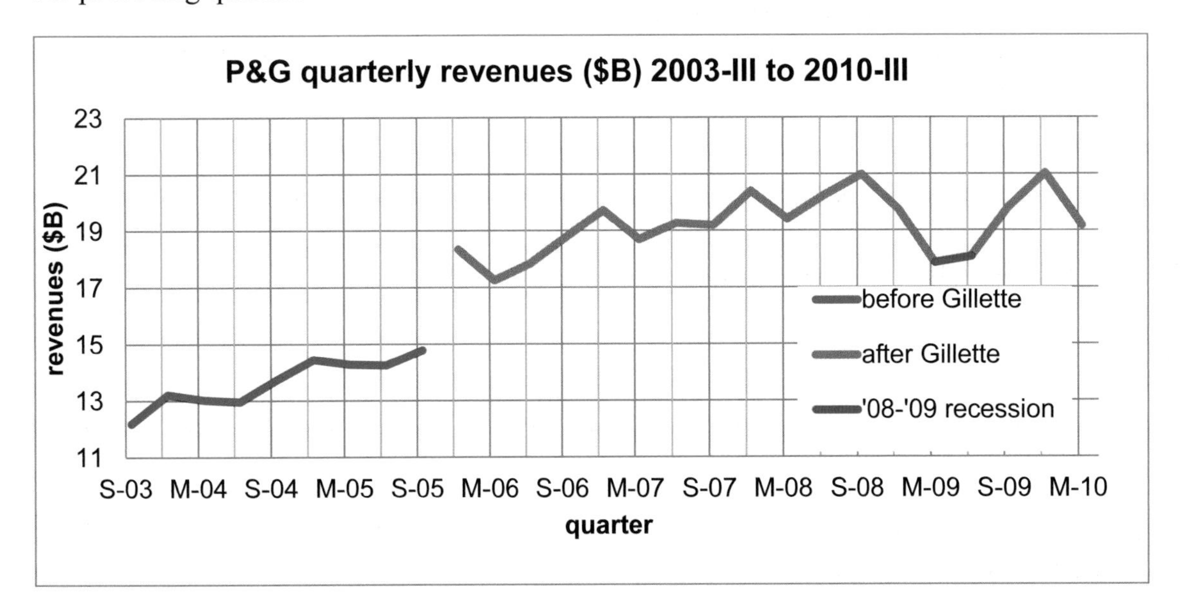

Inertia. Procter & Gamble manufactures and markets packaged goods, many of which could be viewed as necessities. Consumers who choose P&G brands tend to be brand loyal, and choose P&G brands repeatedly. Management would like to know how strong this loyalty response is. For each new dollar in revenues, what proportion can be expected a year later?

Some of the products in P&G's portfolio are seasonal. Laundry detergent, for example, is seasonal, because families wear more layers of clothing in colder seasons. Consequently, quarterly revenues tend to be highest in the fourth quarter each year. One convenient way to account for seasonality is to include an inertia component, such as past year revenues, as a predictor. Past year revenues are, thus, useful for estimating loyalty, and also because they account for seasonality.

'08-'09 Recession. P&G management saw revenues fall below forecasts the first two quarters of 2009. Management believed that the disappointing results were the result of the economic recession of 2008–2009 and the related financial crisis of 2009. In the U.S., GDP actually lost value in the first two quarters of 2009. Once decision makers became aware that the economy was officially in recession, major changes occurred. Companies downsized, and consumers were forced to reduce spending. Management wants to confirm that slowed revenues in quarters I and II of '09 were due to recession, and not some firm specific problem. They also want to confirm that revenues had begun to recover and would regain their previous momentum.

Build a time series model of P&G revenues, including the *Gillette acquisition*, the *'08-'09 recession* and the past year $revenues_{q-4}$.

Add an indicator of the *Gillette boost*, equal to zero in quarters before December 2005 and equal to one in December 2005 and quarters after.

Allow for the impact of the 2009 recession by adding an indicator *'08-'09 recession,* equal to one in the first two quarters of 2009 and equal to zero in other quarters.

Make an inertia component, past year $revenue_{q-4}$, by copying $revenues_q$, and then shifting the lagged inertia column four quarters. Past year revenue from September '02 will appear in the $revenue_{q-4}$ column in the same row as $revenue_q$ in September '03. (Your regression will begin with data from September '03, as a consequence.)

Be sure to exclude the two most recent quarters, December '09 and March '10, to build your model. Then you will be able to test its validity for forecasting.

Assess the Durbin Watson *DW* statistic to decide whether or not your model has accounted for trend, cycles and seasonality in the quarterly data.

Validate your model, then add the two most recent quarters and recalibrate.

Sensitivity analysis to find expected response under alternate scenarios.

Find forecasts with:

i. The *Gillette* indicator set to zero to determine what *revenues* would have been had the acquisition not been made.

ii. The *'08-'09 recession* indicator set to zero to determine what *revenues* would have been had the recession not occurred.

Deliverables.

1. Write your model equations for

   i. The baseline quarters before the *Gillette acquisition*
   ii. Quarters after the *Gillette acquisition*
   iii. Quarters during the *'08-'09 recession*

2. What is the margin of error in your forecasts?

3. What percent of each dollar of new revenue can management expect to come from repeat sales to loyal customers 1 year later?

4. What is the *95 % prediction interval* for revenues in March 2011?

5. What is the expected percent increase in *revenues* in March 2011, relative to revenues in March 2010?

6. Make a table to show

   i. How much the Gillette acquisition has enhanced *revenues* in each of the quarters since December 2005.
   ii. The percent of *revenues* contributed by Gillette relative to what *revenues* would have been without Gillette in each of the quarters since December 2005

7. Make a table to show

   i. *Revenue* lost in each of the first two quarters of 2009 due to the *'08-'09 recession*
   ii. The percent reduction from expected *revenues* had there been no recession

8. Illustrate your model fit and sensitivity analysis with a scatterplot of

   i. *Revenue predictions*, September 2003 through March 2011
   ii. *Actual revenues*
   iii. *Revenue predictions* without the *Gillette acquisition* from December 2006 through March 2011
   iv. *Revenue predictions* without the *'08-'09 recession* from March 2009 through March 2011

CASE 12-2 Store24 (A): Managing Employee Retention* and Store24 (B): Service Quality and Employee Skills**

Work with one or two partners.

Problem. Management needs to know what controllable factors are driving store sales and profits. If management or crew tenure, management or crew skill, or service quality are driving performance, programs to increase tenure, skill or quality will be created.

There has been some grumbling from managers of stores not located in residential neighborhoods, as well as from managers of stores not open 24 h, since sales or profit potential may be more limited in those stores. Data are in **Case 12 Store24.xlsx.**

Your assignment. Identify the drivers of store sales and profits and quantify the influence of each.

Deliverables. Present your results in a memo to management.

Content.

Explain Variations in Store Performance

- Identify the more powerful drivers among the controllable factors, so that resources can be directed to programs which could improve those driver levels.
- Compare their influences on sales and profits
- Quantify the potential performance limitations facing managers of stores located in residential neighborhoods or stores not open 24 h. (Be sure to account for other differences which may work with store location or hours to drive performance differences.)

Implications

- Programs to improve tenure, skill, or quality vary in cost. Suggest a budget for developing programs which ought to be considered.
- Suggest ways in which stores **not** located in residential neighborhoods and stores not open 24 h could be improved to compensate for their possibly lower than average sales or profit potential.

Format

- Refer to graphics to support your presentation and suggestions. Embed one graphic in your memo and attach additional exhibits. Choose axes that make comparisons easy. Be sure to include a stand alone title and label your axes, including units.
- Include your regression equations below the embedded graphic.
- Use single spacing with 12 pt Times New Roman font.

Strict limit of two pages of text.

---

* Harvard Business School case 9602096
** Harvard Business School case 9602097

Chapter 13
Nonlinear Multiple Regression Models

In this chapter, nonlinear transformations are introduced that expand multiple linear regression options to include situations in which marginal responses are either increasing or decreasing, rather than constant. We will explore Tukey's Ladder of Powers to identify particular ways to rescale variables to produce valid models with superior fit. An example will be offered in the context of naïve models built for forecasting, and in Chap. 14, a second example with cross sectional data expands on nonlinear model building and sensitivity analysis.

13.1 Consider a Nonlinear Model When Response Is Not Constant

To decide whether or not to use a nonlinear model, first rely on your logic:

- Do you expect the response, or change in the dependent, performance variable, to be constant, regardless of whether a change in an independent variable is at minimum values or at maximum values? Linear models assume constant response.
- Is the dependent variable limited or unlimited?

Linear models are unlimited. If your dependent variable couldn't be negative, because it is measured in dollars, purchases, people, or uses, a nonlinear model is logically more appropriate. After consulting your logic, plot your data.

13.2 Skewness Signals Nonlinear Response

When either the dependent, performance variable y or a driver x is skewed, performance response to differences or changes in the driver is nonlinear. Rescaling skewed variables to roots or logarithms will linearize response, allowing use of multiple linear regression which will link the rescaled variables.

Increasing marginal response, in which differences or changes in x produce larger and larger responses in y, occurs if

- The dependent variable y is positively skewed, and/or
- The driver x is negatively skewed.

These two possibilities are shown in the upper and left plots in Fig. 13.1.

C. Fraser, *Business Statistics for Competitive Advantage with Excel 2013: Basics, Model Building, Simulation and Cases*,
DOI 10.1007/978-1-4614-7381-7_13, © Springer Science+Business Media New York 2013

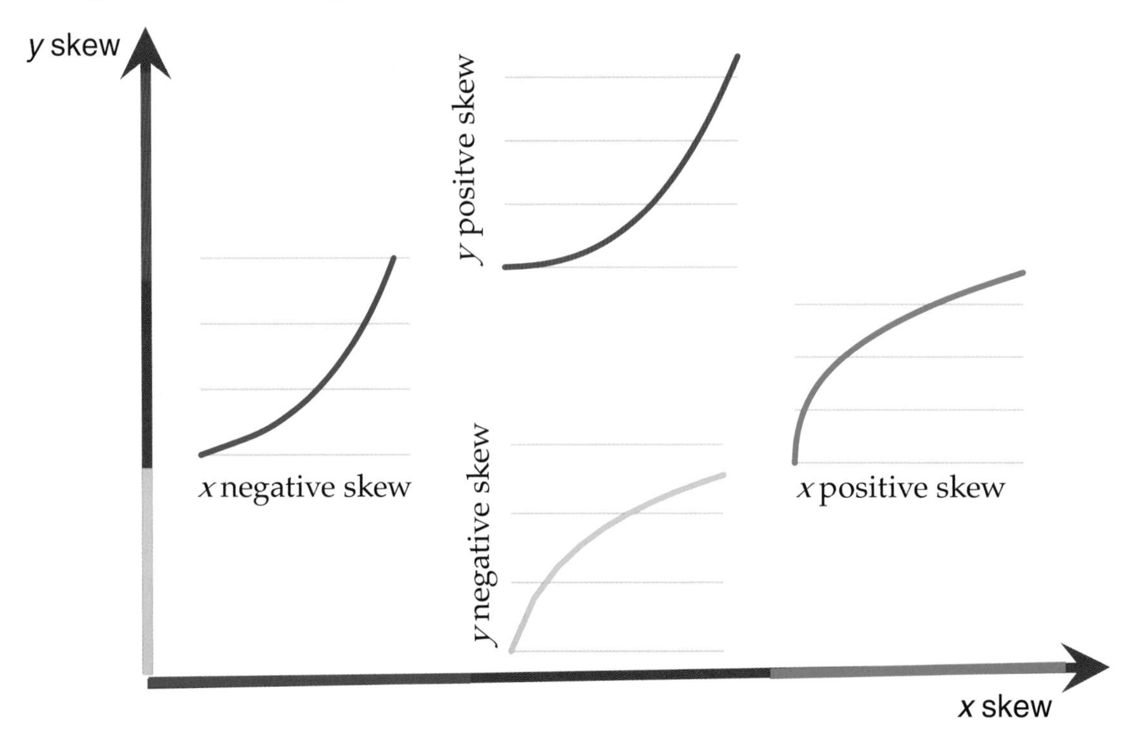

Fig. 13.1 Nonconstant marginal response

Decreasing marginal response, in which differences or changes in x produce smaller and smaller responses in y, occurs if

- The dependent variable y is negatively skewed, and/or
- The driver x is positively skewed.

These two possibilities are shown in the lower and right plots in Fig. 13.1.

To linearize response so that multiple linear regression can be used, the goal is to rescale toward the center of Fig. 13.1. Tukey offered a simple heuristic to quickly suggest ways to rescale variables when residuals from linear regression would be either skewed or hetero-skedastic. Scales are chosen which reduce skewness of both independent and dependent variables. Models built with variables which have been rescaled to reduce skewness will be nonlinear.

If a variable is positively skewed, as is the variable on the left in Fig. 13.2, shrinking it by rescaling in roots, or natural logarithms will *Normalize*. Square roots are higher power, .5, than cube roots, .33, and are less radical. Natural logarithms make a bigger difference than square roots.

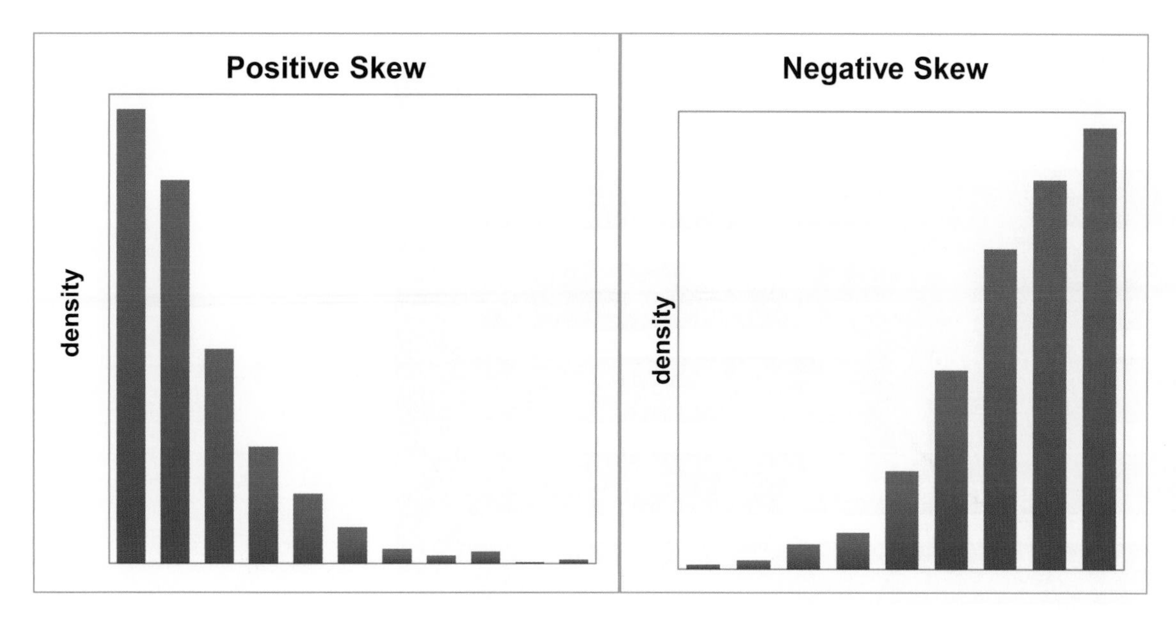

Fig. 13.2 Positively and negatively skewed variables

When a variable is negatively skewed, as is the variable on the right in Fig. 13.2, expanding it by rescaling to squares or cubes will *Normalize*. A higher power, such as cubes, will make a bigger difference.

Moving from the center up or down the *Ladder of Powers,* Fig. 13.3, changing the power more, changes the data and its skewness more. More skewness calls for adjusting more.

The left, blue plot in Fig. 13.4, features a negatively skewed driver. Rescaling driver values to their squares or cubes will produce the linear relationship, moving the plot right to Normal skewness in Fig. 13.4.

The upper, red plot in Fig. 13.4 reflects a positively skewed dependent variable. Rescaling values of that dependent variable to their natural logarithms or roots will linearize the relationship, moving the plot down to Normal skewness in Fig. 13.4.

The lower, gold plot in Fig. 13.4 features a negatively skewed dependent variable. Rescaling dependent variable values to their squares or cubes will linearize the plot, moving it up to Normal skewness in Fig. 13.4.

The right, green plot in Fig. 13.4 reflects a positively skewed independent variable. Rescaling driver values to their roots or natural logarithms will linearize the relationship, moving the plot left to Normal skewness in Fig. 13.4.

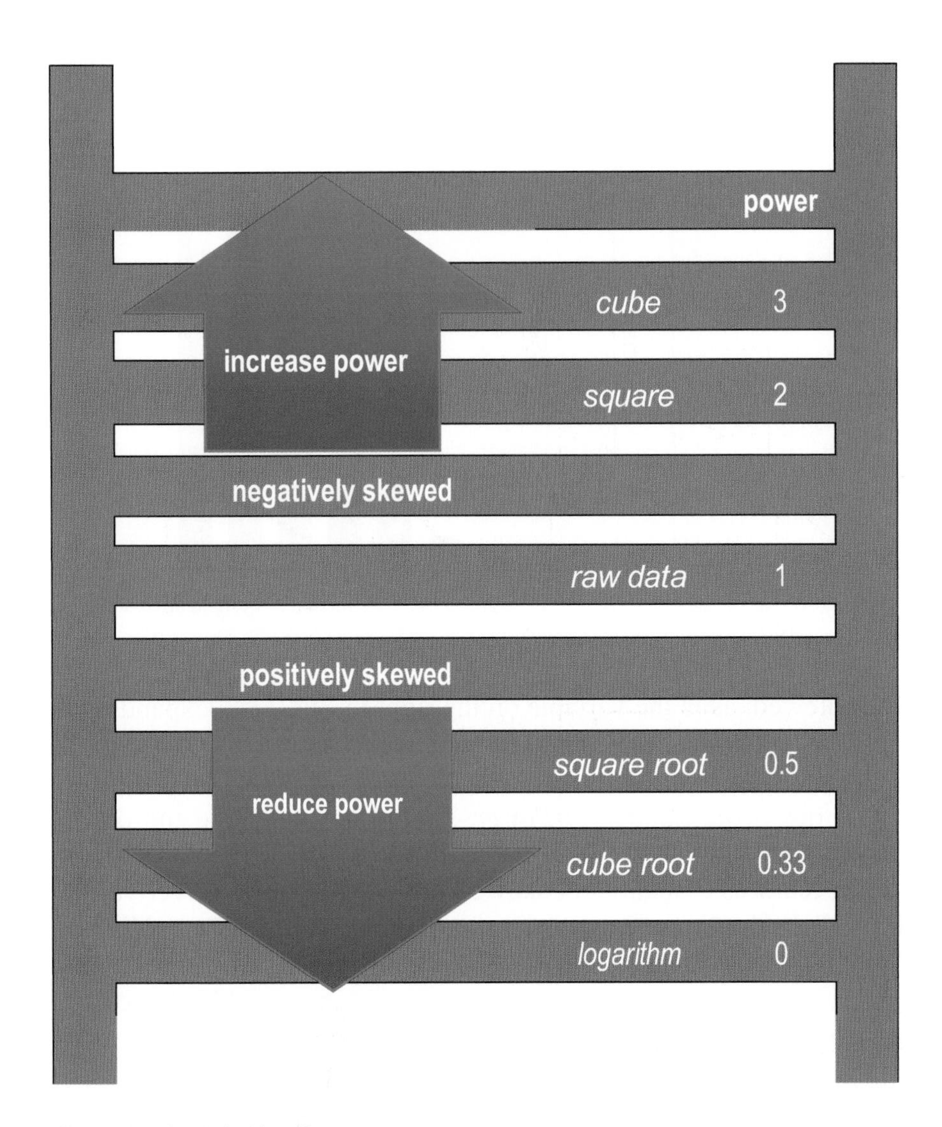

Fig. 13.3 Tukey's ladder of powers

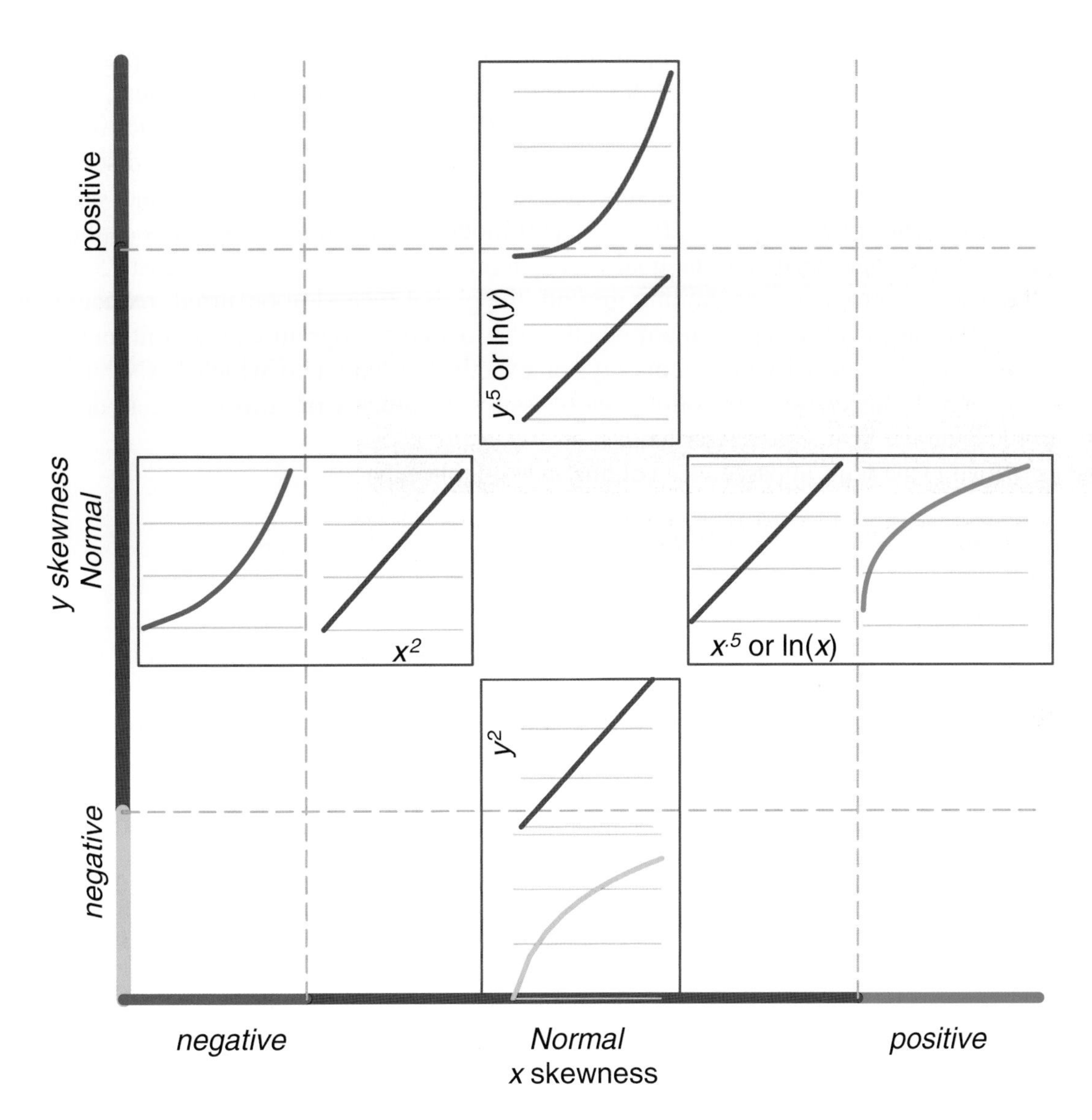

Fig. 13.4 Rescaling to reduce skew linearizes

13.3 Rescaling *y* Builds in Interactions

Jointly, two drivers may make a larger difference than the sum of their individual influences. For example, advertising levels may be more effective when sales forces are larger. The impact of population growth in a country may influence imports more if growth in GDP has been relatively high. When the dependent variable is rescaled, the model becomes multiplicative, which produces interactions between predictors. With this potential benefit of improved fit and validity, comes the cost of transforming predictions in rescaled units back to the original units.

Example 13.1 LAN Airlines in 2011

LAN Airlines, a Chilean multinational, had achieved status as a major global carrier with innovative strategies. Two innovations distinguished LAN. With insightful route planning, the cargo and passenger businesses shared capacity. Passengers might be flown from Chile to

Germany, where cargo was loaded for shipment to the United States. In the States, passengers boarded for flights to Chile, for example. After review of the success of low cost carriers, such as Ryan Air, LAN initiated a two tier fare system in 2007. International flights, which offered premium service, were sold at premium prices. Intra country flights were sold at discounted fares. The low cost intra country fares appealed to price conscious travelers and enabled purchase of more planes, including new, fuel efficient models. Critical to the success of the low cost program was passenger load, and the goal was to achieve passenger load of at least 75 %.

Just as the low cost program was gaining ground, the global recession occurred, reducing air travel. It was not clear to LAN management whether the low cost program would continue to be effective in light of the altered world economy, or exactly the extent to which LAN business would suffer from the recession. To address these issues, a forecast of firm profits is desired, with initial focus on the firm's passenger load.

Passenger load is the ratio of passenger volume to available capacity:

- The volume of passengers carried, *RPKs*, revenue passenger kilometers,
- Available capacity, *ASKs,* available seat kilometers,

Under the direction of Madelaine Kearing, the modeling team built naïve models of the two passenger load components using indicators for the *low cost program.* Model forecasts would be used and integrated in a Monte Carlo simulation to forecast LAN passenger load in 2016.

After hiding the two most recent observations, Madelaine checked skewness of passenger *ASK* and *RPKs* (Fig. 13.5). *ASKs* and *RPKs* were mildly positively skewed, like the upper quadrant of Fig. 13.1, reflecting increasing annual growth in both capacity and volume.

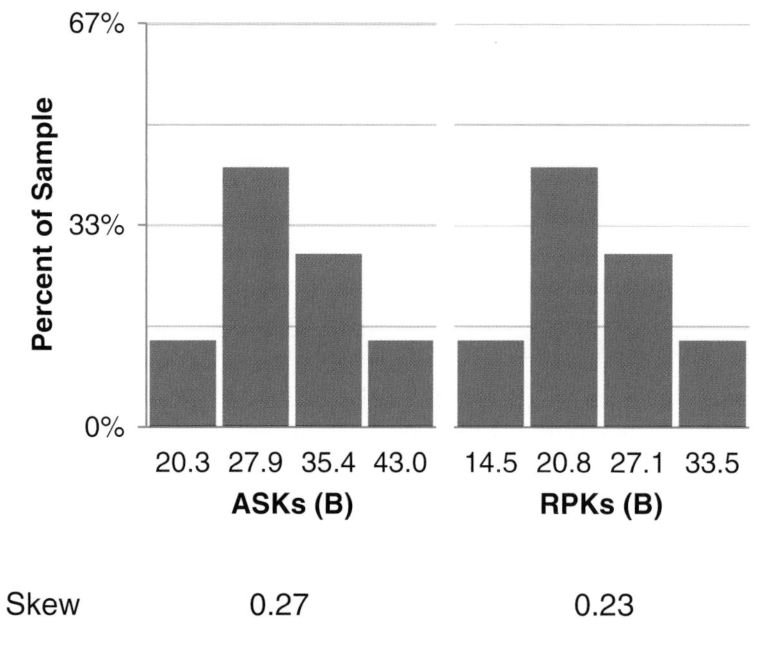

Fig. 13.5 Skewness of passenger revenue components

The distributions of *ASKs* and *RPKs* contained no values more than one standard deviation below the sample mean and more than 33 % within one standard deviation below the sample mean. Using the roots or the natural logarithms of *ASKs* and *RPKs* would increase the chances that the naïve models would be valid for forecasting.

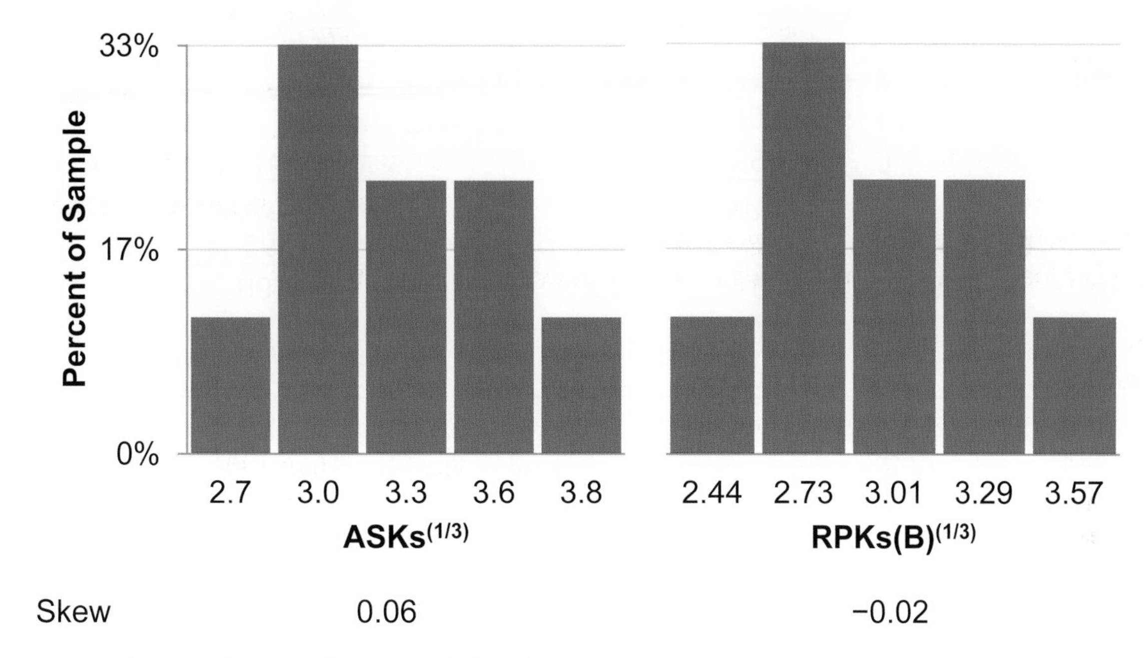

Fig. 13.6 Skewness of ASKs and RPKs rescaled to cube roots

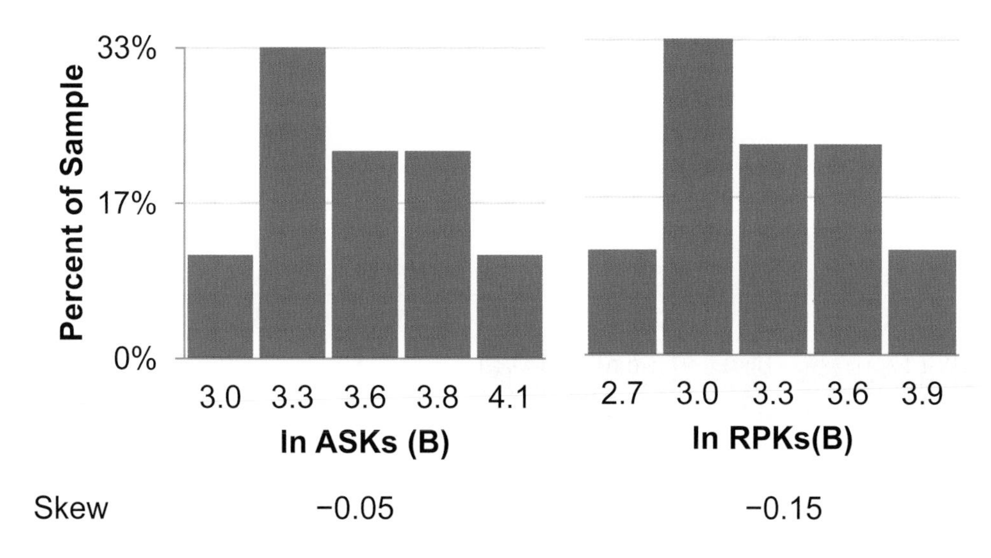

Fig. 13.7 Skewness of ASKs and RPKs rescaled to natural logarithms

The square roots, cube roots, shown in Fig. 13.6, and the natural logarithms, shown in Fig. 13.7, improved skewness. The modeling team chose to use the natural logarithms of *ASKs* and the cube roots of *RPKs*. Regression results are shown in Tables 13.1 and 13.2.

Table 13.1 Regression of ln ASKs

R Square	.998
Standard error	.014
Observations	7

ANOVA

	df	*SS*	*MS*	*F*	*Significance F*
Regression	2	.452	.226	1,114.2	.000
Residual	4	.0008	.0002		
Total	6	.453			

	Coefficients	*Standard error*	*t Stat*	*p value*	*Lower 95 %*	*Upper 95 %*
Intercept	−229	10.8	−21.2	.0000	−259	−199
Year	.116	.005	21.5	.0000	.10	.13
Low cost	.052	.022	2.4	.08	−.01	.11

Observation	*Pred ln pass asks (B)*	*Residuals*	*DW*	*T*	*k*	dL	dU
1	2.93	−0.02	1.63	7	3	.47	1.90

Table 13.2 Regression of cube roots of RPKs

R Square	.998
Standard error	.014
Observations	7

ANOVA

	df	*SS*	*MS*	*F*	*Significance F*
Regression	2	.477	.239	1,263.4	.0000
Residual	4	.001	.0002		
Total	6	.478			

	Coefficients	*Standard error*	*t Stat*	*p value*	*Lower 95 %*	*Upper 95 %*
Intercept	−232	10	−22.3	.0000	−261	−203
Year	.117	.005	22.5	.0000	.10	.131
Low cost	.063	.021	3.0	.04	.004	.121

Observation	*Predicted RPKs* $(B)^{(1/3)}$	*Residuals*	*DW*	*T*	*k*	dL	dU
1	2.35	−0.02	1.97	7	3	.47	1.90

The models were significant, both the low cost indicator and trend were significant, and their influences were in the expected positive direction in both models.

The Durbin Watson statistic indicated the possibility of unaccounted for trend or cycles in *ASKs*. The residual plot revealed lower than expected *ASKs* in the last validation sample year, 2009. The decline in *ASKs* could be related to the global recession, but a global recession indicator was not significant. The models made sense and accounted for nearly all of the annual variation in passenger load components. The modeling team moved on to validate.

Comparing lower and upper 95 % prediction intervals for the two most recent years confirmed that both models had validity for forecasting. Actual *ASKs* and *RPKs* fell within the prediction intervals. The models were recalibrated using all available data:

$$\ln A\hat{S}Ks(B)_t = -215^a + .068^b \times Low\ Cost_t + .11^a \times t$$

RSquare: .998[a]

$$R\hat{P}Ks(B)_t^{(\frac{1}{3})} = -238^a + .053^b \times Low\ Cost_t + .12^a \times t$$

RSquare: .998[a]

[a]Significant at .01 or better; [b]Significant at .05 or better.

The *ASKs* equation is in logarithms. To see the equation in the original scale of billion *ASKs*, use the exponential function to reverse the natural logarithms:

$$e^{\ln A\hat{S}Ks(B)_t} = e^{-215^a + .068^b \times Low\ Cost_t + .11^a \times t}$$

$$A\hat{S}Ks(B)_t = e^{-215^a + .068^b \times Low\ Cost_t + .11^a \times t} \quad (13.1)$$

$$= 2.7\text{E-}94 \times e^{.068 \times Low\ Cost_t} \times e^{.11 \times t}$$

The *RPKs* equation is in cube roots. To see the equation in the original scale of billion *RPKs*, cube both sides, rescaling back:

$$[R\hat{P}Ks(B)_t^{(\frac{1}{3})}]^3 = [-238 + .053 \times Low\ Cost_t + .12 \times t]^3$$

$$R\hat{P}Ks(B)_t = [-238 + .053 \times Low\ Cost_t + .12 \times t]^3 \quad (13.2)$$

The fits and forecasts are shown in Fig. 13.8.

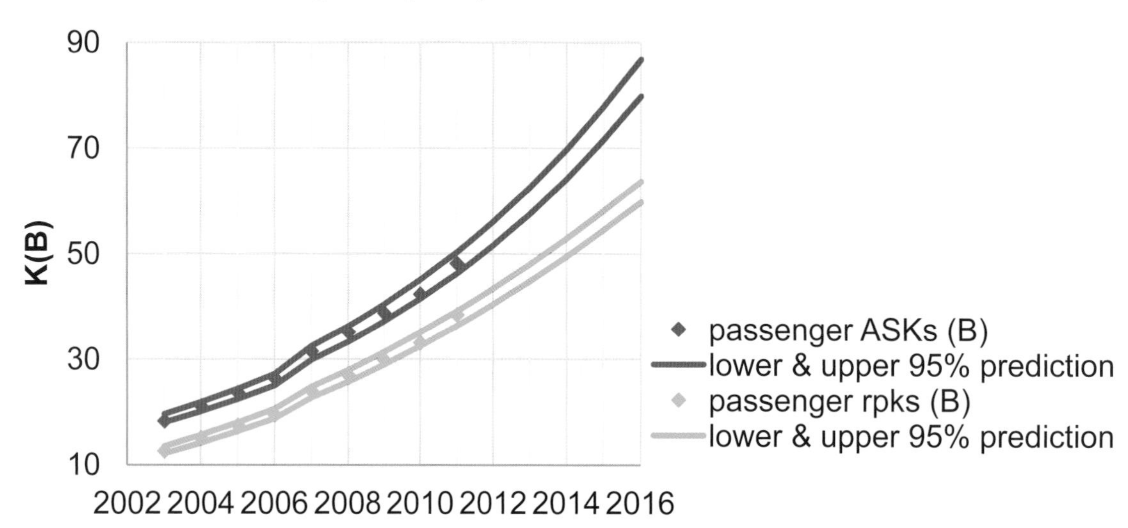

Fig. 13.8 Fits and forecasts for ASKs and RPKs

13.4 The Margin of Error Is Not Constant with a Nonlinear Model

With a nonlinear model, the margin of error is not constant. The margins of error, half the distance between the lower and upper 95 % prediction interval bounds, increases as the forecast date moves further from the present. The 2016 LAN forecast for passenger capacity can be expected to fall no further than 3.5 billion ASKs (= (86.8 − 79.8B)/2) from actual, and the 2016 LAN forecast for passenger volume can be expected to fall no further than 1.9 billion RPKs (= (63.7 − 59.8)/2) from actual.

13.5 Sensitivity Analysis Enables Scenario Comparisons

When a dependent variable is rescaled to build a nonlinear model, the model is multiplicative. The impact of each of the drivers depends on values of all of the other drivers. Expanding the right sides of the equations, the interactions are apparent. (Don't let this expansion scare you! The models would be presented in the forms shown in Eqs. 13.1 and 13.2.)

The capacity equation features an exponential function on the right side, which can be written as the product of the three terms:

$$A\hat{S}Ks(B)_t = e^{-215^a + .068^b \times Low\ Cost_t + .11^a \times t}$$

$$= 2.7\text{E-}94 \times e^{.068 \times Low\ Cost_t} \times e^{.11 \times t}$$

For years before the Low Cost program was initiated, where the *Low Cost* indicator is set to zero, the equation becomes:

$$A\hat{S}Ks(B)_t = 2.7\text{E-}94 \times e^{.068 \times 0} \times e^{.11 \times t}$$

$$= 2.7\text{E-}94 \times e^{0} \times e^{.11 \times t}$$

$$= 2.7\text{E-}94 \times 1 \times e^{.11 \times t}$$

$$= 2.7\text{E-}94 \times e^{.11 \times t}$$

In years after the Low Cost program was initiated, where the *Low Cost* indicator is set to 1, the equation becomes:

$$A\hat{S}Ks(B)_t = 2.7\text{E-}94 \times e^{.068 \times 1} \times e^{.11 \times t}$$

$$= 2.7\text{E-}94 \times e^{.068} \times e^{.11 \times t}$$

$$= 2.7\text{E-}94 \times 1.07 \times e^{.11 \times t}$$

$$= 3.0\text{E-}94 \times e^{.11 \times t}$$

The *low cost* program multiplier of 1.07 multiplies capacity by 107 %. While this multiplier is constant, the low cost impact is greater in years where capacity is greater.

Evaluating the impact of the Low Cost program can be accomplished by comparing capacity with and without the program, the distance between red and orange lines in Fig. 13.9. In 2016, the Low Cost program is forecast to motivate additional capacity of 5.5 billion ASKs.

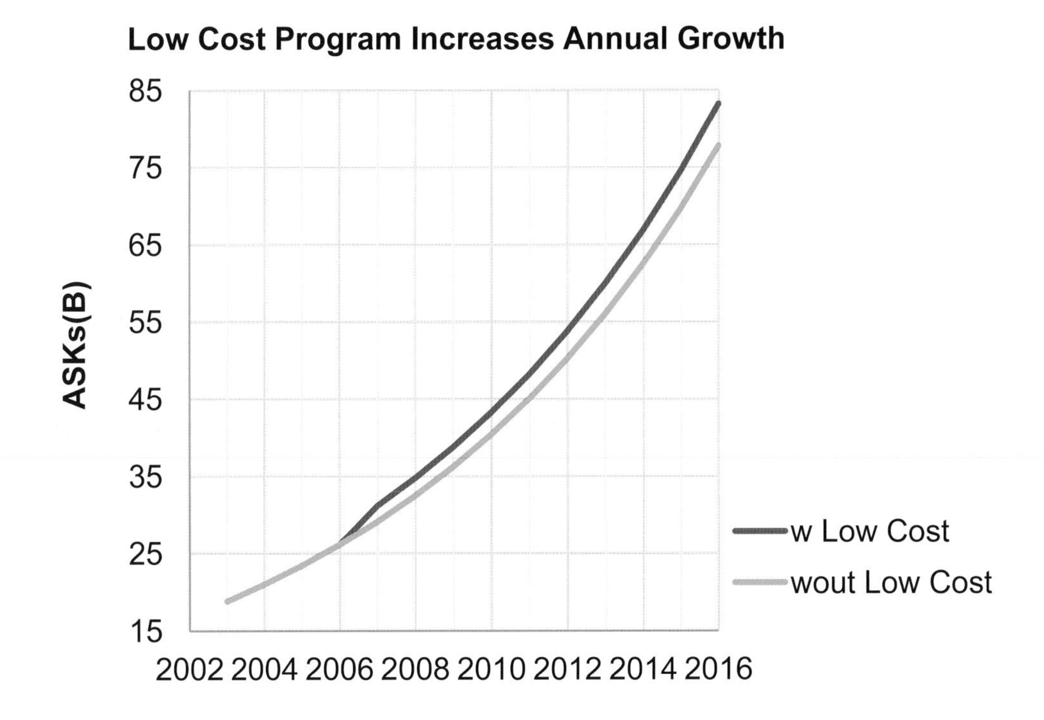

Fig. 13.9 Impact of the low cost program on annual trend

The right side of the volume equation features cubes. Expanding those:

$$R\hat{P}Ks(B)_t = [-238 + .053 \times Low\ Cost_t + .12 \times t]^3$$

$$= (-238 + .053 \times Low\ Cost_t + .12 \times t)$$

$$\times (-238 + .053 \times Low\ Cost_t + .12 \times t)$$

$$\times (-238 + .053 \times Low\ Cost_t + .12 \times t)$$

$$= -238^3$$

$$+3 \times (-238^2 \times .053 \times Low\ Cost_t - 238 \times .053^2 \times Low\ Cost_t^2)$$

$$+.053^3 \times Low\ Cost_t^3$$

$$+3 \times (-238^2 \times .12 \times t - 238 \times .12^2 \times t^2\ ) +.12^3 \times t^3$$

$$+3 \times (.053^2 \times .12 \times Low\ Cost_t^2 \times t + .053 \times .12^2 \times Low\ Cost_t \times t^2)$$

$$+\ 6 \times -238 \times .053 \times .12 \times Low\ Cost_t \times t$$

The Low Cost program positively influences volume, which exhibits a positive annual trend. The annual trend is greater in years following initiation of the Low Cost program.

The impact of this interaction can be seen by splitting the regression equation into two pieces:

- The baseline trend *t*:

$$R\hat{P}Ks(B)_t\ baseline\ trend = b_0^3$$

$$+3 \times (b_0^2 \times b_t \times t + b_0 \times b_t^2 \times t^2)$$

$$+\ b_t^3 \times t^3$$

$$= -238^3$$

$$+3 \times (-238^2 \times .12 \times t - 238 \times .12^2 \times t^2\ )$$

$$+.12^3 \times t^3$$

- Terms with *Low Cost:*

$$R\hat{P}Ks(B)_t\ due\ \ to\ Low\ Cost = 3 \times (b_0^2 \times b_{lc} \times low\ cost_t + b_0 \times b_{lc}^2 \times low\ cost_t^2$$

$$+\ b_t^2 \times b_{lc} \times t^2 \times low\ cost_t$$

$$+ b_t \times b_{lc}^2 \times t \times low\ cost_t^2)$$

$$+ 6 \times b_0 \times b_t \times b_{lc} \times t \times low\ cost_t$$

$$+ b_{lc}^3 \times low\ cost_t^3$$

$$= 3 \times (-238^2 \times .053 \times Low\ Cost_t$$

$$-238 \times .053^2 \times Low\ Cost_t^2$$

$$+ .053^2 \times .12 \times Low\ Cost_t^2 \times t$$

$$+ .053 \times .12^2 \times Low\ Cost_t \times t^2)$$

$$+ 6 \times -238 \times .053 \times .12 \times Low\ Cost_t \times t$$

$$+ .053^3 \times Low\ Cost_t^3$$

Passenger volumes with and without the *Low Cost* program components are shown in Fig. 13.10.

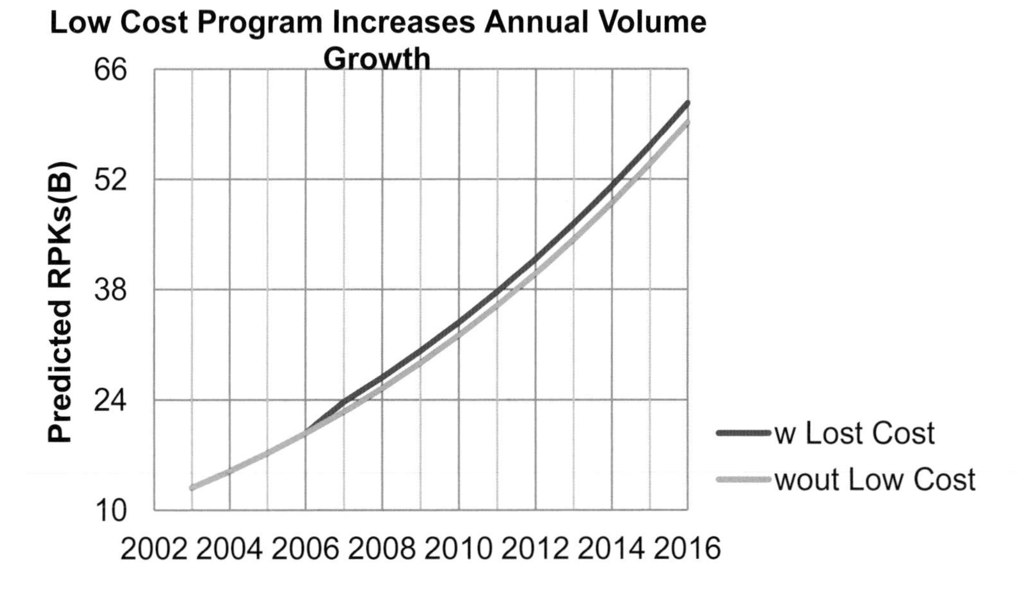

Fig. 13.10 Volume trends with and without low cost program

Comparing 2016 forecasts with and without the Low Cost program, the distance between red and orange lines in Fig. 13.10, the impact on passenger volume of the program is expected to add 2.4 billion RPKs.

Key to the success of LAN's profitable strategy was achieving at least 75 % passenger load. If planes flew at less than 75 % capacity, potential profits were not realized. The low cost program was spurring growth in both passenger capacity and volume. Management needed forecasts of passenger load, as well. Passenger load could be as low as the ratio of the lower 95 % prediction interval bound for volume to the higher 95 % prediction interval bound for capacity, the "worst" case. The "best" case would be the ratio of the upper 95 % prediction interval bound for volume to the lower 95 % prediction interval bound for capacity. Table 13.3 compares these extreme possibilities, which are illustrated in Fig. 13.11.

Table 13.3 "Best" and "worst" case passenger load forecasts

	RPKs (B)			*ASKs (B)*			*Load (%)*		
Year	Actual	Lower 95 %	Upper 95 %	Actual	Lower 95 %	Upper 95 %	Actual	"Worst"	"Best"
2003	12.7	12.2	13.6	18.3	18.1	19.6	69	62	75
2004	15.1	14.2	15.7	21.1	20.2	21.9	72	65	78
2005	17.5	16.4	18.1	23.7	22.5	24.4	74	67	80
2006	19.5	18.8	20.7	26.4	25.1	27.2	74	69	82
2007	24.0	22.7	24.8	31.6	29.9	32.5	76	70	83
2008	27.0	25.8	28.0	35.2	33.4	36.3	77	71	84
2009	29.8	29.0	31.4	38.8	37.2	40.4	77	72	84
2010	33.1	32.5	35.1	42.4	41.5	45.1	78	72	85
2011	38.4	36.3	39.1	48.2	46.3	50.3	80	72	85
2012		40.4	43.4		51.6	56.1		72	84
2013		44.8	48.0		57.5	62.6		72	83
2014		49.5	52.9		64.2	69.8		71	82
2015		54.5	58.1		71.6	77.8		70	81
2016		59.8	63.7		79.8	86.8		69	80

With the margin of error in load forecasts more than 5 % (= (80−69 %)/2), executives could not effectively evaluate the passenger business. Both extremes were the ratio of extreme outcomes for both volume and capacity. The "best" and "worst" case load percentages would occur with less than one tenth of one percent chance (= 2.5 × 2.5 % = .06 %), making the "best" to "worst" load prediction interval a 99.9 % confidence interval. A 95 % confidence interval was needed, instead, which the modeling team could supply with Monte Carlo simulation using the regression model standard errors and predictions as assumptions for expected volume and capacity in 2016.

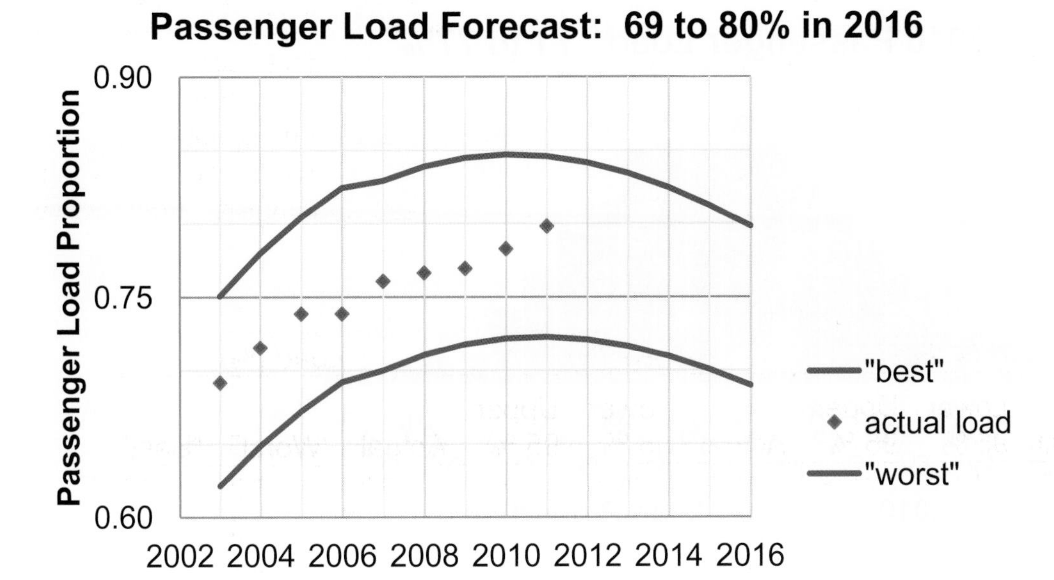

Fig. 13.11 "Best" and "worst" passenger load forecasts

13.6 Nonlinear Models Inform Monte Carlo Simulation

To identify the 95 % prediction interval for passenger load percentage in 2016, the modeling team set up a spreadsheet linking passenger capacity and volume to passenger load. Random samples of possible values for 2016 capacity and volume were generated, using the 2016 forecasts as most likely values and the regression standard errors as measures of dispersion. Since both capacity and volume were positively skewed, a sample of natural logarithms of 2016 ASKs was drawn, and a sample of cubed roots of 2016 RPKs was generated. Those were then rescaled to ASKs and RPKs in billions to find the sample of possible values for passenger load in 2016. The distribution of possible values in 2016 is shown in Fig. 13.12, with 95 % confidence interval 71–77 %. There was a 65 % chance that passenger load would be less than the 75 % goal in 2016 if capacity and volume following existing trends. The margin of error in the 2016 forecast is half the 95 % confidence interval: 3.0 %.

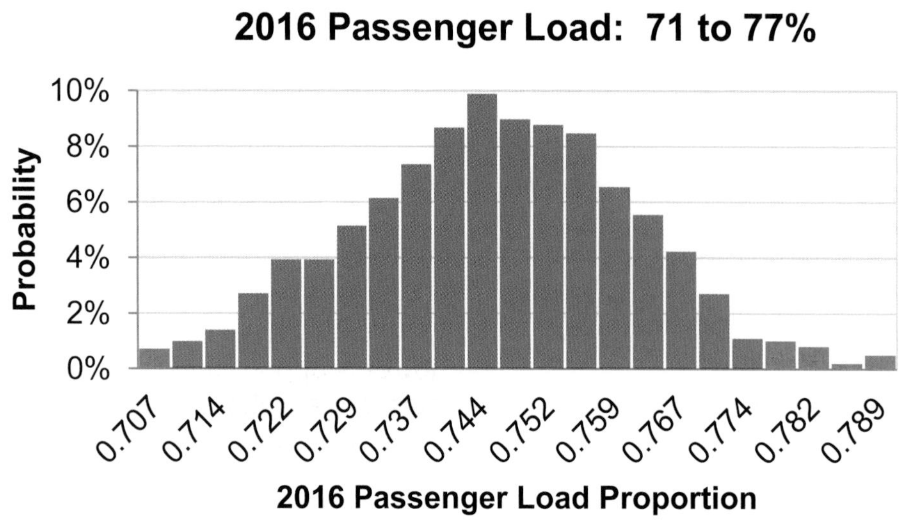

Fig. 13.12 Ninety-five percent confidence interval for 2016 passenger load proportion

13.7 Gains from Nonlinear Rescaling Are Significant

To see the gain from building nonlinear models, compare results with those from simpler linear models. The naïve linear models of passenger capacity and volume, using the same *low cost* indicator and sample, excluding the two most recent observations, are:

$$A\hat{S}Ks(B) = -5{,}880^{a} + 2.48^{b} \times Low\ Cost_t + 2.94^{a} \times t$$

RSquare: $.996^{a}$

$$R\hat{P}Ks(B) = -4{,}920^{a} + 2.11^{b} \times Low\ Cost_t + 2.46^{a} \times t$$

RSquare: $.997^{a}$

^{a}Significant at .01 or better; ^{b}significant at .05 or better.

The linear models provide comparable fits, though neither linear model is valid. Both ASKs and RPKs are underestimated and cannot be reliably used to forecast. Passenger capacity and volume is growing by increasingly larger increments annually, particularly since the Low Cost program was initiated. The linear models assume constant rates of growth, which do not fit historic data (Fig. 13.13).

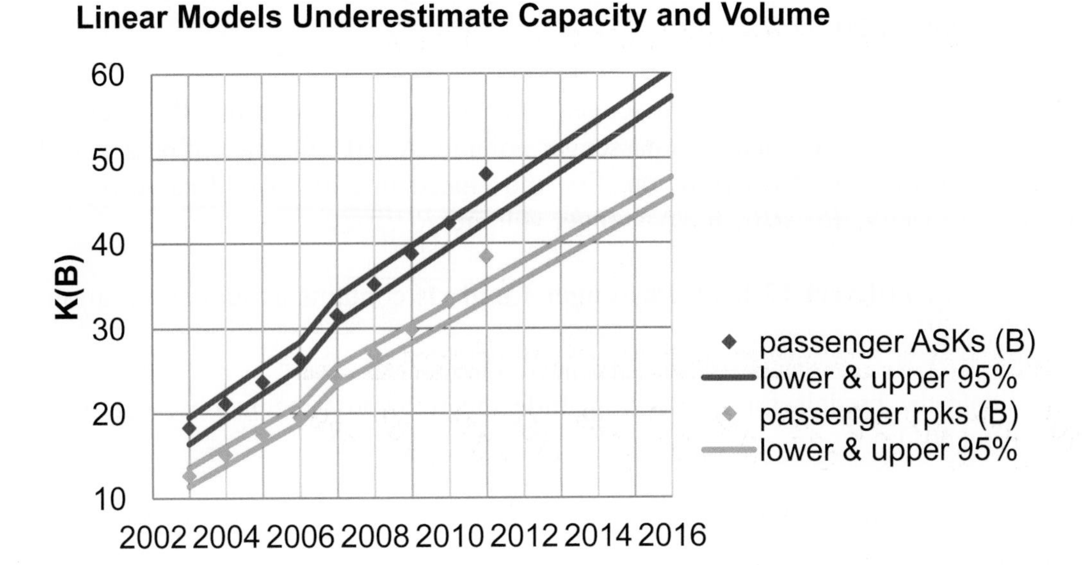

Fig. 13.13 Linear model fits and forecasts are not valid

In addition, the linear models ignore the interactions between the Low Cost program and annual trends, which have accelerated with the Low Cost program in place.

13.8 Nonlinear Models Offer the Promise of Better Fit and Better Behavior

It is a challenge to think of an example of truly linear (constant) response. Responses tend to be nonconstant and nonlinear. The fifth dip of ice cream is less appetizing than the first. Consumers become satiated at some point, and beyond that point, additional consumption is less valuable. Adding the 20th stock to a portfolio makes less difference to diversification than adding the third. A second ad insertion in a magazine enhances recall more than a tenth ad insertion. As a consequence of nonconstant, changing marginal response, nonlinear models promise superior fit and better behaved models, with valid forecasts. Nonlinear models which feature a rescaled dependent variable incorporate interactions between drivers, adding another realistic and useful aspect. Nonlinear models do carry the cost of transformation to and back from logarithms, roots, or squares. In some cases, a linear model fits data quite well and is a reasonable approximation. Thinking logically about the response that you've set to explain and predict, and then looking at the distribution and skewness of your data and your residuals, will sometimes lead you toward the choice of a nonlinear alternative.

Skewness signals nonlinear response. Tukey's Ladder of Powers can help quickly determine the particular nonlinear model which will fit a dataset best. When a variable is positively skewed, rescaling to roots or natural logarithms often reduces the positive skew. Negatively skewed variables are sometimes Normalized by squaring or cubing. The amount of difference corresponds to the power – square roots with power .5 are less radical than logarithms with power 0 and squares with power 2 are less extreme than cubes with power 3.

Excel 13.1 Rescale to Build and Fit Nonlinear Regression Models with Linear Regression

Passenger Load at LAN Airlines. Passenger load is a key measure of efficiency which captures the percent of capacity filled with passenger volume. A 2016 forecast of passenger load is needed to evaluate the Low Cost program, first introduced in 2007. Build naïve models to forecast passenger capacity, *ASKs(B)*, and passenger volume, *RPKs(B)*.

Historical annual data in **Excel 13 LAN Passenger Load.xls** contains annual observations on *ASKs(B)* and *RPKs(B)*.

In order to validate the models for forecasting, hide the two most recent observations with data from 2010 and 2011.

Assess skewness and choose scales. Assess *skewness* of *ASKs(B)*and *RPKs(B),* using the function

=**SKEW(***array***).**

C16 =SKEW(C2:C8)

	A	B	C	D
1	t	low cost $_t$	passenger ASKs(B) $_t$	passenger RPKs(B) $_t$
8	2009	1	38.8	29.8
9	2010	1	42.4	33.1
10	2011	1	48.2	38.4
11	2012	1		
12	2013	1		
13	2014	1		
14	2015	1		
15	2016	1		
16	skew		0.2662154	0.2271244

Add rescaled variables which are roots and natural logarithms of *ASKs* and *RPKs*, using

=*cell*^**.5**

=*cell*^**(1/3)**

=**LN**(*cell*)

E2 =C2^0.5

	A	B	C	D	E	F	G	H
1	*passenger ASKs(B)$_t$*	*passenger RPKs(B)$_t$*	passenger ASKs(B)$_t^{.5}$	passenger ASKs(B)$_t^{(1/3)}$	ln passenger ASKs(B)$_t$	passenger RPKs(B)$_t^{.5}$	passenger RPKs(B)$_t^{(1/3)}$	ln passenger RPKs(B)$_t$
2	18.3	12.7	4.28	2.64	2.91	3.56	2.33	2.54
3	21.1	15.1	4.60	2.77	3.05	3.89	2.47	2.72
4	23.7	17.5	4.87	2.87	3.16	4.18	2.60	2.86
5	26.4	19.5	5.14	2.98	3.27	4.42	2.69	2.97

Find the skewness of the six rescaled variables:

D16 =SKEW(D2:D8)

	A	B	C	D	E	F	G	H	I	J
1	*t*	*low cost$_t$*	*passenger ASKs(B)$_t$*	*passenger RPKs(B)$_t$*	passenger ASKs(B)$_t^{.5}$	passenger ASKs(B)$_t^{(1/3)}$	ln passenger ASKs(B)$_t$	passenger RPKs(B)$_t^{.5}$	passenger RPKs(B)$_t^{(1/3)}$	ln passenger RPKs(B)$_t$
15	2016	1								
16	skew		0.2(154	0.23	0.11	0.06	-0.05	0.05	-0.02	-0.15

Use the scales that produce skewness closest to zero, *ln ASKs*, and *RPKs*$^{(1/3)}$ and run regression with year *t* and the *low cost* indicator. To assess positive autocorrelation in the residuals, find *DW* for both models and compare with the online critical values dL and dU for models with seven observations (T = 7) and three coefficient estimates (K = 3):

	A	B	C	D	E	F	G
1	SUMMARY OUTPUT						
2							
3	*Regression Statistics*						
4	Multiple R	0.999104					
5	R Square	0.998208					
6	Adjusted F	0.997312					
7	Standard I	0.014244					
8	Observati(	7					
9							
10	ANOVA						
11		*df*	*SS*	*MS*	*F*	*ignificance F*	
12	Regressior	2	0.452107	0.226054	1114.232	3.21E-06	
13	Residual	4	0.000812	0.000203			
14	Total	6	0.452919				
15							
16		*Coefficients*	*andard Err*	*t Stat*	*P-value*	*Lower 95%*	*Upper 95%*
17	Intercept	-228.996	10.79133	-21.2204	2.92E-05	-258.958	-199.034
18	t	0.115787	0.005384	21.50759	2.76E-05	0.10084	0.130734
19	low costt	0.051894	0.021757	2.385105	0.075569	-0.00851	0.112302
20							
21							
22							
23	RESIDUAL OUTPUT						
24							
25	*bservation*	*n passenge*	*Residuals*	*DW*	*T k*	*dL*	*dU*
26	1	2.925826	-0.01762	1.626346	7. 3.	0.46723	1.89636

The *ln ASKs* model is significant, and both the *low cost* indicator and trend *t* are significant, positive drivers, together accounting for 99.8 % of the annual variation in passenger capacity. The residuals *may* or *may not* be free from unaccounted for trend, seasons, cycles, shocks and shifts.

To consider the possibility that the residuals contain accounted for trend, seasons, cycles, shocks or shifts, plot them by year:

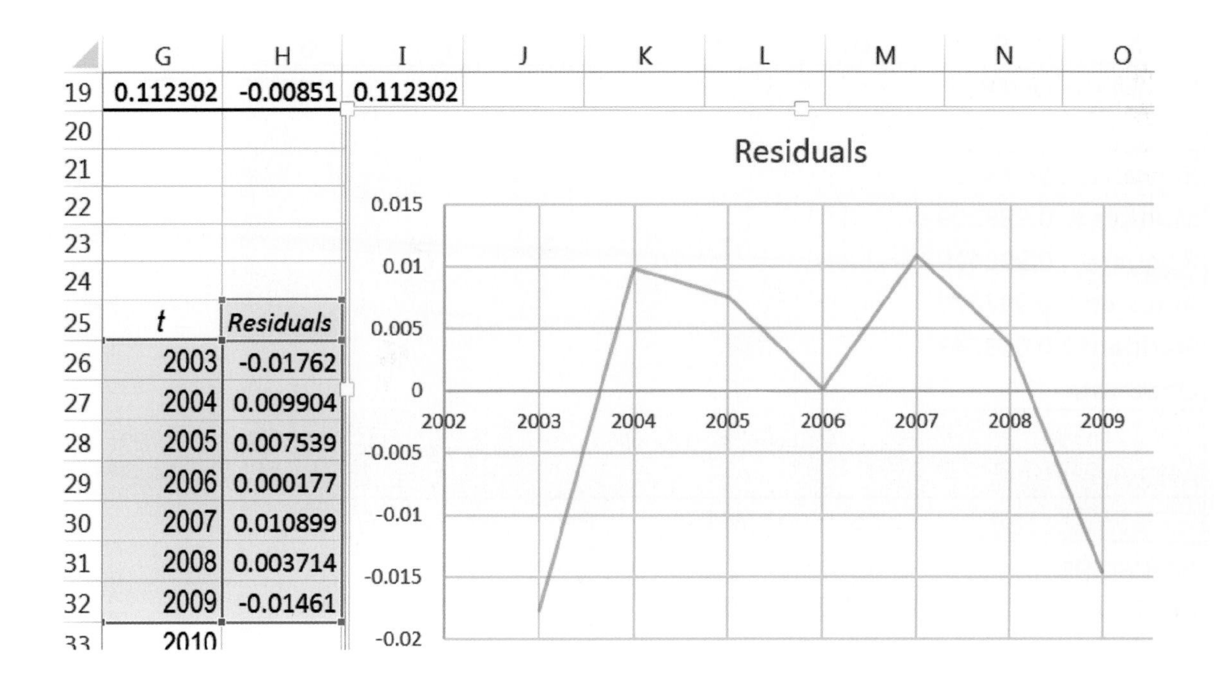

The model fits worst in the first and last year (excluding the two most recent years that were not used to build the model). However, no obvious pattern of trend, cycle, seasonality, shift or shock is apparent.

	A	B	C	D	E	F	G
1	SUMMARY OUTPUT						
2							
3	*Regression Statistics*						
4	Multiple R	0.999209					
5	R Square	0.998419					
6	Adjusted F	0.997629					
7	Standard I	0.013743					
8	Observatic	7					
9							
10	ANOVA						
11		*df*	*SS*	*MS*	*F*	*gnificance F*	
12	Regressior	2	0.477259	0.23863	1263.403	2.5E-06	
13	Residual	4	0.000756	0.000189			
14	Total	6	0.478015				
15							
16		*Coefficients*	*andard Err*	*t Stat*	*P-value*	*Lower 95%*	*Upper 95%*
17	Intercept	-231.797	10.41234	-22.2618	2.41E-05	-260.706	-202.888
18	t	0.116897	0.005194	22.50408	2.31E-05	0.102475	0.131319
19	low costt	0.062668	0.020993	2.985164	0.040531	0.004382	0.120955
20							
21							
22							
23	RESIDUAL OUTPUT						
24							
25	*bservation*	*ssenger RI*	*Residuals*	*DW*	*T k*	*dL dU*	
26	1	2.34755	-0.01622	1.969504	7. 3.	0.46723 1.89636	

The $RPKs^{(1/3)}$ model is significant, and both the *low cost* indicator and trend *t* are significant, with positive coefficient estimates. Together they account for 99.8 % of the annual variation in passenger volume. The residuals are free of unaccounted for trend, cycles, seasons, shifts and shocks.

Assess the validity of both models for forecasting.

For each, find the *critical t* and *margin of error:*

D7 =B7*C7

	A	B	C	D
3	*Regression Statistics*			
4	Multiple R	0.999104		
5	R Square	0.998208		
6	Adjusted R	0.997312	critical t	me
7	Standard E	0.014244	2.776445	0.039546

Copy the ln ASKs critical *t* and margin of error and paste into the RPKs(1/3) regression, reusing formulas:

3	*Regression Statistics*			
4	Multiple R	0.999209		
5	R Square	0.998419		
6	Adjusted R	0.997629	critical t	me
7	Standard E	0.013743	2.776445	0.038158

Copy the data cells and paste without formulas next to *DW* and the regression sheets using

Alt H V S U.

In the *ln ASKs* sheet, delete data cells containing $ASKs^5$, $AKSs^{(1/3)}$ *RPKs* and $RPKs^{(1/3)}$.

	F		G	H	I	J
25	*dL*	*dU*	*t*	*low costt*	passenger ASKs(B)t	ln passenger ASKs(B)t
26	6723	1.89	2003	0	18.3	2.91
27			2004	0	21.1	3.05
28			2005	0	23.7	3.16

In the *RPKs(1/3)* sheet, delete data cells containing *ASKs*, $ASKs^5$, $ASKs^{(1/3)}$, *ln ASKs*, and $RPKs^5$:

	F		G	H	I	J
25	*dL*	*dU*	t	low costt	passenger RPKs(B)t	passenger RPKs(B)t(1/3)
26	6723	1.89	2003	0	12.7	2.33
27			2004	0	15.1	2.47
28			2005	0	17.5	2.60

Using the regression equation, add a column of predicted values, locking cell references to coefficient estimates:

K26 *fx* =$B$17+$B$18*G26+$B$19*H26

	G	H	I	J	K
25	*t*	*low costt*	passenger ASKs(B)t	ln passenger ASKs(B)t	predicted ln passenger ASKs(B)t
26	2003	0	18.3	2.91	2.93
27	2004	0	21.1	3.05	3.04
28	2005	0	23.7	3.16	3.16

Copy the *predicted* column from the *ln ASKs* sheet and paste into the *RPKs(1/3)* sheet to reuse the formulas, producing *predicted* values there:

K26 *fx* =$B$17+$B$18*G26+$B$19*H26

	G	H	I	J	K
25	t	low costt	passenger RPKs(B)t	passenger RPKs(B)t(1/3)	predicted passenger RPKs(B)t(1/3)
26	2003	0	12.7	2.33	2.35
27	2004	0	15.1	2.47	2.46
28	2005	0	17.5	2.60	2.58

Find the *lower* and *upper* 95 % prediction intervals (in natural logarithms for *ln ASKs* and in cube roots for $RPKs^{(1/3)}$) from *predicted* columns and the margins of error *me* in D7, locking the *me* cell references:

L26 =K26-$D$7

	K	L	M
25	predicted ln passenger ASKs(B)t	95% prediction interval bound	95% prediction interval bound
26	2.93	2.89	2.97
27	3.04	3.00	3.08
28	3.16	3.12	3.20

Copy and paste the *ln ASKs lower* and *upper* columns to reuse the formulas in the *RPKs(13)* sheet:

L26 =K26-$D$7

	K	L	M
25	predicted passenger RPKs(B)t(1/3)	lower 95% prediction interval bound	upper 95% prediction interval
26	2.35	2.31	2.39
27	2.46	2.43	2.50
28	2.58	2.54	2.62

Assess validity of each model, comparing predictions with actual values in the two most recent years, 2010 and 2011:

	G	H	I	J	K	L	M
33	2010	1	42.4	3.75	3.79	3.75	3.83
34	2011	1	48.2	3.87	3.90	3.86	3.94

Actual *ln ASKs* values fall within the 95 % prediction intervals for 2010 and 2011, providing evidence of the validity of the *ln ASKs* model for forecasting.

	G	H	I	J	K	L	M
33	2010	1	33.1	3.21	3.23	3.19	3.27
34	2011	1	38.4	3.37	3.35	3.31	3.38

The $RPKs^{(1/3)}$ model also forecasts actual values in 2010 and 2011 accurately, providing evidence of validity for forecasting.

Recalibrate both models, running the regressions with all of the available data:

Copy and paste formulas for *critical t* and the margin of error *me* to find the recalibrated *me*s:

	A	B	C	D	E	F	G
1	SUMMARY OUTPUT						
2							
3	*Regression Statistics*						
4	Multiple R	0.998998					
5	R Square	0.997997					
6	Adjusted F	0.997329	critical t	me			
7	Standard I	0.017075	2.446912	0.041782			
8	Observati(	9					
9							
10	ANOVA						
11		*df*	*SS*	*MS*	*F*	*'gnificance F*	
12	Regressior	2	0.871697	0.435848	1494.829	8.03E-09	
13	Residual	6	0.001749	0.000292			
14	Total	8	0.873446				
15							
16		*Coefficients*	*andard Err*	*t Stat*	*P-value*	*Lower 95%*	*Upper 95%*
17	Intercept	-215.454	8.837566	-24.3793	3.13E-07	-237.078	-193.829
18	t	0.109031	0.004409	24.73002	2.88E-07	0.098243	0.119819
19	low costt	0.067948	0.022909	2.965968	0.025088	0.011891	0.124004

	A	B	C	D	E	F	G
1	SUMMARY OUTPUT						
2							
3	*Regression Statistics*						
4	Multiple R	0.999139					
5	R Square	0.998279					
6	Adjusted F	0.997706	critical t	me			
7	Standard I	0.01688	2.446912	0.041304			
8	Observati(	9					
9							
10	ANOVA						
11		*df*	*SS*	*MS*	*F*	*gnificance F*	
12	Regressior	2	0.991753	0.495876	1740.316	5.1E-09	
13	Residual	6	0.00171	0.000285			
14	Total	8	0.993462				
15							
16		*Coefficients*	*andard Err(*	*t Stat*	*P-value*	*Lower 95%*	*Upper 95%*
17	Intercept	-237.386	8.736415	-27.172	1.64E-07	-258.763	-216.009
18	t	0.119685	0.004358	27.46083	1.54E-07	0.109021	0.13035
19	low costt	0.052685	0.022647	2.326347	0.058935	-0.00273	0.1081

Update predictions: copy and paste the data, predicted column and lower and upper columns from the validation regression sheets to the recalibrated regression sheets to reuse formulas:

	G	H	I	J	K	L	M
25	*t*	*low costt*	passenger ASKs(B)t	ln passenger ASKs(B)t	predicted ln passenger ASKs(B)t	lower 95% prediction interval bound	upper 95% prediction interval bound
26	2003	0	18.3	2.91	2.94	2.89	2.98
27	2004	0	21.1	3.05	3.04	3.00	3.09
28	2005	0	23.7	3.16	3.15	3.11	3.20
29	2006	0	26.4	3.27	3.26	3.22	3.30
30	2007	1	31.6	3.45	3.44	3.40	3.48
31	2008	1	35.2	3.56	3.55	3.51	3.59
32	2009	1	38.8	3.66	3.66	3.62	3.70
33	2010	1	42.4	3.75	3.77	3.73	3.81
34	2011	1	48.2	3.87	3.88	3.83	3.92

	G	H	I	J	K	L	M
25	t	low costt	passenger RPKs(B)t	passenger RPKs(B)t(1/3)	predicted passenger RPKs(B)t(1/3)	lower 95% prediction interval bound	upper 95% prediction interval bound
26	2003	0	12.7	2.33	2.34	2.30	2.38
27	2004	0	15.1	2.47	2.46	2.42	2.50
28	2005	0	17.5	2.60	2.58	2.54	2.62

Predictions are in natural logarithms for *ASKs*. To rescale back to *ASKs* in billions, create three new columns for *predicted, lower* and *upper*, using the exponential function

=**EXP(***cell***)**

N26	=EXP(K26)

	G	H	I	J	K	L	M	N	O	P
25	*t*	*low costt*	passenger ASKs(B)t	ln passenger ASKs(B)t	predicted ln passenger ASKs(B)t	lower 95% prediction interval bound	upper 95% prediction interval bound	predicted passenger ASKs(B)t	lower 95% prediction interval bound (B)t	upper 95% prediction interval bound (B)t
26	2003	0	18.3	2.91	2.94	2.89	2.98	18.8	18.1	19.6
27	2004	0	21.1	3.05	3.04	3.00	3.09	21.0	20.2	21.9
28	2005	0	23.7	3.16	3.15	3.11	3.20	23.4	22.5	24.4
29	2006	0	26.4	3.27	3.26	3.22	3.30	26.1	25.1	27.2

Predictions for *RPKs* are in cube roots. To rescale back to billions of *RPKs*, create three new columns, *predicted, lower* and *upper* using

=*cell*^**3**

N26	=K26^3

	G	H	I	J	K	L	M	N	O	P
24										
25	t	low costt	passenger RPKs(B)t	passenger RPKs(B)t(1/3)	predicted passenger RPKs(B)t(1/3)	lower 95% prediction interval bound	upper 95% prediction interval bound	predicted passenger RPKs(B)t	lower 95% prediction interval bound (B)	upper 95% prediction interval bound (B)
26	2003	0	12.7	2.33	2.34	2.30	2.38	12.9	12.2	13.6
27	2004	0	15.1	2.47	2.46	2.42	2.50	14.9	14.2	15.7
28	2005	0	17.5	2.60	2.58	2.54	2.62	17.2	16.4	18.1

To illustrate the fits and forecasts on a single graph, copy *RPKs(B)* $_{,t}$ ... *lower* and *upper* data cells from the recalibrated *RPKs* regression sheet and paste without formulas into the recalibrated *ASKs* regression sheet using

Alt H V S U

	N	O	P	Q	R	S	T	U	V	W	X
25	predicted passenger ASKs(B)t	lower 95% prediction interval bound (B)t	upper 95% prediction interval bound (B)t	passenger RPKs(B)t	passenger RPKs(B)t(1/3)	predicted passenger RPKs(B)t(1/3)	lower 95% prediction interval bound	upper 95% prediction interval bound	predicted passenger RPKs(B)t	lower 95% prediction interval bound (B)	upper 95% prediction interval bound (B)
26	18.8	18.1	19.6	12.7	2.33	2.34	2.30	2.38	12.9	12.2	13.6
27	21.0	20.2	21.9	15.1	2.47	2.46	2.42	2.50	14.9	14.2	15.7
28	23.4	22.5	24.4	17.5	2.60	2.58	2.54	2.62	17.2	16.4	18.1

Delete data cells in cube roots and *predicted RPKs* which are separating *RPKs* from the *lower* and *upper* data cells:

	M	N	O	P	Q	R	S
25	upper 95% prediction interval bound	predicted passenger ASKs(B)t	lower 95% prediction interval bound (B)t	upper 95% prediction interval bound (B)t	passenger RPKs(B)t	lower 95%	upper 95% p
26	2.98	18.8	18.1	19.6	12.7	12.2	13.6
27	3.09	21.0	20.2	21.9	15.1	14.2	15.7
28	3.20	23.4	22.5	24.4	17.5	16.4	18.1

Move cells not needed for the fit and forecast plot to the end: the *low cost* indicator cells, *ln ASKs(B), predicted, lower* and *upper* cells in natural logarithms, and *predicted ASKs(B)*:

	G	H	I	J	K	L	M	N	O	P	Q	R	S
25	*t*	passenger ASKs(B)t	lower 95% prediction interval bound (B)t	upper 95% prediction interval bound (B)t	passenger RPKs(B)t	lower 95%	upper 95%	*low costt*	ln passenger ASKs(B)t	predicted ln passenger ASKs(B)t	lower 95% prediction interval bound	upper 95% prediction interval bound	predicted passenger ASKs(B)t
26	2003	18.3	18.1	19.6	12.7	12.2	13.6	0	2.91	2.94	2.89	2.98	18.8
27	2004	21.1	20.2	21.9	15.1	14.2	15.7	0	3.05	3.04	3.00	3.09	21.0
28	2005	23.7	22.5	24.4	17.5	16.4	18.1	0	3.16	3.15	3.11	3.20	23.4

Select *t, passenger ASKs(B)$_t$, lower, upper, passenger RPKs(B)$_t$, lower* and *upper* data and insert a scatterplot:

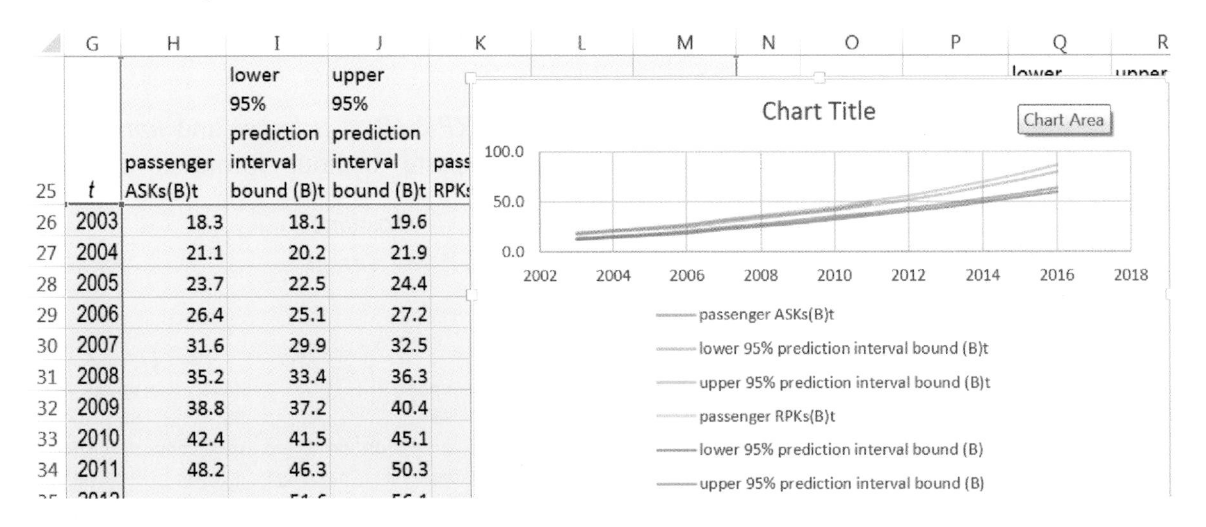

	G	H	I	J
25	*t*	passenger ASKs(B)t	lower 95% prediction interval bound (B)t	upper 95% prediction interval bound (B)t
26	2003	18.3	18.1	19.6
27	2004	21.1	20.2	21.9
28	2005	23.7	22.5	24.4
29	2006	26.4	25.1	27.2
30	2007	31.6	29.9	32.5
31	2008	35.2	33.4	36.3
32	2009	38.8	37.2	40.4
33	2010	42.4	41.5	45.1
34	2011	48.2	46.3	50.3

Add a vertical axis title, chart title. Adjust axes. Change actual observations from a line to markers. Color upper and lower lines the same:

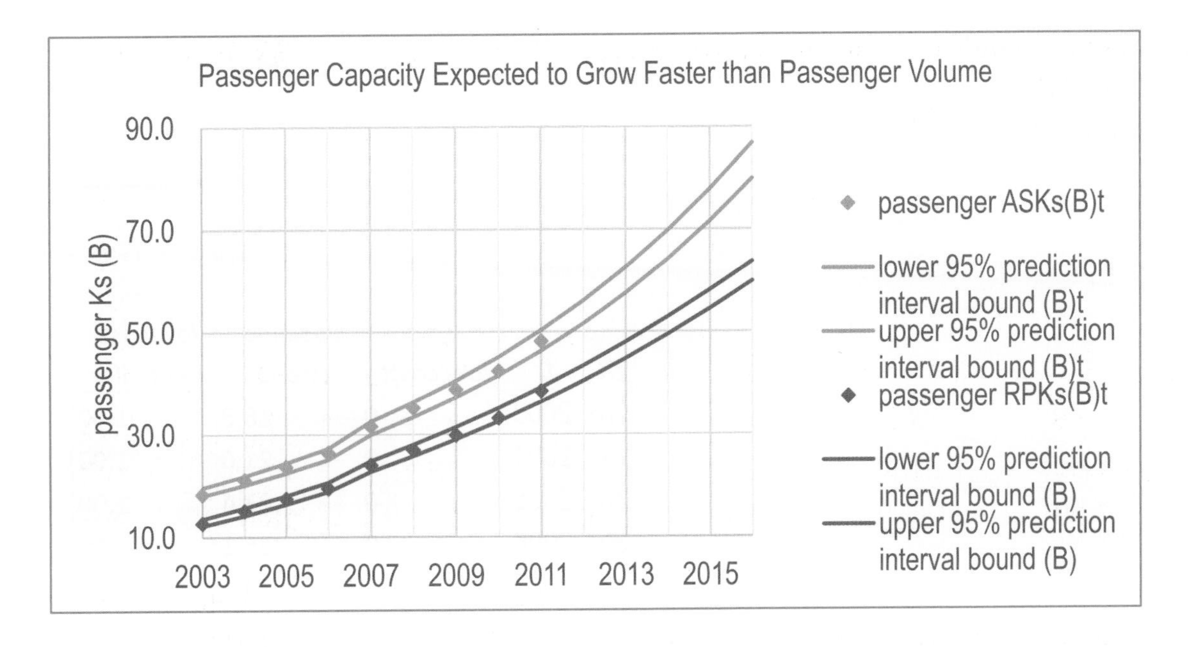

Excel 13.2 Compare Scenarios with Sensitivity Analysis

When a dependent variable has been rescaled, the model becomes multiplicative, and the impact of each driver depends on the values of other drivers. The *ASKs* equation is the product of three components, the intercept, the impact of the *low cost* program, and the trend:

$$A\hat{S}Ks(B)_t = e^{-215^a + .068^b \times Low\ Cost_t + .11^a \times t}$$

$$= e^{.068 \times Low\ Cost_t} \times 2.7\text{E-}94 \times e^{.11 \times t}$$

To see the impact of the *low cost* program, add three columns, the baseline *trend*,

$= \exp(b_0 + b_t \times t)$

(Observations on the *low cost* indicator have been moved back to the column adjacent to year *t* in this screenshot.)

I26 =EXP($B$17+$B$18*G26)

	G	H	I	J	K
25	*t*	*low costt*	*baseline trend*		
26	2003	0	18.8		
27	2004	0	21.0		
28	2005	0	23.4		

and the *low cost multiplier,*

$= \exp\,(b_{low\ cost} \times low\ cost_t$

J26 =EXP($B$19*H26)

25	G	H	I	J
25	*t*	*low costt*	*baseline trend*	*low cost multiplier*
26	2003	0	18.8	1.00
27	2004	0	21.0	1.00
28	2005	0	23.4	1.00
29	2006	0	26.1	1.00
30	2007	1	29.1	1.07

and their product, *predicted ASKs(B).*

K26 =I26*J26

25	G	H	I	J	K
25	*t*	*low costt*	*baseline trend*	*low cost multiplier*	*predicted ASKs (B)*
26	2003	0	18.8	1.00	18.8
27	2004	0	21.0	1.00	21.0
28	2005	0	23.4	1.00	23.4
29	2006	0	26.1	1.00	26.1
30	2007	1	29.1	1.07	31.2

Rearrange data so that *trend* (without the impact of the *low cost* indicator) and *predicted ASKs(B)* follow year *t.* Select data in these three columns and plot to see capacity with and without the *low cost* program:

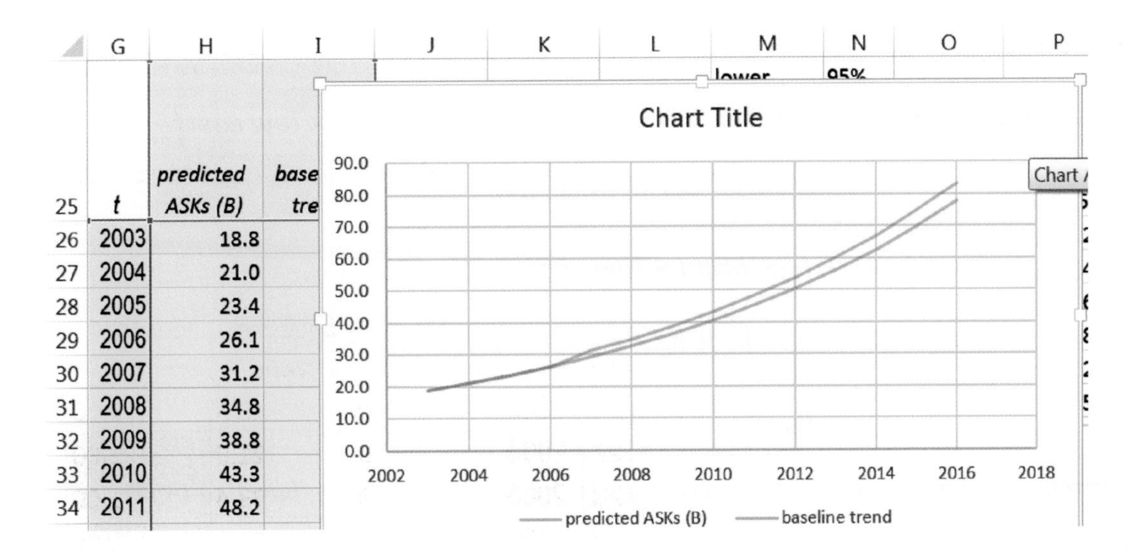

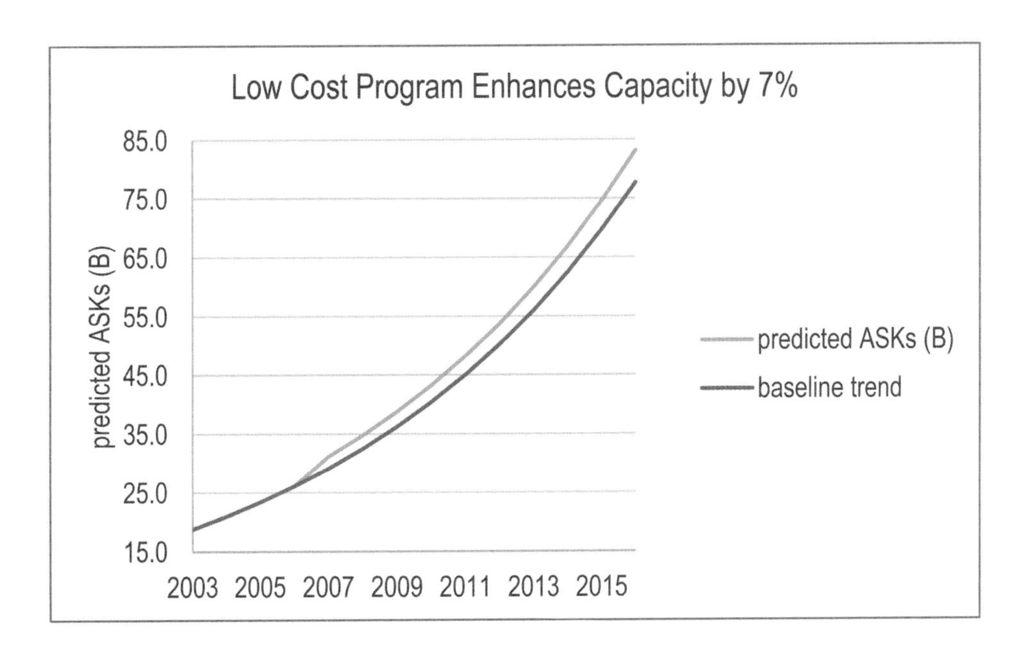

The *RPSs* equation can be split into two parts, that due to trend *t*, with four terms, and that due to the impact of the *low cost* program, with six terms:

- The baseline trend *t*:

$$R\hat{P}Ks(B)_t\ baseline\ trend = b_0^3$$

$$+3 \times (b_0^2 \times b_t \times t + b_0 \times b_t^2 \times t^2)$$

$$+ b_t^3 \times t^3$$

- Terms with *Low Cost*:

$$\hat{RPKs}(B)_t \text{ due to Low Cost} = 3 \times (b_0^2 \times b_{lc} \times low\ cost_t + b_0 \times b_{lc}^2 \times low\ cost_t^2$$
$$+ b_t^2 \times b_{lc} \times t^2 \times low\ cost_t$$
$$+ b_t \times b_{lc}^2 \times t \times low\ cost_t^2)$$
$$+ 6 \times b_0 \times b_t \times b_{lc} \times t \times low\ cost_t$$
$$+ b_{lc}^3 \times low\ cost_t^3$$

Using the formulas shown above, add three columns in the *RPKs(1/3)* sheet, *baseline trend,*

I26 =$B$17^3+3*($B$17^2*$B$18*G26+$B$17*$B$18^2*G26^2)+$B$18^3*G26^3

	G	H	I	J	K	L	M	N	O
25	t	low costt	baseline trend	low cost impact	predicted RPKs (B)t	passenger RPKs(B)t	passenger RPKs(B)t(1/3)	predicted passenger RPKs(B)t(1	lower 95% prediction interval
26	2003	0	12.9			12.7	2.33	2.34	2.3
27	2004	0	14.9			15.1	2.47	2.46	2.4
28	2005	0	17.2			17.5	2.60	2.58	2.5

low cost impact,

J26 =3*($B$17^2*$B$19*H26+$B$17*$B$19^2*H26^2+$B$18^2*$B$19*G26^2*H26+$B$18*$B$19^2*G26*H26^2)+6*$B$17*$B$18*$B$19*G26*H26+$B$19^3*H26^3

	G	H	I	J	K	L	M	N	O	P	Q	R	S	T	U
25	t	low costt	baseline trend	low cost impact	predicted RPKs (B)t	passenger RPKs(B)t	passenger RPKs(B)t(1/3)	predicted passenger RPKs(B)t(1	lower 95% prediction interval	upper 95% prediction	predicted passenge	lower 95% predictio	upper 95% predictio		
26	2003	0	12.9	0.00	12.9	12.7	2.33	2.34	2.30	2.38	12.9	12.2	13.6		
27	2004	0	14.9	0.00	14.9	15.1	2.47	2.46	2.42	2.50	14.9	14.2	15.7		
28	2005	0	17.2	0.00	17.2	17.5	2.60	2.58	2.54	2.62	17.2	16.4	18.1		

and their sum, *predicted RPKs(B)*:

K26 =I26+J26

	G	H	I	J	K
25	t	low costt	baseline trend	low cost impact	predicted RPKs (B)t
26	2003	0	12.9	0.00	12.9
27	2004	0	14.9	0.00	14.9
28	2005	0	17.2	0.00	17.2
29	2006	0	19.7	0.00	19.7
30	2007	1	22.5	1.28	23.8
31	2008	1	25.5	1.39	26.9

Notice that the impact of the *low cost* program increases each year, with the trend.

Move *predicted RPKs(B)t* and the *baseline trend* columns next to year *t* then select data in the three columns and plot to see the impact of the *low cost* program:

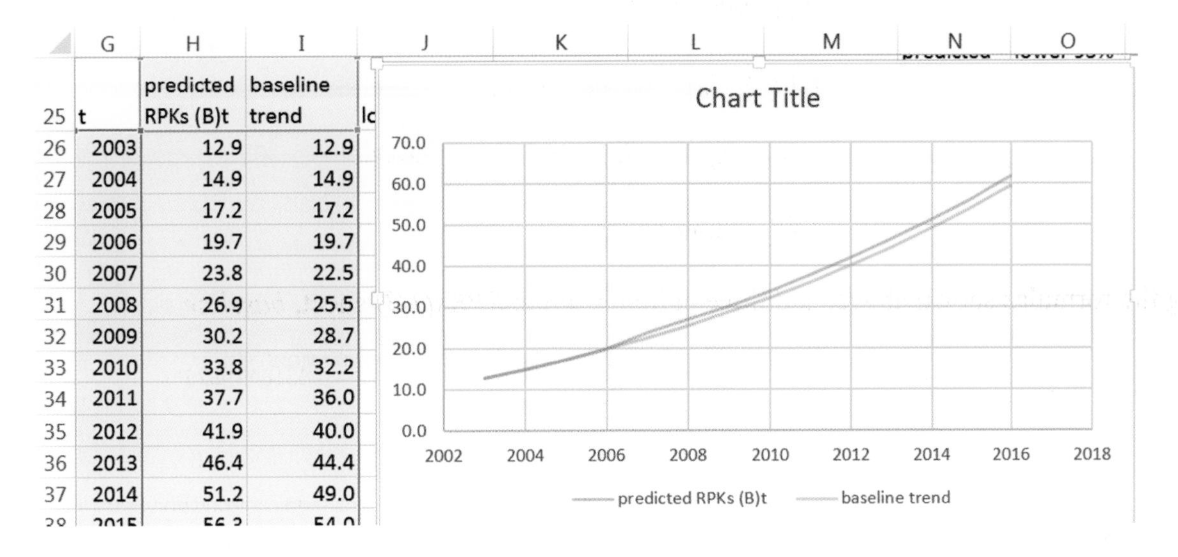

	G	H	I
25	t	predicted RPKs (B)t	baseline trend
26	2003	12.9	12.9
27	2004	14.9	14.9
28	2005	17.2	17.2
29	2006	19.7	19.7
30	2007	23.8	22.5
31	2008	26.9	25.5
32	2009	30.2	28.7
33	2010	33.8	32.2
34	2011	37.7	36.0
35	2012	41.9	40.0
36	2013	46.4	44.4
37	2014	51.2	49.0
38	2015	56.3	54.0

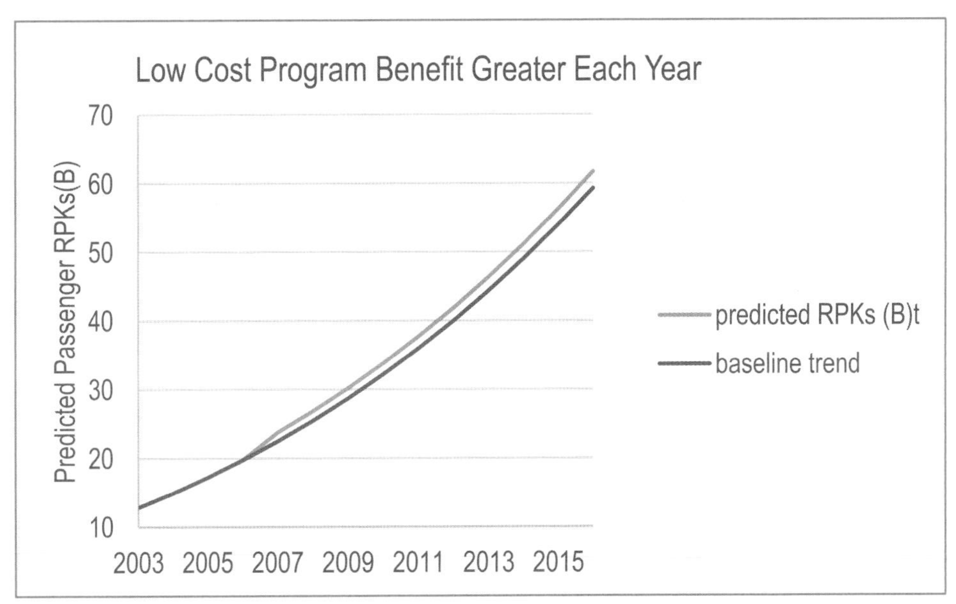

Excel 13.3 Use Nonlinear Regression Estimates with Monte Carlo Simulation

To estimate 2016 passenger load percents, samples of possible values for 2016 passenger capacity, ASKs, and passenger volume, RPKs, are needed. Both ASKs and RPKs were positively skewed in the historical data sample. Generate ln ASKs and $\text{RPKs}^{(1/3)}$, and then rescale to ASKs(B) and RPKs(B), which will then reflect historical data.

On ASKs recalibrated regression sheet, generate 1,000 values for *ln ASKs(B)* using the regression prediction for 2016 as the sample mean and the regression standard error as the sample standard deviation:

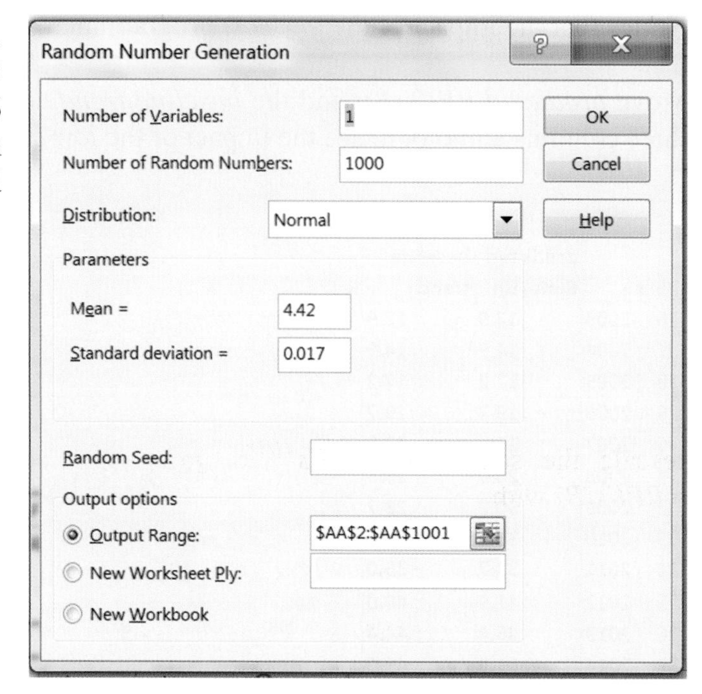

On the same sheet, generate a sample of 1,000 possible values for $RPKs^{(1/3)}$ using the regression 2016 prediction for the sample mean and the regression standard error for the sample standard deviation:

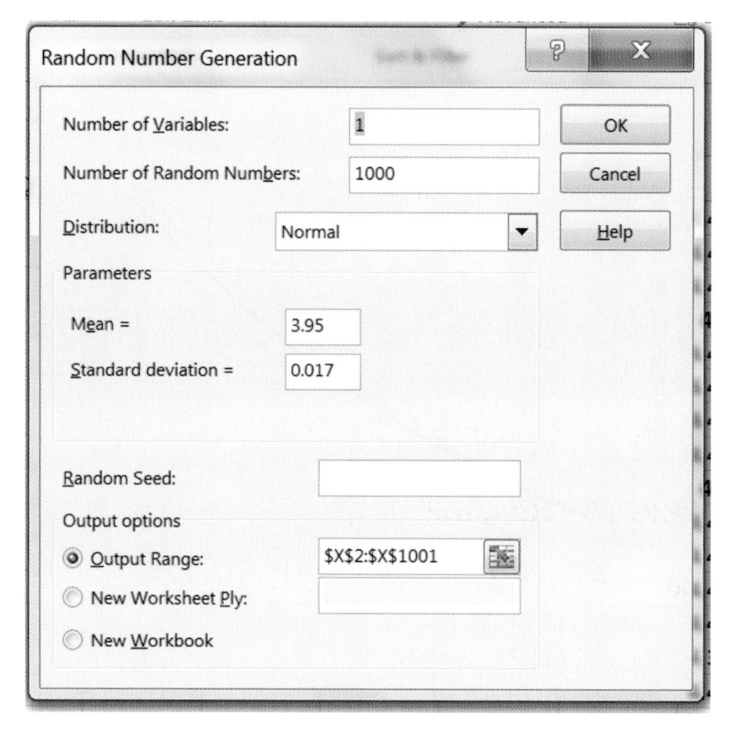

Rescale the sample of *ln ASKs(B)* to *ASKs(B)* using

=EXP(*cell)*

Y2 =EXP(W2)

	W	X	Y
1	ln $ASKs(B)_{2016}$	$RPKs(B)^{(1/3)}{}_{2016}$	$ASKs(B)_{2016}$
2	4.44	3.95	84.7
3	4.44	3.95	84.7
4	4.42	3.96	83.2
5	4.42	3.94	83.1

Rescale the sample of $RPKs^{(1/3)}$ to *RPKs(B)* with

=*cell*^**3**

Z2 =X2^3

	W	X	Y	Z
1	ln $ASKs(B)_{2016}$	$RPKs(B)^{(1/3)}{}_{2016}$	$ASKs(B)_{2016}$	$RPKs(B)_{2016}$
2	4.44	3.95	84.7	61.5
3	4.44	3.95	84.7	61.8
4	4.42	3.96	83.2	61.9

Find the sample of possible values for 2016 *passenger load percent* from the ratio of *RPKs(B)* to *ASKs(B):*

AA2 =Z2/Y2

	W	X	Y	Z	AA
1	ln $ASKs(B)_{2016}$	$RPKs(B)^{(1/3)}{}_{2016}$	$ASKs(B)_{2016}$	$RPKs(B)_{2016}$	$load_{2016}$
2	4.44	3.95	84.7	61.5	0.727
3	4.44	3.95	84.7	61.8	0.729
4	4.42	3.96	83.2	61.9	0.743

Find the 95 % prediction interval for 2016 *passenger load percent* using

=**PERCENTILE**(*array*,.975)

And

=**PERCENTILE**(*array*,.025)

AA1... =PERCENTILE(AA2:AA1001,0.025)

	W	X	Y	Z	AA
1000	4.41	3.94	82.6	61.1	0.740
1001	4.41	3.97	82.4	62.6	0.760
1002				97.50%	0.772
1003				2.50%	0.710

Find the margin of error in the *2016 passenger load percent* forecast by dividing the 95 % prediction interval by 2:

AA1... =(AA1002-AA1003)/2

	Y	Z	AA	AB
1002		97.50%	0.772	
1003		2.50%	0.710	
1004		me	0.031	

To see the distribution of possible 2016 passenger load percents, make a histogram, using the sample range to make 30 histogram bins.

Find the sample max and min, using

=**MAX**(*array)*

and

=**MIN**(*array*)

Divide the range by 30 to find histogram bin width:

AA1... =AA1007/30

	Y	Z	AA
1005		max	0.788
1006		min	0.694
1007		range	0.094
1008		range/30	0.003139

Make 30 histogram bins, starting with the *min,* and then adding the bin width.

(Lock the bin width cell reference and then drag down to fill in bins until the *max* is reached.)

AB1... =AB1002+$AA$1008

	Y	Z	AA	AB
1001	82.4	62.6	0.760	bins
1002		97.50%	0.772	0.694
1003		2.50%	0.710	0.697
1004		me	0.031	0.700
1005		max	0.788	0.703
1006		min	0.694	0.706
1007		range	0.094	0.710
1008		range/30	0.003139	0.713

Produce the histogram of possible 2016 *passenger load percents:*

Reduce decimals, select *bins* and *Frequency*, excluding the **More** row at the end, and insert a PivotTable.

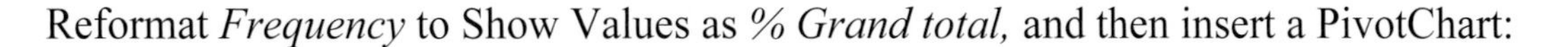

Reformat *Frequency* to Show Values as *% Grand total,* and then insert a PivotChart:

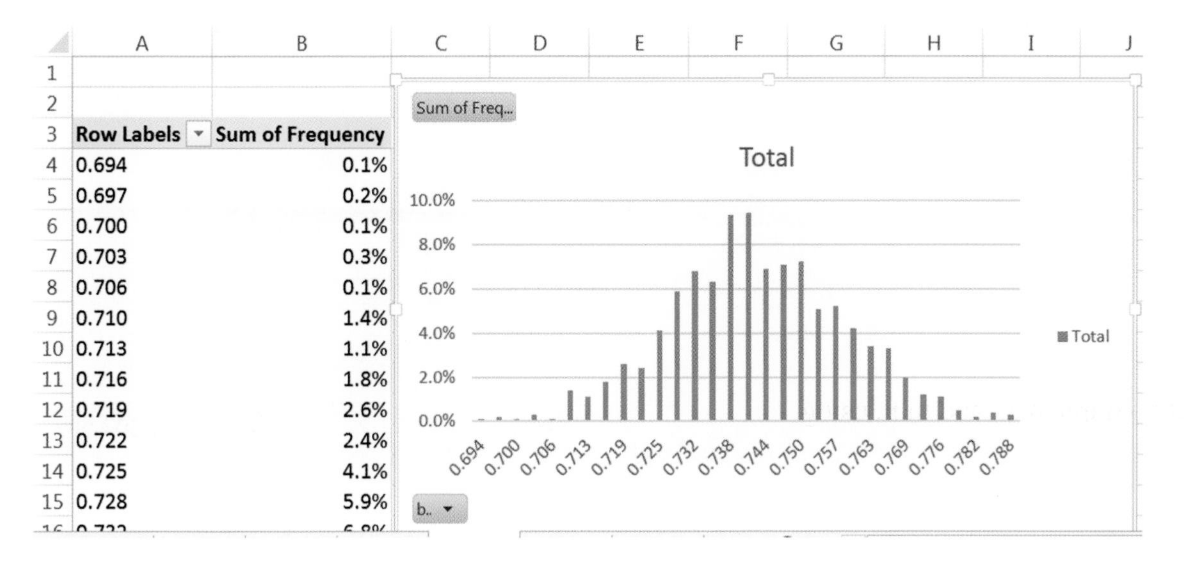

Row Labels	Sum of Frequency
0.694	0.1%
0.697	0.2%
0.700	0.1%
0.703	0.3%
0.706	0.1%
0.710	1.4%
0.713	1.1%
0.716	1.8%
0.719	2.6%
0.722	2.4%
0.725	4.1%
0.728	5.9%

Add axes titles, adjust the vertical axis, chart title and distinguish the 95 % prediction interval with bar colors:

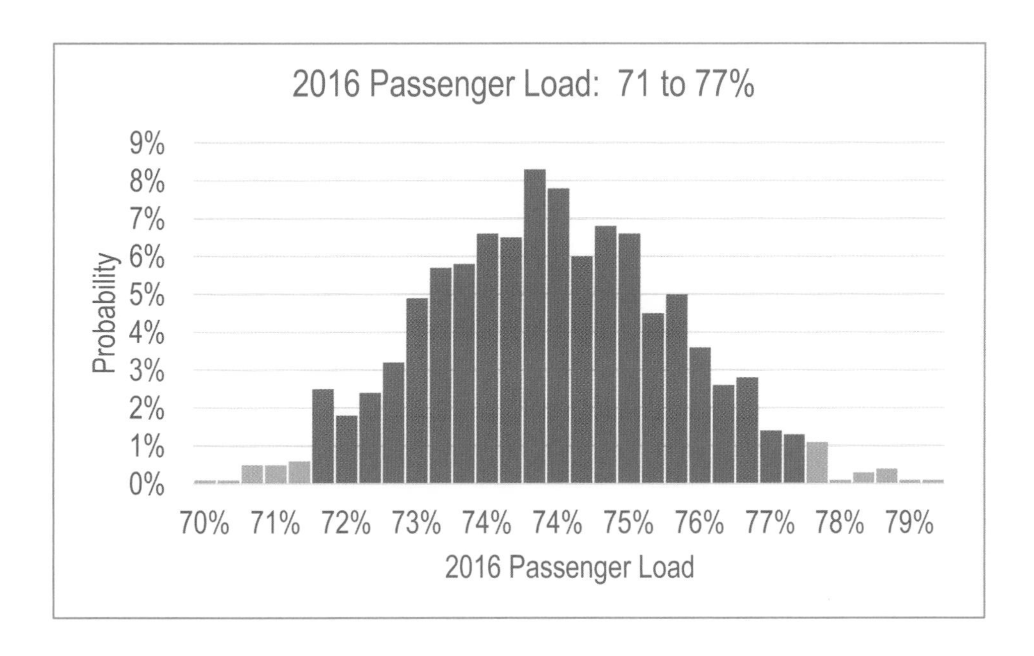

Lab 13 Mattel's Acquisition of Oasys

Despite losses in '08 and '09, Mattel management is claiming that revenue growth has recovered and that record setting revenues will exceed $6B in '12. Mattel management is counting on demand from the growing segment of elementary school children to fuel revenues. In '05, Mattel acquired Oasys, giving them access to enhanced video sales in the U.S. and other global markets. This shift in revenues was first noticeable in the first full year following the acquisition, '06.

Build a *valid* model of Mattel revenues to forecast revenues in '11, '12 and '13 from data in **Lab 13 Mattel.xls**. The dataset contains *Mattel Revenues* (B$) in billion dollars and *U.S. population (M) of under 5, 5–9, and 10–14 year olds* in millions for selected years '91 through '10. Use years '95 through '08 to build your model, holding out revenues in '09 and '10 to validate.

1. Find skewness: *revenues:*_____
2. Rescale and find skewness: *revenues:* _______
3. Which of these age groups drive revenues? ___ under 5 ___ 5–9 yos ___10–14 yos
4. What is the margin of error in your forecasts, B$? _____
5. Illustrate your model fit and forecast graphically.
6. Explain how you know that your model is valid.
7. Write your equation in the original scale of billion dollars:
8. How much did the Oasys business contribute to revenues in '09 and '10? _________
9. How much can the Oasys business be expected to contribute to revenues in '12 and '13?_______
10. How much of the Oasys business is due to 10–14 year olds in '09 and '10? _________
11. How much can 10–14 year olds be expected to contribute to the Oasys business in '12 and '13? ____________
12. Illustrate the impact of 10–14 years olds, the Oasys acquisition, and their interaction.

Lab 13 Nonlinear Forecasting with Naïve Models: LAN Airlines Passenger Revenues

The "low cost" model implemented in '07 included two goals: (1) increase passengers by 40 %, and (2) acquire 20 % more capacity.

Now in place 5 years, the low cost model appeared to be working:

Passenger traffic had increased each year by an average of 21 %, and
capacity had increased each year by an average of 13 %.

However, the lower prices on domestic flights were of some concern, in light of the global recession in 2009. With the recession looming, some fares had been adjusted in 2008, so that Low Cost fare structure had effectively lasted only one year. Passenger revenues in 2009 had fallen below 2008 revenues. The long term impact of the recession on revenues could threaten the low cost program's success.

Information Needed. The Board is asking for evidence that revenue improvements from the low cost model are sustainable. You have been asked to

- Determine the impacts of the low cost program on prices (*revenue per ASK*) and *passenger revenue*
- Determine the impact of the recession on *revenue per ASK* and *passenger revenue*
- Produce a forecast of *passenger revenue*

Considerations. LAN's global business expansion occurred after the '01 industry disruption; data before '03 is not thought to represent well current businesses.

Management contends that all elements under their control will be managed according to the low cost plan presented. Recent growth in performance indicators is considered representative of future growth through '16.

Passenger Revenue Depends on Capacity (ASKs) and Prices (Revenue per ASK)

Data are in **Lab 13 LAN Passenger Revenue.xlsx.** *Assess skewness using only years prior to 2009, since the recession impact will affect skewness and only one year of data affected by the recession is available for model building.* Add a *low cost launch* indicator, equal to 0 in '03 through '06, equal to 1 in '07, and equal to 0 in years 2008 through 2016, and add a *recession* indicator, equal to 0 before '09 and equal to 1 in '09 and later years; then run regressions to find average annual change in each performance indicator, accounting for the low cost and recession shifts. Assess positive autocorrelation using:

T	k	dL	dU
7	3	.47	1.90

1. Is *revenue per ASK* increasing at an ___increasing or ___diminishing rate?

2. Illustrate your fits and forecasts for *Passenger revenue per ASK* in years'03 through '16.

3. Write your equation for *passenger revenue per ASK* in the original units (cents):

4. Estimate the impact of the recession on *passenger revenue per ASK* in '16:
   Impact of recession in '16: ___________________________________

5. Illustrate the impact of the recession on passenger revenues per ASK in years '03 through '16.

6. What is the *margin of error* in your forecast of *revenue per ASK*? ____________

7. Passenger revenue depends on both capacity and prices. Find the "best" and "worst" outcomes in '16, with and without continuing global recession:

	'16 Capacity (ASKs)(B)	*'16 Revenue per ASK (cents)*	*'16 Revenue per ASK (cents) without recession*	*'16 Passenger revenue ($B)*	*'16 Passenger revenue ($B) without recession*
"Worst"					
"Best"					

8. Illustrate "worst," expected, and "best" cases for passenger revenue in years '12 through '16, with and without continuing global recession.

9. How likely is the "best" case scenario? _____ or one in ________

Lab 13 Forecasting with Uncertain Drivers: LAN Passenger Revenues

After reviewing naïve models quantifying the impact of the low cost program and the global recession on passenger revenue, the Board is now pondering those impacts in the passenger business. Over the past 5 years, passenger revenues have grown at an average annual rate of 17.2 %, due, in part of the successful low cost program. Is this level of growth sustainable, given the global recession?

According to the "best case" scenario, with continuing recession, revenues could grow at an average annual rate of 21.9 %. However, in the "worst case," revenues would grow at an average annual rate of only 19.1 % if the recession continued.

If the global economy recovered, in the "best case," revenues could grow at an average annual rate of as much as 24.8 %, and even in the "worst case," average annual revenue growth would be 24.0 %.

The Board has asked you to provide a 95 % prediction interval for passenger revenues in 2016, with and without continuing global recession, to narrow the possibilities to this more likely range.

Data are in **LAN 2016 forecast.xlsx.**

1. Use 2016 predictions for *ASKs* and *revenues per ASK* and standard errors from regression models with Monte Carlo simulation to forecast possible values for *passenger revenues* in 2016, with and without a continuing global recession.

	Passenger revenue ($B)	
	With continuing global recession	Without continuing global recession
Lower 95 %		
Upper 95 %		

2. What average annual growth rate from 2011 passenger revenues is possible?

	Average annual growth in passenger revenue	
	With continuing global recession	Without continuing global recession
Lower 95 %		
Upper 95 %		

3. Illustrate the distribution of possible outcomes for revenue in the passenger business in '16, with and without continuing global recession.

CASE 13-1 Nonlinear Hybrid Sales

Ford executives have asked for a forecast of hybrid sales in the U.S. over the next 24 months, through March 2014. They are particularly interested in knowing whether hybrid sales are driven by higher gas prices, and impacts of the Cash for Clunkers program of July and August 2009 and Toyota's recalls of Prius hybrids in January and February 2010.

*Gas Prices***.** Higher gas prices have traditionally spurred hybrid sales months later. Executives believe that the delay may be as long as 24 months, since car purchases typically involve research, trade ins, and financing arrangements.

*Cash for Clunkers***.** The Cash for Clunkers program was designed to stimulate sales of fuel efficient cars, including hybrids. Ford executives would like to know what the program's boost amounted to in the 2 months when it was available, July and August 2009.

*Plugs Ins***.** The first electric plug ins became available in December 2010. Ford executives weren't sure whether the plug ins would siphon hybrid sales, or motivate hybrid sales, after potential buyers test drove the plug ins.

*Limited Supply following the 2011 Earthquake***.** Following the Japanese earthquake in Spring 2011, hybrid supply was limited in May through October of 2011. Sales seemed to have dipped as potential buyers waited for available cars.

Build a valid model of monthly hybrid sales to provide Ford executives with a 24 month forecast and to quantify the impacts of

- Past gas prices,
- The Cash for Clunkers program,
- The entry of plug ins, and
- The Earthquake drive supply restriction.

Lab 13 Hybrid Shifts & Shocks.xls contains data on *hybrid sales* $(K)_m$, and past *gas prices (cents per gallon)*$_{m-24}$, for months in the period September 2008 through January 2012.

Gas prices have been lagged. (For example, *gas price*$_{m-24}$ in June 2009 is *gas price*$_{June\ 2007}$.)

1. Which of the variables is positively skewed?

2. Illustrate your model fit and forecast.

3. Write your model equation in the original scale of hybrid sales Use standard format with two or three significant digits.

4. In July and August 2009, how much lower would sales have been had there been no Cash for

   Clunkers program? _____________ in July ____________ in August

5. In May through October 2011, how much higher would sales have been had there not been an earthquake in Japan?

May	June	July	August	September	October

6. What impact have plug ins had on sales December 2011 through January 2012?

Difference between actual sales and predicted sales had there been no plug ins	
December	January

7. Gas prices rose \$.60 in the 6 months, October 2006 through April 2007. Find the impact of this price increase on hybrid sales 2 years later, October 2008 through April 2009:

   \$.60 increase in gas prices in 2006–2007 led to ___ an increase ___ a decrease of

   _______________ in hybrid sales in 2008–2009.

8. Gas prices fell \$.63 in the 7 months June 2011 through January 2012. Find the impact of this price decline on expected hybrid sales 2 years later, June 2013 through January 2014:

   \$.63 decline in gas prices in 2011–2012 will lead to ___ an increase ___ a decrease of

   _______________ in hybrid sales in 2013–2014.

9. Why is the recent hybrid response to gas price changes (in 2008) different from past response (in 2007)?

CASE 13-2 Fuel Cost Shock and Shift at LAN Airlines

LAN Airlines executives are satisfied that the Low Cost Program will increase revenues in the future. They are not sure what the impact on costs will be. The program included plans to upgrade aircraft to more fuel efficient models, reducing fuel costs, a major cost component:

*Total fuel costs = fuel cost per total system ATK * total system ATKs*

Since the low cost program was introduced in 2007, the *Commodity Bubble* is thought to have increased fuel costs in 2008.

The *global recession* is thought to have dampened the increase in fuel costs from 2008 through 2010, and slowed capacity expansion (total system ATKs), at least temporarily (in 2009).

Executives are focused on the '16 forecast and are not concerned about explaining variation in the financial components at this time. The Commodity Bubble and recession shocks will affect skewness measures. Assess skewness using data from years 2003 through 2007. Limit rescaling possibilities to square roots, natural logarithms, and squares, and rescale if skewness can be improved.

A. **Forecasting Fuel Cost Components**.

Forecast '16 *total fuel costs*, using company data from 2003 to 2011 in **13 Forecastng LAN costs.xlsx.** Build a valid models of *fuel cost per total system ATKs* and *total system ATKs*.

Durbin-Watson critical values: 5 % significance

T	K	dL	dU
7	2	0.7	1.36
7	3	0.47	1.9
7	4	0.28	2.46

1. Present scatterplots of your fits and forecasts of *total system ATKs* and *fuel cost per total system ATK*. Rescale axes to make good use of space, include axes title with units, and add a title which describes the conclusion which the viewer should see.

2. Write your equations for *total system ATKs* and *fuel cost per total system ATK* in the original scales, with intercept and indicator impact(s) following the equal sign.

3. Explain in a single sentence how you know that your models are free of unaccounted for trend and seasonality:

4. Explain in a single sentence how you know that your models are valid:

5. Explain in a single sentence what it means that *year* is significant in your equations:

6. What is the margin of error in your '16 *total system ATK forecast*?

Account for uncertainty in the two '16 total fuel cost components to generate a sample of 1,000 hypothetical values for *'16 total fuel cost.*

7. Forecast *'16 total fuel cost* with 95 % confidence:

8. Illustrate your forecast of *'16 total fuel costs*, showing the distributions of possibilities. Adjust axes to make good use of space, include axes labels (with units on the x axis), and add a title that describes the conclusion with viewers should see.

9. An intern used naïve forecasts to predict *fuel cost per total system ATKs* and *total system ATKs* in '16. From 95 % prediction intervals for two *total fuel cost* components, the intern forecasts that *'16 total fuel costs* could be

   - As low as $3.13B ("worst"), if both components were at their "worst" levels within 95 % prediction intervals, and
   - As high as $3.55B ("best"), if both components were at their "best" levels within 95 % prediction intervals

   a. What is the probability that *'16 total fuel costs* will be as high as $3.55B?
   b. The chance of *'16 total fuel costs* as high as $3.55B is 1 in ________.

Chapter 14
Sensitivity Analysis with Nonlinear Multiple Regression Models

In this chapter, the insights offered by sensitivity analysis of nonlinear regression models built with multiple drivers and cross sectional data are examined. With nonlinear models, the impact of each driver depends on the values of other drivers. This joint influence adds an element of realism relative to linear regression models with constant response and provides richer insights from sensitivity analysis.

Example 14.1 Marriott Hotel Pricing

Competition in the DC area has intensified among hotels in recent years. Marriott executives believe that the Marriott brand name commands a price premium, particularly for hotels that offer high quality accommodations and amenities. The modeling team is tasked with quantification of the Marriott brand premium. Available data comes from online reservation sites and includes *Guest rating, Star rating, starting room price,* and hotel name for 41 hotels in the DC area.

The sample distribution of starting room prices, shown in Fig. 14.1, is positively skewed. Five percent are *outliers*, more than three standard deviations above the sample mean, shown in red in Fig. 14.1. Those relatively expensive hotels are of particular interest to Marriott, and ought to be included in the regression, if skewness can be reduced so that they are no longer outliers. To Normalize the *Starting Room Prices* for use with linear regression, scales to reduce skewness, square roots and natural logarithms, were considered. The natural logarithms improve skewness and relatively expensive hotels are no longer outliers.

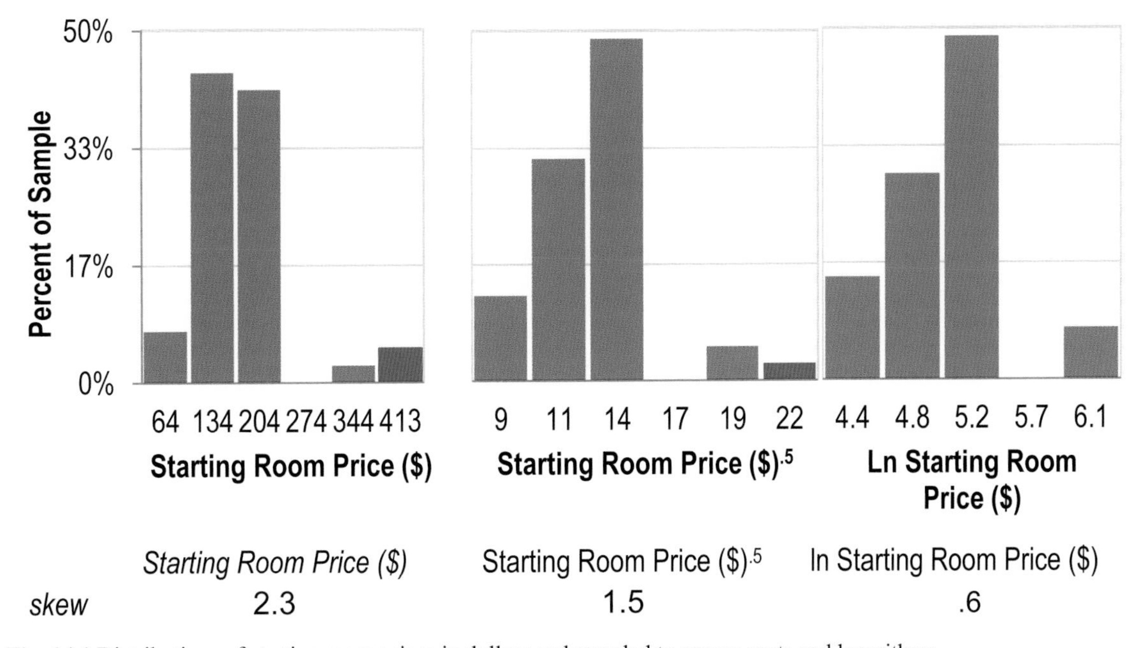

Fig. 14.1 Distributions of starting room prices in dollars and rescaled to square roots and logarithms

C. Fraser, *Business Statistics for Competitive Advantage with Excel 2013: Basics, Model Building, Simulation and Cases*,
DOI 10.1007/978-1-4614-7381-7_14, © Springer Science+Business Media New York 2013

The distribution of *Guest Ratings* is negatively skewed, with more than half above the sample mean, 4.2, but within one standard deviation of the sample mean. Increasing the power will improve skew. The squares are shown in Fig. 14.2, and these produce a more Normal distribution:

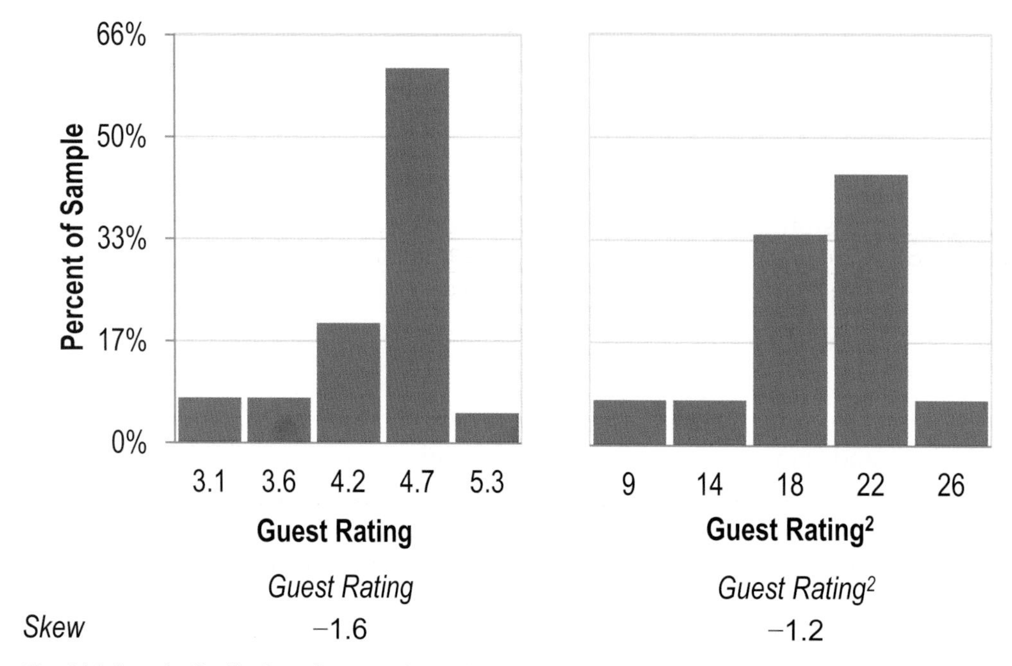

Fig. 14.2 Sample distribution of guest rating and guest rating squares

Stars are mildly negatively skewed, −.2, and are not improved with their squares.

The regression of *Ln Starting Room Prices* with a *Marriott* indicator, the squares of *Guest Rating,* and *Stars* is shown in Table 14.1. The model is significant, and all three drivers are significant, with positive coefficient estimates. Together, quality ratings and the Marriott brand account for 74 % of the variation in DC hotel starting room prices. The model equation is in natural logarithms:

$$Ln\ Starting\ \hat{R}oom\ Price(\$) = 3.24^a + .243^b \times Marriott + .304^a \times Stars$$

$$+ .030^b \times Guest\ Rating^2$$

R Square: $.74^a$
^{a}Significant at .01 or better; ^{b}Significant at .05.

Table 14.1 Regression of Ln starting room prices

Regression statistics	
R square	.742
Standard error	.227
Observations	41

ANOVA

	df	*SS*	*MS*	*F*	*Significance F*
Regression	3	5.49	1.83	35.5	6E−11
Residual	37	1.91	.05		
Total	40	7.40			

	Coefficients	*Standard error*	*t stat*	*p value*	*Lower 95 %*	*Upper 95 %*
Intercept	3.24	.16	20.0	2E−21	2.91	3.57
Marriott	.243	.112	2.2	.04	.015	.470
Stars	.304	.066	4.6	4E−05	.171	.438
Guest rating²	.030	.014	2.2	.04	.002	.058

To express the equation in dollars, the exponential function is used for both sides:

$$e^{Ln\ Starting\ \hat{R}oom\ Price\ (\$)} = e^{3.24+.243\times Marriott+\ .304\times Stars+.030\times Guest\ Rating^2}$$

$$Starting\ \hat{R}oom\ Price\ (\$) = e^{3.24+.243\times Marriott+\ .304\times Stars+.030\times Guest\ Rating^2}$$

The equation can be written to show the multiplicative influences of each of the drivers:

$$Starting\ \hat{R}oom\ Price\ (\$) = 25.6 \times e^{.243\times Marriott} \times e^{.304\times Stars} \times e^{Guest\ Rating^2}$$

The equation can also be written to show the impact of the Marriott brand.

For Marriott hotels, with the *Marriott* indicator set to 1:

$$Starting\ \hat{R}oom\ Price\ (\$) = 25.6 \times 1.27 \times e^{.304\times Stars} \times e^{Guest\ Rating^2}$$

And for other hotels, with the *Marriott* indicator set to zero:

$$Starting\ \hat{R}oom\ Price\ (\$) = 25.6 \times e^{.304\times Stars} \times e^{Guest\ Rating^2}$$

The Marriott brand commands a price that is, on average, 127 % (= $\exp(b_{Marriott}) = 1.27$) the price of other hotels of comparable quality.

14.1 Sensitivity Analysis Reveals the Relative Strength of Drivers

When the dependent variable is rescaled to build a nonlinear model, the model is multiplicative. The impact of each of the drivers depends on values of all of the other drivers. Predicted *starting room prices* can be compared for contrasting scenarios, such as those which are linked to lower prices (lower quality hotels with lower *Star* and *Guest Ratings*) and those which are linked to higher prices (higher quality hotels with higher *Star* and *Guest Ratings*).

As an example, to identify the impact of *Guest* ratings on *Starting Room Prices* for Marriott hotels, the *Marriott* indicator is set to one. Since the *Stars* coefficient is positive, lower *Stars* would reduce *starting room price* and the impact of *Guest* rating will be lower than in the contrasting case in which *Stars* are higher. For both of these contrasting scenarios, find predicted *Starting Room Prices* at varying levels of *Guest Rating* to find its impact. Predicted prices by *Guest* ratings for Marriott hotels, under contrasting *Stars* scenarios, are illustrated in Fig. 14.3:

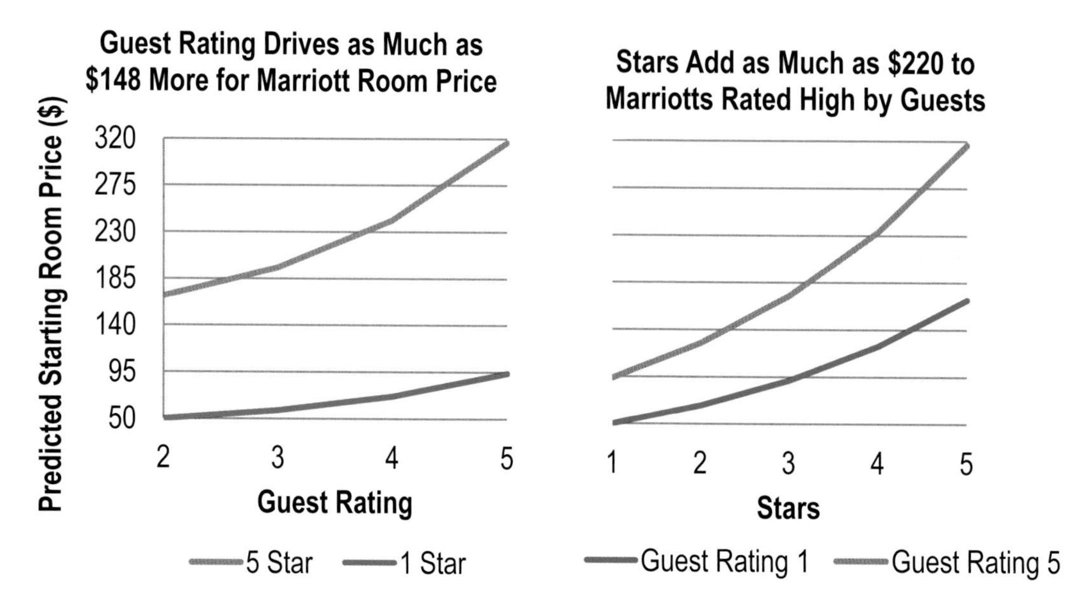

Fig. 14.3 Predicted Marriott starting room prices by quality scenario

To see visually which matters more, *Guest Ratings* or *Stars*, the scenario analysis is repeated, varying *Stars* for contrasting *Guest Rating* scenarios, shown on the right panel of Fig. 14.3.

Five star Marriott hotels benefit more than hotels with fewer Stars from Guest Rating improvements. Starting Room Price response to Guest Rating is increasing at an increasing rate, particularly for higher Star hotels. Similarly, Marriott hotels rated high by guests benefit more from improvements in Star ratings. Stars make a greater difference, justifying prices higher by as much as $220, for hotels rated high by guests.

The impact of the scenario is apparent from the equation. Looking at the impact of Stars on Starting Room Price response to Guest Ratings, holding the Marriott indicator constant at one, with five *Stars*, the equation becomes:

$$Starting\ \hat{Room}\ Price\ (\$) = 32.6 \times e^{.304 \times Stars} \times e^{Guest\ Rating^2}$$

$$= 32.6 \times e^{.304 \times 5} \times e^{Guest\ Rating^2}$$

$$= 32.6 \times 4.58 \times e^{Guest\ Rating^2}$$

And, with one *Star*, the equation becomes:

$$Starting\ \hat{Room}\ Price\ (\$) = 32.6 \times e^{.304 \times 1} \times e^{Guest\ Rating^2}$$

$$= 32.6 \times 1.36 \times e^{Guest\ Rating^2}$$

Guest Rating improvements will be 3.4 (= 4.58/1.36) times more powerful for Five Star Marriott hotels than for One Star Marriott hotels.

14.2 Gains from Nonlinear Rescaling Are Significant

To see the gain from building a nonlinear model, compare results with those from a simpler linear model. The linear model of *Starting Room Prices* using the same variables and sample is:

$$Starting\ \hat{R}oom\ Price(\$) = -78 + 33 \times Marriott + 2.4 \times Guest\ Rating$$
$$+60^{a} \times Stars$$

R Square: 57 %[a]
[a]Significant at .01 or better

Since neither the Marriott indicator nor *Guest Rating* is significant, the final equation would simplified to:

$$Starting\ \hat{R}oom\ Price(\$) = -61^{a} + 60^{b} \times Stars$$

R Square: 54 %[a]
[a]Significant at .05 or better; [b]Significant at .01 or better.

Marriott executives would not learn that the brand name adds value to room prices or that *Guest Ratings* multiply the impact of the brand and of *Star* ratings.

Comparing residuals from the nonlinear and linear models, shown in Fig. 14.4, the nonlinear model residuals are less skewed, approximately Normal, and better satisfy multiple linear regression assumptions:

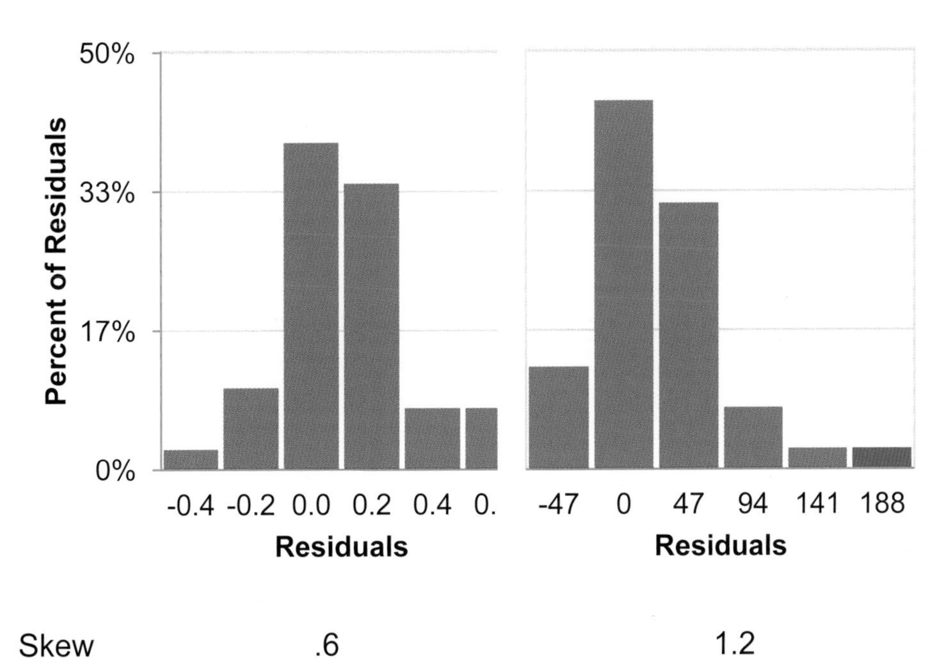

Fig. 14.4 Residuals from the nonlinear model (*left*) are closer to normal

14.3 Sensitivity Analysis with Nonlinear Models Reveals Interactions

Nonlinear models which feature a rescaled dependent variable also feature built in interactions. The impact of one driver depends upon the values of other drivers. Comparing predicted values under hypothetical scenarios allows quantification of the importance of each driver and more accurate inference for alternative scenarios under consideration.

Logarithmic models are multiplicative, with each driver multiplying the impact of others. Models built with roots are additive, but incorporate pairwise and higher order interactions. Either will add realism when modeling performance in business. Rarely do drivers impact performance in isolation. Had the Marriott modeling team been content to use a linear model, they would not have discovered the importance of the brand name, or the joint impact of Star ratings and Guest ratings, which reinforce each other. In addition to producing more believable, useable results, nonlinear models better satisfy the assumptions of regression analysis.

Excel 14.1 Sensitivity Analysis with Hypothetical Scenarios for Nonlinear Models

Pricing of Marriott Hotels

Hotel pricing reflects the quality of accommodations and amenities offered by competing hotels, as well as the value of a chain's brand name. Quantify the impact of hotel quality, reflected in *Star* and *Guest ratings*, as well as the Marriott brand name, with a model of Starting Room Prices in the DC area. Data for 41 hotels, including six Marriott hotels are in Excel 14 Pricing Marriott hotels.xlsx, and include *starting room price*, *Stars*, and *Guest Rating.*

Use Excel's **SKEW(*array*)** function to assess *skewness* of *Starting Room Prices ($)*, *Stars*, and *Guest Rating.*

B43 =SKEW(B2:B42)

	A	B	C	D	E
1	Hotel	Starting Room Price ($)	Marriott	Stars	Guest Rating
40	Sofitel	175	0	4	4.6
41	Westin	149	0	4	4.2
42	Westin Grand	149	0	4	4.2
43	skew	2.276	2.078	-0.177	-1.618

Starting room price is positively skewed, *Stars* is approximately Normal, and *Guest Rating* is negatively skewed. Make two new columns to consider shrinking *Starting Room Prices* with square roots and natural logarithms. Add a third new column to consider the squares of *Guest Rating*, and assess skewness of the rescaled variables:

E43 =SKEW(E2:E42)

	A	B	C	D	E	F	G	H
1	Hotel	Starting Room Price ($)	Marriott	Stars	Guest Rating	Starting Room Price ($)$^{.5}$	ln Starting Room Price ($)	Guest Rating^2
42	Westin Grand	149	0	4	4.2	12.21	5.00	17.64
43	skew	2.276	2.078	-0.177	-1.618	1.459	0.608	-1.182

The natural logarithm of *Starting Room Price* reduces skewness to the approximately Normal range, −1 to +1. The squares of *Guest Rating* improve skew. Use these two, with the Marriott indicator and *Stars* in regression:

	A	B	C	D	E	F	G
1	SUMMARY OUTPUT						
2							
3	*Regression Statistics*						
4	Multiple R	0.861526					
5	R Square	0.742227					
6	Adjusted F	0.721327					
7	Standard I	0.22707					
8	Observatic	41					
9							
10	ANOVA						
11		*df*	*SS*	*MS*	*F*	*gnificance F*	
12	Regressior	3	5.493131	1.831044	35.51244	5.52E-11	
13	Residual	37	1.907743	0.051561			
14	Total	40	7.400874				
15							
16		*Coefficients*	*andard Err*	*t Stat*	*P-value*	*Lower 95%*	*Upper 95%*
17	Intercept	3.240956	0.161654	20.0487	1.82E-21	2.913414	3.568499
18	Marriott	0.242509	0.112474	2.15614	0.037644	0.014616	0.470403
19	Stars	0.304371	0.065807	4.625166	4.46E-05	0.171032	0.437709
20	Guest Rati	0.030035	0.013842	2.169877	0.03651	0.001989	0.058081

The model is significant and all three drivers are significant, with expected, positive coefficient estimates.

To assess the *Normality* of the residuals, find the residual skewness:

C68 =SKEW(C27:C67)

	A	B	C	D
66	40	4.988251	0.015695	
67	41	4.988251	0.015695	
68		skew	0.633238	

The residuals are approximately Normal.

To see *predicted Starting Room Price ($)* values, first find *predicted ln starting room price ($)*.

Copy *Starting Room Price ($)*, *ln Starting Room Price ($)*, *Marriott*, *Stars*, and *Guest Rating*[5] and *Guest Rating*, and then paste into the regression sheet, next to residuals. Use the regression equation to find *predicted ln starting room price ($)*:

J27 =$B$17+$B$18*F27+$B$19*G27+$B$20*H27

	C	D	E	F	G	H	I	J
26	Residuals	Starting Room Price ($)	ln Starting Room Price ($)	Marriott	Stars	Guest Rating ^{2}	Guest Rating	predicted ln Starting Room Price ($)
27	0.623132	169	5.13	0	2.5	16.81	4.1	4.51
28	-0.50533	76	4.33	0	3.5	17.64	4.2	4.84
29	-0.00083	149	5.00	1	3	20.25	4.5	5.00

To rescale back to the original *Starting Room* Price scale million dollars, make *predicted starting room price($)* from the exponential function of *predicted ln starting room price ($)*:

K27 =EXP(J27)

	I	J	K
26	Guest Rating	predicted ln Starting Room Price ($)	predicted Starting Room Price
27	4.1	4.51	91
28	4.2	4.84	126
29	4.5	5.00	149

To isolate the importance of a driver, compare predicted *Starting Room Prices* of hypothetical Marriott hotels.

Determine the difference in Marriott *starting room prices* driven by differences in *Guest Rating* by adding eight new rows to the dataset which describe two sets of four hypothetical Marriott hotels:

- Four with Five *Star* ratings, and
- Four with One *Star* ratings.

Within each set of four, hypothetical hotels are identical, except that they differ only with respect to *Guest Rating*, from two to five.

Rearrange columns so that the indicators and the original variables are adjacent, followed by *Guest Rating*^{2}, *predicted ln Starting Room Price ($)*, and *predicted Starting Room Price($)*.

Focus on *Marriott* hotels, as an example. Set *Marriott* to one in all eight rows.

In the first four rows, set *Stars* to the sample maximum, 5. In the second four rows, set *Stars* to the sample minimum, 1.

For each set of four hypothetical hotels, input *Guest Ratings* from the sample maximum, 5, to the minimum, 1.

	E	F	G	H	I	J	K
26	*ln Starting Room Price ($)*	*Marriott*	*Stars*	*Guest Rating*	*Guest Rating* ^{2}	*predicted ln Starting Room Price ($)*	*predicted Starting Room Price*
67	5.00	0	4	4.2	17.64	4.99	147
68	Stars 5	1	5	5.0	25.00	5.76	316
69	Stars 5	1	5	4.0	16.00	5.49	241
70	Stars 5	1	5	3.0	9.00	5.28	196
71	Stars 5	1	5	2.0	4.00	5.13	168
72	Stars 1	1	1	5.0	25.00	4.54	94
73	Stars 1	1	1	4.0	16.00	4.27	71
74	Stars 1	1	1	3.0	9.00	4.06	58
75	Stars 1	1	1	2.0	4.00	3.91	50

The *Guest Rating*^{2} formula, regression equation formula for *Ln Starting Room price* and formula to rescale Ln *Starting Room Prices* back to dollars will produce predicted prices for the eight hypothetical hotels.

To see this expected *Starting Room Price* response to differences in *Guest Rating*, rearrange columns so that *predicted Starting Room Price* follows *Guest Rating* and then make a scatterplot of the first four hypothetical hotels' *predicted Starting Room Price ($)* by *Guest Rating.*

Add the second series of four hypothetical hotels' *predicted Starting Room Price* by *Guest Rating.*

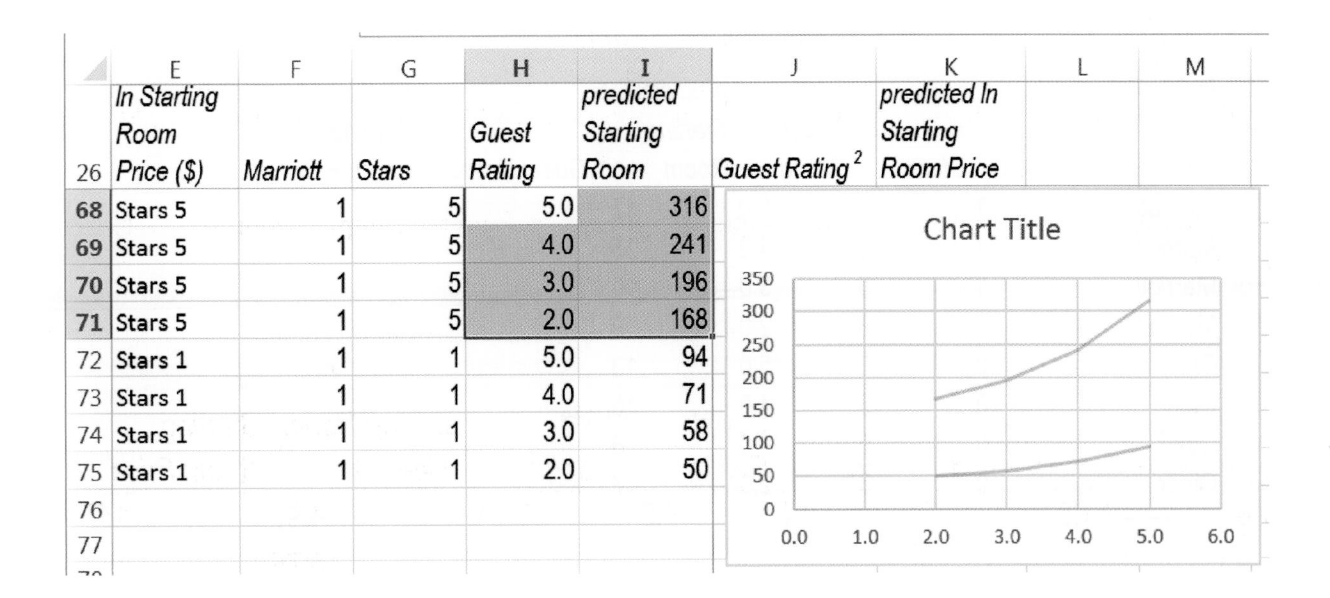

	E	F	G	H	I	J	K	L	M
26	ln Starting Room Price ($)	Marriott	Stars	Guest Rating	predicted Starting Room	Guest Rating ^{2}	predicted ln Starting Room Price		
68	Stars 5	1	5	5.0	316				
69	Stars 5	1	5	4.0	241				
70	Stars 5	1	5	3.0	196				
71	Stars 5	1	5	2.0	168				
72	Stars 1	1	1	5.0	94				
73	Stars 1	1	1	4.0	71				
74	Stars 1	1	1	3.0	58				
75	Stars 1	1	1	2.0	50				
76									
77									

Choose a Style, reset axes minimum, maximum, and major units, and add axes:

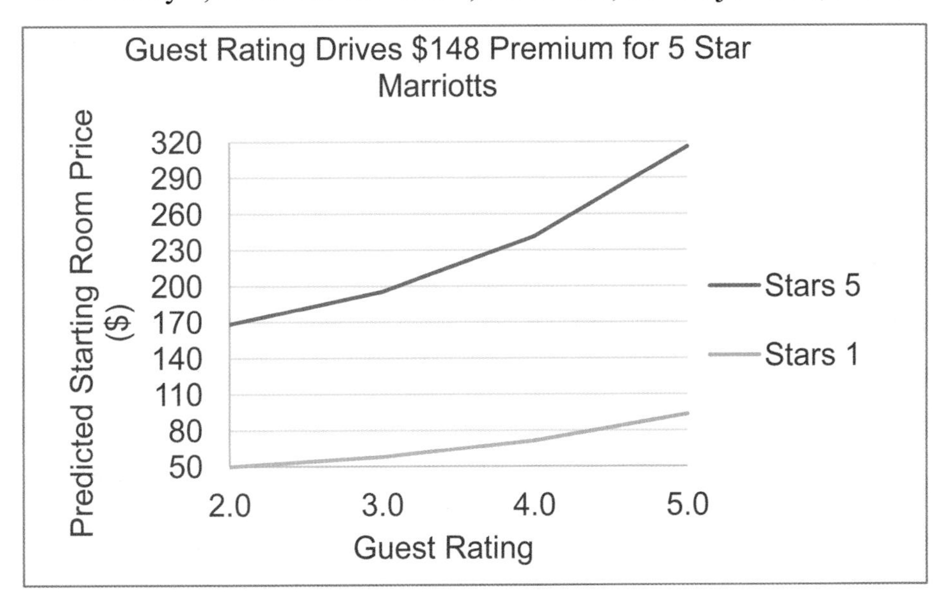

The driver impacts depend on hotel brand, and the Marriott brand indicator multiplies their influences. To see this, compare Marriott and Other hotels.

Add eight new hypothetical hotels, four Marriott and four Other.

For the first set of four, let *Guest Rating* take on values from the sample minimum, 2, to the sample maximum, 5. Set the remaining driver, *Stars*, to the median *Star* rating.

Repeat this process to fill in comparable rows for the four Other hotels.

Drag down *Guest Rating^2*, *ln Starting Room Price*, and *Starting Room Price.*
Plot each of the two series of *Starting Room Price* by *Guest Rating*:

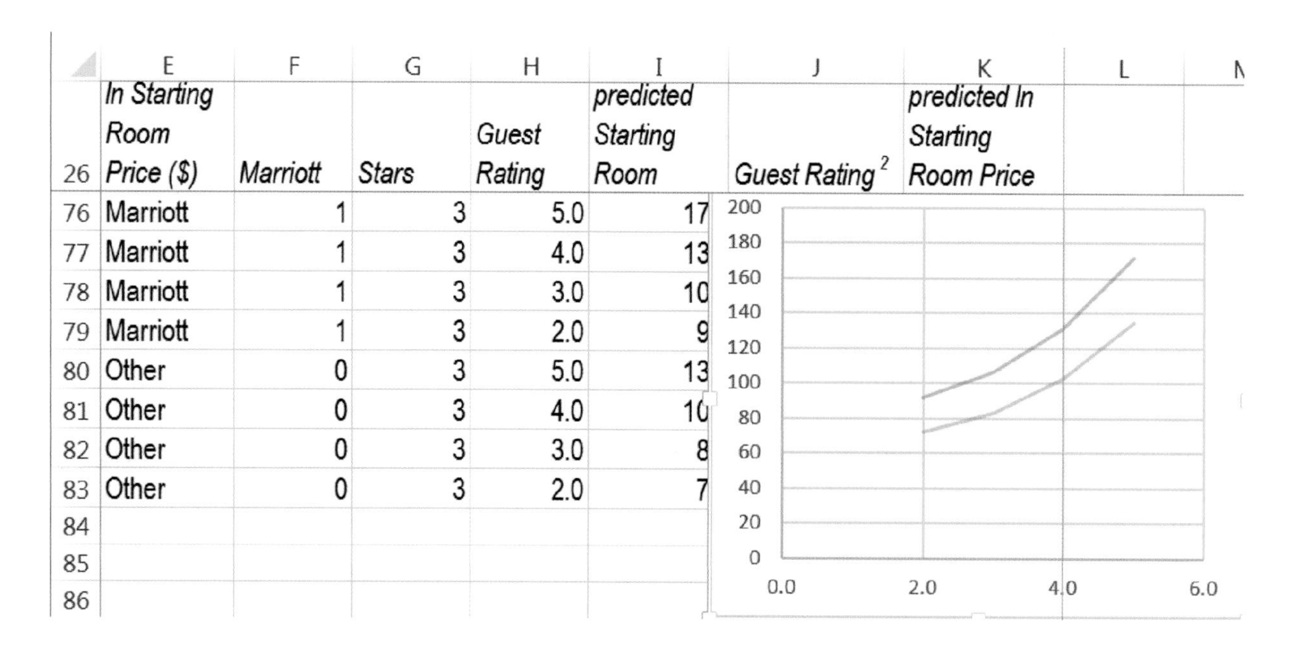

	E	F	G	H	I	J	K	L
26	ln Starting Room Price ($)	Marriott	Stars	Guest Rating	predicted Starting Room	Guest Rating 2	predicted ln Starting Room Price	
76	Marriott	1	3	5.0	17			
77	Marriott	1	3	4.0	13			
78	Marriott	1	3	3.0	10			
79	Marriott	1	3	2.0	9			
80	Other	0	3	5.0	13			
81	Other	0	3	4.0	10			
82	Other	0	3	3.0	8			
83	Other	0	3	2.0	7			
84								
85								
86								

Adjust axes, add axes labels, and chart title:

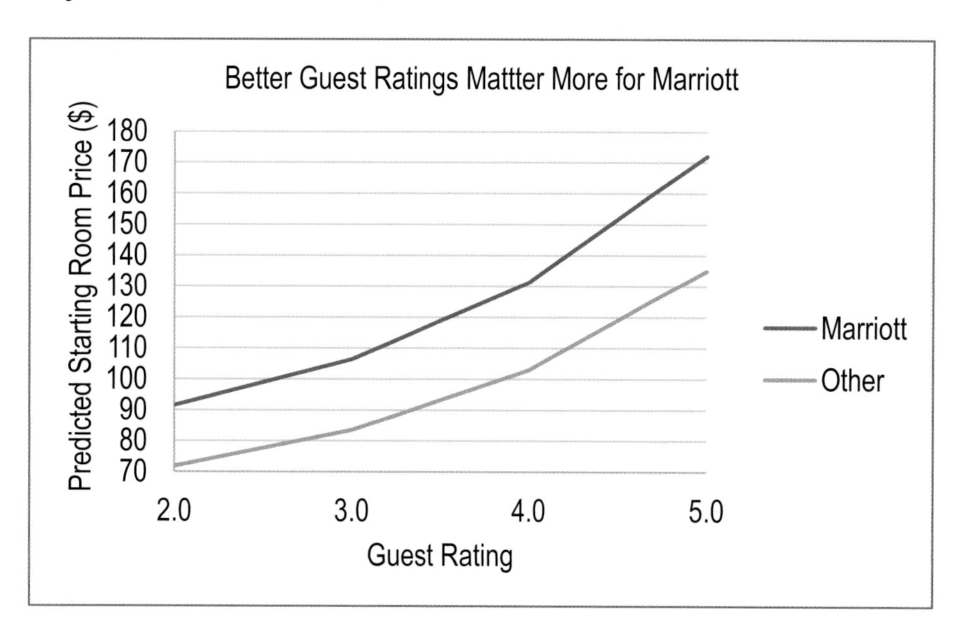

CASE 14-1 Promising Global Markets for EVs

Shiso Motors has designed an inexpensive Electric Vehicle (EV) which targets global segments where air pollution is severe. EVs have not yet been introduced in many countries, and Shiso executives believe that the level of carbon emissions is a good surrogate for potential demand.

It is generally believed that economic productivity, population, and carbon based fuels drive emissions, and Shiso managers would like quantify these impacts, in order to prioritize targeted global segments.

There is disagreement whether GDP or GDP per capita is the stronger driver of emissions.

Some believe that population is a key driver, while others argue that GDP per capita matters more.

It is thought that emissions are highest in the BRIC countries, though rapidly emerging markets that have not yet reached the level of economic development of the BRICs may be attractive.

Build a Model to Explain Emissions Differences Across Countries

Shiso has asked you to build a model to explain differences in emissions across countries.

Shiso.xlsx contains data on *emissions, GDP, population, urban population, GDP per capita, oil production*, and fuel sources used to generate electricity in 32 countries. The 32 include developed nations, the *BRIC* nations, and two emerging market segments, "emerging" and "fast emerging."

Use natural logarithms when rescaling variables would improve skewness:

1. Which potential drivers influence emissions?

___ GDP ___ GDP per capita ___ Population ___ Urban population

___ Oil Production ___ Electricity generated from oil

___ Electricity generated from coal

2. Write your model of emissions (in the original scale of *kt*) in a form which shows the separate impacts of the intercept (first) and each continuous driver. (Show significance levels for continuous drivers.)

3. One of the Shiso managers is convinced that emissions (and potential demand for an EV) are higher in countries where more electricity is produced from oil. Is this manager correct? Explain your answer in two sentences or less:

A. Sensitivity Analysis of Driver Impacts

1. The most influential driver of emissions is:

   (a) Expected emissions in the country where this driver is at the maximum level are

   ___ % of expected emissions in the country where this driver is at the minimum level.

   (b) Make a graph to illustrate the differences in expected emissions in response to differences in levels of this most influential driver, showing driver levels at the 1, 5, 10, 25, 50, 75, 90, 95 and 99th percentiles to illustrate response, *in the original scale (not percentiles)*. Rescale axes to make good use of space, include axes labels with units, and a title that describes what the viewer should see.

   (c) Make a second graph to illustrate the differences in expected emissions in response to differences in levels of a second continuous driver, showing driver levels at the 1, 5, 10, 25, 50, 75, 90, 95 and 99th percentiles to illustrate response, *in the original scale (not percentiles).* Rescale the horizontal axis to make good use of space, and rescale the vertical axis to match the vertical axis that you used in b. Include axes labels with units, and a title that describes what the viewer should see. Show this graph to the right of the graph in b.

   (d) Are differences in emissions across global markets increasing at an *increasing* or *decreasing* rate with increasing values of this driver?

   _____increasing at an increasing rate   _____increasing at a diminishing rate

2. Emissions in South Africa are noticeably higher than emissions in Thailand. Explain in a single sentence why this is the case:

3. If GDP grows by 10 % in both China and Columbia, how much will expected emissions increase in each of the two countries?

   ______% increase in China   ______% increase in Columbia

Erratum

Cynthia Fraser
McIntire School of Commerce, University of Virginia, Charlottesville, VA, USA

Cynthia Fraser (Au.) *Business Statistics for Competitive Advantage with Excel 2013,* Third Edition, DOI: 10.1007/978-1-4614-7381-7, pp xiii © Springer Science+Business Media New York 2013

DOI 10.1007/978-1-4614-7381-7_15

In the original publication of this book information regarding access to the data files is not included in the Preface. The access information is given below.

The Data Files, Solution Files, and Chapter PowerPoints

The data files for text examples, cases, lab problems and assignments are stored on Blackboard and may be accessed using this link:

https://blackboard.comm.virginia.edu/webapps/portal/frameset.jsp

Instructors can gain access to the files, as well as solution files and chapter PowerPoints by registering on the Springer site:

http://www.springer.com/statistics/business%2C+economics+%26+finance/book/978-1-4614-7380-0

Business people can gain access to the files by emailing the author cfg8q@virginia.edu.

The online version of the book can be found at
http://dx.doi.org/10.1007/978-1-4614-7381-7

Index

DOI 10.1007/978-1-4614-7381-7,

Printed in the United States of America